Boston Studies in the Philosophy and History of Science

Founding Editor
Robert S. Cohen

Volume 343

The series *Boston Studies in the Philosophy and History of Science* was conceived in the broadest framework of interdisciplinary and international concerns. Natural scientists, mathematicians, social scientists and philosophers have contributed to the series, as have historians and sociologists of science, linguists, psychologists, physicians, and literary critics.

The series has been able to include works by authors from many other countries around the world.

The editors believe that the history and philosophy of science should itself be scientific, self-consciously critical, humane as well as rational, sceptical and undogmatic while also receptive to discussion of first principles. One of the aims of Boston Studies, therefore, is to develop collaboration among scientists, historians and philosophers.

Boston Studies in the Philosophy and History of Science looks into and reflects on interactions between epistemological and historical dimensions in an effort to understand the scientific enterprise from every viewpoint.

Marius Stan • Christopher Smeenk

Editors

Theory, Evidence, Data: Themes from George E. Smith

 Springer

Editors
Marius Stan
Boston College
Boston, MA, USA

Christopher Smeenk
Western University
Ontario, ON, Canada

ISSN 0068-0346   ISSN 2214-7942   (electronic)
Boston Studies in the Philosophy and History of Science
ISBN 978-3-031-41040-6   ISBN 978-3-031-41041-3   (eBook)
https://doi.org/10.1007/978-3-031-41041-3

This Springer imprint is published by the registered company Springer Nature Switzerland AG
The registered company address is: Gewerbestrasse 11, 6330 Cham, Switzerland

Paper in this product is recyclable.

Introduction

This volume grew out of the conference *On the Question of Evidence: A Celebration of the Work of George E. Smith*, which met at Tufts University in May 2018. It was a venue for presentations by Jody Azzouni, Katherine Brading, Robert DiSalle, Allan Franklin, Michael Friedman, Bill Harper, Teru Miyake, Eric Schliesser, and Chris Smeenk. A number of those presentations evolved into chapters included here. To them, we have added chapters by other scholars and thinkers who have been influenced, directly or indirectly, by George's work. Collectively, these pieces aim to honor that work by engaging with prominent themes from it.

In what follows, we begin with a brief introduction to George's philosophy of science, and to his more prominent and distinctive contributions to doing it. Thereby, we highlight some of the challenges this poses to long-held commitments within philosophy, history, and sociology of sciences. Then we elaborate on key themes in his philosophy as we summarize and introduce the chapters in this volume. We conclude with some remarks on George as a teacher, and on the philosophical impact of his teaching.[1]

George Smith on Evidence in Mature Science

The history of evidential reasoning, if viewed over extended periods rather than in isolated episodes, and regarded as more than illustrations to support philosophical positions, could produce a decisive transformation in our understanding of scientific knowledge. Current views of the nature of scientific knowledge have been forged, in large part, in response to the threat of radical theory change made prominent first by Kuhn, then by subsequent historiography of scientific practice. To fend off that threat, any claim to have achieved a distinctive form of permanent knowledge

---

[1] This volume began as a project jointly co-edited by Smeenk, Schliesser, and Stan. An encounter with long Covid led to Eric Schliesser being unable to remain at the helm as co-editor.

must establish some form of continuity through transitions, such as from Newtonian gravity to general relativity and from classical mechanics to quantum theory. Many scholars have concluded that claims to establish secure knowledge should be dismissed as a misconception, reflecting a false image perpetuated by science pedagogy, for example, rather than genuine insights into what science has achieved. In addition, the program to find a single logic of scientific method or scientific inference seems to have failed—in virtue of accumulating evidence showing that actual scientists are often opportunistic in their practices. In so far as such projects still remain (e.g., Bayesianism or various kinds of causal modeling), they have a highly normative character.

Yet such a dismissive response to claims about the epistemic status of many sciences fails to account for evident progress in understanding the natural world. It is hard to deny that various scientific fields have succeeded in establishing theoretical claims that, in many respects, have also served as effective guides for action, even for those skeptical about whether the theories themselves have any claim to permanence.

George Smith has often described his work as inspired by asking how we came to have high quality evidence in any field. In particular, how have scientists in successful fields turned data from observations or experiments into strong evidence for substantive claims? What is the nature of the knowledge claims that have the best case for being firmly established? And in what sense, if at all, can we take these claims as permanent, stable contributions to scientific knowledge? In pursuing these questions he has developed a striking, distinctive account of the nature, scope, and limits of scientific knowledge. He based his account on historical and philosophical investigations of exemplary evidential reasoning spanning extended lines of research (over decades or centuries) in physics.

George considers extended lines of research for several related reasons (see the paper reprinted in this volume). He largely agrees with historians and sociologists who have argued that the scientific community typically accepts substantive claims based on surprisingly weak evidence. But he reframes the question of high quality evidence as more appropriately posed with regard to the consequences of accepting some claims as a first step in a line of inquiry. Rather than, for example, considering only the case Newton could make in favor of gravity at the time of writing the *Principia*, we should keep in mind the case Simon Newcomb could have made two centuries later. George strikingly disagrees with historians who are tempted to generalize from their apt criticisms of local case histories, often at the *advent* of a line of inquiry, to a general claim that would apply to Newcomb as well as Newton. His approach also contrasts with philosophers of science, who have too often focused exclusively on the kinds of arguments that scientists make in motivating the community to pursue new ideas—such as an emphasis on novel predictions. The strongest evidence in favor of a well-established theory often differs from the kind of case that could have been made for it when it was first introduced; or even when it found widespread acceptance. More importantly, what is epistemically distinctive about science, what enables it to make progress, will be more clearly discerned in how a line of research unfolds subsequently rather than in the initial debates.

This position leads to a distinctive innovation in how George practices philosophy of science—an innovation that has led to a new genre of philosophical scholarship. Many of his studies are working papers or case studies toward the development of what one might call a *longue durée* review article. They can be book-length pieces, as his famous "Closing the Loop" attests. In such a *longue durée* review article, George surveys and evaluates how theory development and a field's evidential practices (which are constitutive of the field) interacted over an extended period of time.

The shift to a long view is only useful in concert with Smith's characteristically thorough assessment of what is actually being put to the test through subsequent research. This means that in practice George often redoes the calculations—and pays careful attention to attempts to replicate earlier results, as well as to the way that scientists assess the evidence in their own time as presented in textbooks and ordinary review articles. This practice is manifest in both his case studies and the more substantial, retrospective review articles. He redoes calculations not just from meticulous care to understand how past scientists would have evaluated the evidence, but, in particular, so as to enable his readers to assess which results or measurements were constitutive or evidentially relevant for particular lines of inquiry.

By contrast, philosophers of science too often rely on an abstract, hypothetico-deductive account of theory-testing: scientists test a theory T by deducing its consequences, then checking these via observation or experiment. According to this view, Newtonian gravity, for example, is tested by deducing predictions for planetary positions given some information about the initial conditions. One of the leitmotifs of Smith's work is that an exclusive focus on successful predictions neglects crucial aspects of scientific practice. Treating theories as a monolith obscures the fact that evidence bears *differently* on specific aspects of the theory. George has emphasized, for example, the stark contrast between ample evidence for the inverse-square variation with distance of the gravitational force, as opposed to the weak evidence for its dependence on the mass of both interacting bodies. The hypothetico-deductive account also misses the role of a different type of claim: in the case of planetary astronomy, calculations of planetary positions are based on the substantive assumption that the masses and forces taken into account in the derivation are complete. But if the comparison with observations requires a claim of this kind, what can we say is actually being tested?

Paying attention to the role that theories play in guiding an extended line of inquiry leads to a richer understanding of the logic of theory-testing, developed in detail for the case of celestial mechanics in George's magisterial "Closing the Loop." Based on surveying nearly 250 years of planetary astronomy, he argues that the main question being pursued was not, as philosophers would have it, whether Newtonian gravity "saves the phenomena." Indeed, throughout the period, he considers there was only a brief time during which the existing models fully fit with available observations of planetary motion. At all other times, there were systematic discrepancies between the calculated and observed motions. Celestial mechanics used these discrepancies to discover further physical features of the solar system.

The main aim of this line of research was to identify robust physical sources for existing discrepancies—such as new planets, but also subtle physical effects such as the changing rotational speed of the Earth—and then "close the loop," by adding this new feature to a more complete model. George once put it as follows:

> Newton's *Principia* forced the test question within orbital astronomy for his theory of gravity to be not whether calculated locations of planets and their satellites agree with observations; but whether robust physical sources can be found for each systematic discrepancy between those calculations and observation—with the further demand of achieving closer and closer agreement with observation in a series of successive approximations in which more and more details of our solar system that make a difference become identified, along with the differences they make.[2]

Pursuit of this iterative approach, adding further details to develop ever more sophisticated models and identifying even more subtle features of the solar system, lead to an enormously rich picture of, as Smith puts it, the configurational details of the solar system and the differences they make to observed motions. The identification of what George calls "second order phenomena" depends on a contrast between the motions, as described at a specific stage of inquiry, and increasingly precise and carefully targeted observations. Success at each stage in discovering more subtle effects that have clear physical sources indirectly confirms the earlier steps in the process. A discrepancy at any given stage only has physical significance, and helps to identify another detail that contributes to observed motions, if the earlier calculations have incorporated the features with larger impacts accurately.

What makes "Closing the Loop" so compelling is George's masterful history of this field, combined with the philosophical and technical acuity needed to recognize and articulate the logic of theory-testing that this history discloses. Part of his insightful analysis pertains to the specifics of celestial mechanics. And George often insists, in print and in person, that particular evidential strategies may not generalize to other fields; that one first has to do the work in reconstructing the *longue durée* evidential practices of them. This reluctance stems from George's recognition that each field faces often highly specific challenges to developing evidence, regarding—to put it roughly—accessibility to the kinds of quantities that allow for fruitful theorizing. This challenge is general, but giving a full account of how to respond to it quickly becomes intricate and domain specific.

Before we articulate (despite George's reticence) some general features of his approach to evidential reasoning, it is worth noting that his detailed work also corrects another tendency common among historians of philosophy. While we have emphasized above that his work can be understood as incorporating and accepting the findings of those focused on scientific practice at a given time, but by shifting toward extended periods (also extended in space) undermines their nominalist

---

[2] This is the first of three principal conclusions stated in the manuscript version of "Closing the Loop," which do not appear in the published text.

and skeptical tendencies, it is also worth noting that—while George reiterates the emphasis on theories familiar from an older historiography, e.g., Koyré—he undermines the older view that theories alone can explain important scientific debates. His work shows that often attitudes toward particular lines of empirical evidence shape how theories are adhered to.

While George is cautious about generalizing from the results of his life-time of research, we think there are also general features of his approach to evidential reasoning, revealed in other detailed case studies—such as his recent monograph (with Seth) on early-twentieth century experimental investigations of molecular reality. Here we will highlight two interrelated aspects of Smith's account, before considering their implications for questions of continuity.

First, physical theories typically re-describe phenomena by introducing novel quantities, such as mass and force in Newtonian mechanics, and law-like regularities that hold among them.[3] This immediately raises a challenge: how do scientists reliably gain access to the proposed fundamental quantities, and justify their use? George's work shows the value of taking *these* questions—rather than obtaining successful predictions—as the central challenge. Scientists usually proceed by exploiting functional relationships between the target quantity of interest and a "proxy" quantity (or quantities) that can be measured (more) directly with high precision. Experimental design focuses on finding measurable proxies with a particularly clear and reliable connection to the target quantities. The role of theory is to establish such functional relationships, to enable *theory-mediated measurements* of the target quantities—that is, to show how the theoretical quantities are manifest in, and can be constrained by, observable phenomena, and can give rise to second-order phenomena. But, crucially, success in theory-mediated measurements provides evidence for the aspects of theory used to derive the functional relationship.

This success can take several forms: stability in the outcomes of repeated measurements of the same type; convergence in the values determined by measurements of different types, based on different functional relationships; and amenability to increasing precision. Smith and Seth (2020) clarify how each of these types of success contribute to justifying the use of novel physical quantities and make a persuasive case that success in all three senses can eliminate the possibility that the target quantity is merely an artifact or useful fiction. On their account, early twentieth century physicists established reliable access to the microphysical realm (of atoms and molecules) through stable, convergent measurements of three crucial parameters: Avogadro's number, N_0; the fine structure constant, α; and the charge of the electron, e. Contemporary physicists now rely on collaborations such as the CODATA group to determine consensus best values for dozens of fundamental parameters. Rather than seeing these as merely values to be inserted to facilitate comparisons with observations, on George's view the stable, consistent determinations of these parameters over time provides evidence for the array

---

[3] A note on our terminology: by "phenomena" we mean roughly what Newton meant by it: a robust empirical regularity accepted by a scientific or expert community.

of physical theories that fix the relevant functional relationships used in these measurements.

To describe the second aspect, we can turn to a feature of Newton's methodology that George has elucidated wonderfully: namely, his response to the challenge of drawing conclusions from phenomena despite their complexity. Smith and Harper (1995) aptly characterized this approach as a "new way of inquiry" in their seminal paper, drawing a sharp contrast between Newton and seventeenth century contemporaries such as Galileo and Huygens. Consider Newton's realization, as he was revising the *De Motu* drafts that would lead to the *Principia*, that the planets do not travel along closed orbits. Due to the mutual gravitational interaction among the planets and the sun, they would instead move around a common center of gravity and follow trajectories too complex to admit of a simple geometric description. Rather than give up on the project of determining the forces from observed motion in light of this remarkable complexity, Newton developed the mathematical framework needed to make inferences regarding the force that hold even if the orbits can be described as only approximately elliptical. The first aspect of Newton's response to complexity is to establish functional relationships between observations and target quantities that are robust, in the sense that an *approximate* description—thus not necessarily exact—of the observed phenomena can still yield an approximate value of the target quantity. Borrowing a Latin phrase that Newton used frequently, George often describes this in terms of "if *quam proxime* (very nearly) then *quam proxime*' reasoning." In the context of reasoning from observed phenomena of motion to forces responsible for them, Newton proves that, if the (observed) antecedent holds very nearly (*quam proxime*), then the (inferred) consequent also holds *quam proxime*.[4] (We will discuss an example of this type of result momentarily.)

This is coupled with a recognition that a specific type of idealized representations play an essential role in guiding further inquiry. Astronomers in the 1670s faced an underdetermination problem: there were several distinct—and inequivalent—ways of representing planetary motions. Newton recognized that the planets only follow *approximately* Keplerian orbits for theoretical reasons, but in the case of the lunar orbit the departure from Kepler's area law was well-established observationally. What then is the status of the Keplerian orbits? As George puts it:

> The complexity of true motions was always going to leave room for competing theories if only because the true motions were always going to be beyond precise description, and hence there could always be multiple theories agreeing with observation to any given level of approximation. On my reading, the *Principia* is one sustained response to this evidence problem.

Newton recognized that the Keplerian orbits (unlike the alternatives) hold in idealized, counterfactual circumstances: if the sun were at rest, and the interactions

---

[4] However, we must note that Newton employs his phrase, *quam proxime*, in several semantic regimes, not just one. As a matter of fact, a closer study of these regimes is part of George's current research on approximation in physical theory construction. Relatedly, see also the chapter by Guicciardini in this volume.

among the planets neglected, then a planet would follow a Keplerian orbit. They thus have a physical status that alternative descriptions lack. Clarity regarding exactly what must be the case for Keplerian orbits to hold—counterfactually—also makes it possible to treat departures from Keplerian motion as second-order phenomena: as a means to identify further physical details relevant to orbital motions. In the case of the moon, despite enormous challenges to developing a satisfactory theory, Newton could make the case that the departures from Keplerian motion were due to the perturbing effects of the sun. More generally, idealizations stating regularities that would be *exact in precisely specifiable circumstances* support the identification of discrepancies. In the best case, the character of these discrepancies further indicate which assumptions of the idealization have to be modified or relaxed.

Although we have focused on George's account of Newtonian methodology to elucidate this approach to complexity, we expect an account along these lines holds for much of modern physics, where it is constrained by robust background theory. His own work highlights how similar considerations apply to the study of molecular reality, and to the structure of molecules (see Smith and Seth 2020 and Smith and Miyake 2021, respectively). More broadly, there is a striking similarity between this account of methodology and the effective field theory approach in modern particle physics, developed by Ken Wilson and others.

Taken as a whole, George's account of methodology leads to a fundamental reframing and response to concerns about permanence and continuity stemming from Kuhn. Even profound transitions at the fundamental theoretical level do not necessarily generate discontinuities in evidential reasoning.[5] Following George's insights, the focus should be on whether the functional relationships presupposed in evidential reasoning continue to hold in light of a new theory—for example, as an approximation within a limited domain. This has a family resemblance to a kind of structural realism that his friend, Howard Stein, has espoused, or the real patterns that his long-time Tufts colleague, Dan Dennett, has defended. But, in George's work, the emphasis is on the challenges to sustaining a particular line of evidential reasoning—not on ontology. In one sense, this is a much weaker demand than what philosophers typically consider, and one that is often imposed by scientists developing new theories, but in another sense it is much more ambitious: a record of success in obtaining stable, convergent measurements, and of identifying further details that make a difference in an iterative approach, in fact make a compelling case that we should demand continuity of evidential reasoning, just as Einstein took recovery of a Newtonian limit as a criteria of adequacy in formulating general relativity. Finally, even though the evidential reasoning itself presupposes and depends directly on (some aspects of) theory, the standards of success do not. Whether a line of research has led to increasingly precise, convergent measurements of fundamental parameters, or continued to identify robust physical sources that can

---

[5] This is not to say that Smith rejects the possibility of Kuhnian discontinuities in evidence; Buchwald and Smith (2002) consider the discontinuity in the treatment of polarization through the transition from ray to wave optics.

be checked by a variety of independent means, can be evaluated in relatively theory-neutral terms.

We hope that this brief précis of one aspect of George's thought is sufficient to show that it has deep ramifications for central debates in history and philosophy of science. But, we also want to highlight a different aspect of his work. As we noted, he approaches historical material with a meticulous attention to detail—perhaps inspired by his working practice as an engineer interested in failure of complex systems (in particular, turbine engines)—often leading to profound reassessments of canonical texts, or lines of argument, that one would expect to have few surprises left to divulge. Several philosophers have, for example, taken Perrin's case in favor of atomism based on his own experimental measurements of Brownian motion, in conjunction with an array of other results, as exemplifying a successful argument in favor of unobservable entities (while disagreeing, of course, about the precise nature of the argument or the correct conclusions to draw from it). Smith and Seth (2020) show that many aspects of the philosopher's lore about this case, even those that are particularly relevant to the case for atomism, are simply incorrect. For example, Perrin's own measurements of N_0 were rejected within a decade after the publication of *Atomes*, due to a substantial and persistent discrepancy with other precision measurements, threatening philosophical arguments that rely uncritically on Perrin's claim of sufficient agreement among the different measurements.

George's most influential contributions along these lines have led to a dramatic re-evaluation of Newton's achievements and the reception of Newton's work. Smith and Schliesser (2000) give an account of the early response to Newtonian gravity that corrects lore about the debate between Newton and his continental critics, in particular Huygens and Leibniz. Those debates are not best understood as merely a clash of metaphysical systems, but as driven, in large part, by the limitations of available empirical evidence. In addition, George's critical reading of Newton has revealed aspects of the argumentative structure of the *Principia* that seem to have escaped notice from Newton's own time. Our favorite example concerns an argument that Newton did *not* make, even though it is often attributed to him: namely, that the inverse-square law of gravity follows from an observed Keplerian ellipse. Even a relatively cursory reading of the *Principia* reveals that Newton did not actually give this argument, but Smith gives an extremely insightful analysis of the argument he actually gave and what it reveals about Newton's methodology. The incorrect argument requires that the antecedent holds exactly: if a body moves on an ellipse with a force directed at one focus, then the force varies as the inverse square of the distance. But Newton also proved that if a body moves on an ellipse with the force directed at the center, then the force varies directly with distance. The planets all move on nearly circular orbits, so it is extremely challenging to draw a distinction between these two cases observationally. Any method that depends on establishing exact results observationally is extraordinarily fragile. By contrast, the argument Newton actually gives relies on a functional relationship that is robust. The apsidal precession theorem establishes a functional relationship between the motion of the apsides (the points on the orbit closest and farthest from the sun) with each orbit and the exponent of the force law (for approximately circular orbits).

For a closed orbit, with no motion of the apsides, the force varies with the inverse square of distance. But, crucially, this result also implies—and this is an instance of "if *quam proxime* then *quam proxime*" reasoning noted above—that for orbits with *nearly* stable apsides the force law is *nearly* inverse-square.

Themes from George Smith: A Synopsis

George's work has made an impact on epistemology, the history of philosophy of science, and Newtonian scholarship. In casting new light on the *Principia*— its argument structure, epistemic support, and long-term posterity—he has changed research agendas and opened new vistas onto old problems and figures. The papers in this volume attest to that broad influence.

Epistemology

In work on Newton and in collaborations on late-classical physics, George had devised and relied on a tripartite division of the epistemic status a proposition can have in the course of research. Specifically, he has distinguished between a proposition counting as a hypothesis; being "taken to be true"; and being established. What underpins this distinction? As we explained above, it is the quality of the evidence for that proposition—at a given time; and over long stretches—and the *constitutive* role assigned to it during ongoing research that involves it.

Knowledge In his paper, Jody Azzouni asks how Smith's taxonomy of epistemic stances dovetails with some central distinctions in traditional epistemology. He argues that propositions taken to be true (by Smith's criteria for that notion) really count as *knowledge*. To make his case, Azzouni distinguishes two concepts of knowledge, viz. ordinary and philosophical. The latter is an artifact of conceptual engineering: it has an inbuilt constraint, namely, that knowledge be infallible. However, Azzouni sides with the ordinary notion—the one behind vernacular uses of "to know." This notion, he explains, does not require infallibility (about the particular proposition P at issue). And so, he concludes, if a community of inquiry has enough evidence for taking P to be true, then that community knows that P.

Azzouni then finishes *à rebours*. From the vantage point of his dual picture, he examines Descartes' notion of knowledge. His verdict is that Descartes endorsed the infallibility condition, notwithstanding the Cartesian distinction between *scientia* and *cognitio*. Moreover, Azzouni argues provocatively, Descartes' above endorsement appears to have contaminated Newton's thought too, in two respects. First, both Newton and Descartes take deduction to be—if suitably carried out— a procedure that cannot fail. Second, both think that a proposition counts as knowledge only if it has been "deduced." There are, of course, key differences between their idea of deduction and our modern concept. Still, Azzouni concludes, Newton shared with Descartes the commitment that "deduction" is the gold standard for knowledge, and knowledge requires infallibility.

This line of argument suggests two directions for further research. It should prompt us to take a deeper look at Descartes' and Newton's respective accounts of deduction, and how they took specific mathematics—frameworks and approaches— to underwrite the presumed infallibility of their "deductions." On a broader note, it encourages us to inquire into when and why the early modern tenet of infallibility (of knowledge) gave way to the current situation in which fallibility is the default position in epistemology.

Strong Evidence George first uncovered the very special role of systematic discrepancies in Newton's work, and their legacy for gravitation theory in the centuries after him. When they are systematic, such discrepancies become very valuable, as second-order phenomena in the sense described above: they drive research forward, by pointing to aspects of the theory that need refining or strengthening; and they yield *strong* evidence for the theory, if it accounts for them from its own resources. This particular insight suggests two lines of further research. Do systematic discrepancies play this role in science more broadly, beyond research in Newtonian gravitation, where Smith first uncovered it? And, did anyone else between Newton and Smith grasp the special value that discrepant phenomena have for research in the exact sciences? The paper by Teru Miyake takes up these questions, and answers them compellingly. In his recounting, it was the natural philosopher John Herschel who in the 1820s explained the evidential value of "residual phenomena," or observed discrepancies—between (first order) predictions from theory and the observed behavior of target objects. Such phenomena further confirm the theory, if it can account for them; or they require—and may even suggest—revisions to theory, if its laws cannot handle those "residues." Though he saw the method as working at its best in physical astronomy (thus foreshadowing, as it were, Smith's magisterial paper of 2014), Herschel seems to have thought the method was more general, potentially useful in many areas of exact science. His lucid case for it would influence key figures of Victorian logic and philosophy of science, as Miyake shows.

At the same time, against the backdrop of George's complex picture of evidence we presented above, we can see more clearly the limits of Herschel's insight into the confirmatory role of "residual phenomena."

Empiricism and Metaphysics A recurring theme in Smith's work is how, in the process of theory building, Newton always took great care to anticipate how he might go wrong, viz. how the evidence available falls short of supporting his physical claims. Robert DiSalle's paper takes up that theme and extends its reach. In particular, he asks whether Newton took the same evidential care in regard to the *philosophical* foundations of his theory, not just the physical details of the solar system. DiSalle argues in the affirmative: he points out how much empirical import attaches to some of the key foundational terms in the *Principia*—absolute velocity, relative motion, true rest, uniform translation, absolute rotation, and the like. In his recounting, Newton gradually found a way to demarcate which of these terms are empirically well-grounded, and which ones are fated to remain 'hypothetical.' The key, DiSalle explains, was Newton's gradual refinement of the relativity of motion, and the resulting insight into how to formulate a concept of inertia compatible with Corollaries V and VI to the laws of motion.

Alongside his philosophical analysis, DiSalle offers a nuanced reconsideration of historical figures like Berkeley and Mach, commonly known as critics of absolute space and time. He suggests that their grasp of "Newtonian relativity" was really deeper than we have appreciated so far. More broadly, we can see in his paper a plea, inspired by George's picture of Newtonian evidence, to reflect on the various ways in which a metaphysics could be empiricist—by way of looking at the empirical credentials of the metaphysical ingredients proffered as foundations for physical theory.

Measurement and Evidence A very important theme from Smith is the evidential value of measurements. In his reconstruction, when measured values of a key parameter converge over time, they count as evidence (for the theory in which the parameter is embedded).

That evidence is even stronger when that particular ingredient—the expression that connects a theory-bound quantity with a measured parameter—can be further refined. As we explained above, behind such expressions, or functional dependencies, are tacit idealizing assumptions. If these assumptions get relaxed (with the functional expression refined accordingly); and if new, further measurements bear out the newer, refined formulas linking theory to metric proxies—then that counts as strong evidence for the theory.

Allan Franklin's study brings this theme to the fore. His diachronic survey highlights how attempts to measure a certain parameter (the mean density of the earth) yielded more exact values in the long run: from Cavendish to the late nineteenth century. At the same time, all such experimental attempts have to grapple with the challenge of disaggregation. Namely, the need to identify factors—temperature gradients, friction damping, material-specific properties—that might lead to discrepant measurements; and to screen them off, or at least to estimate their effects, so as to subtract them from discrepant phenomena.

In this respect, Franklin's paper dovetails with the studies by Miyake, and Biener and Domski in this volume. They too single out for reflection Smith's emphasis on discrepant phenomena, and to what extent we can marshal such discrepancies in support for theory.

Evidence in the Sciences of the Past Certain key elements in Smith's account of strong evidence—for instance, the emphasis on systematic discrepancies, and on converging values of a measured parameter—suggests that such evidence is available just in advanced, strongly mathematized science. Which implies, inter alia, that the sciences of the past—and disciplines that study one-off events, in particular—are at an evidential disadvantage. That is because their specific domain exhibit neither repeatable patterns (from which we could discern *systematic* discrepancies) nor *serial* values of measured parameters.

In recent years, however, Carol Cleland and others have argued that the historical sciences are not at an evidential disadvantage, relative to the physical sciences above. Craig Fox's paper takes a critical look at the reasons for their epistemic optimism. In their account, these sciences rely on a common pattern of confirmation. Specifically, they infer from present traces of past events—artifacts, material remnants, physical leftovers, and byproducts of extinct processes—to some common cause responsible for those traces. Underwriting this generic pattern of inference is a key assumption, which these authors call the "Common Cause Principle." Fox inquires into the warrant for this assumption. That warrant, he explains, is a claim further upstream, namely, that the past is overdetermined. That claim is really two ideas. One is epistemological, and says that, for us to infer reliably to some past event, we do not need all (causal) traces of that event as evidence for it—just *some* traces are enough. The other regards ontology and says that past events always leave at least *some* causal traces into the present.

These authors think their key claim—that the present overdetermines the past— follows from David Lewis' analysis of causation in terms of possible worlds. However, Fox shows, the analysis does *not* support their metaphysical premise— because in his account Lewis excluded certain sets of possible worlds by fiat. In particular, Lewis' analysis ignored "backtracking counterfactuals." But that move, as Fox points out, in effect excluded—illegitimately—the possibility that our present is causally and nomically compatible with many, *different* pasts. And so, their key premise (that the historical past is overdetermined) holds by stipulation, not in virtue of metaphysical argument. Thus Fox's incisive scrutiny undermines (or challenges) one case for optimism about evidential reasoning in the historical sciences—and leaves the important task of developing a more compelling case open to further work.

Fox's piece is on the cutting edge of recent studies that investigate the structure of confirmation in science about the past. That enlarges and sharpens our picture of the descriptive import and explanatory power of George's account of evidence.

Newton Scholarship

George's research has drawn renewed attention to Newton's methodology—a generous term that covers both heuristics, or guidance for "reasoning more securely" in natural philosophy, with empirical results appropriately guiding inquiry; and also logics of confirmation, viz. patterns of evidential reasoning, and constraints on admissible inferences.

Newton on Methodology In regard to the former—heuristics—Monica Solomon in her paper seeks to uncover a new aspect of Newton's thought on that topic. She argues that a key element in the famous Scholium on space, time, and motion is best understood as a piece of methodology. In particular, Solomon claims, Newton meant his example of the two globes (connected by a cord, rotating around each other in empty space) as a terse blueprint for setting up, and tackling, problems in orbital dynamics. As she explains, the globes are an epitome of the type of dynamical system that Newton studies in *Principia*. The example requires us to think of the globes as having the quantitative properties Newton sets down in his definitions—properties that covary in accordance with his laws of motion—and as being sufficiently far away from perturbing factors that we can treat the globes as a quasi-isolated system (a supposition that Newton at the end of his book, in the General Scholium, seems to reaffirm).

Thereby, Solomon breaks with a long tradition (going back to Mach) that saw Newton's globes example as making a point about the metaphysics and epistemology of true motion. In effect, her paper makes a case for further study of Newton's heuristic resources—a topic that harmonizes nicely with George's analyses, and which is sure to reward further scholarship on it.

Newton, Method, and Optics The paper by Howard Stein brings to the fore, as does much of Smith's work, the precision and discipline with which Newton evaluated claims in natural philosophy. His richly historical account unfolds along two strands of inquiry. One clarifies Newton's distinctive treatment of the levels of certainty that can be attained in scientific inquiry. Stein here sharply contrasts Newton's capacity for "multiple vision," an ability to assess the differing roles and epistemic status of components of a proposed theory, with the more one-dimensional views of his contemporaries. Hooke, Huygens, and other critics often assumed that Newton relied upon, but failed to acknowledge, some form of mechanical hypotheses or metaphysical posits. The contrast reflects the novelty of Newtonian inquiry, a theme that resonates with Smith's work. But Stein further takes contemporary commentators to task for misreading Newton's methodology and reveals the rewards of a more sophisticated reconstruction of Newtonian empiricism.

The second strand addresses the status of metaphysics, given Newton's claims to establish specific claims regarding the nature of light and gravity without needing to "mingle conjectures with certainties." By contrast with the dominant Cartesian tradition, for Newton metaphysical notions ultimately have an *a posteriori* character. Newton introduced or adopted those metaphysical elements in response to the foundational and explanatory needs of physical theory, not before (and independently of) them. And in addition, Newton regarded his metaphysics as essentially revisable—not at will, of course, but always in reaction to the changing evidential fortunes of the *empirical* theories. Here Stein in particular reviews, briefly, the transformative steps Newton took in formulating his optical theories, and later in revising traditional metaphysical categories of space, body, and force in the course of writing the *Principia*. Stein reads Newton's assessment as again illustrating a capacity for multiple vision, in this case treating the doctrines of space and time as having a more secure status than the more conjectural account of the nature of body. Furthermore, as Stein emphasizes, far from entirely eschewing hypotheses, such as the corpuscular theory of light, Newton specifically acknowledged their value—not as something presupposed as part of his experimental reasoning but as a guide to further inquiry.

The *Principia* and Hypotheses From his earliest work in optics and the disputes it prompted, Newton always tried to demarcate results he had established from "mere hypotheses." Much of George's work is a reconstruction of how Newton developed a method to "reason more securely" in natural philosophy—by contrast with the explicitly hypothetical methods of his contemporaries. Huygens, for example, clearly endorsed a version of reasoning from hypotheses. The chapter by Zvi Biener and Mary Domski revisits the broad idea that, at least in mathematical mechanics, Newton did not reason from, nor did he endorse, hypothetical assertions.

For the case of the "mathematical" Book I of the *Principia* (and its application in Book III to gravitational phenomena), Newton's rejection of hypotheses appears unimpeachable. However, when they turn to Book II—the mathematics of motion in resisting media, plus Newton's experimental basis for it—Biener and Mary Domski think that his rejection of hypotheses looks shakier. For one, Newton there starts from assumptions about the dynamics of a resisting medium: *suppositions* about which of its properties are causally relevant; and about their kinematic effects (on a solid moving in that medium). For another, it is hard, they argue, to find in Book II the key elements that allowed Newton to rise above hypothetico-deductivism: robust theorems (for inferring force laws from motion phenomena); convergent values of a measured parameter; or a mathematical treatment that allows him to disaggregate the respective contributions (of various physical causes and mechanisms) to a complex motion effect, e.g., the decay of pendulum swings in water. But, if those sources of evidential reasoning are lacking, Biener and Domski conclude, isn't Newton's pattern of confirmation in Book II closer to Galileo's and Huygens' hypothetical approaches than we have thought so far?

The Architectonic of the *Principia* The argument structure of Newton's treatise has long been an elusive puzzle. Katherine Brading in her chapter sheds new light on this difficult topic. The *Principia* unites two disciplines, or areas of inquiry, she argues. Books I and II establish results in rational mechanics—the applied mathematics of orbits induced by central forces of interaction. In contrast, Book III turns to physics: a theory of a force (gravity) seated in bodies, its effects on them, and its quantitative relations to other properties of material bodies, like mass, shape, and volume. According to Brading, in the *Principia*, these two disciplines are not simply juxtaposed; they are connected by a conceptual bridge, as it were. That bridge is the definitions and "axioms, or laws of motion" that Newton placed at the outset, before his three books. These elements serve a dual function in the treatise: they make up the axiomatic basis of his mathematical mechanics; and they are the nomic basis of his quantitative physics of gravity.

This framework allows Brading to elucidate the diachronic context of his theory, not just its synchronic makeup. In particular, she explains, with the *Principia* Newton continued, while transforming very drastically, a program that goes back to Descartes. It was the program of combining rational mechanics with physics into a "philosophical mechanics," as Brading calls it (because physics then counted as a branch of philosophy). He expanded the scope of rational mechanics well beyond what anyone had done by 1700. And, he showed how the physics of gravity is amenable to mathematization. Specifically, well-established empirical methods of measurement give us a quantitative handle on the effects of gravity. From those effects, Newton's rational mechanics lets us infer the strength and direction of the gravitational forces responsible for them; and to treat these forces as endowed with a measure, or algebra. Thereby, Brading's study dovetails with George's well-known emphasis on Newton treating forces as quantities, viz. actions endowed with a ratio structure, or measure.

Mathematical Methods The *Principia* is famously geometric: geometric objects (e.g., lines, plane curves, areas) stand for properties of motion and force; and geometric methods (e.g., auxiliary constructions, diagrammatic reasoning) are among its key vehicles for proof in rational mechanics. Newton's reliance on this geometric framework—its scope of representation, heuristic merits, inferential limits, and equivalence to other frameworks then available—have been a topic for much scrutiny. Debates began already in his time, and scholarship in the last half-century has greatly advanced our understanding of Newton's formal methods. The chapter by Niccolò Guicciardini makes a major contribution to that understanding, by shedding light on a difficult, elusive, and little studied topic. Broadly described, that topic is integration in Newton's mathematical thought: its meaning, scope, key techniques, and comparative virtues. Guicciardini establishes a number of novel and important results. One is that, overall, Newton was a good deal more of an "algebraist" than the *Principia* might lead us to expect. At least in matters of integration, he clearly favored proof procedures that rely on the rule-governed manipulation of algebraic formulas—rather than inferring from inspection of a geometric object. Another major result is that Newton had an incipient notion of differential equation and had worked out ways to solve some classes—primarily, by expansion into power series; by numeric approximation, where feasible; and by change of variable, or substitution.

This particular result is subtle, but matters greatly, in two respects. For one, as Guicciardini notes, Newton's approaches to "fluxional," or differential, equations differ from our modern approach (which goes back to Leibniz and his disciples) that seeks analytic, closed-form solutions for them. We may be tempted to think that difference counts as a weakness of Newton's fundamental concept of "fluxion." But the reason lies elsewhere, as Guicciardini explains: many of Newton's intended applications—for the techniques above—are in the dynamics of perturbed systems. In general, those systems are *not* integrable in closed form. His approaches (numeric integration, and expansion into infinite series) are excellent approximations of those generally unavailable exact solutions. And so, what prima facie looks like a weakness is in fact a source of strength. For another, it helps us see that the gap between Newton's mathematical methods and modern formulations (of gravitational dynamics) is not as great as a casual look at the *Principia* might suggest. Thanks to Euler and Lagrange, mechanics settled into the form familiar to us late-moderns, viz. of a connected set of differential equations (of motion) derived from dynamical laws. In that regard, the *Principia* with its geometric language looks separated by a gulf from our versions of mechanics. Guicciardini's study reveals that to be an exaggeration, ultimately. Though he lacked a function concept (and the Leibnizian notation that eases their calculus so greatly), Newton after all did have *the* key ingredient of modern mechanics: describing the motion (over an instant) by way of a differential expression, which then must be integrated. Guicciardini's paper resonates with another theme from George Smith, among whose breakthroughs has been the careful study of Newton's mathematical methods at the more advanced stages of theory construction in the *Principia*.

History of Philosophy of Science

George's work has opened new lines of research on the diachronic aspects of foundations for mechanics, and on how major philosophical figures responded to Newton's achievement. A number of chapters in this volume explore these new lines.

The Status of Gravity A widespread view has it that Newton was averse to metaphysical inquiry into the objects and results of the *Principia*, e.g., the ontology of gravity; and that when competitors took issue with his physics on metaphysical grounds, they were ultimately ill-advised to do so. In his paper, Andrew Janiak gives good reasons to resist this picture. His starting point is Newton's own attempts to make sense of his key result, in Book III, proposition 7, that "gravity is in all bodies universally." Namely, to elucidate its real semantic content, underlying ontology, and the broader metaphysical framework that supports them. As Janiak shows, this was no easy task, and Newton grappled with it at length. Nor was the answer clear and uncontroversial—not even to his fellow travelers like Roger Cotes, let alone to antagonists like Leibniz. Newton's struggles with this central question—what does it really mean to assert that gravity is universal—required him to make forays into the very metaphysics that he allegedly eschewed. Janiak's study has another, implicit benefit (in addition to turning the tables on the received view). In particular, it breaks a new path to better situating other, important figures (such as Emilie du Châtelet and Kant) in their own efforts to elucidate the meaning of gravity being universal.

Kant and Newton George's careful work on Newton's approach to evidence has yet another important benefit: it has opened a new vista on eighteenth-century philosophers' dialogue with the *Principia*.[6] Michael Friedman's paper takes a further step

---

[6] Studies of that dialogue are old, to be sure. However, before Smith they tended to be rather one-sided, or narrowly focused on the inertial-kinematic basis of the *Principia*: its metaphysics of space, time, and eighteenth-century reactions to it. Representative for that work are Earman 1989, Friedman 1992, and Rynasiewicz 1995.

on this new ground, by looking at Kant's response to these aspects of Newton. He thinks there are two significant aspects of this response. Kant strongly endorsed Newton's idea—which George was the first to highlight for us—that theory-mediated measurement is a very significant source of confirmation in gravitational dynamics. However, Friedman claims, Kant favored a notion of phenomena that is thicker than Newton's analogous concept. Specifically, Newtonian phenomena of motion were exclusively kinematic: patterns of planet- or satellite motion over time, equilibrium configurations, and the like. The notion Kant preferred, however, is that of phenomena as "involving causal and dynamical information" as well, not just kinematic content. Friedman argues in favor of Kant's stronger notion, because he thinks it does useful work in contemporary philosophy of science: it can help us chart a middle path in the disputes between realism and instrumentalism (about the relation of theory to its target objects).

The paper by Katherine Dunlop unfolds in the same register as Friedman's investigation above. Thereby, our image of Newton's reception by the philosophers after him becomes clearer. At a critical juncture in his argument for universal gravity, Newton in Book III relies on his third law of motion. That key move won him few friends, it seems. On the Continent, some objected that it asserts bodies to interact without contact, which they dismissed as unintelligible. Others, like Roger Cotes—and also Euler, later in the century—demurred that Newton lacked enough evidence to claim the law applies to actions at a distance. In her study, Dunlop uses Kant so as to cast a new, and more positive, light on Newton's move.

Kant endorsed action-at-a-distance early in his career, and so he had none of the Continentals' qualms about it. Accordingly, his reaction to Newton's *Lex Tertia* was different. For one, Dunlop explains, he argued that Newton relies on that law well before the master argument in Book III. In fact, he needs it already in Book I; specifically, in section 11, where Newton shows how to reduce the two-body problem (of two particles interacting as they orbit around each other) to two separate, more easily tractable one-body problems, viz. of a mass in Kepler motion around a fixed center of force. Kant thinks that Newton needs to assume the third law for his approach to go through, but that he did not acknowledge it as an explicit premise.

For another, Dunlop adds, Kant's natural philosophy helps elucidate some important but otherwise baffling claims by Newton. Kant distinguished between 'dynamical' and 'mechanical' treatments of force. The former is quantitative, or mathematical, but it has causal import—it regards forces as causes (of acceleration) that a source could exert even while stationary. From that vantage point, Dunlop argues, Newton's treatment of force in section 11 counts as "dynamical," hence causally relevant. That would resolve the apparent tension between his well-known claims that he treated gravity merely "mathematically," and yet that his treatment shows gravity to "really exist," and "suffice for" celestial and terrestrial motions.

Foundations of Classical Mechanics   The three laws of motion that Newton asserted purport to be general: they apply to forced motion beyond the relatively narrow class of gravitational effects. Many theorems in Book I—stating relations between accelerations and their corresponding orbits—are about forces other than inverse-square and placed at a focus (which gravity is). And, Newton in a famous passage conjectured that many things led him "to have a suspicion that *all* phenomena may depend on certain *forces*," and hoped his theorems and methods will help his posterity discover those further forces (Newton 2004, 60; emphasis added). To be sure, many after him did continue his agenda. Still, a mere century after the *Principia*, Lagrange created a genuine alternative to Newton's foundation. His book, *Mécanique Analytique*, unified statics and dynamics from a dual basis—a principle of virtual work, and a postulate known as "D'Alembert's Principle." In some respects, Lagrange's framework is more powerful than Newton's. The chapter by Sandro Caparrini takes a closer look at *Mécanique Analytique*. It shines a light on the layered structure of that key treatise, and on the early growth of the theory it contains. In Lagrange's lifetime, the book went through two editions; in the interim, French mathematicians extended greatly the reach of mechanics, to novel and difficult phenomena. Caparrini shows convincingly how Lagrange was able to incorporate those new advances into his framework, turning it into an even more formidable competitor for the tradition of theorizing that came out of the *Principia*.

Caparrini's study resonates with an important theme from George Smith, though not explicitly. In Katherine Brading's terms above, the theory in Lagrange's book is a rational mechanics, not a physics. Warrant for its results cascades downstream: from its basic, most general principles, by deductive reasoning alone. In *Mécanique Analytique*, empirical facts are very scarce, adduced mostly as illustration, not confirmation. That raises the weighty question, where does empirical evidence—especially strong evidence, of the kind that Smith has so fruitfully explored so far—enter Lagrange's "analytic" mechanics? The question is far from easy, to be sure, but it is to be hoped that scholars will take it up to grapple with it. Thereby, another theme from Smith would come to the fore, namely, that confirmation in advanced, strongly mathematized theories is *diachronic*: it is temporally extended, and rests on a historical record of accumulating, and increasingly stronger evidence for the theory.

Early Scientific Cosmology The mathematical astronomy in Ptolemy's *Almagest* rests inter alia on a small number of extra-mathematical assumptions—about what is at rest, what moves, and how far the stars are from us. Ptolemy there merely gestures at argument for these assumptions, or "hypotheses," as he calls them. His proffered support for them in that book is cursory and rash; which feeds the suspicion that the *Almagest* is a collection of simulation software, as it were: mere algorithms for predicting or retrodicting ephemerides and select orbital parameters—not a genuine "system of the world." The basis for that system would come from physics, we may think. In particular, from Aristotle's philosophical physics, which—thanks to its doctrines of natural place, motion, and five elements—easily entails that the earth is at rest in the world center while stars and planets revolve around it.

In his chapter, the late Noel Swerdlow subverts this received wisdom. He does so by way of a synoptic study of *Planetary Hypotheses*, an important tract by

Ptolemy. That work, Swerdlow argues, contains a theory of cosmology, or system of the world. But, it is not based on metaphysical premises from Aristotle. In fact, it is properly scientific. Specifically, it is quantitative: it makes claims about celestial distances and orbital parameters. Likewise, it is empirical: the inputs for theory building are empirical givens, e.g., long-term observational data or patterns of perception; and the evidence for the theory is empirical as well. And, Swerdlow suggests, it is supported by physical assumptions, e.g., about the causal mechanisms of planetary motion, and about the physical consequences of counterfactual setups. Thus, Swerdlow concludes, Ptolemy's *Planetary Hypotheses* has every right to count as a scientific cosmology; indeed, it was the first one of its kind. To help the reader, he ends with a synopsis of the complicated transmission and reception of Ptolemy's theory above.

There is a broader lesson here, and it lies at the confluence of two strands of thought. For one, George's painstaking work has shown inter alia not just how subtle Newton's methods for gathering evidence were—but also how easy it was for many figures after him to miss those methods. For another, the papers above show how unclear—and far from obvious or uncontroversial—the foundations of Newton's theory were, in the century after him: its ontology, basic semantics, and generic methods. Together, these two strands suggest a revisionist conclusion that challenges an interpretive consensus going back to Kuhn's *Structure*. In that influential book, Kuhn had argued that when paradigm-making work transforms its field (into an arena for "normal" science), it *ends* previous controversies (about the ontology and methods suited for that domain), and it produces consensus (about the basic objects, acceptable methods, legitimate problems, and criteria for solutions). Kuhn counted Newton's *Principia* among the epitomes for his case (Kuhn 1996, 11, 17). But, the chief results in the papers by Friedman, Solomon, Janiak, and Dunlop cast increasing doubt on Kuhn's picture of the *Principia*'s role in the history of exact science. Thereby, these results help clear the field for a new generation of scholars to step in and determine what Newton's book really did to physics and its philosophy—in effect, how eighteenth-century philosophers and their successors answered "Newton's challenge to philosophy" (Schliesser 2011).

This volume ends, appropriately, with George E. Smiths' reflections on the two crafts—the philosophy and history of exact science—that he has cultivated and fostered so admirably. The occasion for his reflections is the theme of revisiting accepted science. More specifically, the diachronic process whereby pieces of theory—once they become "accepted," or used constitutively to carry out further research—get tested over and over again, often with increased stringency; and the long-term outcomes of such "revisiting." He illustrates this process with examples, discussed in exquisite detail, from gravitation theory and late-classical physics. These examples, and others, support his concluding message—really, a dual lesson for students of science. Philosophers who investigate the epistemic aspects of science ought to pay close attention to the diachronic side of its credentials as knowledge: for any given theory, they must study the history of the evidence for it, in Smith's memorable phrase. And, historians of science ought to avoid dogmatic allegiance to the idea that social-group dynamics holds the master key to understanding the birth and growth of scientific knowledge. Rather, they would do well to pay attention to the extended record of testing and revisiting the epistemic credentials of that knowledge as it grew.

To both communities above, in sum, George emphasizes the crucial importance of *longue durée*, fine-grained study of the confirmation processes behind the production of scientific knowledge. These processes begin when a theory has been accepted, not *before*. And so, a corollary of his lesson is that we ought to revisit—and be prepared to drastically revise—Kuhn's old picture of normal science. Thereby, his lesson resonates with the dominant note of the papers in this volume.

De magistro

George's work is unique as well in a respect that makes it hard to present synoptically in any introduction, not just this one. He has conveyed much of his philosophy of science by a route that goes beyond the standard of our time, viz. the journal article or book chapter qua discrete, printed units of research. In particular, that route has been his legendary two-semester course on Newton's *Principia*— really, a master class in the history and philosophy of evidential reasoning developed at Tufts University but offered at a number of other institutions (including Stanford, Notre Dame, and Duke). Roughly, the first semester puts the student in a position to read the *Principia* by studying seventeenth-century primary texts (including Galileo, Kepler, Huygens, and pre-*Principia* Newton). The second is a close reading of the *Principia*, theorem by theorem; and then an overview of Newton's impact on mechanics after him.

The course is pitched to undergraduate students, but often the auditors include graduate students and faculty. For many years, it was offered on Wednesday evenings with a three-hour time slot interrupted by modest breaks. George *lectured* by partially reading from amazing lecture notes and using the blackboard when necessary. (The lecture notes would be made available after class, and after further careful editing.)

What made George's lectures mesmerizing was that he took all the students at all levels seriously as genuine interlocutors in the shared adventure of understanding Newton's method. And what made the whole point even more remarkable: many sessions would start with his excitement of his latest discovery—sometimes an overnight discovery—of the evolution and development of Newton's thought.

The paper assignments for students taking the course for a grade were all clearly designed to foster a collective endeavor to understand the evidential status of particular works at a given time. An assignment could read: "what was the status of Kepler's laws in 1680?" This could open the door to more metaphysical papers on what exactly the nature of a Keplerian law was in the late seventeenth century; or to examinations of seventeenth-century discussions of Kepler in astronomical texts of the period.

George's pedagogic methods center on very high expectations from his students by assigning challenging primary texts (and a lot of them) without flipping the classroom. What he does do, and he does this amazingly well, is prepare the student *qua* student to be a co-equal in his astounding intellectual adventure. He makes sure they acquire all the technical background, one firm step at the time, and then puts them in the position to contribute to active research, if they so wish. (Many students end up writing term papers that could be the basis of a journal article.) Subsequently, George would often come to co-author with students, by drawing on their specialized mathematical, linguistic, or research skills. In turn, his lectures along the years become enriched by what he learns or discovers while collaborating with them or grappling with their questions. Along the way, he invites them out on frigid winter nights to stare through a telescope, so as to experience what Galileo might have felt when he turned one toward the Moon. (George makes sure to let them try to see *anything* with the magnification that Galileo had available.) Through his course, which he has taught for nearly three decades, George has reached and influenced some four generations of research, from senior luminaries to current graduate students. Inter alia, the chapters in this volume attest to the enduring influence of his teachings.

Marius Stan
Chris Smeenk
Eric Schliesser

References

Buchwald, J.Z., and G. Smith. 2002. Incommensurability and discontinuity of evidence. *Perspectives on Science* 9: 463–498.

Earman, J. 1989. *World enough and space-time*. MIT Press.

Friedman, M. 1992. *Kant and the exact sciences*. Harvard University Press.

Kuhn, Th. 1996. *The structure of scientific revolutions*, 3rd ed. University of Chicago Press.

Miyake, T., and G.E. Smith. 2021. Realism, physical meaningfulness, and molecular spectroscopy. In *Contemporary scientific realism*, ed. T.D. Lyons and P. Vickers, 159–182. Oxford University Press.

Newton, I. 1999. *The* Principia: *Mathematical principles of natural philosophy*, ed. and trans. I.B. Cohen with A. Whitman. University of California Press.

Newton, I. 2004. *Philosophical writings*, ed. A. Janiak. Cambridge University Press.

Rynasiewicz, R. 1995. 'By their properties, causes and effects.' Newton's scholium on space, time, place and motion – II. The context. *Studies in History and Philosophy of Science* 26: 295–321.

Schliesser, E. 2011. Newton's challenge to philosophy: A programmatic essay. *HOPOS: The Journal of the International Society for the History of Philosophy of Science* 1: 101–128.

Publications by George E. Smith

Several engineering papers and more than 140 engineering reports issued by various government agencies, and by NREC and Turbomachinery Solutions, Inc.

1979a. [with M.L. Kean] Issues in core linguistic processing. *Behavioral and Brain Sciences* 2 (3): 469–70.

1979b. [with S.M. Kosslyn, S. Pinker, and S.P. Shwartz] On the demystification of mental imagery. *Behavioral and Brain Sciences* 2(4): 535–48.

1980. [with S.M. Kosslyn] An information-processing theory of mental imagery: a case study in the new mentalistic psychology. *PSA 1980*, Vol. 2, 247–66. University of Chicago Press.

1983. [with N. Daniels] The plasticity of human rationality. *Behavioral and Brain Sciences* 6(3): 490–1.

1986. The dangers of CAD. *Mechanical Engineering* 108(2).

1994a. [with S. Cohen, G. Smith, R. Chechile, and B. Cook] Designing curricular software for conceptualizing statistics. *Proceedings of the First Scientific Meeting of the International Association of Statistics Education*, eds. L. Brunelli and G. Cicchitelli, 237–45. University of Perugia.

1994b. [with S. Cohen, R. Chechile, F. Tsai, and G. Burns] A method for evaluating the effectiveness of educational software. *Behavior Research Methods, Instruments, and Computers* 26(2): 236–41.

1995. [with W. Harper] Newton's new way of inquiry. *The Creation of Ideas in Physics: Studies for a Methodology of Theory Construction*, ed. J. Leplin, 113–66. Kluwer/Springer.

1996a. [with S. Cohen and R. Chechile] A detailed, multisite evaluation of curricular software. *Assessment in Practice*, ed. Trudy W. Banta *et al.*, 220–2. San Francisco: Jossey Bass.

1996b. [with S. Cohen, R. Chechile, and G. Burns] Impediments to learning probability and statistics from an assessment of curricular software. *Journal of Educational and Behavioral Statistics* 21(1): 35–54.

1996c. [lead author; with L. Brown, R. Chechile, S. Cohen, R. Cook, J. Ennis, D. Garman, S. Lewis] *ConStatS: Software for Conceptualizing Statistics*. Englewood Cliffs: Prentice-Hall.

1996d. Chandrasekhar's *Principia*: an essay review. *Journal for the History of Astronomy* 27: 353–62.

1996e. [with E. Schliesser] Huygens' 1688 report to the VOC. *De Zeventiende Eeuw* 12: 196–214.

1997a. [with J. Z. Buchwald] Thomas S. Kuhn, 1922–1996. *Philosophy of Science* 64: 361–76.

1997b. J. J. Thomson and the electron, 1897–1899: an introduction. *The Chemical Educator* 2(6): 1430–71.

1998a. [with W. Harper] Isaac Newton. *Encyclopedia of Philosophy*. London: Routledge.

1998b. Newton's study of fluid mechanics. *International Journal of Engineering Science* 36: 1377–90.

1999a. The achievements of Book 2. I. Newton, *Mathematical Principles of Natural Philosophy*, trans. I.B. Cohen and A. Whitman, 188–194. University of California Press.

1999b. The motion of the lunar apsis. I. Newton, *Mathematical Principles of Natural Philosophy*, trans. I.B. Cohen and A. Whitman, 257–64. University of California Press.

1999c. Planetary perturbations: the interaction of Jupiter and Saturn. I. Newton, *Mathematical Principles of Natural Philosophy*, trans. I.B. Cohen and A. Whitman, 211–17. University of California Press.

1999d. Newton and the problem of the Moon's motion. I. Newton, *Mathematical Principles of Natural Philosophy*, trans. I.B. Cohen and A. Whitman, 252–57. University of California Press.

1999e. A puzzle in Book 1, Prop. 66, Coroll. 14. I. Newton, *Mathematical Principles of Natural Philosophy*, trans. I.B. Cohen and A. Whitman, 265–68. University of California Press.

1999f. How did Newton discover universal gravity? *The St. John's Review* XLV (2): 32–63.

2000a. Fluid resistance: why did Newton change his mind? *Foundations of Newtonian Scholarship*, eds. R. Dalitz and M. Nauenberg, 105–36. Singapore: World Scientific.

2000b. [with D. Mindell] The emergence of the turbofan engine. *Atmospheric Flight in the Twentieth Century*, eds. P. Galison and A. Roland, 107–55. Dordrecht: Kluwer.

2001a. The Newtonian style in Book 2 of the *Principia*. *Isaac Newton's Natural Philosophy*, eds. J.Z. Buchwald and I.B. Cohen, 249–97. MIT Press.

2001b. [with I.B. Cohen, A. Whitman, and J. Budenz] Newton on fluid resistance in the first edition: English translations of the passages replaced or removed in the second and third editions. *Isaac Newton's Natural Philosophy*, eds. J.Z. Buchwald and I.B. Cohen, 299–313. MIT Press.

2001c. J.J. Thomson and the electron, 1897-1899. *Histories of the Electron: The Birth of Microphysics*, eds. J.Z. Buchwald and A. Warwick, 21–76. MIT Press.

2001d. La scoperta dell'elettrone [Discovery of the electron] *Storia della scienza*, Vol. 7, *La scienza dell'Ottocento*. Istituto della Enciclopedia Italiana.

2001e. Review of J.R. Christianson's "On Tycho's Island: Tycho Brahe and His Assistants, 1570–1601." *Journal of Interdisciplinary History* 32: 130–1.

2001f. Comments on Ernan McMullin's "The impact of Newton's *Principia* on the philosophy of science." *Philosophy of Science* 38: 327–38.

2001g. [with J.Z. Buchwald] An instance of the fingerpost: review of M. Christie, "The Ozone Layer: A Philosophy of Science Perspective." *American Scientist* 89(6): 546–9.

2002a. From the phenomenon of the ellipse to an inverse-square force: why not? *Reading Natural Philosophy*, ed. D. Malament, 31–70. La Salle: Open Court.

2002b. The methodology of the *Principia*. The *Cambridge Companion to Newton*, eds. I.B. Cohen and G.E. Smith, 138–73. Cambridge University Press.

2002c. [with I.B. Cohen] *The Cambridge Companion to Newton*. Cambridge University Press.

2002d. [with I.B. Cohen] Introduction. *The Cambridge Companion to Newton*, eds. I.B. Cohen and G.E. Smith, 1–32. Cambridge University Press.

2002e. [with I. B. Cohen] Newton and the lunar motion. Review of N. Kollerstrom, *Newton's Forgotten Lunar Theory. Journal for the History of Astronomy* 33: 212–3.

2002f. [with J. Z. Buchwald] Incommensurability and discontinuity of evidence. *Perspectives on Science* 9: 463–98.

2003. Review of "Meanest Foundations and Nobler Superstructures: Hooke, Newton, and 'The Compounding of the Celestiall Motions of the Planetts.'" *Physics Today* 56(9): 61–2.

2005a. [with S.R. Valluri, P. Yu, and P.A. Wiegert] An extension of Newton's apsidal precession theorem. *Monthly Notices of the Royal Astronomical Society* 358: 1273–84.

2005b. Was wrong Newton bad Newton? *Wrong for the Right Reasons*, eds. J.Z. Buchwald and A. Franklin, 127–160. Springer.

2006. The *vis viva* dispute: a controversy at the dawn of dynamics. *Physics Today* 59(10): 31–6.

2007a. Isaac Newton. *The New Dictionary of Scientific Biography*, ed. Noretta Koertge, 48–53. Farmington Mills, MI: Charles Scribner's Sons.

2007b. Isaac Newton. *Stanford Encyclopedia of Philosophy*. (http://plato.stanford.edu)

2007c. Newton's *Philosophiae Naturalis Principia Mathematica. Stanford Encyclopedia of Philosophy*. (http://plato.stanford.edu)

2008. [with J. Maienschein] What difference does history of science make, anyway? *Isis* 99: 318–21.

2009. [with J. W. Dauben and M. L. Gleason] Seven decades of history of science: I. Bernard Cohen (1914–2003). *Isis* 100: 4–35.

2010. Revisiting accepted science: the indispensability of the history of science. *The Monist* 93: 545–79.

2012. How Newton's *Principia* changed physics. *Interpreting Newton*, eds. A. Janiak and E. Schliesser, 360–95. Cambridge University Press.

2013. On Newton's method. Book symposium on W. Harper's "Isaac Newton's Scientific Method." *Metascience* 22: 215–246.

2014. Closing the loop: testing Newtonian gravity, then and now. *Newton and Empiricism*, eds. Z. Biener and E. Schliesser, 262–351. Oxford University Press, 2014.

2016a. [with R. Iliffe] eds. *The Cambridge Companion to Newton*, second edition. Cambridge University Press.

2016b. [with R. Iliffe] Introduction. *The Cambridge Companion to Newton*, second edition, 1–32. Cambridge University Press.

2019. Newton's numerator in 1685: a year of gestation. *Studies in History and Philosophy of Modern Physics* 68: 163–77.

2020a. [with Raghav Seth] *Brownian Motion and Molecular Reality: A Study in Theory-Mediated Measurement.* Oxford University Press.

2020b. Experiments in the *Principia*. The *Oxford Handbook of Newton*, eds. E. Schliesser and Chr. Smeenk. Oxford University Press.

2020c. The *Principia*: from conception to publication. The *Oxford Handbook of Newton*, eds. E. Schliesser and Chr. Smeenk. Oxford University Press.

2020d. Newton's laws of motion. The *Oxford Handbook of Newton*, eds. E. Schliesser and Chr. Smeenk. Oxford University Press.

2021a. [with T. Miyake] Realism, physical meaningfulness, and molecular spectroscopy. *Contemporary Scientific Realism and the Challenge from the History of Science*, ed. T. Lyons and P. Vickers, 159–80. Oxford University Press.

2021b. [with the assistance of J.M. Musca] Du Châtelet's commentary on Newton's *Principia*: an assessment. *Epoque Emilienne*, ed. R. Hagengruber, 255–310. Springer.

Forthcoming, a. Newtonian relativity: a neglected manuscript and understressed corollary. *Newtonian Relativity*.

Forthcoming, b. [with E. Schliesser] Huygens' 1688 report to the directors of the Dutch East India Company on the measurement of longitude at sea and the evidence it offered against universal gravity. *Archive for History of Exact Sciences.*

In preparation, a. *Isaac Newton's De Motu Corporum, Liber Secundus: a Variorum Translation of the Manuscript and Related Manuscripts from 1685*, trans. G. E. Smith and A. Whitman, with the assistance of R. Strobino, and with commentary by S. Hesni and G.E. Smith, in preparation.

In preparation, b. [with Chr. Smeenk]. Newton on constrained motion: a commentary on Book 1, Section 10 of Newton's *Principia*. *Archive for History of Exact Sciences.*

Contents

Editors and Contributors

About the Editors

Marius Stan is Associate Professor of Philosophy at Boston College. He works in history and philosophy of science, with an emphasis on eighteenth century physics. He is co-author, with Katherine Brading, of *Philosophical Mechanics in the Age of Reason* (Oxford), and author of *Kant's Natural Philosophy* (Cambridge).

Christopher Smeenk is Professor of Philosophy and Director of the Rotman Institute at Western University, Ontario. He works in general philosophy of science, philosophy of cosmology, and history of physics. In addition to numerous articles, he is co-editor of the *Oxford Handbook of Newton* (with Eric Schliesser) and co-author of *The Aim and Structure of Cosmological Theory* (with James Owen Weatherall).

Contributors

Jody Azzouni Tufts University, Medford, MA, USA

Zvi Biener University of Cincinnati, Cincinnati, OH, USA

Katherine Brading Department of Philosophy, Duke University, Durham, NC, USA

Sandro Caparrini Politecnico di Torino, Chestnut Hill, MA, USA

Robert DiSalle Department of Philosophy, Western University, London, ON, Canada

Mary Domski University of New Mexico, Albuquerque, NM, USA

Katherine Dunlop University of Texas at Austin, Austin, TX, USA

Craig W. Fox Edelstein Center for History and Philosophy of Science, Technology and Medicine, Hebrew University of Jerusalem, Jerusalem, Israel

Allan Franklin University of Colorado, Boulder, Boulder, CO, USA

Michael Friedman Stanford University, Stanford, CA, USA

Niccolò Guicciardini Università degli Studi di Milano, Milano, Italy

Andrew Janiak Department of Philosophy, Duke University, Durham, NC, USA

Teru Miyake School of Humanities, Nanyang Technological University, Singapore, Singapore

Eric Schliesser University of Amsterdam, Amsterdam, the Netherlands

George E. Smith Tufts University, Medford, MA, USA

Monica Solomon Department of Philosophy, Faculty of Humanities and Letters, Bilkent University, Ankara, Turkey

Howard Stein Department of Philosophy, University of Chicago, Chicago, IL, USA

N. M. Swerdlow University of Chicago, Chicago, IL, USA
California Institute of Technology, Pasadena, CA, USA

Chapter 1
Smith, Smith and Seth, and Newton on "Taking to Be True"

Jody Azzouni

1.1   Introduction

Taking (a proposition) to be true is an epistemic theme appearing throughout George E. Smith's work; this includes his marvelous new book with Raghav Seth, *Brownian motion and molecular reality: A study in theory-mediated measurement* (2020; hereafter Smith and Seth). They use this notion to categorize changes in scientific perspectives both towards the ontological claim that molecules exist and towards molecular-kinetic theory. They illustrate the shift in viewpoint occurring over the successive editions of Ostwald's and Nernst's respective textbooks. Nernst writes, in his 6th edition, that, "the theory begins to lose its hypothetical character"; later he writes, "... we may well acknowledge that the theory has lost its hypothetical character" (Smith and Seth, 3). Ostwald does an "about-face on atomism" (2): In earlier editions he characterizes the "atomic conception" as a "convenient mode of representation," as having "enormous heuristic value" (2). In the 4th edition (1909) he writes (2),

> I am now convinced that we have recently become possessed of experimental evidence of the discrete or grained nature of matter, which the atomic hypothesis sought in vain for hundreds and thousands of years.... [The most cautious scientist is now justified in] speaking of the experimental proof of the atomic nature of matter. The atomic hypothesis is thus raised to the position of a scientifically well-founded theory, and can claim its place in a text-book intended as an introduction to the present state of our knowledge of General Chemistry.

He writes more strongly in 1909 about molecular reality (Smith and Seth, 336): "... it appears that *the final proof of the grained or atomistic-molecular nature of*

J. Azzouni (✉)
Tufts University, Medford, MA, USA

matter has been obtained, after a fruitless search during a whole century" (Italics in the original).[1]

Smith and Seth characterize the about-faces as shifts from these propositions having a hypothetical status to being *taken to be true*. This, in scientific contexts, is a "standing" that can be temporary because further testing of such propositions can quickly move them—although it needn't—into a different, more entrenched, standing.[2]

They draw a *tripartite* epistemic distinction: *mere-hypothesis-status, being taken to be true,* and lastly, progressive "entrenchment." Smith and Seth (382) write (emphasis theirs):

> The [taken-to-be-true] standing . . . is at the center of the second distinction that we think may be of some benefit in the debate between realism and instrumentalism. It is a threefold distinction among (1) a proposition, whether constitutively presupposed in ongoing research or not, continuing to have the status of a hypothesis; (2) a proposition being taken to be true and granted a presumption of inviolability in ongoing research predicated on it; and (3) a proposition that is accruing evidence and thereby becoming increasingly entrenched as a result of continuing to be tested, *en passant* yet often stringently, in conjunction with research predicated on it. The latter half of the distinction provides a response to historians and sociologists who challenge the epistemic authority of the sciences by stressing how often their most fundamental principles have been accepted on the basis of little or no evidence.[3]

Smith and Seth (2020) is largely restricted to the historical period from around 1903 until the 1911 Solvay conference. It thus officially covers the period during which (according to Smith and Seth) the standing of molecules and molecular-kinetic theory shifted from *hypotheses* to being *taken to be true*. Their primary focus is on the role in this episode of Perrin's Brownian motion research; the book only briefly indicates the explosion of experimental results that (presumably) subsequently sealed their status as entrenched.[4] Throughout (and with great subtlety)

---

[1] His remarks on the kinetic theory of heat are more measured (Smith and Seth, 335): "If the movements of a small particle suspended in a liquid are calculated on the basis of this [kinetic theory] hypothesis, the agreement with the movements actually observed is so close that we are compelled to regard this agreement as a fairly satisfactory proof of the kinetic nature of heat."

[2] See Smith and Seth, 321. "Fleeting" is how they describe the moment "at the end of 1912 and beginning of 1913" during which "the molecular and molecular-kinetic hypotheses had just the standing [taken to be true] we are attributing to them and nothing beyond it . . . ." (358).

[3] Smith's suggestion isn't that "taking to be true" is a special *semantic* status for propositions—in between truth and falsity. It's a special *epistemic* status.

[4] Ironically (as Smith and Seth show), it quickly emerged that Perrin's Brownian-motion values for Avogadro's number deviated from 5% to 10%, or more, from the results of Millikan, Planck, Rutherford and Geiger. His values weren't used—notably by Bohr—in subsequent research. Smith and Seth (277) mention that Perrin's 1911 paper for the conference doesn't—italics theirs— "even note the *systematic* difference between his values for Avogadro's number from Brownian motion and all the comparably well-founded values with which he directly compared his in the paper. Instead he compares them one-by-one, never raising the somewhat obvious question, once the systematic difference is noted, whether it might be pointing to some systematic error in his results." Smith and Seth (131) distinguish between N_0, "the number of molecules in a mole of

Smith and Seth raise questions like: Why didn't the theory of Arrhenius being confirmed end "the hypothetical standing of that theory?" And they answer (72):

> Insofar as what atoms amount to and how they combine to form enduring molecules was still almost entirely a matter of conjecture, to an even greater extent was how an atom could be or become electrically charged.

In this paper, I worry about distinguishing "taking to be true"—both Smith's notion and Newton's original that Smith takes himself to be borrowing—from the *ordinary* notion of knowledge. I stress the word "ordinary," in "ordinary notion of knowledge" because the historically shifty and wavering *philosophical* notion of knowledge isn't the same as the contemporaneous ordinary (folk) notion of knowledge. The ordinary notion (the one we use, and necessarily use, when we speak) derives from the natural-language word "know." This word is a lexical universal that's (additionally) rigid in its properties: it's immune to conceptual engineering (modifications); see Azzouni (2020), especially § 11.4. The philosophical notion, therefore, not inaccurately, should be described as a historically-changing sequence of misunderstandings of the ordinary notion or (more charitably) attempted conceptual engineerings of it that, for lexical reasons, fail. I'll also speculate a little about how the philosophical notion—especially the (sometimes wavering, or inconsistently held) view that knowledge requires infallibility or certainty—influenced Descartes and Newton as well as Smith and Seth.[5]

The key issue is fallibility: Are knowers fallible (as most epistemologists think today) or infallible, as—arguably—Descartes thought, and as those both influenced by him (and opposed to him) probably did in the immediate couple of centuries after he wrote?[6] By knowers being "infallible," I mean: it isn't possible for knowers to

---

gas at standard conditions—and the symbol N, for the number of granules needed for their mean translational energy to sum to 3RT/2—that is, for what Perrin called 'Avogadro's number' in his statement of the results of his Brownian motion experiments." We *still* don't know why Perrin's values deviate; it's (Smith and Seth, 132, 323–4, 325–6, and elsewhere, suggest) likely because N_0 and N don't match—probably because of macroscopic features of granules (and/or their relatively sparse interactions between one another, and/or because of the effects of the viscosity of the fluid on the surfaces of the granules) all of which differ significantly from how the molecules themselves interact. See, in particular, Smith and Seth's discussion (177, note 5), where the hypothesis is found wanting that inaccuracies in Perrin's measurements of the sizes of the granules is the source of the deviations. Perrin provided an experimentally good measure of N, but (alas) not of N_0!

[5] When philosophers (historically and today) talk about "know(s)" and "knowledge" they mostly take themselves as speaking in the vernacular (rather than having coined specialized terminology as they self-consciously do with other words); however, they place conditions on "know(s)," and the like, that these words don't have in ordinary usage—necessary conditions like awareness of the knowledge had or infallibility of the knowing agent towards what they know. Philosophers are thus unaware of how much their usage deviates from the vernacular: they think these really are necessary conditions on these (ordinary) words. Again, see Azzouni (2020) for details about this and for other aspects of "know(s)."

[6] This is controversial: I'll make some further remarks later.

be wrong.[7] If knowers are fallible then, I'll show, we can't distinguish presumed *knowledge* from Smith's "*taking* to be true." I'll speculate, further, that if Newton originally used phrases equivalent to "taking to be true," as Smith claims, to indicate a special epistemic status apart from knowledge, then that's because—like Descartes (and maybe *because* of Descartes)—he thought knowledge required infallibility, and he thought the propositions established in science *apart from those that can be "deduced" from observations* weren't established infallibly.[8]

1.2   Smith (and Smith and Seth) on *Taking to Be True*

Smith and Seth (346), and Smith elsewhere, attribute "take" to Newton, as used in his fourth rule of reasoning (I quote their translation):

> In experimental philosophy, propositions gathered from phenomena by induction should be taken to be true, either exactly [*accuraté*] or very nearly [*quamproximé*], notwithstanding any contrary hypotheses, until yet other phenomena should render such propositions either more exact [*accuratiores*] or liable to exceptions.[9]

---

[7] There are cognitive traps in the ordinary word "possible." Since knowing *p* implies *p*, knowers can't be wrong insofar as: if they know *p* then *p* is true. What's meant is that the methods by which knowers come to know what they know aren't (potentially) misleading (Azzouni (2020), section 10.2).

[8] Eric Schliesser (email, April 24, 2018) notes that scales of epistemic confidence were common in the early modern period, e.g., (in Hume) *probability*, *proof*, and finally *demonstration*, where "proofs" always involve causal claims. And for Huygens, "probability bordering on certainty"—see Sect. 1.4—is "moral certainty" which is distinguished from demonstrative certainty and mere probability. I speculate that these epistemic distinctions would have all been uncontroversially described as types of "knowledge" had not the infallibility of knowledge not been so influential an assumption. I have an alternative hypothesis for why Newton adopted the Latin phrase that Smith and Seth translate as "take to be": to avoid Moorean-style infelicities. See note 37.

[9] "Taken to be" isn't in the translation of the 3rd edition of the *Principia* by I.B. Cohen and Whitman. Instead (796): "*In experimental philosophy, propositions gathered from phenomena by induction should be considered either exactly or very nearly true notwithstanding any contrary hypotheses, until yet other phenomena make such propositions either more exact or liable to exceptions.*" This formulation with "consider," perhaps, can be taken to come to the same thing as the formulation with "taken to be." Smith (2014, 272) puts Rule 4 this way: "*In experimental philosophy, propositions gathered from phenomena by induction should be regarded as either exactly, or very, very nearly true notwithstanding any contrary hypothesis, until yet other phenomena make such propositions either more exact or liable to exceptions.*" Smith adds (2014, 273), "that the main verb in [this rule] is 'should be regarded'—in Latin, a form of the verb *habere*, "to hold." Smith's thought: These rules aren't saying that 'propositions gathered from phenomena by induction' *are* exactly or very, very nearly true universally, but that they should be nevertheless *treated* as so. Arnold Koslow writes—email, April 27, 2018—that the Motte translation uses, instead of "considered to be true," "esteemed to be true"; and—email, April 29, 2018—that in the Motte translation as revised by Cajori (1947), we read "In experimental philosophy we are to look upon propositions . . . as accurately or . . . ."

Smith and Seth then write (346):

> The propositions to be taken to be thus true as of 1913 were: (1) All moles of any substance contain the same definite number of chemically demarcated molecules, which to current approximation is around 64×10^{22}; (2) the molecules in all gases have the same definite mean translation kinetic energy per degree of absolute temperature, which to current approximation is around 1.95×10^{-16} ergs per degree Kelvin; and (3) there exists a definite fundamental, universal unit of negative charge, which to current approximation is around 4.5×10^{-10} esu, and it is always joined with a definite rest mass, the combination constituting the *electron*, with the number of them required to liberate a mole of molecules of any monovalent substance in electrolysis . . . always the same.

Statements like "I take A to be true," or "scientists now take A to be true," *in the vernacular*, aren't transparently interpretable as Smith and Seth do. They're most naturally understood as exclamatory renditions of "A is true" or "scientists have established A," or the like. Smith and Seth (383) disagree. They write: "The phrase 'take to be true' alone is enough to indicate that the evidence required for a proposition to acquire this intermediate standing is something short of evidence that it is true."

Consider, however, this use of 'take to be true' by Smith and Seth themselves:

> Granted, authorities were telling the scientific community that the molecular hypothesis should now be taken to be true, and hence criticizing it and proposing alternatives to it were no longer in order. (7)

It's hard not to read this use of "take" as describing an authoritative *command*, rather than as indicating that a distinctive non-knowledge epistemic status for a hypothesis has emerged—so that "taken to be" is operating as "regarded as" would be. Natural reading: Scientific authorities were saying at that time: This is true, and (if you're a sensible member of the scientific community) you should act and believe accordingly![10]

This is why Smith and Seth (and Smith elsewhere) need to explain *carefully* what they take "take to be true" to have meant to Newton and what we're to take it to mean to us if we're to apply it to historical shifts in the perspectives of scientific disciplines towards scientific propositions as Smith and Seth mean us to. (In the previous sentence, I further illustrated the natural use of "take," but with respect to "mean." I'll similarly use/illustrate "take" later in this paper).

What follows are descriptions of some of the important properties of "taking to be true," as Smith, Smith and Seth (and Newton, so they claim) understand it.

---

[10] Compare: "You should take it to be true that there is a God," said to someone (say) by an official of the Inquisition." Notice that if we were to urge (or command) someone to believe in this way, or if we talk about what he "should" believe, "take" is unavoidable. "Look at this evidence! You should really take yourself to know he's the murderer." Otherwise, we have say things like: "Come on—you *know* he did it." The surface semantic flavor of "considered" (see the second paragraph of note 9) suggests even less an avoidance of a commitment to something as a known truth—this is also true of the archaic "esteemed" and the phrases "look upon," and "should be regarded." (I can't judge the original Latin, although what I say is true of the literal "held as" in English).

First: "Taking to be true" is generic, applying in principle to any scientific proposition (351–9). I presume Smith and Seth understand the phase as syntactically unrestricted, and so (in principle) applicable to propositions as widely as "believe(s)" or "know(s)" are, or as the unadorned adjectives "true" and "false" are.

Second: Professionals in a scientific field taking a proposition to be true isn't a matter of those professionals *themselves* believing something specific about it— thinking they don't know (or believe) it, for example, but thinking that it's now acceptable to use it in further research. Smith (2010, 546) writes: "Being accepted, as I use the term, has nothing as such to do with what individual scientists happen to believe." He writes, further, "Being accepted, as I use the term here, has only to do with what the community is prepared to take for granted, perhaps tacitly, in ongoing research, and not with how individuals in the community thought about the matter." This is also true of the mere-hypothesis-status of scientific propositions.[11]

Third, taking a scientific proposition to be true licenses its use in experimental design *constitutively*.[12] Smith and Seth write:

> For us, a proposition enters constitutively into a research result just in case either the statement or the evidential reasoning supporting the result indispensably presupposes it. Put another way, the result ceases to be warranted by the putative evidence without the proposition in question. Correlatively, a proposition does not enter constitutively into a result if it can be rejected or eliminated while leaving the result and the evidence for it intact. (378)

One achievement of Smith and Seth's book is that they show molecular-kinetic theory and the hypothesis of molecules—despite being employed by Perrin at crucial points in his evidential *reasoning* (e.g., when describing the design of one of his experiments)—are *dispensable* in the sense quoted above: they aren't (despite their presence in Perrin's reasoning) playing a constitutive role in the design of any of his experiments. (See, specifically, Smith and Seth's section 4.3). To my

---

[11] Smith and Seth (73) write, "Although we have not surveyed the literature to verify it, we are confident that a large majority of the research community of physicists had ceased questioning the atomic and molecular composition of matter, if not before, then surely after the unqualified experimental demonstration of van't Hoff's claim that the gas law extends to liquid solutions." They write later (73–74): "Granting that such a large fraction of the relevant research communities had ceased questioning the reality of molecules by 1900, however, obligates us to end the chapter with a response to the question, *By what authority did the reality of atoms and molecules and their relation to heat still have only the status of conjectured hypotheses in 1900?*" (Their italics). They answer this question by adopting the stance of a book-length review article (written at that time) that would have objectively evaluated the evidence available for these hypotheses (26–27). This fruitful historical strategy is their invention (as far as I know).

[12] The quotation from Smith and Seth (346) that I gave above about the propositions taken to be true as of 1913 continues thus (italics theirs): "These propositions are to be presupposed as fully established in ongoing research—that is, they are licensed to enter *constitutively* into the design of experiments and rules for measurement—until yet other phenomena should render them either more exact or liable to exceptions."

knowledge, the evidential independence of Perrin's Brownian motion results from the molecular hypothesis and from molecular-kinetic theory is made clear for the first time, in their book.[13]

Fourth, and finally, taking a proposition to be true grants it a protective "innocent until proven guilty" status (350), or (as Smith and Seth also put it (352)), a "presumption of inviolability." Smith and Seth write, contrasting the status of a proposition taken to be true with one that has the status of a mere hypothesis:

[W]hen the research is predicated on a claim that still has the status of a hypothesis, its continuing failure puts the hypothesis itself into dispute. When, by contrast, it is predicated on a claim for which evidence has provided adequate grounds for it to be taken to be true, it does not. It then has the status like the legal presumption of innocence until proven guilty beyond reasonable doubt. Decisive evidence is required even to raise doubts about its truth. This, of course, was the whole point of Newton's introducing the phrasing, *take to be true, either accuraté or quamproximé, not withstanding any contrary hypotheses*, in his fourth rule. (350; their italics)

Smith and Seth quote Euler with approval, taking what he says to be an illustration of "taking to be true":

... since M. CLAIRAUT has made the important discovery that the movement of the apogee of the Moon is perfectly in accord with the Newtonian hypothesis ..., there can no longer remain the least doubt about this proposition .... One can now maintain boldly that the two planets Jupiter and Saturn attract each other mutually in the inverse ratio of the squares of their distance, and all irregularities that can be discovered in their movements are infallibly caused by this mutual action .... *And if the calculations that one claims to have drawn from this theory are not found to be in good agreement with the observations, one will be always justified in doubting the correctness of the calculations, rather than the truth of the theory* (350f.; their emphasis).[14]

This fourth property raises worries about how the notion of "taking to be true" differs from, say, "known," "taking to be known," "taken to be known to be true," and the like. The worry—especially if "taking to be true" induces a "presumption of inviolability"—is that it looks like a presumption of knowledge can only be distinguished from "taking to be true" if knowledge requires, beyond inviolability, *infallibility*. If knowledge is fallible, nothing in the above four conditions distin-

---

[13] It's not obvious that *scientists* realized this at the time, *or later*.

[14] Significantly, Euler doesn't just confirmationally sideline *mathematical calculation*; he does the same to metaphysics in a passage (apparently) brought to attention by Arnold Koslow in the 1970s. See Schliesser (2011, 104–105), on Euler, and on the Newtonian challenge to metaphysics that began to be posed by about 1700 on the basis of what was seen as the *knowledge* that was encapsulated in Newton's mechanics.

guishes "taking to be true" from *knowledge—tout court*.[15] I turn to exploring this next.

1.3 Descartes' *Clarity and Distinctness*: "Deduction" (for Him) Is a Gold Standard for Knowledge

It grossly understates what Descartes thought he'd discovered to call the criterion of *clarity and distinctness* a decision procedure for short surveyable inferences. One reason for this is that built into his criterion is the transparency of mental states (or of some mental states, anyway): one isn't only aware of the topics focused on by one's mental state (e.g., that $1 + 1 = 2$) but of the mental state itself (one's recognition of the clarity and distinctness of one's grasp of this mathematical proposition).[16] Leave aside this metacognitive requirement—it's important but doesn't bear directly on the issues of this paper. Another reason that calling the criterion a decision procedure understates what Descartes thinks he has found is that decision procedures (think of truth tables) even for short inferences are open to executional failures by *anyone*—regardless of their competence and/or despite their presence of mind—and for reasons unrecognized at the time of execution. What, more strongly, Descartes offers is a "rule for the direction of mind" that *can't go wrong* if it's employed sincerely by anyone who isn't insane (who doesn't think, for example, that he's a gourd).

The same is therefore thought true of consistency—our (competent) recognition of short and surveyable consistent propositions: There is (it follows) a *decisive test* for the recognition of contradiction—at least for short surveyable propositions. Hume invokes just this criterion to establish his celebrated result about induction. I quote from *An Enquiry Concerning Human Understanding* (22):

> All reasonings may be divided into two kinds, namely demonstrative reasoning, or that concerning relations of ideas, and moral reasoning, or that concerning matter of fact and existence. That there are no demonstrative arguments in the case, seems evident; since it implies no contradiction, that the course of nature may change, and that an object, seemingly like those which we have experienced, may be attended with different or contrary effects. May I not clearly and distinctly conceive that a body, falling from the clouds, and which, in all other respects, resembles snow, has yet the taste of salt or feeling of fire? Is there any more intelligible proposition than to affirm, that all the trees will flourish in DECEMBER and JANUARY, and decay in MAY and JUNE? Now whatever is intelligible, and can

---

[15] Chris Smeenk asks: "[Smith] wants to accord certain hypotheses the status of "taken to be true," while also acknowledging fallibility. Newton's characterization of gravity and the entire framework of the *Principia* are still "working hypotheses" that may be incorrect on his account; yet Euler's epistemic attitude of taking the laws to be true is still ... appropriate. Is it your position that there's simply no middle ground for Smith (or Newton, or ...) to occupy?" My answer: Yes, if knowledge is fallible then this middle ground is occupied by *knowledge*. As I'll discuss with respect to Kripke's dogmatism paradox in section 5, gathering more evidence for what we nevertheless know isn't merely permissible; it's *rational*. It's rational even in the case of *deductions*.

[16] Gaukroger (1995, 350).

be distinctly conceived, implies no contradiction, and can never be proved false by any demonstrative argument or abstract reasoning *a priori*.

If transparently available *recognition* of clarity and distinctness—or something similar—is a standard for knowledge, then infallibility follows.[17] We can't be wrong if we competently take ourselves to know something. Descartes doesn't establish this corollary via what "know(s)" means (that is, he doesn't establish it by considerations of ordinary usage). Rather, it follows from a supposed fact about a method he thinks we have that's imposed as a standard: deductions *executed in a certain way* suffice for knowledge. This is how I explain, interpreting the reasoning in the above passage, why Hume thinks we never know any generalizations based on induction: we never know more than what we've directly experienced of the instances of those inductions (e.g., so many risings of the sun).[18]

Caveat: We mustn't understand the early modern notion of "deduction" as we contemporaries understand *our* notion of it. For Descartes (as *we* have to put it) *premises* can be *deduced*: the *cogito* is an example. The gold standard for knowledge I'm attributing to Descartes isn't purely inferential, in the contemporary sense—e.g. inferences *from* B *to* C. The same is true of Hume's idea of the recognition/deduction of mathematical/logical truths, as the quotation from him makes clear; and I'll claim next that the same is true of Newton too. This helps explain why Newton *literally thinks* that, using secure mathematics, generalizations can be *deduced* from phenomena.

1.4 Taking to Be True, Deduction from Phenomena, and Newton

Newton famously disdained hypotheses. Smith and Seth (351) tell us (also Smith (2010), and other work) that "the principle distinction Newton was making with his fourth rule was between 'hypotheses', for which he had long had a low regard (to say the least), and 'propositions gathered from phenomena'."[19] They write:

---

[17] It follows, anyway, if immediate experience is the other (and only other) gold standard for knowledge. I omit further discussion of this.

[18] Hume's result about induction survives the rejection of there being a general *test* (Church's theorem) for contradictoriness (requiring a general recognition procedure for *consistency*) because, although there's no general test for contradictoriness, there are many successful recognition procedures for *specific* sets of consistent propositions—e.g., the set of sentences containing the instances $Pa_1, \ldots Pa_n, \ldots$ and $\neg(x)Px$. Eric Schliesser—email, April 29, 2018—points out that my above description is anachronistic: what's at issue for Descartes and Hume is the inspection of ideas and not methods of induction/deduction. That's *right*; and this difference *really matters*: See the very next paragraph, and (for that matter) the next section.

[19] Smith and Seth (351) point out that "the hypothesis [Newton] had in mind was the view that planets are carried around in their orbits by fluid vortices, hypothesized to provide a contact mechanism to counteract their centrifugal tendency," on the one hand, and his (inductively

[Newton] had come to use the phrase, "take to be true," between the first and second editions of the *Principia*, initially in licensing the inductive leap from the evidence that his law of gravity holds for the planets to taking it to hold universally for all matter. Two considerations appear to have led him to do so: the realization that in any such leap to a sweeping generalization, one is always simply *taking* the general claim to be true; and his coming to appreciate the potential for much stronger evidence for his theory of gravity than he could muster in the *Principia* if others in the future were to use it, instead of relying on observation, to pin down the enormous systematic complexities of planetary motion that result from their mutual interactions. (351; italics theirs)

I detect three claims here about Newton's methodology. The first claim is that he draws a distinction between kinds of *inductive generalizations*.[20] Some are mere hypotheses. There are no (or few) indications of how they can be evidentially supported by subsequent research against competitor hypotheses. Other inductive generalizations aren't mere hypotheses: they have "the potential for much stronger evidence." The second claim is that, for Newton, the second kind of inductive generalization involve a *taking* of the general claim to be true." Mere hypotheses have no potential for graduating to this status—there can be no "gathering from phenomena" for *them*. The other kind of inductive generalizations have this potential—in Smith and Seth's words, "if others in the future were to use it, instead of relying on observation, to pin down the enormous systematic complexities of planetary motion that result from their mutual interactions."

The third claim is that knowledge of a proposition *p*—*at a time*—isn't had if subsequent "much stronger evidence" might subsequently be available for it. This claim can only be connected to the second claim by the thought that if the possession of knowledge of *p* yields infallibility, then of course, wanting or needing *additional* evidence for *p*, given knowledge of it, is senseless.

The rest of this paper discusses this purported taking-to-be-true status; but I want to use the remainder of this section to lay out a *hypothesis* (hopefully not a *mere* hypothesis) about the source of Newton's distinction between the mere hypotheses that he refuses to feign and the better generalizations "gathered from phenomena" that he embraces. I claim the distinction is rooted in his view of deduction—*one he shares with Descartes*. The first textual point supporting this is from the quotation from the General Scholium in note 20: his phrase "deduce from phenomena." Claim:

---

established) law of gravity, on the other. Newton also apparently didn't think his corpuscular view of matter was a hypothesis in this sense (see Shapiro 2002, 249). This distinction fuels his irritation towards Pardies, and his attitude towards Hooke. (See Westfall (1980, 242), and later pages).

[20] What follows the statement of the fourth rule—Newton (1999, 796)—is: "This rule should be followed so that arguments based on induction may not be nullified by hypotheses." *Alternative* hypotheses are meant. Further, Newton adds at the end of the General Scholium the famous phrase (I quote this from Smith (2002, 139)—it's in Newton (1999, 943): "I have not yet been able to deduce from phenomena the reason for these properties of gravity, and I do not feign hypotheses. For whatever is not deduced from the phenomena must be called a hypothesis...." Given that the words "hypothesis," "theory," etc. were used loosely in this period (see Shapiro, 2000, 144–154), Newton is *stipulating* a meaning for the word "hypothesis"; he's not relying on how it's commonly understood—some of the targets of his ire (note 19) are thus *innocent*.

We're to understand him *literally*. He means "deduce," exactly as I've described Descartes and Hume as understanding it. When a hypothesis (even an inductive generalization!) is *deduced* from phenomena, *it has met the gold standard for knowledge!*[21]

There's more to say (which is speculative—part of my *hypothesis*) about how *anyone* could think *any* inductive generalization can *ever* be deduced from phenomena. On this, we contemporaries labor under the burden of knowing too much about deduction and mathematics—and, in particular, about how *those* cause underdetermination of "theory by data." Part of our recognition of this is due to our living in a post-Hume period; but that's *not* the whole story about underdetermination. Part of the reason is also that we (contemporaries) don't include in demonstrative reasoning or inference, or whatnot, content that's not conditional—we don't (in a phrase, include mathematical *premises*).[22] These two *parts* dovetail nicely, preventing us (so I claim) from seeing that for Newton if a hypothesis is appropriately mathematized[23] then it admits of deduction from the phenomena because no alternative hypotheses, *given the use of that mathematics,* can be deduced instead.[24]

Smith and Seth, among others, treat underdetermination of theory by data as generating alternative threatening hypotheses in a way, I'm claiming, that Newton thought the mathematics he was employing *sometimes* excluded.[25] To support

---

[21] Notable is how "deduce" and "deduce from phenomena," and related phrases ("corollary" and "infer") show up in the *Principia*, e.g. Rule 3 (795) "We know by experience that some bodies are hard. Moreover, because the hardness of the whole arises from the hardness of its parts, we justly infer from this not only the hardness of the undivided particles of bodies that are accessible to our senses, but also of all other bodies." Equally notable is where they *don't* show up—places where Newton seems to think he's not succeeded in *deducing* from phenomena the results he's striving for. Consider that he never describes his two solutions for the moon's gravitational pull on the earth's oceans along with the gravitation and rotation of the earth as deductions from phenomena. I'm confining these remarks to a footnote because they require far more textual support than I can give now. I'll add that I'm opposing Newton here to Locke and aligning him with Kant (2004). See van Leeuwen (1970, 136–137) for a discussion of Locke and Newton that attributes to both of them the same view about "appearances" and what can and can't be deduced from them.

[22] A couple of messy qualifications I set aside: We do have one-line proofs that presuppose whatever logic we use; relatedly, we don't know how to conditionalize reasoning on the choice of logic (Lewis-Carroll considerations). And, there are still philosophical attempts, even in the contemporary setting, to design successor notions to a priori *truths*, or to tuck into logic (via higher-order means) mathematical content. By "conditionalization" I'm *not* speaking of what Smith and Seth (384), and Smith in earlier work, are speaking of when they describe Newton as using "accuraté" and "quamproximé" to qualify the clauses they appear in.

[23] "*Appropriately* mathematized, i.e. by using *secure* mathematics such as geometry instead of *insecure* mathematics, such as the calculus (or even algebra). I can't get further into this now, but see note 25.

[24] Therefore, I'm here disagreeing with Stein (1990, 219) on what Newton means by "deduce."

[25] The irony is that how new mathematics *itself* can generate alternative hypotheses didn't become clear until some centuries after Newton's calculus notation had been deserted for Leibniz's; that is, until it was appreciated how differential equations—allowing the *expression* of functions that amounted to alternative hypotheses (that heretofore couldn't be expressed)—*could be written*

Newton's distinction between mere hypotheses and generalizations gathered from phenomena, therefore, Smith and Seth rely on *empirical evidence* as enabling the exclusion of alternatives—and they take Newton as thinking the same thing. They write:

> One way to lessen worries about an inference to an only plausible explanation is to marshal evidence that the phenomena in question, in and of themselves, determine certain specific features of the proposed explanation that are then so strongly supported to amount to a constraint on any possible alternative explanations. (21)

Smith and Seth give Newtonian gravity as an example. I claim empirical considerations can't exclude alternative hypotheses this way: More theoretical scaffolding, however tightly constrained by empirical evidence, can't eliminate cleverly designed theoretical alternatives.[26]

Newton's relationship to Descartes' work was "tortured"—to put it mildly.[27] Notable (and relevant) is his eventual hostility to Descartes' sloppy *mathematics*. Similarly explained, I think, is Newton's evident hostility to those who fail to see the importance of his distinction between mere hypotheses and what he's done with respect to gravity and elsewhere—e.g. Pardies and Hooke: they're underestimating how empirical results when securely tethered to the appropriate mathematics exclude alternatives. Newton's disdain was surely in part due to his low opinion of his opponents' mathematical abilities: they couldn't see how the mathematics was excluding alternative hypotheses because their grasp of the mathematics was so bad!

Ironically, Descartes and Newton are on the same page about this.[28] Descartes tries to justify *all* our empirical means of learning about the world via *deduction*. Huygens, I hypothesize, doesn't agree with Descartes and Newton: he's *anticipating* Hume. Smith quotes Huygens (1690):

> One finds in this subject a kind of demonstration which does not carry with it so high a degree of certainty as that employed in geometry; and which differs distinctly from the method employed by geometers in that they prove their propositions by well-established

---

down. Crucial to understanding why Newton took his distinction to be principled, I'm claiming, is recognizing that Newton didn't think secure mathematics allowed alternative hypotheses to be cogent—if one had actually deduced a hypothesis from the phenomena using that mathematics. (Further details about this can only be given in future work).

[26] That's really a twentieth century insight: Goodman's (1983) grue and the like (Although, perhaps, Henri Poincaré saw it first).

[27] One place (apart from Newton's published and unpublished work) to get a feel for this is to read Westfall (1980)—especially 379–380, and later discussion.

[28] Another hypothesis: Descartes and Newton were somewhat alone in think that reasoning—specifically mathematical reasoning—could carry us so far into the empirical realm. Most at the time thought reasoning was much more restricted in its scope. Shapiro (2000, 46) writes, for example, "Believable statements in the field of history as in law were recognized to be different from the truths of logic and were by their very nature incapable of achieving mathematical certainty or metaphysical truth." Another irony: Hume shrank the scope of reasoning and mathematics into contours that most intellectuals of the period already saw it to have, except insofar as they (Locke, for example) thought Newton had achieved more.

and incontrovertible principles, while here principles are tested by the inferences which are derivable from them. The nature of the subject permits of no other treatment. It is possible, however, in this way to establish a probability which is little short of certainty. This is the case when the consequences of the assumed principles are in perfect accord with the observed phenomena, and especially when these verifications are numerous; but above all when one employs the hypothesis to predict new phenomena and finds his expectations realized. (Smith 2002, 139–40)

Smith (2002) takes himself to be illustrating Huygens' adherence to hypothetical-deductive methods. I'm instead stressing Huygens' apparent recognition of the gap between the "demonstrations" of geometers and the "demonstrations" of empirical scientists. For Huygens, demonstrations yielding a probability "little short of certainty" are, I think, still insufficient for knowledge.

Consider this (indirect) exchange between Newton and Huygens (via Oldenburg's correspondence. Newton writes:

> You know the proper method for inquiring after the properties of things is to deduce them from experiments; and I told you that the theory, which I propounded, was evinced to me, not by inferring 'tis thus because not otherwise, that is, not [by] deducing it only from a confutation of contrary suppositions, but deriving it from experiments concluding positively and directly. The way therefore to examine it is by considering whether the experiments . . . do prove those parts of the theory, to which they are applied, or by prosecuting other experiments, which the theory may suggest for its examination.

And Huygens writes Oldenburg:

> What you have put in your late Journals from Mr. Newton confirms still further his doctrine of colors. Nevertheless the thing could very well be otherwise, and it seems to me that he ought to content himself if what he has advanced is accepted as a very likely hypothesis. (Shapiro 2000, 154–5)

This debate, and similar ones, don't really make sense if the (implicit) issue isn't knowledge but instead some other epistemic "standing."

Going forward, I take it that Newton distinguished two kinds of generalizations, those open to further evidence-gathering—deductions from phenomena—and mere "hypotheses" that shouldn't be taken to be true, or used constitutively in empirical research; those deduced from phenomena instead play this role. Further, if we're infallibilists about knowledge, then because whatever is inductively established (but not deduced from phenomena) is open to further evidential tests, that standing isn't one of knowledge. Thus, the evidence that Newton's laws of gravity holds for the planets has been deduced from phenomena; the inductive generalization that the law holds universally of all matter has not.[29] That Newton's laws hold for the planets is knowledge; that it holds of all matter isn't.

---

[29] I, here, assume this because Smith (and Smith and Seth) do. A case—I think—can be made that Newton thinks he's deduced the general law of gravitation from phenomena via the mathematics enabling him to go from point masses to masses-in-general. (See note 21 for the style of reasoning that establishes this, in his view). Making this case definitively goes beyond this paper.

1.5   Fallibilism in Contemporary Epistemology and in Ordinary Discourse

Let's return to fallibilism. Most contemporary epistemologists are fallibilists. I don't know exactly when, in the twentieth century, fallibilism became the default position for philosophers; but Gettier presumes it. He writes, making a *preliminary* point, that,

> ... in the sense of 'justified' in which S's being justified in believing P is a necessary condition of S's knowing that P, it is possible for a person to be justified in believing a proposition that is in fact false. (Gettier 1963, 121)

Kripke's "dogmatism paradox," and that it's seriously regarded, also illustrates how the fallibility of knowledge is now taken to be the default position about knowledge; it also illustrates how further evidence bears on already-in-place knowledge. Kripke's "paradox" is an argument, using apparently incontestable principles (e.g., known closure, and other simple inferential principles) that:

> If I know something now, I should, as a rational agent, adopt a resolution not to allow any future evidence to overthrow it,

> that is, if an agent A knows *p*,

> A should resolve not to be influenced by any evidence against *p*. (Kripke, 2011, 43)

As Kripke points out: "... this does not seem to be our attitude towards statements that we know—nor does it seem to be a rational attitude" (2011, 43).

I mention the paradox not to resolve it (or even dwell on it)[30] but to acknowledge that Kripke is right: the attitude the reasoning of the paradox apparently requires is widely rejected: it doesn't seem to be regarded as an attitude that would be rational towards *anything* we know. As mentioned, this general reception of Kripke's dogmatism paradox is further evidence of the contemporary view that fallibility is the contemporary default position in characterizations of knowledge.[31]

Why is fallibility the default position of contemporary epistemologists? What's been noticed[32] is that we (ordinarily) attribute all sorts of *empirical knowledge* to ourselves and to one another, and we do so in numerous contexts in which it's clear the agents possessing this knowledge (ourselves and others) have no *hope* of infallibility.[33] This empirical knowledge isn't restricted to immediate

---

[30] I dwell on it, and presume to resolve it, in Azzouni (2020, 357–362)

[31] That knowledge is fallible isn't *universally* accepted by contemporary and near-contemporary philosophers. Lewis (1996, 419) writes that "[I]t seems as if knowledge must be by definition infallible. ... To speak of fallible knowledge, of knowledge despite uneliminated possibilities of error, just *sounds* contradictory." His (only) stated reason for this claim is superficial, however: the oddity of Moorean expressions. See section 6 for discussion of them.

[32] By, among many, Moore (1925), Harman (1973), and, indeed, Lewis (1996, 418).

[33] There is a burgeoning literature in cognitive ethology, almost all twenty-first century, and *all of which* explores animal *knowledge*. "Know," and its cognates are *uncontroversially* applied to

perceptual knowledge (however thick or thin that's taken to be); it includes many generalizations—empirical ones. Related to this is that (contemporary) philosophers accept that empirical generalizations are propositions we can know.

Other kinds of usage evidence for the fallibility of knowers that (I'd like to think) have been noticed are these: We speak naturally (and wistfully) of thinking we knew something or other that we've subsequently discovered we were wrong about.[34] Similarly, we can (and do) describe ourselves and others as having, respectively, *taken* ourselves or *taken* themselves, to know something and later discovering otherwise; we can similarly *speak in the present tense* as *taking* ourselves (or others) to know something, and thus allowing ourselves to be open to discovering that we're wrong later.[35] On infallibility views of knowers, this distinction between "taking oneself to know *p*" and "knowing *p*" collapses—for to take someone to know something (because that someone can't be wrong) *is* to attribute knowledge to that someone.[36]

1.6 Accepting the Ordinary Notion of Knowledge Is Fallible Retains Important Insights About Scientific Practice That Smith and Seth Stress

I've just claimed that the use of "take," in phrases like, "we take ourselves to know *p*," or "I take myself to know *p*," doesn't indicate a different epistemic status from that of "I know p"; rather, it's *sometimes* used as a tool for acknowledging the fallibility of knowers; otherwise—recall from Sect. 1.2—it's an emphatic way of expressing knowledge. Fallibility is hard to acknowledge by straightforwardly using the word "know" because its factivity—*A* knows *p* implies *p*—impedes this. To say, "I know *p*, but I might be wrong," in most circumstances is to say something

---

animals in both the scientific and popular literature. That these animals (including honeybees and other insects) are infallible isn't an option.

[34] Wittgenstein (1969, 3e, 12), writes, "One always forgets the expression 'I thought I knew'."

[35] This is a use that "take" can *sometimes* be put to—recall the observations about it in section 2.

[36] I'm focusing on contemporary usage. What about the use of the word "know" in, for example, Newton's time? I claim—recall note 5—that their ordinary word "know" didn't require infallibility either, and indeed, that it's the same as our contemporary one. The reader should read (as I've done) fiction from that time, e.g., *Gulliver's Travels*, or the writings of Daniel Defoe, to verify that the usage claim of fallibility that I've made about the ordinary contemporaneous use of "know" were in place in British English in the seventeenth and eighteenth centuries. (I'm *not* claiming that the *philosophical views* expressed explicitly and tacitly by these authors about knowledge presupposes fallibility. I'm claiming that their novelistic depictions of ordinary speech—e.g., when quoting characters—show that uses of "know(s)" and related expressions in the vernacular are fallibilistic. This parenthetical remark was prompted by a remark of Eric Schliesser in an email on 4/24/2018).

unacceptable.[37] This is because "I know p," implies p, as I just noted, and *that* rules out that *in the circumstances of utterance,* $\neg p$. "I might be wrong (about p)," however, denies *in the circumstances of utterance* that $\neg p$ can be ruled out.[38] Notice: I'm claiming the (Moorean) awkwardness of "I know p, but I might be wrong," is due only to the factivity of "know" and not to anything stronger (like infallibility). On the other hand, "I take myself to know p, but (of course) I might be wrong," doesn't sound similarly awkward; *this is because this use of "take to" in "take to know" cancels the factivity of "know."*[39]

"Cancel," as just used, isn't *Gricean* "canceling." If there *are* necessary conditions on an agent knowing p, that p is true is surely one of them. Rather, this use of "take to," in "taking [oneself] to know p," or in "taking p to be true," expresses an "iterated cognition": an assertion by the knowing agent of her *knowing p* or *p being true*. This assertion of iterated cognition *isn't* factive: it isn't required (for an agent to take herself to know p or to take p to be true) that she actually know p or that p be true.[40] But only the factivity of know(s) is deleted. In particular, there's no diminution of the evidence on the basis of which one takes p to be true or takes oneself to know p.

So, to acknowledge fallibility without stumbling over the factivity of "know," we utilize "take to know" instead. But, importantly, this isn't to introduce a new epistemic category in between knowledge and mere hypothesizing. Smith, as well as Smith and Seth, impound "take to be true," as *terminology* to provide an intermediate epistemic status between mere hypotheses and knowledge because of this role, already in place in natural language, of "take to." If Newton did the same—if he wasn't, that is, merely choosing a Latinate form of speech that avoids Moorean awkwardness—then (I'm hypothesizing), he introduced this

---

[37] Consider Smith's translation of Newton's 4th rule: "In experimental philosophy, propositions gathered from phenomena by induction should be taken to be true, either exactly or very nearly notwithstanding any contrary hypotheses until yet other phenomena should render such propositions either more exact or liable to exceptions." Suppose, instead, we have, "In experimental philosophy, propositions gathered from phenomena by induction are true [or, alternatively, "known to be true"], either exactly or very nearly notwithstanding any contrary hypotheses until yet other phenomena should render such propositions either more exact or liable to exceptions." Either formulation is Moorean-unacceptable, a tautology, or worse—something to the effect of: p is true unless it's not; p is true until it's not; p is known unless it's not, etc. Notice that replacing "take to be" with "considered," "esteemed to be," "looked upon," or "be regarded as" also circumvents otherwise unpleasant Moorean phrasing. (I'm abusing use and mention, in the above; or I'm tacitly employing a kind of variable, p, that functions as needed within quotations. Or whatever).

[38] "I might be wrong that p," here, isn't the expression of the metaphysical possibility of $\neg p$, or anything (weird) like that. Rather, it's a characterization of the specific circumstances that the speaker takes herself to be in: *She* can't guarantee that p holds in those circumstances. That's why she can't couple "I might be wrong about p" with "I know p"—the factivity of "know" *does* present her as guaranteeing that p holds in those circumstances if she asserts knowledge of p. (See Azzouni (2020), section 10.3, for details and arguments).

[39] Again, further remarks about usage are needed to establish this. See Azzouni (2020), section 10.3.

[40] Again, this is so if "take to know p" isn't meant to emphasize one's knowledge of p.

terminological innovation because he continued to think, as Descartes did, that knowledge requires infallibility.[41]

If we appreciate that knowledge is fallible, we appreciate that knowledge, in general, is open to, and may even require, further evidence (this is what we recognize by thinking the conclusion of Kripke's dogmatism "paradox" is *unacceptable*): testing one's knowledge in light of further evidence often makes sense (and is often reasonable). Thus, the acknowledgement of the fallibility of knowledge is compatible with an important insight Smith and Seth, and Smith (elsewhere, e.g., Smith 2010) stress: when evaluating the evidence for a scientific proposition, one can't just look at the moment the proposition is adopted (and considered empirically verified) by scientists, but at the much stronger evidence for the proposition that emerges subsequently.

Smith and Seth write:

> Smith has elsewhere invoked Newton's "take to be true" phrasing to contrast the standing the law of gravity had on the basis of the evidence presented in the *Principia* with the increasing entrenched standing that it came to have over the course of the next two centuries as a consequence of evidence accruing to it from the research that presupposed it. This contrast is no less apropos in regard to the reality of molecules. Anyone asking now what the evidence is that molecules exist is misconceiving the way evidence works in physics if they focus on the time before 1913. This question can be meaningfully answered only through a critical review of *all* the evidence bearing on molecular reality that has been, and is still being, generated since 1913. (359; their italics)

I can put the compatibility of the fallibility of knowledge with Smith and Seth's claim another way: If knowledge is fallible, then relying on knowledge involves risk—and risk comes in degrees. Thus, the decision to rely on what one *knows*—to employ it constitutively in experimental design and scientific research— is to nevertheless (often) take chances. This important observation (of Smith and Seth and Smith 2010) remains intact as well—even when foregoing Smith's interpretation of the phrase "take to be true": using *known* scientific propositions constitutively in experimental design and research can be risky. One has to evaluate when there's enough evidence for the propositions in question to make doing so worth it.

---

[41] *Does* Descartes think "knowledge" requires infallibility? There's controversy because it's thought "knowledge" isn't univocal for him: he's got both "scientia" and "cognitio" (see, e.g., Pasnau (2017) on this); scientia requires infallibility, perhaps, but cognitio needn't. I claim: One of Descartes' gold standards for knowledge is appropriately-managed deduction—whether that involves empirical matters or not—and that entails infallibility. (So, I claim, both "scientia" and "cognitio" require infallibility—their difference is based on something else, systematicity, say). The ordinary notion of knowledge, however (rendered verbally in living languages, one way or the other), continues to affect everyone—however they try to conceptually engineer their epistemic notions; and *its* usage is compatible with fallibility. Thus there's waffling on everyone's part—in print and in their thinking—e.g., when Descartes describes his vortex hypothesis, magnetism, etc., as *cognitio*. Again: to make this case adequately requires more textual evidence than I'm allowed to stuff into this paper. My thanks to Chris Smeenk for pressing me on this, and for citing email exchanges between Alan Nelson and Eric Schliesser, where Nelson presses the nonunivocality claim for Descartes that I'm responding to.

As I mentioned earlier, Smith and Seth note that (most) professionals were convinced of molecular reality by 1900—perhaps on the basis of van't Hoff's results.[42] If knowledge is something that's more or less well supported by evidence—even good evidence—this is compatible with professionals still thinking that the existence of molecules, even if already known, still needed to be *definitively* established.

This seems—as a matter of the sociological reality—to be the situation vis-à-vis molecules (and electrons) in the time period studied in Smith and Seth's book. It explains why, for example, Perrin did so much to try to establish that it was *his work* that definitively and finally established the existence of molecules—even though he must have realized that everyone (at the Solvay conference, for example) was *already* convinced of their existence. Perrin wanted ontological credits, *nevertheless*, as definitively establishing what everyone else *already knew*.[43] Millikan, too, wanted *ontological credit* for having "shown" the existence of the electron (as opposed, presumably, to Thomson having done so, *despite our knowing* on the basis of Thomson's work of the existence of electrons).[44] In both cases, I suggest, the change in respective status of propositions about the existence of, respectively, molecules and electrons isn't one from some earlier intermediate state of hypothesizing the existence of these things to one of taking the respective propositions to be true. That's because knowledge of the existence of these things was already taken to be had; what was needed was more, as Smith and Seth indicate: reasons to take the risk of using this knowledge constitutively in experimental design and in research. In very large measure (as Smith and Seth show) this *additional* knowledge turned on supplementing the sheer fact of molecular reality with detailed numerical results about the properties of molecules that could be subsequently used in research and experimental design—for example (but hardly the *only* example), a measure of Avogadro's number that could (finally) be *trusted*.

Acknowledgments I'm grateful for years of insightful conversation on this (and many other topics) with my colleague George Smith. I'm similarly grateful for conversational insights—sporadic at times—with Eric Schliesser. My thanks to Chris Smeenk for valuable pushback and for citing additional resources that I helped myself to during the last rewrite of this paper. I'm,

---

[42] Smith and Seth (65) express puzzlement about why Ostwald didn't take those results "as providing some sort of application of kinetic theory to the liquid state," and therefore the success of this application as evidence for the theory.

[43] Smith and Seth (24) write about Perrin's 1909 monograph that "it provided Perrin with his own priority claim, namely the first to present in systematic comprehensive form the evidence derived from research on Brownian motion in support of molecular reality ....." And, by virtue of the convergence of the values of Avogadro's number, Perrin is quoted as saying (293), "the real existence of the molecule is given a probability bordering on certainty." (This echoes Huygen's phrasing, cited in section 4. This is another typical locution contemporary speakers sometimes adopt to square knowledge claims with simultaneous admissions of fallibility. It doesn't work).

[44] Smith and Seth (286) write: "Millikan always insisted that he had been the one to finally show that electricity consists of discrete unit charges ...." We *often* know things that haven't yet been "definitively shown": this happens all the time (by the way) in mathematics.

lastly, grateful to the organizers of "On the question of evidence: A celebration of the work of George E. Smith"—a conference on May 11–12, 2018, where I was invited to give the first version of this paper.

References

Azzouni, J. 2020. *Attributing knowledge: What it means to know something*. Oxford University Press.
Gaukroger, S. 1995. *Descartes: An intellectual biography*. Oxford University Press.
Gettier, E.L. 1963. Is justified true belief knowledge? *Analysis* 23: 121–123.
Goodman, N. 1983. *Fact, fiction, and forecast*. Harvard University Press.
Harman, G. 1973. *Thought*. Princeton University Press.
Hume, D. 1977. *An enquiry concerning human understanding*. Hackett Publishing Company.
Kant, I. 2004. *Metaphysical foundations of natural science*. Trans. M. Friedman. Cambridge University Press.
Kripke, S. 2011. On two paradoxes of knowledge. *Philosophical Troubles: Collected Papers* 1: 27–51.
Lewis, D. 1996. Elusive knowledge. In *Papers in metaphysics and epistemology*, Vol. 2 (1999), 418–445. Cambridge University Press.
Moore, G.E. 1925. A defense of common sense. In *Philosophical papers* (1962), 32–59. Routledge.
Newton, I. 1999. The *Principia*. Trans. I.B. Cohen and A. Whitman. University of California Press.
Pasnau, R. 2017. *After certainty: A history of our epistemic ideals and illusions*. Oxford University Press.
Schliesser, E. 2011. Newton's challenge to philosophy: A programmatic essay. *HOPOS: The Journal of the International Society for the History of Philosophy of Science* 1: 101–128.
Shapiro, B.J. 2000. *A culture of fact: England, 1550–1720*. Cornell University Press.
Shapiro, A.E. 2002. Newton's optics and atomism. In *The Cambridge companion to Newton*, ed. I.B. Cohen and G.E. Smith, 227–255. Cambridge University Press.
Smith, G.E. 2002. The methodology of the *Principia*. In *The Cambridge companion to Newton*, ed. I.B. Cohen and G.E. Smith, 138–173. Cambridge University Press.
———. 2010. Revisiting accepted science: The indispensability of the history of science. *The Monist* 93: 545–579.
———. 2014. Closing the loop: Testing Newtonian gravity, then and now. In *Newton and empiricism*, ed. Z. Biener and E. Schliesser, 262–351. Oxford University Press.
Smith, G.E., and R. Seth. 2020. *Brownian motion and molecular reality: A study in theory-mediated measurement*. Oxford University Press.
Stein, Howard. 1990. 'From the phenomena of motions to the forces of nature': Hypothesis or deduction? *PSA 1990*: 209–222.
van Leeuwen, H.G. 1970. *The problem of certainty in English thought 1630–1690*. 2nd ed. Martinus Nijhoff.
Westfall, R.S. 1980. *Never at rest: A biography of Isaac Newton*. Cambridge University Press.
Wittgenstein, L. 1969. *On certainty*, ed. G.E.M. Anscombe and G.H. von Wright. Harper & Row.

Chapter 2
'To Witness Facts with the Eyes of Reason': Herschel on Physical Astronomy and the Method of Residual Phenomena

Teru Miyake

2.1 Introduction

One of the distinctive features of George Smith's work on celestial mechanics is his emphasis on the role of what he calls "second-order phenomena" in the production of high-quality evidence.[1] On Smith's view, these gaps between theoretical predictions and observations can, under certain circumstances, be a source of evidence far stronger than that achievable through the hypothetico-deductive method. The practice of examining gaps between predictions and observations for the purposes of discovery and testing is commonplace in certain sciences such as seismology, and has played an important role in their development. I use the term *reasoning from residuals* as a general term for this practice.[2] I think it is worth investigating examples of this set of practices from the history of science, in order to understand the different ways in which reasoning from residuals is done, under what situations it is done, and how it contributes to the growth of knowledge in certain sciences.

For a long time, I thought that the only philosopher of science who has written in any detail about such gaps between theory and observation is Smith, until a few years ago, when I read the work of John Herschel, a nineteenth century British natural philosopher who has views that are, in many ways, remarkably similar to

---

[1] See, for example, 'Closing the Loop' (Smith, 2014).

[2] To be clear, this term is not intended to refer specifically to Smith's view about the particular practice he identified in celestial mechanics. Rather, I intend it to encompass a diverse set of practices from a variety of different sciences, among which is the practice Smith describes.

T. Miyake (✉)
School of Humanities, Nanyang Technological University, Singapore, Singapore
e-mail: TMiyake@ntu.edu.sg

© The Author(s), under exclusive license to Springer Nature Switzerland AG 2023
M. Stan, C. Smeenk (eds.), *Theory, Evidence, Data: Themes from George E. Smith*, Boston Studies in the Philosophy and History of Science 343, https://doi.org/10.1007/978-3-031-41041-3_2

Smith's. I first became aware of the work of Herschel when I read the following passage from William Thomson and Peter Guthrie Tait's landmark *Treatise on Natural Philosophy*:

> A most important remark, due to Herschel, regards what are called *residual phenomena*. When, in an experiment, all known causes being allowed for, there remain certain unexplained effects (excessively slight it may be), these must be carefully investigated, and every conceivable variation of arrangement of apparatus, etc., tried; until, if possible, we manage so to isolate the residual phenomenon as to be able to detect its cause. It is here, perhaps, that in the present state of science we may most reasonably look for extensions of our knowledge; at all events we are warranted by the recent history of Natural Philosophy in so doing. (Thomson and Tait 1883, 443)

When I read the passage from Thomson and Tait, the notion of "residual phenomena", attributed to Herschel, seemed to me to be the sort of thing that Smith calls "second-order phenomena". It did not escape my attention that Thomson and Tait thought that residual phenomena played a central role in extensions of our knowledge in the nineteenth century—a claim that seemed very encouraging for my interest in investigating reasoning from residuals.

It turns out that the term "residual phenomena" and the related terms "residue" and "residual" (used as a noun) have a rather complicated history. As far as I know, the term was coined, as Thomson and Tait suggest, by Herschel, in the *Preliminary Discourse on the Study of Natural Philosophy* (1830). The term then underwent an evolution in the work of subsequent writers on scientific methodology. In Volume 2 of the *Philosophy of the Inductive Sciences* (1840), Herschel's friend William Whewell describes the "Method of Residues" as one of a series of "Special Methods of Induction Applicable to Quantity", attributing the idea to Herschel. The "Method of Residues" also appears in Volume 1 of John Stuart Mill's *System of Logic* (1843) as one of the famous "Four Methods of Experimental Inquiry". To illustrate the Method of Residues, Mill relies entirely upon two very long passages from Herschel's *Preliminary Discourse*. In the late nineteenth century, books on the methodology of science or inductive logic often contained long discussions of either residual phenomena or the method of residues (see, for example, William Stanley Jevons's *Principles of Science*, or Thomas Fowler's *Elements of Inductive Logic*). Despite wide recognition of the importance of residual phenomena for understanding the methods of science, the concept underwent a sharp decline around the turn of the twentieth century.

What are residual phenomena, exactly? And why might someone like Thomson, writing in the 1880s, think they are so important for the progress of science? We might begin by examining how Herschel takes residual phenomena to fit into the proper method of natural philosophy. The problem is that even getting clear on what Herschel takes to be the proper method is complicated. Interpreters of Herschel[3] tend to separate into two camps: those who take him to be a Baconian inductivist,

---

[3] The philosophy of science literature on Herschel's methodology goes back to Ducasse (1960), and it includes Olson (1975), Jain (1975), Agassi (1981), Oldroyd (1986), and the vastly underappreciated work of Good (1982, 1987), Bolt (1998), and Cobb (2012a, b).

and those who take him to be an early advocate of the hypothetico-deductive method. Bolt (1998, 10) takes Herschel to be a hypothetico-deductivist who did not want to present the appearance of running counter to the "inductive" philosophy of Newton. The result, according to Bolt, is that the *Preliminary Discourse* presents, on the surface, an inductive methodology for the lay reader, while also containing a more advanced, hypothetico-deductive methodology for the enlightened (Bolt 1998, 286–7). On the other hand, Cobb (2012a, b) defends the inductivist interpretation of Herschel's methodology based on a study of Herschel's experimental research on electromagnetic phenomena. Complicating matters is that Herschel's own statement in the *Preliminary Discourse* of how the natural philosopher ought to arrive at and test general laws in fact describes *three* ways to do so. In addition to induction and the hypothetico-deductive method, Herschel mentions a third method "combining the advantages of both without their defects" (Herschel 1830, 199). Herschel takes the third method to be the most powerful, but unfortunately, he is not entirely clear on exactly what this method involves.

One thing that *is* clear throughout Herschel's writings is that Herschel takes physical astronomy to be an exemplar for the methods of natural philosophy. In trying to understand Herschel's views about the method of natural philosophy and the role of residual phenomena in that method, I will thus turn to an article on "Physical Astronomy" that he wrote for the *Encyclopaedia Metropolitana*, and completed by 1823, prior to writing the *Preliminary Discourse*. Despite the promise of this article for better understanding Herschel's views about the methodology of natural philosophy, there is no previous writing on Herschel that examines the "Physical Astronomy" in any detail. Marvin Bolt's (1998) dissertation is the only work I know of that contains more than a passing mention of the article. This might be because it was eclipsed by Herschel's later writings on astronomy, in particular the *Treatise on Astronomy* (1833) and the *Outlines of Astronomy* (1864). As we shall see, however, the "Physical Astronomy" is a completely different work. While the latter books were written for a popular audience, the "Physical Astronomy" was written for those who had mastered the elements of mathematical analysis. This is important, because, as we shall see, the whole key to understanding Herschel's views about the methods of natural philosophy is understanding the role that mathematics plays in the investigation of nature.

The aim of this chapter is to argue for the following claim. In the "Physical Astronomy", Herschel takes the whole difficulty of physical astronomy to be dealing with the complexity of the planetary motions, and describes a method for dealing with these complexities that is neither inductive nor hypothetico-deductive. In this method, the primary use of theory is as a tool that allows the skilled practitioner to separate out the various effects that together make up the complex planetary motions, and identify details in the solar system that give rise to them. Although Herschel does not use the term "residual phenomena" in the "Physical Astronomy," these complexities in the orbital motions are prime examples of what he would later come to call residual phenomena. I think a further argument can be made that the methodological view presented in the *Preliminary Discourse* is an extension and modification of the methodological views Herschel presents in the "Physical Astronomy." More specifically, it is an attempt to be more explicit about the methodology, extend it to sciences other than physical astronomy, and make it understandable to a wide audience—one that has no familiarity with the mathematical tools employed in the "Physical Astronomy." I will not attempt to make a full argument for this view here. I shall merely show how some of the most striking features of Herschel's views in the *Preliminary Discourse* can be traced back to the "Physical Astronomy."

The remainder of this chapter consists of six further sections. Section 2.2 provides some background to the "Physical Astronomy". Section 2.3 examines the Introduction to the "Physical Astronomy", showing that Herschel saw the article as presenting a method for dealing with the complexity of the planetary motions. Section 2.4 discusses the parts of the "Physical Astronomy" dealing with perturbations, going over two examples in detail. Section 2.5 summarizes Herschel's conception of research in physical astronomy. Section 2.6 argues that many of the methodological views in the *Preliminary Discourse* have their origins in the "Physical Astronomy." The concluding Sect. 2.7 briefly touches on parallels between the views presented by Herschel in the "Physical Astronomy" and the view of George Smith in his paper "Closing the Loop."

2.2   The "Physical Astronomy"

The "Physical Astronomy" is one of three articles that Herschel wrote for the *Encyclopaedia Metropolitana*, the other two being "Light" and "Sound."[4] To call these works articles for an encyclopedia is rather misleading, for they go into much more rigor and detail than typical articles written for modern encyclopedias. The article on "Light," written by 1827, was 245 pages long, and it was at the time the best-available account in English of both Jean-Baptiste Biot's emissionist theory of polarization and Fresnel's wave theory of diffraction and polarization.[5] Copies of

---

[4] There is some confusion about the dates for these articles. Many secondary sources, such as the biography of Herschel by Buttman (1970), give 1845 as their date of publication. I take the dates from Bolt (1998) to be more reliable. He lists the dates of publication as 1829 for "Physical Astronomy," 1830 for "Light," and 1830 for "Sound" (Bolt 1998, 417). Herschel himself, in an 1861 article for *The Mathematical Monthly*, gives the following as the dates of completion (not publication) of these articles: 1823 for "Physical Astronomy," 1827 for "Light," and 1829 for "Sound." Throughout this chapter, I will use 1829 as the publication date for the "Physical Astronomy," keeping in mind that it was probably written by 1823.

[5] See Buchwald (1989, 291) and Good (1982, 1). The article was an important resource for optical research among the younger generation of English-speaking natural philosophers in the 1830s, among them Baden Powell and William Rowan Hamilton. Powell wrote in 1832: "The publication of Sir J. Herschel's Treatise on Light forms an epoch in the history of the science, and has given a material impulse to the study of it . . . " (Powell 1832, 433). Hamilton recounts in a letter that he kept the article under his pillow (along with Coleridge's *Aids to Reflection*) in early 1832, the year he later discovered conical refraction (Graves 1882, 515).

this article were widely circulated in England after 1827, and an influential French translation of it appeared on the Continent. The article on "Sound," written by 1829, was not quite as influential, but Lord Rayleigh states in the Preface to his great work *Theory of Sound* (1877) that "since the well-known Article on Sound in the *Encyclopaedia Metropolitana*, by Sir John Herschel, no complete work has been published in which the subject is treated mathematically" (Rayleigh 1877, v). Such are the standards to which Herschel held himself when he wrote these articles.

The "Physical Astronomy," written by 1823, meets those lofty standards. The *Treatise on Astronomy* (1833) and the *Outlines of Astronomy* (1864) are much better-known works of astronomy by Herschel, but they were written for a popular audience. Neither book presupposes any more knowledge on the part of its readers than basic geometry and trigonometry, and they contain no equations. The "Physical Astronomy," on the other hand, presupposes familiarity with French analytic methods that had only recently been introduced into England, and it contains hundreds of equations throughout its 83 pages.[6] Herschel intended for the work to provide readers with sufficient preparation for reading the works of the great French analysts, in particular Laplace's *Celestial Mechanics*.

As we shall see, the work lays out a research program for physical astronomy, in the sense that it provides examples of the kinds of research questions that can be investigated, and an introduction to the tools needed to carry out such investigations. Most of these research questions have to do with complexities in astronomical phenomena, such as deviations of the planetary motions from Keplerian motion. Theory provides the set of tools needed to carry out these investigations. We shall now examine some parts of the "Physical Astronomy" in detail, beginning with the Introduction, in which Herschel presents in general terms the picture of the primary concern of physical astronomy as dealing with the complexity of the orbital motions.

2.3 Human Limitations and Complexity in Physical Astronomy

Herschel starts off the "Physical Astronomy" by emphasizing the limitations of the human intellect in the face of the complexity of natural phenomena. Natural philosophers should "arrange and classify facts and phenomena, with a view to trace the agency of their remote, or, at least, their proximate causes, and ascend as high as the imperfection of human means of observation, and the limited powers of the human intellect will allow us in the scale of generalization" (Herschel 1829, 647). Herschel contrasts intellectually limited humans with ideal beings for whom "much of that complication we observe in natural phenomena would disappear:

---

[6] Many of the mathematical references in the article are to Lacroix's *An Elementary Treatise on Differential and Integral Calculus*, which had been translated into English by Herschel and his Cambridge friends Charles Babbage and George Peacock in 1816.

many effects, which seem to us independent of each other, and linked by no natural connection, would be in their eyes collateral results of one and the same principle" (Herschel 1829, 647).[7]

The following passage makes it clear that these angels are really Herschel's way of expressing the power of mathematical analysis for investigating natural phenomena:

> The man who has learnt to regard the fall of a leaf and the precession of the equinoxes as results equally certain and unavoidable of a law capable of being stated in three lines, and understood by a child of ten years old, has made already a considerable step in this way— the patient exercise of his natural reason has stood him in the stead of a sharper intellect; and secrets which an angel might penetrate perhaps at a glance, become revealed to man by the slow, yet sure, effects of persevering thought. (Herschel 1829, 647)

Natural philosophers can now, through the power of mathematical analysis, show how seemingly disparate phenomena like the fall of a leaf and the precession of the equinoxes are both the result of the law of gravity. Although humans do not have the ability to see these connections at a glance, they can, through the slow but sure application of mathematical analysis, obtain cognitive powers approaching those of such angels.

Supposing that mathematical analysis gives natural philosophers the means to show that utterly disparate phenomena are the result of one and the same principle, such as the law of gravity, we might yet ask how the underlying principles are to be found in the first place. Herschel does not have an answer, but he takes physical astronomy to be unique among the sciences in this regard. He believes it is the one science for which the inference to the underlying principle was done relatively easily, and all at one go:

> [The] one feature in physical astronomy which renders it remarkable among the sciences . . . . . . is this—that the fundamental law embracing all the minutiae of the phenomena so far as we yet know them, presents itself at once, on the consideration of broad features and general facts, deduced by observations even of a rude and imperfect kind, in such a form as to require no modification, extension, or addition when applied in minute detail. In other sciences, when an induction of a moderate extent has led us to the knowledge of a law which we conceive to be general, the further progress of our inquiries frequently obliges us either to limit its extent or modify its expression. (Herschel 1829, 647)

The inference to the law of gravity was made, according to Herschel, from "rude and imperfect" observations, but it has since required no modifications. He contrasts

---

[7] Herschel likely takes the idea of such ideal beings from the Scottish Common Sense philosopher Thomas Brown. Richard Olson (1975) has written extensively on the influence of Brown on the *Preliminary Discourse*, pointing out that many of the illustrative examples come from Brown's *Lectures on the Philosophy of the Human Mind* (1824). The influence is already there in the earlier "Physical Astronomy." In a discussion of the power of reasoning, and in mathematics in particular, Brown refers to "races of beings . . . . . . able to feel, in a single comprehensive thought, all those truths, of which the generations of mankind are able, by successive analysis, to discover only a few, that are, perhaps, to the great truths they contain, only as the flower which is blossoming before us, is to that infinity of future blossoms enveloped in it, with which, in ever renovated beauty it is to adorn the summers of other ages." (Brown 1824 [vol 2], 198–9).

physical astronomy with sciences like chemistry or physical optics, where general laws, once arrived at, had to be modified or restricted in response to new discoveries. For example,

> the Cartesian law of refraction when applied to the extraordinary ray in crystallized media, and even to the ordinary, if the reports of some recent experiments are to be relied on— together with innumerable other laws, simple, natural, and resting on extensive inductions, have all been either overset, extended, or materially modified by the progress of science. (Herschel 1829, 647–8)[8]

Herschel states in a bit more detail how he takes the law of gravity to have been arrived at, and how it was subsequently tested.[9] He takes Newton to have started with the assumption that the Moon moves uniformly in a circle around the Earth. From this assumption, Newton inferred the "true law of gravity" and the claim that the domain over which the law holds extends from the Earth to the Moon. Next, the law was "confirmed" and its domain extended to the boundaries of the solar system through the observation that the planets, under the assumption of uniform circular motion, exhibit motion in accordance with the 3/2 Power Rule of Kepler.[10] Since then, the law of gravity has been undergoing continual testing through a set of more complicated phenomena:

> Every thing more refined than this—the elliptic motions of the planets and satellites—their mutual perturbations—the slow changes of their orbits and motions, denominated secular variations—the deviation of their figures from the spherical form—the oscillatory motions of their axes, which produce nutation and the precession of the equinoxes—the theory of the tides, both of the ocean and the atmosphere, have all in succession been so many trials for life and death in which this law has been, as it were, pitted against nature; trials, whose event no human foresight could predict, and where it was impossible even to conjecture what modifications it might be found to need. (Herschel 1829, 648)

The overall method through which the law of gravity was arrived at, and has been subsequently tested, according to Herschel, is through an initial inference to the law from a highly idealized characterization of the phenomena, and then the testing of it through an examination of how the law of gravity can account for a great number of subtler phenomena in the solar system.

Although Herschel takes the law of gravity to have undergone stringent testing through the observation of the planetary motions, he is also careful to note that there

---

[8] This is a very brief reference to recent work on double refraction, about which Herschel gives a longer exposition in the *Preliminary Discourse* (Herschel 1830, 29–33). The "recent experiments" are a reference to work by Fresnel that showed that in double refraction the ordinary ray does not precisely follow Snell's Law.

[9] There is a much more detailed account of the inference in Part I, pages 649–51.

[10] This account of Newton's inference to the law of gravity is quite different from that given in Book 3 of the *Principia*. It appears to be based on a famous description of Newton's thought leading up to the *Principia* that is given in the Preface to Henry Pemberton's *A View of Sir Isaac Newton's Philosophy* (1728). I have not found any references to Pemberton in Herschel's published work, but the preface was well-known to Herschel's contemporaries, and Herschel surely knew of it. See, e.g. Whewell (1837, 158).

are other distance scales at which it has not been tested so severely: "The cautious philosopher however will still regard it as worthy inquiry, whether, at enormous distances, like those of the fixed stars, or at such comparatively microscopic intervals as those we are ordinarily conversant with on the surface of our planet, the rigorous law of a force as the inverse square of the distance may not suffer some modification." (Herschel 1829, 648) He notes in particular that the possibility that terrestrial gravity deviates from the Newtonian law is not incompatible with the testing that has been done thus far: "The experiments of Maskelyne and Cavendish, which may perhaps be adduced as supporting its rigorous application, are far too gross, and differ too widely in their results, to be cited in so delicate a matter, besides which, their results, as applied to such an inquiry, are affected with an unknown element, the mean density of the earth." (Herschel 1829, 648).[11] This passage suggests that Herschel is very conscious of the possibility that the law of gravity might not hold exactly, especially at certain distance scales, and could require modification after all, much as Snell's Law required modification in the case of physical optics.

The Introduction, then, makes the following points in setting up the subsequent presentation of the methods of physical astronomy:

1. Much of the seeming complexity in natural phenomena is due to human limitations. If the human intellect were such that humans could, at a glance, trace out the mathematical consequences of underlying principles such as the law of gravity, they would see many more connections between disparate phenomena than they are cognizant of now.
2. Physical astronomy is unique among the sciences in that the underlying principle (the law of gravity) was arrived at relatively quickly and it has never undergone any modifications, but there is no guarantee that it will not require any modifications in the future.
3. If any modifications to the law of gravity are required, they will be found through examining the subtler phenomena in the solar system, such as the perturbations, and through extending the domain of observation throughout which the law has been tested.

These points suggest that now that the law of gravity has been inferred and shown to account broadly for the phenomena, the major task that lies ahead for the physical astronomer is to examine the subtler phenomena of the solar system, such as the perturbations, in order to find out whether modifications to the law of gravity are required. The rest of the article provides examples of the kind of research that

---

[11] Herschel here is referring to attempts by Maskelyne to measure the attraction of Schiehallion, a Scottish mountain, and torsion pendulum experiments done by Cavendish, which could potentially decide the question, but they were limited by their use of an unreliable estimate of the mean density of the earth. Usually, in fact, these experiments are taken to be measurements of the earth's mean density, under the assumption that the law of gravity holds exactly. See, e.g., Poynting (1894) and Bullen (1975); cf. also Allan Franklin's chapter in this volume.

is carried out by astronomers under this conception of physical astronomy, and introduces readers to the tools that are needed to carry out such investigations.

2.4 Herschel on Perturbations

The remainder of Herschel's article consists of three parts, most of it devoted to perturbations. Part I, taking up 24 pages, is on the inference to the law of gravity, the attraction due to spherical bodies under the inverse square law, the two-body problem, and the determination of orbital elements from observations. Part II, taking up 41 pages, is on the planetary perturbations. It first lays out a general theory of perturbations, then investigates the perturbations under the assumption of circular orbits, and finally investigates the effects on the perturbations when the orbits are not circular. Part III, taking up 19 pages, is on the perturbations of the moon.

My focus in this chapter will be on Parts II and III, but there is much of interest in Part I. In particular, what is striking is Herschel's showing how, even though the motions of the planets under the two-body assumption are simple, the determination of orbital elements for the planets is anything but, because observations are being made from a moving platform—the earth.[12] The impression one gets in reading Part I is that determining the motions from observation is an extremely difficult problem—even under the two-body assumption!

Herschel starts off his discussion of the planetary perturbations in Part II, Section I with the comment that up until that point in the article, the planetary motions were investigated under the assumption that only two attracting bodies existed, in which case the bodies undergo motion in elliptical orbits in accordance with Kepler's laws of planetary motion. Planetary tables calculated under this assumption agree with observations to such an extent that they "leave no doubt of the correctness of the principles from which they have been deduced" (Herschel 1829, 673). There are, however, small discrepancies, which Herschel describes as follows:

> 1st. The ellipse in which each planet moves, must be conceived to change its eccentricity and position in space, by exceedingly slow gradations—so slow indeed, as to be insensible in a single revolution of the planet, and only discoverable by a comparison, such as we have described, of its nature in past and present ages.
> 2dly. The planet itself is not always found exactly in its place in the ellipse even when so varied; nor indeed is it always found in the exact periphery of the ellipse at all. But if we conceive an imaginary point to describe this ellipse according to the rigorous laws

---

[12] More specifically, Herschel shows that the problem of determining the elements of a planet's orbit from observations can be reduced to a system of nine equations in six unknowns (Herschel 1829, 664). He states: "The complication of the relations in question precludes all idea of a direct solution of the problem, and to apply them to any particular case indirect and approximative ones have been invented, but with every assistance from such simplification; and, after all the force of analysis has been exercised upon it, it still remains a very difficult problem, and one which our limits will by no means permit us to enter upon in its full extent" (1829, 665).

> of elliptic motion, the real planet will never be very far distant from it, but will oscillate or revolve round it like a secondary about its primary in an orbit of extremely small dimensions, yet according to laws of a very complicated nature, too complicated indeed for observation alone to unravel. (Herschel 1829, 673)

At this point, then, research becomes focused not so much on the motions of the planets themselves, but their deviations from "the rigorous laws of elliptic motion"—i.e., motion in accordance with the three laws of Kepler. These deviations consist of two types. In the first type of deviation, the planet is conceived to still be undergoing Keplerian motion, but the orbital elements that define the ellipse are slowly changing. The second type of deviation involves extremely complicated motions that are not capturable in this way.

Almost all of these deviations are too subtle to be detected solely from observation, without actively seeking them out. The first type of deviation takes place so slowly that astronomers must look far into the past in order to bring them into comparison with observations. The second type of deviation is extremely complex, and is the combined result of several different causes working at once. This makes it extremely difficult to identify what is giving rise to the deviations, and in particular, determining whether the deviation is an indication that a modification is needed to the law of gravity. According to Herschel, the major task of the physical astronomer is to investigate these deviations, and to determine what is giving rise to them, for "deviations like these must have some positive and decided cause; as much so as the elliptic motion itself; and if they are to be accounted for on the theory of universal attraction, it is manifest that the same mechanical principles applied in the same way to the case of several bodies abandoned in free space to their mutual attractions, ought to lead us to their explanation" (Herschel 1829, 673). Evidently, when Herschel talks of finding "causes" for such deviations here, he typically means providing an account of how such deviations may arise from the action of the law of gravity on the bodies of the solar system.

What does this imply for the methods of physical astronomy? For starters, it implies that physical astronomers cannot proceed by pure induction—the laws governing these deviations are "too complicated indeed for observation alone to unravel". On the other hand, the method that is used here is not straightforwardly hypothetico-deductive either, in the sense that physical astronomers are usually not in the business of testing the law of gravity by examining the agreement between theory and observation, although the possibility that the law of gravity might require modification is always in the background. Rather, the focus is on examining the deviations between theory and observation to see how they can be accounted for by the action of the law of gravity on the bodies of the solar system. But these deviations can be extraordinarily complex, for they often involve the combined effects of various details in the solar system. Of crucial importance is the ability to separate out these effects and assign different causes to them—a process that I will refer to as *decomposition*. On the view of physical astronomy presented by Herschel, the primary role for theory is its use in decomposition. The way in which theory is used is best seen by examining some examples, which I will discuss in the next two sections.

2.4.1 Example 1: Planetary Perturbations Under the Assumption of Circular Orbits

In Part II, Section I, Herschel starts off by showing that second-order differential equations of a particular type can be integrated under certain circumstances, resulting in solutions that consist of an infinite series of terms. In Section II, he then starts with the equations of planetary motion, including perturbational terms, and shows how to manipulate them into the integrable type of second-order differential equation. In Section III, he shows how the integration may be carried out under the assumption of circular orbits and neglecting the inclination of the orbits relative to the ecliptic. Much of the analysis consists in an investigation of the "perturbative function", which represents, in the equations of motion, the perturbative force due to the disturbing body, and the difference it makes to the solutions when the equations are integrated. Through this analysis, he derives expressions for the perturbations in the radius vector and the longitude. Inspecting these expressions (Herschel 1829, 686–7), he concludes that the effect of perturbation due to a third body, under the assumption of circular orbits, is to (1) permanently alter the mean value of the radius vector, (2) the radius vector and the longitude acquire an additional small periodic motion that can be expressed as an infinite series of sine or cosine functions with coefficients, and (3) the values of these coefficients can be calculated given the mass of the disturbing planet, and the mean radii and mean motions of both the disturbed and disturbing planet.

Herschel points out that since the masses of Jupiter and Saturn are known, their perturbational effect (neglecting the eccentricity of their orbits) on other planets such as the earth can be calculated. He then notes that, given this means of calculation, the unknown masses of planets such as Mars and Venus can be inferred from observed perturbational effects—that is, to use modern terminology, the equations can be *inverted* to determine planetary masses. He then further suggests that, in fact, a system of equations can be set up where the only unknown quantities are the masses of the planets, and the masses of all the planets can be determined by simultaneously solving for the unknowns. Here, Herschel is well aware of the difficulties presented by observational error in such a procedure. He takes a brief detour to discuss the effects of errors of observations, and then refers the reader to Laplace's *Theorie Analytique des Probabilites* for further details. Regarding this method of determining the masses of the planets, Herschel writes:

> This method is laborious, certainly, but considering the perfection of modern observations, the great multitude of them which may be brought to bear upon this point, and the considerable degree of exactness which the theory of the planetary perturbations has now attained, it is not impossible that it may one day be made to render the best service in determining the masses even of those planets which have satellites. At all events, it is highly desirable that it should be applied for that purpose, as its results would lead us to judge how far the latter method can be depended on in cases like that of Jupiter and Saturn, where the great deviation from sphericity of the central body renders the application of [$T = (2\pi \, a^{3/2}) / (M + m)^{1/2}$] somewhat liable to error. (Herschel 1829, 688)

Herschel here is pointing out that this method of determining planetary masses gives the astronomer two different ways of determining the masses of Jupiter and Saturn. The first is the method used since Newton of applying the formula[13] $T = (2\pi a^{3/2})/(M + m)^{1/2}$, where T is the orbital period of a planet, a is the length of its semi-major axis, M is the mass of the planet, and m is the mass of a satellite. The second is the method just mentioned of solving a system of simultaneous equations, taking account of observational error. Now that there are two methods, the second method can be used as a check on the accuracy of the first method, given that the first method is not exact due to the departure from sphericity of these planets.

In addition, Herschel notes that if the two methods fail to result in agreeing values for the mass of a planet, this could be an indication that further investigation is needed:

> This alone, however, will not account for the great difference which Mr. Gauss has lately found between the mass of Jupiter, as obtained from observations of its satellites, and that deduced from the perturbations of the small planets intermediate between Jupiter and Mars [i.e., Ceres, Pallas, Juno, and Vesta], so that the subject must be regarded as open to further investigation, should the calculations of the last named eminent geometer be found to coincide with a more extensive series of observations of those interesting bodies than the shortness of time they have been known has hitherto allowed. (Herschel 1829, 688)

Note here that failure of the two values to converge is not taken by Herschel to be an indication of modifications being needed to the law of gravity. It is rather an indication that further investigation of the motion of the small planets is needed. The failure to converge could be due to further disturbing bodies, some unknown aspect of the mass distribution of Jupiter, including departures from sphericity, and so on. A modification to the law of gravity is always a possibility to be considered, but not before such further investigations are carried out.

2.4.2 Example 2: Planetary Perturbations Due to Eccentricities of Orbits

Sections IV and V of Part II are on the effects on the perturbations due to the eccentricity of the planetary orbits. Following on from the investigation of the perturbative function in Section III, Herschel provides a further analysis of the effects of eccentricity on the perturbative function. He shows that the perturbations in the radius vector and the longitude due to the eccentricities of the planetary orbits will be an infinite series of terms, each of which is a cosine or sine function multiplied by a coefficient and a power of e or e', where e is the eccentricity of the disturbed planet, and e' is the eccentricity of the disturbing planet. Thus, there is a set of terms that has either e or e' as a factor. These terms are said to be the *first-order terms*, and the effects on planetary motion due to these terms are called

---

[13] Herschel provides a derivation of this formula on page 655.

first-order perturbations. There is another set of terms that has e^2, e'^2, or both e and e' as a factor. These are *second-order terms*. Higher-order terms have higher powers of e or e' as factors.

Because the values of e and e' are small, most terms of higher order have little effect on the radius vector or longitude. But there could be some higher-order terms whose coefficients are very large, which could make a significant difference to the planetary motions. The important question here is: *which terms will make a significant difference?* Naturally, one might examine the contribution due to each term one by one, starting with the lowest-order terms. However, the number of terms at each order rapidly increases, making such an investigation impractical:

> Such is the immense number of terms, or rather of series of terms, branching out in all directions, of which the perturbations consist, that it is manifestly in vain to attempt to take account of them all. It is therefore of the highest consequence to have some guiding principle to direct us in our choice of the terms to be retained or neglected. (Herschel 1829, 698)

The guiding principle is provided by an examination of the coefficients of these terms. Herschel shows that the coefficients are fractions that have expressions such as the following in the denominator:

$$(i \pm k)\, n \pm l\, n'$$
$$\left\{(i \pm k)\, n \pm l\, n'\right\}^2$$
$$\left\{(i \pm k)\, n \pm l\, n'\right\}^2 - n^2$$

Here, i, k, and l are positive integers (i can be any positive integer, while there are restrictions on k and l), while n and n' respectively represent the mean motions of the disturbed and disturbing planet. Herschel notes that the denominators for some terms could take on a value of nearly zero when n and n' have certain values, in which case the term would have an outsized influence on the planetary motions:

> . . . should there be an approach to commensurability in the periodic times of the two planets, (as, for instance, should five times the mean motion of the disturbed planet ($5n't$) be very nearly equal to twice that of the disturbing, ($2nt$))[14] this circumstance will render some one of their factors ($5n' - 2n$) very small. In consequence, all the divisors into which this factor enters will become very small, and the inequalities affected by them will, in consequence, acquire from this cause an unnatural magnitude (if we may use such an expression) and must be retained, even though of such an order as would otherwise authorize their rejection. (Herschel 1829, 698)

The numbers 5 and 2 here were chosen by Herschel as a lead-in to his discussion of the so-called "great inequalities" of Jupiter and Saturn, which immediately follows this passage. He explains that when modern observations of Jupiter and Saturn were compared to ancient ones, it appeared to indicate that Saturn had slowed down, while Jupiter had sped up. These observations were "long a difficulty in

---

[14] Herschel makes a slight error here. In the notation he uses, the primed letters always stand for parameters of the disturbing body, so he has switched n and n'.

the way of the theoretical astronomer, and even a stumbling block in the way of the Newtonian philosophy" (Herschel 1829, 698-9). Astronomers had attempted to account for this by examining the perturbations of first and second order due to the eccentricities, but could not find an effect with a long enough period. The terms of higher order had been neglected until Laplace decided to examine the terms of third order:

> Here he immediately encountered the argument $(5n - 2n't + \text{const.})$; and the mean motion of Jupiter being to that of Saturn nearly in the proportion of 5 to 2, if we suppose n' to correspond to Jupiter's and n to Saturn's motion the co-efficient $5n - 2n'$ is very small, and the corresponding period is found on calculation to amount to 918 years. The resulting inequality has also $5n - 2n'$ for its divisor, and its magnitude is thus increased as well as its period lengthened. (Herschel 1829, 699)

Here we see that theory allowed physical astronomers to separate out, from the perturbations of Jupiter and Saturn, the contributions of some of the terms of the infinite series, and effectively search through them to identify the ones that might be giving rise to the observed deviation. Let us call the contribution of a single term to the perturbations a *component*. By calculating the period of the component, they could then examine observations stretching into the distant past in attempting to verify that the cause—the near-commensurability of the orbits of Jupiter and Saturn, in combination with the eccentricities of their orbits—has been correctly identified.

2.5 Research in Physical Astronomy

I have given above two of the more prominent examples in the "Physical Astronomy" of the use of theory for decomposition, but there are many more throughout the article. What we see in these examples is that Herschel typically first develops in detail the mathematical theory needed to deal with the investigations in question. In the case of the planetary perturbations, he starts with the equations of motion, including terms that represent the perturbative forces, and then shows how the equations may be integrated to obtain solutions. He then shows how certain terms in these solutions are the result of the terms representing perturbative forces in the original equations of motion. Usually, some further mathematical development is needed in order to determine from these solutions expressions for the perturbations in observable parameters such as planetary longitudes. Once this is carried out, the components of the perturbations due to each of these terms may in principle be precisely determined, including the relative magnitudes of the effects, the periods of the effects, and so on.

The development of theory in the century after Newton had thus given physical astronomers an important new capability by the time of Herschel. They could start with subtle details of the solar system, such as the eccentricities of the planetary orbits, and calculate precisely how they would make a difference to the observations. This is how, through the use of theory, "secrets which an angel might penetrate perhaps at a glance, become revealed to man by the slow, yet sure, effects of persevering thought."

This new capability has major consequences for research in physical astronomy. I shall focus here on two of them. The first consequence is that it greatly increases the number of phenomena that can be brought to bear on the law of gravity. Each of the subtle components that together make up the complex motions of the planets are in principle available for comparison with theory. Of course, this can be done only if these components can be properly separated out—indeed, properly separating out the individual subtle components of complex motion is one of the primary methodological challenges in the new physical astronomy. In many cases these subtle components show up initially as deviations between theory and observation. The investigation of the great inequalities of Jupiter and Saturn, discussed above, is one such example.

Another example that Herschel treats in Part III is on the motion of the lunar apsides. According to Herschel, the value that Newton calculated for this motion from theory was only half that observed, and it "became so great a stumbling-block in the way of succeeding geometers, as to shake their faith in the theory of gravity" (Herschel 1829, 720). As he did with the inequalities of Jupiter and Saturn, Herschel starts with the equations of motion for the moon including a perturbative force, and carries out an analysis showing how terms arise in the solutions that are the result of this force. He shows that when only terms of first order in the perturbative force are included, the motion of the lunar apsides will be only half that observed, but when terms of second order are included, the deviation could be accounted for. He credits Clairaut with taking the analysis to the second order, and thereby transforming the motion of the lunar apsides from a stumbling-block to a triumph for Newtonian theory:

> ...and the fact, so far from being an objection against gravity, is thus converted into a most cogent argument in its favour—its effects being thus shewn to correspond not merely to general results and first approximations, but to the refinements and niceties of theory. (Herschel 1829, 724)

The examples just mentioned are ones where analysis of the subtle components are taken to provide some verification for the law of gravity. If, on the other hand, the law of gravity is taken to hold exactly, then an investigation of these subtle components can potentially be used to make inferences about details of the solar system. A good example of this is the suggestion by Herschel that the perturbations under the assumption of circular orbits could be used to make determinations of the masses of planets with no satellites, such as Venus or Mars. Herschel also described work by Gauss in which the mass of Jupiter was determined by its perturbational effects on the planetoids between Mars and Jupiter. In this case, there was a lack of agreement between this new method of measuring the mass of Jupiter and the old method. This lack of agreement could of course be due to any number of reasons. It could merely be a calculational error, or a mistake in a derivation. But it could also be due to a previously unknown detail of the mass distribution of Jupiter, or an undiscovered small planet. It could even be due to the failure of the law of gravity to hold exactly. Further investigations would have to be carried out to find out which of these is the reason for the lack of agreement. But such investigations could potentially lead to the discovery of previously unknown details about the solar system, or even a modification to the law of gravity.

The second consequence of the new capability is that it allows astronomers effectively to use a method that I will refer to as the *method of residual phenomena*. As we have seen, theory gives astronomers the ability to start with particular details such as the eccentricity of the planetary orbits, and precisely characterize the component of planetary motion that arises due to that detail. Such components in the planetary motions that have known causes can then effectively be *subducted out* from the complex planetary motions. So one can, in principle at least, start with the planetary motions in their full complexity, and subduct out each of the subtle components with known causes. Suppose there is a significant remainder left after this process is carried out. Let us refer to the remainder as a *residual phenomenon*. If the subduction process has been done properly, the residual phenomenon might be due to a subtle component, or a sum of subtle components, whose cause has yet to be identified. The identification of this cause will then become an open question for further research—can some detail of the solar system be identified that gives rise to the residual phenomenon? The pursuit of this question would involve trying to look for patterns in the residual phenomena—e.g., Is it secular or periodic? If the latter, what is its period?—and then trying to match up this pattern with some detail in the solar system. There are complications, of course—the remainder could be due, not to some previously unknown detail of the solar system, but simple errors in observation, errors in theoretical calculations, mistakes in derivations, and so on. But nevertheless, one of the foci of research in the new physical astronomy is the pursuit of such residual phenomena.

2.6 Physical Astronomy and the Methodology of the *Preliminary Discourse*

Herschel does not explicitly describe the method of residual phenomena in the "Physical Astronomy"—he merely describes the tools that astronomers could use to carry it out. It would no doubt occur to anybody familiar with these tools, however, that this would be one way to proceed in astronomical research. In any case, Herschel *does* describe the method in the *Preliminary Discourse*. He presents a more general version of the method in a passage where he introduces the notion of residual phenomena:

> Complicated phenomena, in which several causes concurring, opposing, or quite indepen-
> dent of each other, operate at once, so as to produce a compound effect, may be simplified
> by subducting the effect of all the known causes, as well as the nature of the case permits,
> either, by deductive reasoning or by appeal to experience, and thus leaving, as it were, a
> *residual phenomenon* to be explained. It is by this process, in fact, that science, in its present
> advanced state, is chiefly promoted. Most of the phenomena which nature presents are very

complicated: and when the effects of all known causes are estimated with exactness, and subducted, the residual facts are constantly appearing in the form of phenomena altogether new, and leading to the most important conclusions. (Herschel 1830, 156)

This more general version of the method of residual phenomena allows for subduction to be done not just by calculating subtle components of the phenomena from established laws such as the law of gravity, but from "appeal to experience"— in some cases the subduction can be done through experimental investigation.[15]

Herschel clearly takes physical astronomy to be the exemplar for the method of residual phenomena. In fact, a good case can be made that the methodological view presented in the *Preliminary Discourse* is an extension and modification of the methodological views Herschel presents in the "Physical Astronomy". To be sure, it also includes many of the lessons Herschel learned in his work, both firsthand and in the writing of encyclopedia articles, in other areas of natural philosophy, such as chemistry, electromagnetism, optics, and the theory of sound. The methods that are suitable for physical astronomy are not necessarily suitable for these other areas, and the methodology presented in the *Preliminary Discourse* certainly reflects these differences.[16] Nevertheless, there is no doubt that many of the most striking features of the methodology of the *Preliminary Discourse* come from his consideration of the methods of physical astronomy.

Let us recall the three points that Herschel makes in the Introduction to the "Physical Astronomy," which I shall summarize as follows: (1) the great challenge of physical astronomy is the complexity of the planetary motions, (2) physical astronomy is unique in that the law of gravity was arrived at from a relatively rough characterization of the phenomena but has never required modification, and (3) any modification to the law of gravity will be found through the investigation of the subtler phenomena in the solar system.

Herschel surely has physical astronomy on his mind when he discusses the process by which a law is "verified" once it is arrived at by induction. Herschel downplays the importance of the initial inductive step. When we make an induction to a law, "we must not . . . . . . be scrupulous as to how we reach to a knowledge of such general facts: provided only we verify them carefully once detected, we must be content to seize them wherever they are to be found" (Herschel 1830, 164). More important, for Herschel, are the subsequent verifications of the induced law. These verifications must be as thorough as possible:

Whenever, therefore, we think we have been led by induction to the knowledge of the proximate cause of a phenomenon or of a law of nature, our next business is to examine

---

[15] This is the case in the second example Herschel gives after this passage, on experiments Arago did on a magnetic needle suspended over a plate of copper.

[16] Good (1982, 1987) and Cobb (2012a, b) show how some of the methodological views Herschel presents in the *Preliminary Discourse* originate from his work on, respectively, optics and electromagnetism. I do not wish to underplay the role of Herschel's work on sciences other than physical astronomy in the formation of his methodological views. What I do want to emphasize, however, is that many of the ideas in the *Preliminary Discourse* originate in his earlier work on the "Physical Astronomy".

deliberately and seriatim all the cases we have collected of its occurrence, in order to satisfy ourselves that they are explicable by our cause, or fairly included in the expression of our law: and in case any exception occurs, it must be carefully noted and set aside for re-examination at a more advanced period, when, possibly, the cause of exception may appear, and the exception itself, by allowing for the effect of that cause, be brought over to the side of our induction ... (Herschel 1830, 165).

The curious phrasing here—any exceptions must be "set aside for re-examination at a more advanced period" until, possibly, they are "brought over to the side of our induction"—makes perfect sense in light of the methods described in the "Physical Astronomy." The motion of the lunar apsides, for example, was a "stumbling-block" for the theory of gravity until Clairaut took the analysis to the second order, upon which it was "converted into a most cogent argument in its favour."

Most important, for Herschel, are cases where "the cause or law to which we are conducted be one already known and recognised as a more general one, whose nature is well understood, and of which the phenomenon in question is but one more case in addition to those already known" (Herschel 1830, 165). These are cases where the natural philosopher ought to be looking for residual phenomena, for they often lead to new discoveries:

Now, this is precisely the sort of process in which *residual phenomena* may be expected to occur. If our induction be really a valid and a comprehensive one, *whatever* remains unexplained in the comparison of its conclusion with particular cases, under all their circumstances, *is* such a phenomenon, and comes in its turn to be a subject of inductive reasoning to discover its cause or laws. It is thus that we may be said to witness facts with the eyes of reason; and it is thus that we are continually attaining a knowledge of new phenomena and new laws which lie beneath the surface of things, and give rise to the creation of fresh branches of sciences more and more remote from common observation. (Herschel 1830, 166)

Of particular note here is the analogy Herschel draws to vision—"to witness facts with the eyes of reason"—for it calls to mind the angels that Herschel refers to in the Introduction to the "Physical Astronomy," who have the ability to see, at a glance, the connections between the law of gravity and subtle components within the planetary motions. And sure enough, Herschel's primary example of this process of verification and discovery is physical astronomy:

Physical astronomy affords numerous and splendid instances of this. The law, for example, which asserts that the planets are retained in their orbits about the sun, and satellites about their primaries, by an attractive force, decreasing as the square of the distances increases, comes to be verified in each particular case by deducing from it the exact motions which, under the circumstances, ought to take place, and comparing them with fact. This comparison, while it verifies in general the existence of the law of gravitation as supposed, and its adequacy to explain all the principal motions of every body in the system, yet leaves some small deviations in those of the planets, and some very considerable ones in that of the moon and other satellites, still unaccounted for; *residual phenomena*, which still remain to be traced up to causes. By further examining these, their causes have at length been ascertained, and found to consist in the mutual actions of the planets on each other, and the disturbing influence of the sun on the motions of the satellites. (Herschel 1830, 166–7)

Herschel then goes on to say that a law of nature, such as the law of gravity, gains the highest degree of verification through "extending its application to a wide variety of cases that were not originally contemplated; in studiously varying the circumstances under which our causes act, with a view to ascertain whether their effect is general; and in pushing the application of our laws to extreme cases" (Herschel 1830, 167). This whole process crucially involves the investigation of residual phenomena, for residual phenomena often present cases that were not originally contemplated.

The following chapter discusses what Herschel calls the "higher degrees of inductive generalization."[17] For Herschel, the lower degrees concern the induction of proximate causes and empirical laws from the phenomena. At the next level, these causes and laws are then taken to be a new set of phenomena, which must be accounted for by a more general theory. The theory of universal gravity is Herschel's primary example of such a higher-level theory. How should we go about formulating such a high-level theory? Herschel says that there are three ways to do it:

> 1st, By inductive reasoning; that is, by examining all the cases in which we know them to be exercised, inferring, as well as circumstances will permit, its amount or intensity in each particular case, and then piecing together, as it were, these *disjecta membra*, generalizing from them, and so arriving at the laws desired; 2dly, By forming at once a bold hypothesis, particularizing the law, and trying the truth of it by following out its consequences and comparing them with facts; or, 3dly, By a process partaking of both these, and combining the advantages of both without their defects, viz. by assuming indeed the laws we would discover, but so generally expressed, that they shall include an unlimited variety of particular laws ;—following out the consequences of this assumption, by the application of such general principles as the case admits;—comparing them in succession with all the particular cases within our knowledge; and, lastly, on this comparison, so modifying and restricting the general enunciation of our laws as to make the results agree. (Herschel 1830, 198–9)

The first method is straightforward induction, while the second method is what we would now call the hypothetico-deductive method. The third method, which Herschel calls a mixture of the first two, involves a process in which the consequences of a theory are deduced, and if there is a gap between the consequences and observations, a modification and restriction of the general enunciation of the laws is made. Herschel takes all three of these processes to be applicable to the investigation of nature, depending on the circumstances. But he notes that the third method is "that which mathematicians (especially such as have a considerable command of those general modes of representing and reasoning on quantity, which constitute the higher analysis,) find the most universally applicable, and the most efficacious." (Herschel 1830, 199). It is striking here that Herschel takes the third method to be particularly effective in the hands of those who have mastered advanced mathematics.

Herschel thus believes that the best way to arrive at a general theory is by being led to it by particular inductions, and then verifying it by tracing out all of

---

[17] The terminology here calls to mind Herschel's reference to a "scale of generalization" in the Introduction to the "Physical Astronomy": Natural philosophers are to "ascend as high as the imperfection of human means of observation, and the limited powers of the human intellect will allow us in the scale of generalization" (Herschel 1829, 647).

its consequences under a wide variety of circumstances, some of which may not be obvious antecedently. The tools for carrying out this verification process are to be provided by mathematical analysis. Herschel's example here, again, is physical astronomy:

> To return to our example: particular inductions drawn from the motions of the several planets about the sun, and of the satellites round their primaries, etc. having led us to the general conception of an attractive force exerted by every particle of matter in the universe on every other according to the law to which we attach the name of gravitation; when we would verify this induction, we must set out with assuming this law, considering the whole system as subjected to its influence and implicitly obeying it, and nothing interfering with its action; we then, for the first time, perceive a train of modifying circumstances which had not occurred to us when reasoning upwards from particulars to obtain the fundamental law; we perceive that *all the planets* must attract *each other*, must therefore draw each other out of the orbits which they would have if acted on only by the sun; and as this was never contemplated in the inductive process, its validity becomes a question, which can only be determined by ascertaining precisely how great a deviation this new class of mutual actions will produce. To do this is no easy task, or rather, it is the most difficult task which the genius of man has ever yet accomplished: still, it *has* been accomplished by the mere application of the general laws of dynamics; and the result (undoubtedly a most beautiful and satisfactory one) is, that all those observed deviations in the motions of our system which stood out as exceptions, or were noticed as residual phenomena and reserved for further enquiry, in that imperfect view of the subject which we got in the subordinate process by which we rose to our general conclusion, prove to be the immediate consequences of the above-mentioned mutual actions. As such, they are neither exceptions nor residual facts, but fulfillments of general rules, and essential features in the statement of the case, *without* which our induction would be invalid, and the law of gravitation positively untrue. (Herschel 1830, 201–2)

The peculiar thing about physical astronomy being the primary example of the third method of inductive generalization is that, although the third method is supposed to involve "modifying and restricting" the induced law on comparison with a wide variety of particular cases, physical astronomy is the one example where the law itself had never been modified. This shows that "modifying and restricting" is not an essential part of the third method—what is important is that the remotest consequences of the law, and the widest variety of particular cases, are investigated. Many of these cases, if not most of them, involve residual phenomena.

2.7 Conclusion

We have seen that the "Physical Astronomy" starts off with an emphasis on the complexity of the planetary motions, and human cognitive limitations in grasping this complexity. Mathematical analysis is the tool through which these limitations are overcome. Mathematical analysis gives astronomers the capability of tracing out the subtle effects of details of the solar system, such as the eccentricities of the planetary orbits, on the complex motions of the planets and the moon. This capability greatly increases the empirical content of astronomy, in the sense that subtle effects of the planetary motions are now subject to comparison with theory. It also gives astronomers the means for making discoveries about previously unknown details of the solar system. Physical astronomy, under this conception, does not proceed by induction. Neither is it, however, hypothetico-deductive, in the sense that astronomers are typically not in the business of comparing observations with theory. They are instead in the business of using mathematical analysis to make discoveries about details in the solar system. Far from testing the law of gravity, this method proceeds on the assumption that the law of gravity holds exactly. Only the persistent failure of this method in finding a cause for a deviation between theory and observation would result in overturning the law of gravity.

This understanding of the methodology of physical astronomy has much in common with that described by George Smith in his 2014 paper "Closing the Loop". Smith claims that "the primary aim in comparing calculation with observation in gravity research is to discover which details of the physical world in point of fact make a difference and what differences they make" (Smith 2014, 285).[18] I have little doubt that Herschel would agree with this characterization, if applied to the physical astronomy of his day.

It is perhaps not surprising that the views of Herschel and Smith would agree, given that both of them arrived at their views through a close examination of the methods of physical astronomy after Newton, especially the work of the great Continental analysts of the eighteenth century. A more interesting question is why the idea of residual phenomena dropped out of discussions of the methodology of science in the late nineteenth century. One factor surely is that only those who were familiar with the tools of mathematical analysis truly understood the significance of residual phenomena. In the *Preliminary Discourse*, Herschel mentions, but does not discuss, any of the mathematical tools that are essential for carrying out the method of residual phenomena. There were probably very good reasons for this—the *Cabinet Cyclopaedia*, for which Herschel wrote the *Preliminary Discourse*, was intended for a mass market (Bolt 1998, 254), which prevented the inclusion of anything too mathematically demanding. The decision not to include discussions of the mathematical tools was unfortunate, however, since without such a discussion, much of what Herschel says about residual phenomena loses its appeal. For those reading the *Preliminary Discourse* without a background in mathematics, it's simply not clear how the method of residual phenomena is supposed to work—how, in particular, the "effects of all known causes are estimated with exactness, and subducted", or what, exactly, is meant by an "exception" to a law, and how such exceptions might be re-examined "at a more advanced period". Only those who were familiar with such mathematical tools—such as Thomson and Tait—understood the power of the method. This might be one reason why the method of residual phenomena failed to take root, and became obscure by the end of the nineteenth century.

---

[18] My repeated use of the terms "details in the solar system" and "differences they make" in this chapter are, as the reader might surmise, a conscious attempt to draw connections to Smith (2014).

Acknowledgment The research for this chapter was supported by the Ministry of Education, Singapore, under its Academic Research Fund Tier 1 Grant, No. RG156/18-(NS).

References

Agassi, J. 1981. Sir John Herschel's philosophy of success. In *Science and society: Studies in the sociology of science*, 388–420. Dordrecht: D. Reidel Publishing Company.

Bolt, M. 1998. *John Herschel's natural philosophy: On the knowing of nature and the nature of knowing in early-19th-century Britain*. Ph.D. Dissertation, University of Notre Dame.

Brown, T. 1824. *Lectures on the philosophy of the human mind*. Vol. II. Philadelphia: John Grigg.

Buchwald, J. 1989. *The rise of the wave theory of light: Optical theory and experiment in the early 19th century*. University of Chicago Press.

Bullen, K.E. 1975. *The earth's density*. Chapman and Hall.

Buttman, G. 1970. *The shadow of the telescope: A biography of John Herschel*. Lutterworth Press.

Cobb, A.D. 2012a. Inductivism in practice: Experiment in John Herschel's philosophy of science. *HOPOS: The Journal of the International Society for the History of Philosophy of Science* 2: 1: 21–54.

———. 2012b. Is John Herschel an inductivist about hypothetical inquiry? *Perspectives on Science* 20 (4): 409–439.

Ducasse, C.J. 1960. John F. W. Herschel's Methods of Scientific Inquiry. In *Theories of scientific method: The renaissance through the nineteenth century*, ed. Edward H. Madden, 153–182. Gordon and Breach.

Good, G. 1982. *J. F. W. Herschel's optical researches: A study in method*. Ph.D. Dissertation, University of Toronto.

———. 1987. John Herschel's optical researches and the development of his ideas on method and causality. *Studies in History and Philosophy of Science* 18: 1–41.

Graves, R. 1882. *Life of Sir William Rowan Hamilton*. Vol. 1. Hodges, Figgis, and Co.

Herschel, J.F.W. 1829. Physical Astronomy. *Encyclopaedia Metropolitana*, 2nd Division: Mixed Sciences 1: 647–729.

———. 1830. *A preliminary discourse on the study of natural philosophy*. Cambridge University Press.

———. 1864. *Outlines of astronomy*. Cambridge University Press.

Jain, C. Lal. 1975. *Methodology and epistemology: An examination of Sir John Frederick William Herschel's philosophy of science with reference to his theory of knowledge*. Ph.D. Dissertation, Indiana University.

Lacroix, S.F. 1816. *An elementary treatise on the differential and integral calculus*. Deighton and Sons.

Mill, J.S. 1843. *A system of logic*. Vol. 1. Cambridge University Press.

Oldroyd, R. 1986. *The arch of knowledge: An introductory study of the history of philosophy and methodology of science*. Methuen.

Olson, R. 1975. *Scottish philosophy and British physics, 1750–1880*. Princeton University Press.

Pemberton, H. 1728. *A view of Sir Isaac Newton's philosophy*.

Powell, B. 1832. Further remarks on experiments relative to the interference of light. *Philosophical Magazine* 1 (Third Series): 433–438.

Poynting, J.H. 1894. *The mean density of the earth*. Charles Griffin and Company Limited.

Rayleigh, L. 1877. *The theory of sound*. Vol. 1. Cambridge University Press.

Smith, G.E. 2014. Closing the loop. In *Newton and empiricism*, ed. Zvi Biener and Eric Schliesser, 263–345. Oxford University Press.

Thompson, William, and P.G. Tait. 1883. *Treatise on natural philosophy*. Vol. 1. Cambridge University Press.

Whewell, W. 1837. *History of the inductive sciences*. Vol. 2. Cambridge University Press.

———. 1840. *Philosophy of the inductive sciences*. Vol. 2. Cambridge University Press.

Chapter 3
Newton on the Relativity of Motion and the Method of Mathematical Physics

Robert DiSalle

3.1 Introduction

The work of George Smith has illuminated how Newton's scientific method, and its use in constructing the theory of universal gravitation, introduced an entirely new sense of what it means for a theory to be supported by evidence. This new sense goes far beyond Newton's well known dissatisfaction with hypothetico-deductive confirmation, and his preference for conclusions that are derived from empirical premises by means of mathematical laws of motion. It was a sense of empirical success that George was especially well placed to identify and to understand, through his experience as an engineer specializing in failure analysis. For Newton, to understand how well his theory was supported by evidence, he had to anticipate, as far as possible, all the ways in which it might be wrong. This paper explores how Newton's empirical method shaped his thinking about space, time, and the relativity of motion.

Newton's dedication to this practice, as we've learned from George, was an essential part of what made the theory of universal gravitation so fruitful as an account of gravity, as a foundation for celestial mechanics, and as a model for the future development of theoretical physics. George's account of this particular practice, and its role in Newton's methodology as a whole, is well known (see, for example, Smith 2002a, b). My purpose is only to point out some further aspects of it that deserve more attention. For the empiricist methodology that made universal gravitation an empirically successful theory is also deeply intertwined, more than has been generally acknowledged, with aspects of Newton's work that seem to be

R. DiSalle (✉)
Department of Philosophy, Western University, London, ON, Canada
e-mail: rdisalle@uwo.ca

© The Author(s), under exclusive license to Springer Nature Switzerland AG 2023
M. Stan, C. Smeenk (eds.), *Theory, Evidence, Data: Themes from George E. Smith*, Boston Studies in the Philosophy and History of Science 343, https://doi.org/10.1007/978-3-031-41041-3_3

more purely philosophical. This particularly applies to Newton's theory of space, time, and motion.

For much of its history, of course, and especially during much of the twentieth century, Newton's theory was viewed as quite separate from, and even as an embarrassment to, his stated empiricist method. According to Newton's laws, the physically meaningful quantities are force, mass, and acceleration, while velocity is relative; his conception of "absolute space," however, implies that there is a meaningful distinction in principle between uniform motion and "absolute rest" in space, even though they are physically indistinguishable. This seemingly naive metaphysical appendage to an empirically well-founded physics earned Newton a low reputation as a philosopher among twentieth-century empiricists. Only Stein (1967) eventually convinced philosophers of science to take seriously, at least, the empirical motivations of Newton's theory of space, time, and motion, and its close connection with the physics of motion, in spite of the superfluous aspects of absolute space. I propose to take a further step in the same direction, one particularly inspired by George Smith's work: Newton's philosophical account of absolute space, time, and motion, with all its flaws, was an integral part of his empiricist methodology. It was integral to his effort to anticipate all the ways in which his causal account of celestial motion, through universal gravitation, might be wrong. To see this, we have to see the development of this theory in connection with Newton's developing understanding of the relativity of motion, through the preliminary drafts and successive editions of the *Principia*.

3.2   Absolute Space and the Principle of Relativity

The internal flaw in Newton's theory of absolute space was first made clear, in a published work at least, by Berkeley (1721). As Newton well knew, motion with respect to space itself is unobservable, and any treatment of motion must begin with a relative space identified by relatively fixed empirical markers: the surface of the earth, the inner walls of a moving ship, the fixed stars. For Newton's application of the laws of motion, all of the relevant phenomena were motions of small bodies with respect to the earth, and displacements of celestial bodies with respect to the fixed stars. The empirical frame of reference for his theory, therefore, along with the theory's evidentiary basis, was essentially the one that astronomers had relied upon for centuries, though the accuracy of the evidence had (beginning with Tycho Brahe) increased dramatically in the century preceding Newton's *Principia*. By interpreting these phenomena through the laws of motion, Newton could infer the forces at work from the apparent motions, and thus describe the solar system as a system of interacting masses moving about a common center of gravity, calculated from their relative masses and positions. This analysis, in turn, led to an account of "true motion" in a certain restricted, yet decisive, sense. It led to a principled account of the true "frame of the system of the world": only the center of gravity of such a system can be its true fixed center (at rest or in uniform motion), and since the Sun has most of the mass of the system, the Sun, though "agitated" by its interactions with the planets, can never be far from the center of gravity. In other words, the decision between the "two chief world systems" could be settled not by the most plausible hypothesis, but by reasoning from the phenomena with the help of the laws of motion. Newton himself noted, of course, that this solution was independent of whether the center of gravity of the system is at rest, or moving uniformly.

It was Berkeley who, embracing Newton's reasoning thus far, drew what now seems to be the obvious inference: that "absolute space" in Newton's sense is superfluous to Newton's theory of motion. Newton had succeeded in finding an empirical frame of reference in which the forces at work within the system can be known, and the quasi-heliocentric structure of the system can be calculated. But it certainly does not follow from this success that there is a truly resting frame of reference—at rest in absolute space—with respect to which the center of gravity (or any body at all) has its true velocity. Berkeley also pointed out, more clearly than Newton's other contemporaries, just how far Newton himself had gone to incorporate the relativity of motion into his theory, and to make it, as we would say, Galilean-invariant. The most obvious example is Corollary V to the laws of motion:

> Corollary V: The motions of bodies enclosed within in a given space are the same among themselves, whether that space is at rest, or moving uniformly in a straight line without circular motion. (Newton 1687b, 19.)[1]

This was Newton's version of the relativity principle that had been used by Galileo and made precise by Huygens, and that is generally known as the principle of "Galilean relativity."[2] Berkeley turned this relativity principle directly against the very idea of absolute motion:

> As it is clear that, according to the principles of those who introduce absolute motion, it cannot be known by any mark whether the entire frame of things is at rest or moved uniformly in a right line, it is evident that no absolute motion of any body can be known. (Berkeley 1721, section 65)

Beyond this, Berkeley recognized a more subtle way in which Newton had integrated the relativity principle within his theory: the re-definition of inertia as an essentially relativistic concept. The "principle of inertia" as we know it was first introduced by Huygens, though unclear or incomplete forms of it had already been asserted by Galileo, Descartes, and Gassendi:

> Hypothesis I: Any body, once moved, if nothing opposes it, will continue to move perpetually with the same velocity in a straight line. (Huygens 1656, 31)

Kepler had defined "inertia" as a body's natural tendency to stay at rest; the core of the newer idea was that a body's natural tendency is to maintain its velocity until acted upon by an external cause. In the thinking of Galileo and Huygens,

---

[1] Translations are my own, unless otherwise noted.

[2] In what follows, "relativity principle" and "relativity of motion" refer to the Galileo-Huygens principle, and "relativistic" is to be understood in in the sense of this principle.

this tendency was essential to the relativity principle, as it formed the dynamical basis for the indistinguishability of uniform motion from rest: since bodies naturally persist in uniform motion, any accelerations among bodies will be simply composed with any uniform motion that they share, and so the bodies will behave in a shared uniform motion exactly as they would in a shared state of rest. Galileo appealed to this principle in explaining the fall of a stone from a tower as the earth moves, and Huygens in explaining collisions of bodies inside a moving canal boat (Galileo 1632, 9f.; Huygens 1656, 31).

Before the *Principia*, however (and for some time after), the persistence of motion was generally understood in a manner incompatible with the relativity principle. For it was generally assumed that a body's power to persist in motion, and its power to change the motion of other bodies in virtue of its own motion, were distinct from its property of resisting changes in its own motion. This implies a difference between the state of rest and the state of uniform motion that, strictly speaking, should have been rejected by all those who claimed to embrace the relativity principle. But it proved difficult to formulate the concept of inertia in a relativistic way. Leibniz, for example, always distinguished between the "active power" of a body to change the motion of another, and its "passive power" to maintain its own state (e.g. 1695, 146; 1699, 170). Perhaps the most useful example is Newton himself, who, in the manuscript *De Gravitatione et aequipondio fluidorum* (1684a), made the same distinction:

Definition 7. Impetus is force in so far as it is impressed on another.
Definition 8: Inertia is the internal force of a body, so that its state may not be easily changed by an external force. (Newton 1684a)

Sometime between writing *De Gravitatione* and beginning the first drafts of the *Principia*, Newton came to see that these Definitions divided into distinct concepts what was a single concept, the "vis insita" or "innate force," seen from different points of view; he explicitly blamed his former division on a failure to take into account the relativity principle. His first statement of the relativity principle appeared in the manuscript "On the motion of spherical bodies in fluids" (1684b), as a "Law" rather than a Corollary (as it became in the *Principia*).

Law 3: The motions of bodies included in a given space are the same among themselves, whether that space is at rest or moves uniformly in a straight line without circular motion. (1684b, p. 40r)

Neither the explicit principle, nor any acknowledgement of it, appears in *De Gravitatione*. After introducing it, however, Newton gradually revised his conception of inertia toward the relativistic notion that eventually appeared in the *Principia*.[3] Every body has a degree of mass; whether this appears as the resistance

---

[3] Newton's development of his mature, relativistic conception of inertia was, evidently, an important part of his development of the concept of mass as presented in the *Principia*. The latter development is central to the history of the conceptual structure of the *Principia*. But it has received a thorough historical and philosophical treatment only recently, in work by Fox (2016).

to motion, or a power proportional to its motion, depends on whether it is regarded as moving uniformly or at rest. The "inertia of the mass" is the same regardless of the velocity attributed to it. A subsequent draft, "De motu corporum" (1685b) contains a relativistic definition of inertia nearly identical to Definition III of the *Principia*:

Definition 3: The internal force of matter is the power of resistance by which any body persists in its state of rest or of moving uniformly in a straight line: it is proportional to the body and does not differ from the inertia of matter except in our mode of conceiving it. A body truly exerts this force only in a change of its state brought about by another force impressed upon it, and the exercise of this force is both resistance and impetus, which are distinct from one another only relatively: resistance in so far as the body, to maintain its state, opposes the impressed force; impetus insofar as the same body, yielding only with difficulty to the force of a resisting obstacle, endeavors to change the state of that obstacle. Resistance .is commonly attributed to resting bodies and impetus to moving bodies; but motion and rest, as commonly understood , are only relatively distinguished from each other; and bodies commonly seen as resting are not always truly at rest. (Newton 1685b, 315)

In remarking that the internal force differs from the inertia of matter only " in our mode of conceiving it," and that impetus and resistance differ "only relatively," Newton was clearly correcting the views he had expressed in *De Gravitatione*.[4] The most critical innovation was the assertion that "a body exerts this force only in a change of state brought about by another force impressed upon it": this explicitly rejects the pre-relativistic idea of a force that is required to maintain a body in its motion, because motion and rest are in fact only relatively distinguished. Newton may have confused some readers of the *Principia* by referring to the innate force as the "force of inertia" or "force of inactivity." In explicating it as he did, however—specifying that a body exerts it only when a force is impressed upon it by another body—Newton showed that he had cleared away the last remnant of the medieval impetus theory, that is, the notion of a special power by which bodies persist in their motions. This conceptual achievement was not lost on Berkeley:

Leibniz confounds impetus with motion. According to Newton, impetus is in truth the same as the force of inertia . . . . (Berkeley 1721, section 16)

Inert body acts just as body moved acts, if the matter is truly examined. This is what Newton acknowledges, where he asserts that the force of inertia is the same as impetus . . . (*Ibid.*, section 26)

In other words, Berkeley saw that Newton had defined a concept of inertia that expressed the profound connection between the principle of inertia and the relativity of motion, on which even a professed advocate of relativity such as Leibniz remained confused.

Generally, Berkeley, more clearly than other early critics of absolute space, saw that Newton had thoroughly integrated the principle of relativity into his theoretical

---

[4] The date of *De Gravitatione* is unknown. However, the development of Newton's conception of inertia toward his mature view, in the drafts of *De Motu*, definitively places *De Gravitatione* before those drafts. (See DiSalle 2020a, b).

physics. This was precisely the ground of his objection to absolute space: there was no need for it in a theory that had spectacularly solved the fundamental problems of terrestrial and celestial motion without any appeal to it.

> The laws of motions and all their effects, and the theorems containing the calculations of the same for different figures of the paths, as well as for accelerations and various directions, and for more or less resistant media, all these hold without the calculation of absolute motion. (Berkeley 1721, section 65)

If we replace metaphysical hypotheses into the nature of things with Newtonian mathematical principles, avoid abstractions such as absolute space, and reject any but sensible measures of motion, according to Berkeley, we will

> leave untouched all those celebrated theorems of the mechanical philosophy, by which the recesses of nature are brought out and the system of the world is subjected to human calculation: And the study of motion will be freed from a thousand minutiae, subtleties, and abstract ideas. (Berkeley 1721, section 66)

The "celebrated theorems" include, evidently, the derivation of universal gravity and of the structure of the solar system, which Newton had achieved by the use of attraction as a "mathematical hypothesis." That is, Newton had brought all observable motions under the rule of the laws of mechanics and the law of universal gravitation. The idea that there was another sense in which bodies could be said to be moving, with respect to a space that cannot be represented in any empirical problem addressed by Newton's mathematical methods, was indeed an "abstract idea" in Berkeley's sense. Berkeley's critique therefore raises, more cogently than anyone else had done, the question, why did Newton ever think that he required the concept of absolute space? We will return to this below. For now, it suffices to recall that in Berkeley's view, Newton began from a misguided belief that bodies, and the space in which they exist and move, exist independently of being perceived. So Berkeley was approaching this question with a particular aim: to show that not even the achievements of Newton's science could justify belief in a material world outside the mind. For anyone who maintained that belief, eliminating absolute space from Newton's physics was more challenging than it was for Berkeley.

It should be recalled here that to the problem of absolute space, within Newton's theory, there is no corresponding problem with the concept of absolute time, or corresponding motive to eliminate the concept from Newton's theory. Absolute time does not imply more than is strictly required by the laws of motion, and therefore does not invoke relations for which the laws do not provide, in principle or to some degree of approximation, empirical measures. Absolute time incorporates two principles: absolute simultaneity, and absolutely "equable flow," that is, absolute equality of time intervals. The first is evidently presupposed by the notion that there are objective spatial distances, and therefore objective relative motions; indeed, this seems to have been presupposed almost universally before the advent of special relativity. None of the historically distinguished advocates for the relativity of time, such as Leibniz or Mach, ever doubted that there is an objective measure of spatial relations at a given time, and successive spatial configurations of bodies—on this assumption rested the entire classical theory of the relativity of motion. Newton made this assumption explicit in the Scholium to the Definitions: all things are located "in time with regard to order of succession, and in space with regard to order of situation" (1687b, p. 7); and again in the General Scholium: "every particle of space is *always*, and every indivisible moment of time is *everywhere*" (1726, p. 528). The laws of motion provide, in principle, ways of determining when events are simultaneous, to some level of approximation. Given the finite velocity of light, which was well known in Newton's time, any such determination would necessarily be retrospective.

For the principle of equable flow, too, empirical content is provided by The laws of motion. The laws determine a physically distinguished state of uniform rectilinear motion, so that any freely moving body must move equal spatial distances in equal intervals of time. Such a body obviously represents an ideal case, not to be encountered where gravity is ubiquitous; Newton noted that, in general, there may be no truly equable movement of any actual body in the universe (Newton 1687b, 7). Yet the laws of motion provide, in principle, empirical measures of how well any motion approximates an inertial motion, and practical comparisons among motions to determine the best available approximation to an absolutely equable measure. For example, an ideal rigid sphere, left to rotate freely, will rotate through equal angles in equal times, but the rotation of the non-ideal earth generally varies; therefore, the day as defined by the actual rotation of the earth must be compared with sidereal motions, as in the traditional "equation of time" (*ibid.*). Ideally, the more closely any such motions approximate equable motion, the more they will tend toward agreement with each other, i.e. toward equal or mutually proportional intervals of time. The equable flow of time, then, is a theoretical concept for which the laws of motion determine a method of approximation. Newton could justly be confident that measurements of time for celestial motions, though imperfect, were ameliorable. For the concept of absolute velocity in absolute space, in sharp contrast, there is no method of approximation, that is, no sensible measure of velocity from which a true measure could be gathered. This is why absolute space could be eliminated on the basis of an internal analysis of Newton's mechanics (cf. below), while absolute time was not displaced until Einstein proposed an alternative theory.

3.3 The Relativity of Motion and Newton's Method

Newton's introduction of the relativity principle (1684b) marked the beginning of what I have called "Newton's theory of relativity."[5] For here Newton began to work on a broader task, namely, to explore the conceptual consequences of embracing the relativity principle: the review and replacement of existing theoretical concepts in accord with the principle, and the development of a new dynamical account of systems of bodies in states of motion that may be indistinguishable from rest.

---

[5] See DiSalle (2020b) for further discussion.

We have already discussed an instance of the first point, in Newton's gradual development of a relativistic conception of inertia. A striking illustration of the second is Newton's novel application of the relativity principle to the traditional question between the heliocentric and the geocentric accounts of the solar system, which gave the question a new meaning. His treatment began with four laws:

> Law 1: By its innate force alone, a body perseveres in uniform motion in a straight line if nothing hinders it.
> Law 2: The change in a body's state of motion or rest is proportional to the force impressed and acts along the straight line in which that force is impressed.
> Law 3: The motions of bodies included in a given space are the same among themselves, whether that space is at rest or moves uniformly in a straight line without circular motion.
> Law 4: By the mutual actions between bodies their common center of gravity does not change its state of motion or rest. (Newton 1684b, 40r)

Newton's essential step forward was to apply the third and fourth principles to the motion of the solar system as a whole. He now made it clear that the motion of the system could be considered, not as having a state of motion in space itself, but as contained in a space of its own, "the whole space of the planetary system," i.e., a space encompassing all of the planets whose motion can be traced with respect to the fixed stars. Then the configuration of the system can be determined from the actions of the bodies among themselves: by Law 3, these actions will be the same, whether the space is at rest or in uniform motion; by Law 4, these actions will not alter the state of motion or rest of the center of gravity, which may be in uniform motion or rest along with the entire system. Newton concluded:

> Moreover the whole space of the planetary heavens either rests (as is commonly believed) or moves uniformly in a straight line, and hence the common center of gravity of the planets (by Law 4) either rests or moves along with it. In both cases (by Law 3) the relative motions of the planets are the same, and their common center of gravity rests in relation to the whole space, and so can certainly be taken for the still center of the whole planetary system. (*Ibid.*, 47r)

The solution to the problem of the system of the world thus becomes the problem of using the laws of motion to find the resting center of the system, which can only be its center of gravity. To solve this problem, it suffices to begin with what Newton would later call a "relative space."

It followed that the traditional question—"which body is at rest in the center of the planetary system?"—rested on an unjustified supposition. In a system of interacting bodies, only their common center of gravity will remain unaccelerated. So the nearest equivalent to the traditional question is, "which body is closest to the system's center of gravity?" By "Law 3," the motions of the bodies in the system will be the same, whether its center of gravity is at rest or in uniform rectilinear motion. In explicitly asserting the dynamical equivalence of "whole spaces" that may moving uniformly or at rest, Newton made it clear that the solution to the problem of "the system of the world" is the same with respect to any such moving space as it is with respect to immobile space.

This discussion provides a further example of Newton's conceptual progress beyond *De gravitatione*. By the same token, it connects Newton's progress in thinking about relativity with his concern about ways in which his reasoning might be wrong. In *De Gravitatione*, Newton had criticized Descartes's conceptions of matter, space and motion; in particular, he noted the incoherence of the concept of matter as nothing but extension, and of motion as ("in the philosophical sense") nothing but the displacement of a body with respect to immediately contiguous bodies. Both concepts were contrary, not only to common sense ideas, but also to the Cartesian explanation of planetary motion, which depended, in spite of the philosophical account of motion, on the supposition that body follows a privileged trajectory in space (a straight line) unless acted upon by external causes. Newton incorporated some of the essential points of his criticism into the Scholium to the Definitions in the *Principia*, though without mentioning Descartes by name.

One particular argument against Descartes, however, was not included in the *Principia*, even though, in retrospect, it seems as if it might have proved a compelling one: Descartes' definitions introduced arbitrariness and uncertainty into the very idea of a trajectory, and therefore undermined the very possibility of a law of motion such as that on which Cartesian physics was founded.[6] If such a law is to be applicable, there must be a way to identify a path for any body, and to characterize its deviation from a rectilinear path. But this is impossible if space is a gyrating fluid vortex, and positions of bodies are defined only by the particles to which they are contiguous.

> [N]o one can assign the place according to Descartes at which the body was in the beginning of the motion undergone, or rather he has not said from where it is possible for a body to be moved. And the reason is that, according to Descartes, it is not possible to define and assign the place except from the position of the surrounding bodies, and that after any motion having been undergone, the position of the surrounding bodies is no longer the same as it was before . . . .It follows that Cartesian motion is not motion, for it has no velocity, no determination, and there is no space or distance that it traverses. Therefore it is necessary that the definition of places, and so of local motion, be referred to some immobile thing, such as extension alone, or space insofar as it is viewed as truly distinct from bodies. (Newton 1684a, 9–11)

Stein identified the essential point that Newton was trying to make: that dynamics as then understood, founded on the principle of inertia, required just that connection of space with time implied by the privileged status of uniform rectilinear motion; moreover, Newton's contemporaries were in no reasonable position to deny this, as they all adopted such a principle unquestioningly as the foundation for their program for explaining the planetary motions (cf. Stein, 1967).

Yet one can also see why Newton abandoned this argument, once having fully embraced the relativity principle. For the conclusion that he had drawn was too strong: there was no need to refer places and motions to "some immobile thing." Given the relativity principle and the center of gravity principle, Newton could solve the problem of the system of the world without knowing how the system

---

[6] On Newton's criticisms of Descartes, see Stein (1967, 2002). For further discussion and comparison of Newton's criticisms in *De Gravitatione* and the Scholium, see DiSalle (2006, chapter 2), and DiSalle (2020b).

as a whole might be moving with respect to space itself. The decision between the heliocentric and geocentric models was no longer a matter of choosing the more plausible hypothesis, but a matter of calculation using the laws of motion and astronomical evidence.

> Hence truly the Copernican system is proved a priori. For if the common center of gravity is calculated for any position of the planets it either falls in the body of the Sun or will always be very close to it. (Newton 1684b, 47r)

By "a priori," Newton obviously did not mean anything like the usual philosophical meaning, that is, "from first principles," or independent of experience, since the "a priori" proof explicitly appeals to empirical facts. He seems to have meant, rather, a proof from what is previously established: empirical facts about the motions and magnitudes of the planets, combined with the established physical laws that permit such a calculation. This was in stark contrast to the a posteriori argument for the heliocentric view maintained by his mechanistic contemporaries: that a heliocentric theory is a more likely basis for a mechanistic explanation, such as a vortex theory in the Cartesian vein. Newton's conclusion dispensed with the idea of a true central body, but rigorously derived a quasi-heliocentric account from physical principles and phenomena. And he showed that our knowledge of it could not be affected by our ignorance of the motion of the whole system in immobile space.

3.4   Newton's Introduction of Absolute Space

Given Newton's clear-sighted understanding of the principle of relativity, and his commitment to empirical methods for deciding theoretical questions (whenever possible), it is a subtle problem to understand the place of his conception of absolute space within his broader theoretical outlook. There are three points about this conception that should be more widely appreciated. First, the conception was, in fact, Newton's own: he coined the terms "absolute space" and "absolute time," and there is no established usage of these terms before Newton used them. There was, of course, a history of distinguishing "absolute motion" from "relative motion," and Barrow's use of this terminology (1685) was doubtless familiar to Newton. Moreover, Newton was evidently not the first to think of space as infinite, immobile, and homogeneous, or time as flowing equably. To refer to "absolute space" and "absolute time" as Newton's own theoretical terms, therefore, is not to deny that they denoted ideas that had much in common with those of previous philosophers. It is only to acknowledge something that ought to be obvious: that he introduced these terms in order to tell his readers exactly how he meant them to be understood. Newton never asserted that "space is absolute" or that "time is absolute," as if "absolute" were a predicate with an established philosophical significance in this context. Instead, he introduced the terms "absolute space" and "absolute time," along with clear explications of their meanings.

This first point is reinforced by the second, namely, that Newton first intro-duced "absolute space, time, place, and motion" explicitly as "Definitions," in an unpublished draft titled *De motu corporum in mediis regulariter cedentibus* ("On the motion of bodies in regularly yielding media," 1685a). The treatise begins with eighteen definitions, of which the first four distinguish absolute from relative time and space:

> Def. 1. Absolute time is that which by its own nature without relation to anything else flows uniformly. Such it is whose equation Astronomers investigate, and by another name is called Duration....
>
> Def. 2. Time looked at relatively is that which from something some other sensible passage or another flow or passage is measured in respect to the flow or passage of any sensible thing is considered as uniform....
>
> Def. 3. Absolute space so-called is that which by its own nature and unrelated to any other thing whatsoever always remains immobile. As the order of the parts of time is immutable, so also is that of the parts of space....
>
> Def. 4. Relative space is that which is considered immobile with respect to another any sensible thing: such as the space of our air with respect to the earth.... (Newton 1685a)

Granting that Newton was influenced by contemporary ideas about space,[7] it is clear that he wished the reader of his treatise on motion to understand no more or less by these terms than he specified.

In the *Principia*, Newton used more or less the same words, speaking in the same definitional mode, to distinguish absolute from relative space, time, and motion. It might seem puzzling, therefore—not to mention adverse to my point—that he did not include them among the named Definitions (now only eight) with which the book begins. But the puzzle is solved by comparing Newton's statements of the role of definitions. In the earlier treatise, the definitions are to ensure "that the reader, freed from certain common prejudices and imbued with distinct conceptions of mechanical principles, may agree to what follows..." (Newton 1685a). This statement differs little from one that follows the Definitions in the *Principia*: "Hitherto I have laid down the definitions of such words as are less known, and explained the sense in which they are to be understood in what follows" (Newton 1687b, 5). But one important difference stands out: in the *Principia*, the logical function of definitions in the axiomatic structure—to explain how the terms "are to be understood in what follows"—excludes the definitions of absolute space, place, time, and motion. They were crucial to his account of the philosophical context of his theory of motion, and, especially, to his view of the problem of the true structure of the solar system—"the frame of the system of the world"—and how the physics of the *Principia* would solve it. They were not, however, "terms used in the following treatise" in a logical sense: no subsequent reasoning in the book attempts to establish the absolute motion of any actual body in absolute space.[8]

---

[7] Rynasiewicz (1995a, b) offers a thorough account of Newton's immediate antecedents.

[8] Interestingly, the nearest approach to such a claim occurs in Proposition LXIII, in his discussion of motions with respect to the center of gravity of a uniformly moving relative space. He adds the hypothetical statement that "adding to this motion the uniform progressive motion of the entire

Therefore these definitions no longer belonged to the logical structure of the work, as they had in the first draft; instead, they are relegated to the Scholium, where Newton explicated his distinctions and their empirical content, and exposed the philosophical "prejudices" that must be removed.

The third important point about Newton's definitions concerns "absolute space" specifically: the theory of absolute space is, in fact, a space-time theory. This was already noted by Stein (1967), in explicating Newton's theory in four-dimensional terms. But it should be emphasized that Newton's own definition is explicitly spatio-temporal. The term "space-time" is anachronistic, of course, but it is certainly not anachronistic to observe that Newton incorporated time into his definition of absolute space: it "*remains always* similar and immovable" (Newton 1687b, 5; my emphasis). Newton did not think of space-time, but he did think of absolute space as something essentially connected with time. Predicates such as "same position at different times," or "same moment of time at different places," and "same velocity at different times," evidently refer to spatio-temporal relations that were commonly understood before the twentieth century. And when Newton spoke of the "absolute places" as constituting the proper reference-frame for absolute motion, he characterized them as those places that "from infinity to infinity maintain given positions with respect to one another" (Newton 1687b, 8f.). In short, rather than claiming that "space is absolute"—which would not have had an obvious meaning for his readers—Newton claimed that space has a peculiar connection with time, and "absolute space" was his term for this conception of space.

My purpose here is not to belabor a point that seems obvious as soon as one reads Newton's Scholium with care. It is, rather, to provide the right context for the earlier-mentioned question, why did Newton maintain the theory of absolute space in a physical theory that had no use for it? Briefly: Because he understood that physics, in his time, presupposed a certain connection between space and time. This was strictly implied by the accepted principle that the motion of any free body is uniform and rectilinear, and that any change in that motion requires a causal explanation. Neither he nor any of his contemporaries, nor anyone else for the next two centuries, proposed a more appropriate spatio-temporal account of that connection than Newton's own theory. Both Huygens (cf. Stein 1977) and Berkeley (cf. above) were aware that Newton's concepts of absolute acceleration and absolute rotation could dispense with the concept of absolute velocity in absolute space. But neither was in a position to imagine a spatio-temporal structure in which the distinction between uniform motion and acceleration, and between rotation and non-rotation, would be physically meaningful, while absolute velocity would not. Unlike Berkeley, Newton was convinced that the laws of motion, if they were true at all, were true of the universe itself, independent of anyone's perceiving it—and that such laws described motion in absolute space and time.

---

system of the space and the bodies revolving in it, we will then have the absolute motion of the bodies in immovable space." (1687b, pp. 168–69.) He does not, of course, suggest any method by which the "uniform progressive motion" in immovable space might be known.

Only in the twentieth century was the appropriate space-time structure defined, in which all and only the objective relations of Newtonian physics could be represented, without the excess structure represented by absolute space, and philosophers became aware of this mainly after Stein (1967).[9] In the later nineteenth century, however, physicists had already begun to see that absolute space could be dispensed with and replaced with an equivalence class of relative spaces—the "inertial systems" or "inertial frames"—any one of which is as good as any other for the description of a dynamical system. In effect, this development brought out the true import of Newton's use of Corollary V (or earlier, Law 3): there is no fact of the matter about whether a particular system is uniformly moving or at rest, and therefore no need to suppose that there is an encompassing immobile space with respect to which any such system has its true velocity. But this insight was hard won and only slowly digested.[10] In light of this, it is not so remarkable that Newton believed in absolute motion in immobile space. More remarkable is that he endeavored to secure his dynamical reasoning from any doubt that might be thought to arise from our ignorance of the absolute motions.

Even Ernst Mach, known as one of Newton's sternest critics, came to a more sympathetic view of Newton's effort after the emergence of the concept of inertial frame (cf. DiSalle 2002). Absolute space had seemed to him a metaphysical answer to an empirical question: when we state the principle of inertia, relative to what do we describe the motion of a body as uniform and rectilinear? Mach's celebrated answer was that the law has no meaning except as a description of motion relative to the fixed stars and the earth's rotation. But he saw that the concept of inertial frame had placed this question in a different light. The laws of motion do not need a reference-frame relative to which they are meaningful; rather, they are themselves the principles by which an appropriate frame (an inertial frame) can be determined. Having identified one such frame, we know that any other frame in uniform motion with respect to the first is an equally suitable frame. Mach came by this means to appreciate how well Newton had grasped the physical equivalence of such frames, in spite of his belief in absolute space.

> It is very much the same whether we refer the laws of motion to absolute space, or express them abstractly, without express indication of the system of reference. The latter course is unproblematic and practical, for in treating particular cases the student of mechanics looks for a suitable system of reference. But owing to the fact that the first way, whenever there was any actual issue at stake, was nearly always interpreted as having the same meaning as the latter, Newton's error was much less dangerous than it would otherwise have been.
> ....Let us again emphasize that Newton's oft-mentioned Corollary V, which alone has scientific value, makes no mention of absolute space. (Mach 1933, 242)

---

[9] Stein (1967) distinguished the four-dimensional affine structure required by Newton's laws from Newton's absolute space. That Newtonian mechanics corresponds to a four-dimensional space-time structure like the Minkowski space-time of special relativity, only with the Galilean symmetry group instead of the Lorentzian, was spelled out by Minkowski himself (1908). But the precise notion that "the [Newtonian] world is a four-dimensional affine space," as described by Stein, first appears in Weyl (1918, 130).

[10] For the history of the concept of inertial frame, see DiSalle (2020a).

Yet the remark, though illuminating, may be misleading. For it seems to suggest that it was mere good fortune that Newton's theory of space and motion did not compromise the success of his physics. On the contrary, the success of his theory of motion, in spite of the assumption of absolute space, was the result of deliberate efforts by Newton himself. It was Newton who—aware of the empirical problem posed by absolute space—found the means to insulate his physical reasoning against it.

We can see this by comparing Mach's remark to Newton's actual reasoning. Surely Newton supposed that a body not subject to forces moves uniformly with respect to absolute space and time. But he knew as well as Mach did that absolute space does not serve as a reference-frame. As far as celestial motion was concerned, the empirical use of the laws of motion required reference to the fixed stars. This included, evidently, determining the "frame of the system of the world" by calculating its center of gravity. It might appear to be a rather transparent bluff to claim to have distinguished, by reasoning on such a basis, the "true motions" from the apparent. Yet Newton recognized that to find a "suitable frame of reference" was a theoretical exercise for which the laws of motion provided empirical criteria. Taking the fixed stars to be at rest was only provisional; the dynamical analysis of planetary motion determined just how suitable a reference frame they are. If they were not a suitable frame—that is, not in some uniform state of motion—then the effects of its non-uniform motion would have to reveal themselves, in principle, to sufficiently precise measurement. By the third law of motion, every acceleration of any planet would have to be balanced by an equal and opposite reaction within the system. Any acceleration or rotation of the reference frame, therefore, ought to result in unbalanced forces, just as centrifugal forces exhibit the rotation of a frame of reference fixed to the earth. Newton himself discovered a more practical criterion: he could show that the orbits of the outer planets, as far as observation could determine, did not precess as, for example, the orbit of Mercury was later observed to do. That is, their apsides are approximately stable with respect to the fixed stars. But a relative space with respect to which these apsides are sufficiently stable is, Newton showed, a sufficient approximation to one that is at rest or in uniform motion (cf. Book III, Proposition XIV, 1687b, 420). On Mach's proposed empiricist foundation for the laws of motion, it would be meaningless to ask whether the fixed stars comprised a dynamically distinguished frame of reference; on Newton's abstract conception, it was an empirical question.

3.5 Newton's Extended Theory of Relativity

Though he was unable to incorporate his relativity theory into an appropriate theory of space and time, Newton used the relativity principle to ensure that his reasoning about the forces at work in the system would not be undermined by ignorance about the state of motion of the system with respect to absolute space. In giving empirical criteria for establishing that the system is in (approximately) uniform motion or at rest, he came as close as anyone did to articulating the idea of an inertial frame, without giving up the idea of absolute rest altogether. Thus his theory of relativity was the most advanced account of physically equivalent states of motion until the nineteenth century, in spite of his conviction that such empirically indistinguishable states may be genuinely inequivalent with respect to absolute space.

Yet, sometime before the *Principia*, Newton saw the need to extend his theory, or more precisely, his treatment of physically equivalent systems, to consider physically inequivalent systems that may be practically indistinguishable—more precisely, uniformly accelerated systems that may be treated as if they are at rest or in uniform motion. This peculiar kind of accelerated system was described in the remarkable novel principle that became Corollary VI:

> Corollary VI: If bodies are moved in any way among themselves, and are urged by equal accelerative forces along parallel lines, they will all continue to move among themselves in the same way as if they were not acted on by those forces. (Newton 1687b, 20)

A system of bodies acted upon by such a set of forces is in a dynamical state that is, necessarily physically inequivalent to that of a uniformly moving system. For practical purposes, however, it may be treated as equivalent to the latter since, locally, it will be empirically indistinguishable. Of course, Newton was not especially interested in the ideal case described in the Corollary, of forces acting exactly equally and in parallel directions on all parts of a system. His interest was Galileo's discovery, carefully corroborated by Newton himself, that gravity actually does behave very much like such an accelerative force.

Newtonian gravity generally will not act exactly equally, or in parallel directions, on all parts of any interacting system of bodies: the forces will naturally vary with distance, and they will not be parallel, but converging on the centers of gravitational attraction. In principle, however, and in real examples, the actions of gravity may approximate the conditions of Corollary VI as nearly as may be imagined. Jupiter and its moons, for example, are not equally accelerated by the gravitational pull of the Sun, given their varying distances, and their accelerations are not parallel, but converging on the center of the Sun. And Jupiter's elliptical orbit around the Sun does not, as a whole, approximate a uniform rectilinear motion. Yet the Jovian system may be treated as if it is in nearly uniform motion, over limited periods of time, because its immense distance from the Sun renders those differential accelerations negligible. Thus the system approximates the conditions of Corollary VI as closely as observation can determine. It should be emphasized, therefore, that Corollary VI as Newton understood it is strictly not an extended relativity principle. It does not enlarge the class of systems that are in principle equivalent. Any system undergoing such accelerations is, necessarily (by Newton's third law of motion), involved in an interaction with some other system, which must experience an equal and opposition reaction. Corollary VI is, however, an extension of Newton's theory of relativity: it extends his power to make inferences about the forces acting within a system of bodies, even when we may be ignorant of their states of motion in a larger context. In particular, it protects our reasoning about those forces from our ignorance of possible larger interacting systems of which our local system may be a part. In the case of Jupiter and its moons, we may be sure that their accelerations toward the Sun affect their actions among themselves only negligibly. But by the same reasoning, we may trust our analysis of entire solar system, even if it is similarly bound in orbit around some distant and unknown gravitational source.

Newton himself drew these remarkable implications in *De motu Corporum liber secundus* (1687a).[11] He showed that his analyses of the forces at work among the Sun and the planets were not undermined by the possibility that the system as a whole might be accelerated toward some other system.

> It may be imagined that the sun and planets are impelled by some other force equally and in the direction of parallel lines; but such a force (by Cor. VI to the Laws of Motion) would not change the situation of the planets among themselves, nor would produce any sensible effect; but we are concerned with the causes of sensible effects. Therefore let every force of this kind be rejected as being precarious and having nothing to do with the phenomena of the heavens; then all the remaining force by which Jupiter is urged will tend (by prop. 3, coroll. 1) toward the center of the sun. (Newton 1687a, article 13)

The last sentence indicates the methodological thought behind Newton's application of Corollary VI. Newton was explaining his calculation of the force acting on Jupiter, in order to show that it is directed toward the center of the Sun—a calculation that was a crucial part of his argument for a quasi-heliocentric structure. Corollary VI allowed him to assert that such calculations could yield secure theoretical conclusions, even if the entire solar system is acted upon by some external force.

Newton's theory of relativity is integrated with his method of mathematical physics. It concerns not only equivalence of certain states of motion, based on physical quantities that remain invariant across such states; it also concerns the foundation of physical equivalence in the mathematical composition of motions. Galileo and Huygens had understood that the motions within a system could be composed with a common inertial motion of the whole system. Only Newton, however, extended this idea to the composition of accelerative forces. This then enabled him to grasp the idea of approximately equivalent states of motion, approaching by degrees the ideal case of a force acting precisely equally on all bodies in a system. Such an approach is exemplified by the passage from (first) a system such as that of the earth and its moon, in which differential accelerations toward the sun are significant enough to create a difficult three-body problem; to (second) a system like the Jovian system, in which differential accelerations are small enough for us to ignore them and to treat the system as practically isolated; to (finally) an ideal system such as is described in Corollary VI. The latter is essentially a limiting case of a system like Jupiter's, as the size of the system becomes small compared to its distance from the source of the external force (cf. Book I, Prop. LXV, Case 2). Thus Corollary V and Corollary VI belong together, as describing, first, in principle indistinguishable systems of bodies, and second, dynamically

---

[11] This was Newton's original draft a concluding section for the *Principia* written in a "popular" style, posthumously published as *The System of the World* (Newton 1728).

distinct systems that may be indistinguishable to any degree of approximation. Indeed, it is likely that this fact moved Newton to change the relativity principle from an independent assumption to a Corollary: historically, this change coincides with his first use of Corollary VI.

3.6 Hypotheses, Rules, and Phenomena

A final significant change in Newton's thinking appears in the reorganization of Book III, which connects Newton's methodology with the relativity of motion in a particularly striking way. The modern reader of the *Principia*, typically of Newton's third edition, might be surprised to learn that in the first edition, Book III begins with nine hypotheses: how is this to be reconciled with the later remark that "hypotheses have no place in experimental philosophy"? That remark first appears in the General Scholium, added to the second edition (1713, 484), but it can hardly be seen as a late development, since Newton had expressed something like it throughout his career. A change of principles is also implausible, since the principles designated as "hypotheses" in the first edition of Book III remain in the later editions, under different designations. Rather, the probable explanation is the simplest one: Newton was using the word "hypothesis" in two different senses. This explanation, moreover, is straightforward to document and to account for historically. When we consider its philosophical implications, however, and its effects on the arguments of Book III, we see that Newton's revision reflects important aspects of his thinking on absolute and relative motion.

A minor part of the history of Newton's revision is the publication, between 1702 and 1710, of John Harris's *Lexicon Technicum* (1708). This contained a noteworthy definition of "Hypothesis":

> When for the Solution of any Phenomena in Natural Philosophy, Astronomy, etc. some Principles are supposed as granted, that from thence an Intelligible and Plausible account of the Causes, the Effects of the proposed Phenomena may be given, the laying down or supposing such Principles to be granted, is called Hypothesis.... Wherefore an Hypothesis is a Supposition of that which is not, for that which may be; and it matters not whether what is supposed be true or not, but it must be possible, and should always be probable. (Harris 1708)

The point of mentioning this definition is not that it influenced Newton's thinking. Indeed, the influence went primarily the other way, as Harris was a supporter of Newton's natural philosophy, and consulted him on scientific topics discussed in the Lexicon (see, for example, the entry on "Newtonian Philosophy"). Moreover, this definition of Hypothesis is quite consistent with one of Newton's long-established uses of the term. The important point is that this definition does not apply equally to all of the hypotheses in the first edition of Book III, for they are principles of several different kinds.

The nine hypotheses of Newton's first edition fall into four categories. The first two are the canons of inductive reasoning that, in the second edition, would become the first two "Regulae Philosophandi" (along with the new third Rule, with a fourth Rule added in the third edition). The last five assert regularities observed by astronomers, chiefly that the planets in their orbits about the sun, and the known satellites in their orbits around their respective planets, obey (to a good approximation) Kepler's second and third laws of planetary motion; these would be designated "Phaenomena" in the second and third editions. Hypothesis III was a specific hypothesis about the nature of matter: "Every body can be transformed into a body of any other kind and successively take on all the intermediate degrees of qualities" (Newton 1687a, b, 402). In the later editions, this principle is not stated as a fundamental hypothesis, but is absorbed in the text as part of the argument for Proposition VI, Corollary 2. Among the nine Hypotheses, only Hypothesis IV remains a hypothesis in all editions: "That the center of the system of the world is at rest. This is conceded by all, while some contend that the Earth, others that the Sun, rests in that center" (1687a, b, 402). But its presentation changes significantly. Not only is it now Hypothesis I, by default; it is also moved from the beginning of Book III to just before Proposition XI. The shift emphasizes that this Hypothesis is no longer presented as a presupposition of Book III generally; rather, it is introduced specifically for the sake of the argument that follows, namely, the argument that determines "the frame of the system of the world." Indeed, Newton adds a sentence to the original Hypothesis that specifies its dialectical purpose: "Let us see what will follow from this" (1713, 373).

Evidently, then, in describing all nine hypotheses as hypotheses, Newton was using the term in its logical sense: these were the principles assumed in the ensuing arguments. In revising the list, Newton changed his emphasis from their common logical role to their diverse sources of warrant. This does not suggest that Newton changed his mind about the hypothetical character of the claim that the center of the system of the world is at rest, or about the status of "hypotheses, metaphysical or physical," in general. The revisions emphasize, rather, that Newton continued to regard Hypothesis IV as hypothetical, in a sense in which almost all of the other hypotheses were not. This is what likely resulted from Newton's reading of the Harris dictionary entry; in subsequent editions of the *Principia*, Newton ceased to use the word "hypothesis" in two senses. Instead, he used it exclusively to refer to principles assumed for the purpose of a particular argument, but, at best, more plausible than other possible alternatives. Newton was clearly aware that both hypotheses, that the center of the world system is at rest and that it is in uniform motion, were possible. By Corollary V, either was compatible with all of his mathematical reasoning from the phenomena. Therefore neither hypothesis could be established as something more than a hypothesis, and he could defend the former hypothesis only as the more plausible of the two. In the *Liber secundus*, he had characterized the hypothesis that the system is in uniform motion as "hard" (1687a, section 28; or, in Motte's translation, "hardly to be admitted"—1728, 50). In the *Principia*, he says no more against it than that it is "against the fourth Hypothesis" (1687a, b) or "against the Hypothesis" (1713).

Contrast this case with the cases of Hypotheses I and II, and V–IX. The former may perhaps be called hypothetical in a certain sense: we can do no more than suppose that nature is sufficiently simple and uniform to allow us to reject superfluous causes, or to assign the same effects to the same causes. At the same time, however, neither principle is, like Hypothesis IV, just one of two more or less plausible alternatives, of which the selection of one can make little difference to the progress of the subsequent reasoning. Evidently Newton renamed them "Regulae philosophandi," precisely because they guide the entire project of inductive reasoning about causes. Hypothesis I even has the form of a command: "no more causes... should be admitted than are true and sufficient to explain their effects" (1687a, b, 402.) Hypothesis II is more nearly in the form of a hypothesis: "Therefore of effects of the same kind, the causes are the same" (ibid.), and remained so when renamed "Rule II" in the second edition. But in the third edition, Newton restated the second Rule as a command like the the the first: "Therefore to the same effects, the same causes are to be assigned, as far as possible" (1726, 387). This agrees with the forms of Rules III and IV: "the properties of bodies... are to be taken...," and "propositions drawn from the phenomena by induction... should be taken..." (1726, 387–9.) The Rules are, in short, instructions on how to infer general features of nature from observation. Without such instructions, Newton's arguments from evidence could hardly proceed.

In the case of Hypotheses V–IX, the methodological distinction from Hypothesis IV is even more obvious, and Newton expressed it clearly enough by renaming them "Phaenomena." Again, they function logically as hypotheses, insofar as they are presupposed in subsequent reasoning. But their similarity to Hypothesis IV goes no further than this. They are clearly propositions derived from astronomical observation, indeed by Newton's comparison of results from several sources. Despite their similar logical function, the Phenomena evidently have an epistemic basis unlike any that could be claimed, even by Newton, for Hypothesis IV. The reorganization of the beginning of Book III, therefore, has a greater methodological significance than first appears. For, even though the contents of the original nine hypotheses remain in the later editions, and function more or less as they had in the first edition, their separation into distinct categories reveals their distinct epistemic foundations and methodological roles. In recasting most of his original Hypotheses in Book III, he emphasized their sources of strong warrant, distinguishing them radically from a hypothesis (Hypothesis I) that could be no more than merely plausible.

In light of all this, it is a sign of Newton's discernment that, in recasting his Hypotheses from the first edition, he maintained Hypothesis IV as a hypothesis. In presenting it as Hypothesis I, Newton was not only making it clear that he considered it to be hypothetical in a way that the Rules and the Phenomena were not. He also made it very explicit that it was being assumed for the purpose of argument, specifically in the argument that begins with Proposition XI, to determine "the frame of the system of the world." Newton spoke as clearly and carefully on this subject, and moreover with the same concern for possible sources of error, as he did on the other central aspects of his theory of gravity. His extraordinary clarity, in retrospect, is precisely what has drawn the attention of readers from Berkeley onwards to Newton's error regarding absolute space. He made it very clear that no physical argument depended on a knowledge of absolute velocities, or even on the existence of a physical distinction between uniform velocity and rest. The *Principia* contains no claim regarding the absolute velocity of any actual thing. That the system of the world—or any other physical system—is at rest, and not in uniform motion, can be no more than a matter of hypothesis, and either hypothesis is compatible with Newton's dramatic conclusion: that the center of the system is its center of gravity, and that it is nearly heliocentric only because the mass of the Sun, relative to the masses of the other planets, determines that it can never recede far from the center of gravity. That is, neither the dialectical use of Hypothesis I, nor its character as a hypothesis, weakens the warrant for Newton's conclusion regarding the world-system. Against this background, however, the error stands out equally clearly: Newton had no physical or empirical grounds to regard either of the two hypotheses as any more plausible than the other—in retrospect, no grounds to suppose that there is a fact of the matter either way.

3.7   Conclusion

The theory of absolute space should be seen as an essential part of Newton's effort to protect his physical theory against arbitrariness and error. The notion of a background immobile space, with respect to which every body had a definite trajectory, seemed to give unambiguous meaning to the idea of true motion, as far removed as possible from the arbitrariness and uncertainty of Descartes' "motion in the proper sense." But to understand the theory of absolute space is to understand its position in the evolution of Newton's thinking, and therefore its role in the evolution of Newton's "theory of relativity." For an essential part of the latter was Newton's recognition that absolute space was itself a potential source of uncertainty, since motion with respect to space itself was inherently unknowable. Hence Newton's consistent use of the relativity principle to develop physical concepts, and a way of treating physical interactions, that could not be undermined by the problem of absolute space. Relativity for Newton was a physical principle that enabled him to treat the physical properties of any system of interacting bodies without regard to the motion of the whole system in space, except insofar as the system might exhibit rotation or non-uniform acceleration—either of which could be detected, in principle, by the physical means at Newton's disposal. The "general" principle of relativity advocated by Leibniz conferred a certain arbitrariness on the question of the system of the world; Newton's theory of relativity provided a methodological safeguard against arbitrariness. This is because for Newton, the principle of relativity was in fact a physical rather than a philosophical principle. Its basis was not epistemological equivalence, but the physical equivalence of states of motion, as identified by the laws of motion. Moreover, as a physical principle, it necessarily has an approximate dimension: it allows for states of motion that are nearly indistinguishable, and provides for a quantitative assessment of just how nearly indistinguishable they are. If philosophers have overlooked this aspect of Newton, it is perhaps because of the tendency of the last century to set Newton's "absolutism" against all of the philosophical insights that we associate with Einstein's theory of relativity. By understanding how Newton's conceptions of space, time, and relativity evolved together, we begin to see that, not unlike Einstein, Newton undertook a profound critical examination of the physical concepts with which he was working, seeking to separate their true physical content from what is merely relative, apparent, or arbitrary.

This leads me to a closing remark about George Smith. George has been a model for many people of the combination of philosophical rigor with attention to scientific and historical detail, and he has been among the most generous and inspiring colleagues and teachers that I have known. Beyond that, however, he has helped me to appreciate an aspect of Newton that I had not appreciated: that Newton's profound and unprecedented work of conceptual analysis was inseparable from his extraordinary attention to the details of evidential reasoning, and his adherence to the strictest standards for the grounding of theory in reliable evidence. I have tried to show how thoroughly Newton's thoughts about absolute and relative motion exemplified this empiricist methodology.

References

Barrow, I. 1860. *Lectiones mathematicae*, Lecture XII (1685). In *The mathematical lectures of Isaac Barrow*, ed. W. Whewell, 183–198. Cambridge University Press.

Berkeley, G. 1871. *De motu* (1721). In *The works of George Berkeley*, ed. Fraser Campbell, 73–100. Oxford: Clarendon Press.

DiSalle, R. 2002. Reconsidering Ernst Mach on space, time, and motion. In *Reading natural philosophy: Essays in the history and philosophy of science and mathematics*, ed. D. Malament, 167–191. Chicago: Open Court.

———. 2006. *Understanding space-time: The philosophical development of physics from Newton to Einstein*. Cambridge University Press.

———. 2020a. Space and time: Inertial frames. In *The Stanford encyclopedia of philosophy*, 3rd ed. revised, ed. Edward N. Zalta. https://plato.stanford.edu/archives/sum2020/entries/spacetime-iframes/.

———. 2020b. Absolute space and Newton's theory of relativity. *Studies in History and Philosophy of Modern Physics* 72: 232–244. https://doi.org/10.1016/j.shpsb.2020.04.003.

Fox, C. 2016. The Newtonian Equivalence Principle: How the relativity of acceleration led Newton to the equivalence of inertial and gravitational mass. *Philosophy of Science* 83: 1027–1038.

Galileo. 1632. *Dialogo sopra i due massimi sistemi del mondo—Tolemaico e Copernicano*. Reprint, Milan: Mondadori (1996).

Harris, J. 1708. *Lexicon Technicum: or, an universal English dictionary of arts and sciences: Explaining not only the terms of art, but the arts themselves*. London.

Huygens, C. 1929. De motu corporum ex percussione. (1656) *Oeuvres complètes de Christian Huygens* XVI: 30–91. The Hague: Martinus Nijhoff.

Leibniz, G. 1695. Specimen dynamicum, Part I. *Acta Eruditorum*: 145–157.

Leibniz, G.W. 1960. Letter to B. de Volder. (1699) Gerhardt, C., ed. (1960). *Die philosophischen Schriften von Gottfried Wilhelm Leibniz*, ed. C. Gerhardt, 168–170. Reprint, Hildesheim: Georg Olms.

————. 1933. *Die Mechanik in ihrer Entwickelung, historisch-kritisch dargestellt.* Reprint of the 9th edition. Leipzig: Brockhaus.

Newton, I. 1684a. De gravitatione et aequipondio fluidorum. MS Add. 4003, Cambridge University Library, UK. From *The Newton Project*, Cambridge University. http://www.newtonproject.ox.ac.uk/.

————. 1684b. "De motu sphæricorum corporum in fluidis". MS Add. 3965.7, ff. 55–62*, Cambridge University Library, UK. From *The Newton Project*, Cambridge University. http://www.newtonproject.ox.ac.uk/.

————. 1685a. De motu corporum in mediis regulariter cedentibus. MS Add. 3965.5, ff. 25r-26r, 23r-24r, Cambridge University Library, Cambridge, UK. From The Newton Project, Cambridge University. http://www.newtonproject.ox.ac.uk/.

————. 1685b. De motu corporum. MS Add. 3965.5, ff. 21, 26, Cambridge University Library, UK. In *The Background to Newton's Principia*, ed. J. Herivel, 315–317. Oxford: Clarendon, 1965.

————. 1687a. *De motu Corporum Liber Secundus.* MS Add. 4003. Cambridge University Library, UK. From *The Newton Project*, Cambridge University. http://www.newtonproject.ox.ac.uk/.

————. 1687b. *Philosophiae naturalis principia mathematica.* London.

————. 1713. *Philosophiae naturalis principia mathematica.* 2nd ed. Cambridge.

————. 1726. *Philosophiae naturalis principia mathematica.* 3rd ed. London.

————. 1728. *The system of the world.* Trans. A. Motte. London: Fayram.

Rynasiewicz, R. 1995a. "By their properties, causes and effects": Newton's Scholium on time, space, place and motion. Part I: The text. *Studies in History and Philosophy of Science* 26: 133–153.

————. 1995b. "By their properties, causes and effects": Newton's Scholium on time, space, place and motion. Part II: The context. *Studies in History and Philosophy of Science* 26: 295–321.

Smith, G.E. 2002a. The methodology of the *Principia*. In *The Cambridge companion to Newton*, ed. I.B. Cohen and G. Smith, 138–173. Cambridge University Press.

————. 2002b. From the phenomenon of the ellipse to an inverse-square force: Why not? In *Reading natural philosophy*, ed. D. Malament, 31–70. Chicago: Open Court Press.

Stein, H. 1967. Newtonian space-time. *Texas Quarterly* 10: 174–200.

————. 1977. Some philosophical prehistory of general relativity. In *Foundations of space-time theories*, Minnesota studies in the philosophy of science, ed. J. Earman, C. Glymour, and J. Stachel, vol. 8, 3–49. University of Minnesota Press.

————. 2002. Newton's metaphysics. In *The Cambridge companion to Newton*, ed. I.B. Cohen and G. Smith, 256-3-7. Cambridge University Press.

Weyl, H. 1918. *Raum-Zeit-Materie.* Berlin: Springer.

Chapter 4
Henry Cavendish and the Density of the Earth

Allan Franklin

Contrary to the views expressed in many introductory physics textbooks, Henry Cavendish did not measure G, the gravitational constant contained in Newton's Law of Universal Gravitation, $F = G\, m_1 m_2 / r^2$. As the title of his paper states, Cavendish conducted "Experiments to Determine the Density of the Earth (1798)." As discussed below, one can use that measurement to determine G, but that was not Cavendish's intent. In fact, the determination of G was not done until the latter part of the nineteenth century.

Newton had stated that the gravitational force between two objects was proportional to the product of their masses and inversely as the square of the distance between their centers, but he did not use the Universal Gravitational Constant, G.[1,2]

> Gravity exists in all bodies and is proportional to the quantity of matter in each....If two globes gravitate toward each other, and their matter is homogeneous on all sides in regions that are equally distant from their centers, then the weight of either globe toward the other will be inversely as the square of the distance between the centers. (Newton 1999, 810–811)

Cavendish provided no reasons for his measurement of the density of the Earth other than the availability of a suitable experimental apparatus. One may speculate that he felt that this was unnecessary because there had been earlier measurements of that density. These earlier measurements, which had involved considerable effort,

---

[1] I am grateful to George Smith for helpful discussions on this point.

[2] G did not appear in an equation for the force until the late nineteenth century. See the discussion of Boys below.

A. Franklin (✉)
University of Colorado, Boulder, Boulder, CO, USA
e-mail: allan.franklin@colorado.edu

© The Author(s), under exclusive license to Springer Nature Switzerland AG 2023
M. Stan, C. Smeenk (eds.), *Theory, Evidence, Data: Themes from George E. Smith*, Boston Studies in the Philosophy and History of Science 343, https://doi.org/10.1007/978-3-031-41041-3_4

showed the importance of the experiment. These measurements were performed by Bouguer (two) and by Maskelyne.[3,4]

4.1 The Experimental Apparatus

Cavendish's method was to measure the gravitational force between two masses using a torsion balance. As discussed below this can be used to determine the density of the Earth. The apparatus is shown in Fig. 4.1, in detail in Fig. 4.2, and schematically in Fig. 4.3. Briefly, the gravitational force between the small sphere and the large sphere is measured by measuring the twist of the wire suspending the beam holding the two small masses using the period of oscillation. Cavendish remarked that the method had been invented by the late Reverend John Michell, who had passed away before he could complete the apparatus and perform any experiments with it.[5] Cavendish described his apparatus as follows,[6]

> As no more force is required to make this arm turn round on its centre, than what is necessary to twist the suspending wire, it is plain that if the wire is sufficiently slender the most minute force, such as the attraction of a leaden weight a few inches in diameter, will be sufficient to draw the arm sensibly aside . . . .
>
> In order to determine from hence the density of the earth, it is necessary to ascertain what force is required to draw the arm aside through a given space. This Mr. Michell intended to do, by putting the arm in motion and observing the time of its vibrations, from which it may easily be computed. (469–470).[7,8]

---

[3] For an excellent account of these measurements as well as many conducted in the later nineteenth century see Mackenzie 1900.

[4] Judging by the large number of further measurements of both the density of the Earth and of "G" during the nineteenth century, Cavendish's speculation, had it occurred, would have been correct.

[5] Cavendish noted that a similar apparatus had been used by Coulomb. "but Mr. Michell informed me of his intention of making this experiment, and of the method he intended to use, before the publication of any of Mr. Coulomb's experiments (1798, p. 470)."

[6] Hereafter, numbers between 469 and 526 in parentheses denote pages in Cavendish 1798.

[7] The method of calculation used by Cavendish is quite detailed and very complex. A modern and simpler version is given below.

[8] The large weights were lead spheres, 12 inches in diameter and weighing 350 pounds. The smaller lead weights were 2 inches in diameters and weighing about 1 pound 10 ounces. Cavendish labelled the direction of the arm in which the weights were in the position shown in Fig. 4.3 as the positive position of the weights. The midway position was where the weights were perpendicular to the beam, and in which no net gravitational force of the two large spheres was expected.

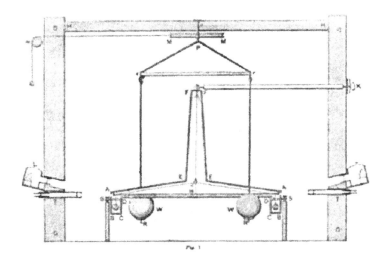

Fig. 4.1 Sketch of Cavendish's experimental apparatus. (Source: "The Cavendish Experiment, Wikipedia, https://en.wikipedia.org/wiki/Cavendish_experiment)

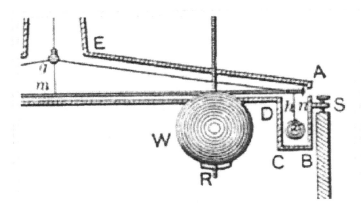

Fig. 4.2 Details of the Cavendish experimental apparatus. (Source: "The Cavendish Experiment, Wikipedia, https://en.wikipedia.org/wiki/Cavendish_experiment)

4.2   Taking Data

Cavendish was extremely careful in attempting to avoid possible confounding effects. He regarded possible temperature gradients in his apparatus as the most important of these. He had found that

> the disturbing force most difficult to guard against, is that arising from the variations of heat and cold; for if one side of the case is warmer than the other, the air in contact with it will be rarefied, and, in consequence, will ascend, while that on the other side, will descend,

Fig. 4.3 Schematic diagram
of the Cavendish experiment.
(Source: "The Cavendish
Experiment, Wikipedia,
https://en.wikipedia.org/wiki/
Cavendish_experiment)

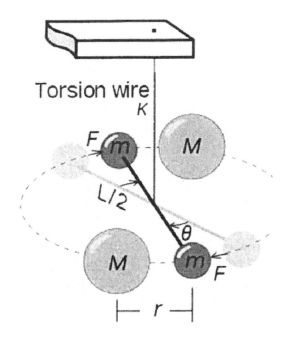

and produce a current which will draw the arm sensibly side.[9] As I was convinced of the
necessity of guarding against this source of error, I resolved to place the apparatus in a room
which should remain constantly shut, and to observe the motion of the arm from without,
by means of a telescope; and to suspend the leaden weights in such manner, that I could
move them without entering the room. (470–471)

This required making changes in Michell's original apparatus.

Cavendish also worried about the effects of air resistance on his torsion balance.
Because the balance would not only be deflected by the gravitational force
between the small and large lead spheres, but also set into oscillation, he used the
following method to find how far the balance arm was moved; and he adjusted his
measurements to deal with any such confounding effects (see also Fig. 4.4).

---

[9] In a footnote Cavendish noted that Cassini had found, in his study of the variation compass,
an effect caused by Cassini standing near his apparatus, which "drew the needle sensibly aside;
which I have no doubt was caused by this current of air (471)." The possibility of human effects on
experimental results was also remarked on in the study of the Fifth Force, a proposed modification
of Newtonian gravity. A composition effect was seemingly observed in the Eötvös experiment on
the equality of gravitational and inertial mass. The effect depended on a local mass inhomogeneity.
Alvaro De Rujula quipped that, "Although malicious rumor has it that Eötvös himself weighed
more than 300 pounds, unspecific hypotheses are not, *a priori*, particularly appealing (De Rujula
1986, 761)." De Rujula's quip is quite unfounded. Eötvös was a mountaineer and appears quite slim
in photographs. There is a mountain in the Dolomites in northern Italy, Cima di Eötvös, named for
him.

Weights moved to positive position.

Extreme points.	Divisions.	Time.			Point of rest.	Time of mid. of vibration.		Difference.	
		h.	$'$	$''$		h.	$'$	$''$	$'$	$''$
3¹,3										
	25	11	10	25 ⎫	–	11	10	40		
	23		11	3 ⎭						
17,1	–	–	–		24,02	–	–		7	3
	22		17	· 6 ⎫	–		17	43		
	23			26 ⎭						
30,6	–	–	–		24,17	–	–		7	1
	25		24	33 ⎫	–		24	44		
	23		25	17 ⎭						
18,4	–	–	–		24,32	–	–		7	5
	23		31	21 ⎫	–		31	49		
	25		32	9 ⎭						
29,9	–	–	–		24,4	–	–		6	59
	25		38	39 ⎫	–		38	48		
	23		39	31 ⎭						
19,4	–	–	–		24,5	–	–		7	6
	23		45	16 ⎫	–		45	54		
	25		46	12 ⎭						
29,3										

Motion of arm on moving weights from

midway to pos.   – $= 3,1$

pos. to neg.   – $= 6,18$

neg. to pos.   – $= 5,92$

Time of one vibration in neg. position   – $= 7' 1''$

pos. position   – $= 7 3$

Fig. 4.4 Data from Cavendish experiment IV. (Source: Cavendish 1798)

To do this, I observed three successive extreme points of a vibration, and take the mean between the first and third as the extreme point of vibration in one direction, and then assume the mean between this and the second extreme as the point of rest; for as the vibrations are continually diminishing, it is evident, that the mean between the two extreme points will not give the true point of rest. (474).

Cavendish next needed to determine the time of vibration. He did this simultaneously with his determination of the midpoints of oscillation. He also recorded the time at which the beam arm arrived at two divisions, on either side of the midpoint (see Fig. 4.4).

The midpoint, or point of rest, calculated earlier is the mean of the first and third extremes and the second extreme, which is 24.65. The arm passed division 25 at 10h23'4" and division 24 at 10h23'57". It therefore passed 24.65 at 10h23'23". A similar calculation found the midpoint for his seventh measurement found the midpoint at 11h5'22". Therefore "six vibrations were performed in 41'59" or one vibration in 7'0". (476)

When Cavendish began his experiments, he found that the stiffness of the wire suspending the beam was such that the period on vibration was approximately 15 min. "I immediately found, indeed, that it was not stiff enough, as the attraction of the weights drew the balls so much aside, as to make them touch the sides of the case; I, however, chose to make some experiments with it, before I changed it (p. 478)." Cavendish replaced the wire in his Experiment 4, and thereafter. The increased stiffness changed the period of vibration to approximately 7 min.

The rods holding the large weights were made of iron and Cavendish worried that there might be confounding magnetic effects. He checked this by removing the large spheres to see if there were any effects due to the rods alone. He found small effects, and replaced the iron rods with ones made of copper: "... the result was, that there still seemed to be some effect of the same kind, but more irregular, so that I attributed it to some accidental cause, and therefore hung on the leaden weights, and proceeded with the experiments (479)."

Cavendish performed three experiments with the initial wire. "These experiments are sufficient to shew, that the attraction of the weights on the balls is very sensible, and are also sufficiently regular to determine the quantity of this attraction.... (484)." From Experiment 4 onwards Cavendish used a stiffer wire. In his paper Cavendish presented the data for all of his experiments. Typical data records are shown in Figs. 4.4 and 4.5.

4.3   The Final Result

Cavendish was now in a position to calculate the density of the Earth. Rather than report his original method of calculation, which is quite unfamiliar to modern readers, I will use a contemporary method. As shown in Fig. 4.3 the two light balls are a distance L/2 from the axis of rotation. A force F, the gravitational force between the small and the large spheres, is exerted on each mass at a distance L/2. The wire was assumed to obey Hooke's Law so

$$\kappa\theta = FL/2 + FL/2 = FL$$

We can find the torsional constant κ from the measurement of the period T of the vibration or oscillation of the beam.

$$\kappa = 4\,\pi^2\,I/T^2,$$

Weights moved to negative position.

15						
	17	19 25 }		10 20 31		
	19	20 41 }	—			
22,4	—	— —	18,72	— —	7	
	20	26 45 }		27 31		
	19	27 22 }	—			
15,1	—	— —	18,52	— —	6 57	
	19	35 1 }		34 28		
	20	48 }	—			
21,5	—	— —	18,35	— —	7 23	
	20	40 23 }		41 51		
	19	41 18 }	—			
15,3	—	— —	18,22	— —	6 48	
	18	48 36 }		48 39		
	19	49 24 }	—			
20,8	—	— —	18,1	— —	6 58	
	19	54 45 }		55 37		
	18	55 45 }	—			
15,5						

Fig. 4.5 Continuation of data from Fig. 4.4. (Source: Cavendish 1798)

where I is the moment of inertia of the beam and the two small weights. Assuming initially, as Cavendish did that the beam was weightless

$$I = mL^2/4 + mL^2/4 = mL^2/2,$$

where m is the mass of the small weight. θ is found from the displacement of the beam, where $L\theta/2$ is approximately the displacement. Thus, we can determine F.

To get the density of the earth one compares the gravitational force, F, between the spheres to W, the weight of the small ball. We assume, as Cavendish did, Newton's Law of Universal Gravitation. In modern notation

$$F/W = \left[GmM_{Large}/r^2\right] / \left[GmM_{Earth}/R^2\right], \text{ where r is the distance between}$$

centers of the small and large spheres and R is the radius of the Earth.

$$F/W = \left[M_{Large}/r^2\right] / \left[M_{Earth}/R^2\right] = \left[M_{Large}/r^3\right] / \left[M_{Earth}/R^3\right] \times r/R$$
$$= \left[M_{Large}/r^3\right] / Density_{Earth}\right] \times r/R$$
$$Density_{Earth} = \left[M_{Large}/r^3\right] \times r/R \times W/F$$

Each of the terms on the right side is known, or has been measured, so the density of the Earth can be calculated.

Cavendish remarked that before a final result could be obtained certain corrections had to be applied:

> first, for the effect which the resistance of the arm to motion has on the time of vibration: 2d, for the attraction of the weights on the arm: 3d, for their attraction on the farther ball: 4th, for the attraction of the copper rods on the balls and arm: 5th, for the attraction of the case on the balls and arm: and 6th, for the alteration of the attraction of the weights on the balls, according to the position of the arm, and the effect which that has on the time of vibration. (511)

He noted that "None of these corrections, indeed, except the last, are of much signification, but they ought not entirely to be neglected." (511) Cavendish then proceeded to make very detailed calculations of these corrections and applied them to his final results. His final results are shown in Fig. 4.6. He further noted that the average values for the density of the Earth for both the experiments performed with wires of different stiffness were the same and used that average value as his final result, 5.48. Using the maximum and minimum values that he had found for the density of the Earth in the last 23 measurements, 5.1 and 5.85, a difference of 0.75, Cavendish estimated the uncertainty in that value of 0.38, half the difference between the maximum and minimum values. In modern terms we might report the result as 5.48 ± 0.38.[10] "Therefore the density is seen to be determined hereby, to great exactness." (521)

Cavendish thought that the spread in his final values for the density of the earth was too large to be accounted for by errors in observation. He worried that this might be due to differences in the temperature between parts of the experimental apparatus.

> As to the difference in the motion of the arm, it may very well be accounted for, from the current of air produced by the difference of temperature; but whether this can account for the difference in the time of vibration is doubtful. If the current of air was regular, and of the same swiftness in all parts of the vibration of the ball, I think it could not; but, as there will most likely be much irregularity on the current, it may very likely be sufficient to account for the difference . . . .
>
> It, indeed may be objected, that as the result appears to be influenced by the current of air, or some other cause, the laws of which we are not well acquainted with, this cause may perhaps act always, or commonly, in the same direction, and thereby make a considerable error in the result.[11] *But yet, as the experiments were tried in various weathers, and with considerable variety in the difference of temperature of the weights and air, and with the arm resting at different distances from the sides of the case, it seems very unlikely that this*

---

[10] Using a standard deviation calculation for the last 23 measurements, Cavendish's result is 5.48 ± 0.19. As we can see, Cavendish was quite conservative in his estimate of his experimental uncertainty. I note that the standard deviation was first used in writing by Karl Pearson in 1894. Gauss also used mean deviation after the publication of Cavendish's paper.

[11] Cavendish noted another possible objection, the fact that the law of gravity might be different at the small distances of his experiments. "With a view to see whether the result could be affected by this attraction [the force of cohesion], I made the 9th, 10th, 11th, and 15th experiments in which the balls were made to rest as close to the sides of the case as they could; but there is no difference to be depended on, between the results under that circumstance, and when the balls are placed in any other part of the case." (522)

The following Table contains the Result of the Experiments.

Exper.	Mot. weight	Mot. arm	Do. corr.	Time vib.	Do. corr.	Density.
1 {	m. to +	14,32	13,42	, ʺ	–	5,5
	+ to m.	14,1	13,17	14,55	–	5,61
2 {	m. to +	15,87	14,69	–	–	4,88
	+ to m.	15,45	14,14	14,42	–	5,07
3 {	+ to m.	15,22	13,56	14,39	–	5,26
	m. to +	14,5	13,28	14,54	–	5,55
4 {	m. to +	3,1	2,95		6,54	5,36
	+ to –	6,18	–	7,1	–	5,29
	– to +	5,92	–	7,3	–	5,58
5 {	+ to –	5,9	–	7,5	–	5,65
	– to +	5,98	–	7,5	–	5,57
6 {	m. to –	3,03	2,9	–	–	5,53
	– to +	5,9	5,71		–	5,62
7 {	m. to –	3,15	3,03	7,4 by mean.	6,57	5,29
	– to +	6,1	5,9			5,44
8 {	m. to –	3,13	3,00	–	–	5,34
	– to +	5,72	5,54	–	–	5,79
9	+ to –	6,32	–	6,58	–	5,1
10	+ to –	6,15	–	6,59	–	5,27
11	+ to –	6,07	–	7,1	–	5,39
12	– to +	6,09	–	7,3	–	5,42
13 {	– to +	6,12	–	7,6	–	5,47
	+ to –	5,97	–	7,7	–	5,63
14 {	– to +	6,27	–	7,6	–	5,34
	+ to –	6,13	–	7,6	–	5,46
15	– to +	6,34	–	7,7	–	5,3
16	– to +	6,1	–	7,16	–	5,75
17 {	– to +	5,78	–	7,2	–	5,68
	+ to –	5,64	–	7,3	–	5,85

Fig. 4.6 Cavendish's final results. (Source: Cavendish 1798)

cause should act so uniformly in the same way, as to make the error of the mean result nearly equal to the difference between this and the extreme; and, therefore, it seems very unlikely that the density of the earth should differ from 5.48 by so much as 1/14 of the whole. (521–522)

Cavendish also remarked that his value for the density of the earth differed markedly from the value found earlier by Maskelyne.

According to the experiments made by Dr. MASKELYNE, on the attraction of the hill Schehallien, the density of the earth is 4 ½ times that of water; which differs rather more from the preceding determination than I should have expected. But I forbear entering into any consideration of which determination is most to be depended on, till I have examined more carefully how much the preceding determination is affected by irregularities whose quantity I cannot measure. (522)

Cavendish did not comment on the values obtained by Bouguer of 4.7 times the density of Cordilleras and 6–7 times the density of Chimborazo.[12] Nevertheless, given the increased accuracy and precision of Cavendish's result when compared to those of Bouguer and Maskelyne, we may agree with Poynting's comment that, "... he made the experiment in a manner so admirable that it marks the beginning of a new era in the measurement of small forces"[13] (Poynting 1913, 63).

4.4   The Constant G

Cavendish, as we have seen, measured the density of the Earth, not the Universal Gravitational Constant, "G." In 1894, C.V. Boys made a passionate argument for formulating Newton's Law of Gravitation as Force $= G$ (Mass$_1$ × Mass$_2$)/Distance^2. He remarked that "**g**," the acceleration due to gravity on the surface of the Earth is

Eminently of a practical and useful character; it is the delight of the engineer and the practical man; it is not constant, but that he does not mind ....G, on other hand represents that mighty principle under the influence of which every star, planet and satellite in the universe pursues its allotted course ....Owing to the universal character of the constant G, it seems to me to be descending from the sublime to the ridiculous to describe the object of this experiment as finding the mass of the earth, or less accurately, the weight of the earth." (Boys 1894, 330).[14]

From the density of the earth we may calculate G as follows.

$$m\mathbf{g} = G \, mM_{Earth}/R_{Earth}^2, \; G = gR_{Earth}^2/M_{Earth}$$
$$= \mathbf{g}R_{Earth}^2 / \left((4/3) \, \pi \, R_{Earth}^3\right) \rho_{Earth}$$

---

[12] Cordilleras and Chimborazo are mountains in South America.

[13] Cavendish was the first to give an estimate of the experimental uncertainty in his result.

[14] Boys (1895) himself made such a determination, as did Poynting (1892). For other references see Mackenzie (1900).

And so, $G = 3g/4\pi R_{Earth}\rho_{Earth}$, where ρ_{Earth} is the density of the Earth. Using Cavendish's value of 5.48 ± 0.38 for that density, we find

$$G = (6.71 \pm 0.47) \times 10^{-11} \, m^3 \, kg^{-1}s^{-2},$$

which is in agreement with the 2014 CODATA value of (6.67408 ± 0.00031) $10^{-11} \, m^3 \, kg^{-1} \, s^{-2}$.

4.5 A Criticism of Cavendish

Francis Baily, Esquire, Fellow of the Royal Society, and Vice President of the Royal Astronomical Society[15] was an early critic of Henry Cavendish's experiments on gravity. Baily did not, in fact, believe that Cavendish had performed a serious measurement, but had merely demonstrated an excellent method of measuring the density of the earth.

> He is of the opinion that Cavendish's object in drawing up his memoir was more for the purpose of exhibiting a *specimen* of what he considered to be an excellent method of determining this important inquiry, than of deducing a result, at that time, that should lay claim to the full confidence of the scientific world.[16] (Baily 1842, 111)

He further criticized Cavendish for performing only 23 experiments,[17] whereas he, Baily, had made 2153 such measurements. This is an exaggeration.

> Baily adopted the method of Reich for reducing the time required to make the number of turning-points requisite for calculating the deviation and the period; that is the masses were moved quickly from one near position to the other and the last turning-point on one series served for the first of the next. Three new turning-points were observed at each position of the masses, and each group of 4 was called an experiment.[18] (Mackenzie 1900, 117)

Had Cavendish used Baily's counting method he would have reported many more experiments.

One possible confounding effect, which Cavendish had also discussed, was the need to keep the temperature in the room constant and to avoid temperature gradients. Cavendish arranged to have his apparatus in a sealed room; he manipulated the masses by means of a series of pulleys, and made his observations using a telescope located outside the room (see Fig. 4.1). This was insufficient for Baily.

---

[15] That is the way he listed himself in the paper. Baily was also president of the Royal Astronomical Society four times. He is best known for his discovery of Baily's beads, an optical phenomenon visible during a total eclipse of the sun.

[16] Given the detailed corrections that Cavendish made to his calculations, this seems highly unlikely.

[17] Cavendish actually reported 29 experiments.

[18] See discussion below for some of the problems of Baily's method.

Cavendish chose an out-house[19] in his garden at Clapham Common; and, having constructed his masses *within* the building, he moved the masses by means of ropes passing through a hole in the wall, and observed the torsion-rod, by means of a telescope fixed in an ante-room on the *outside*. The general temperature of the interior was therefore probably uniform during the time that he was occupied in any one set of experiments: but it is scarcely to be expected that a building of this kind, and in such a situation, would preserve the same uniform temperature for twenty-four hours; especially at the season which he selected for his operations.[20] (Baily 1842, 117)

Baily performed experimental checks to determine if temperature effects could have a significant effect on his final result. He found that there were such effects and discarded his early experimental results. He also made changes to his apparatus to guard against such effects by gilding the weights and covering the box which contained the apparatus with felt.[21] Baily felt that these changes minimized any temperature effects. As I explain below, this was not correct. There were, in addition, other difficulties with Baily's experiment.

Baily's final result for the density of the Earth was 5.67, where the density of water is 1. He noted, I believe incorrectly, that his result disagreed with the value reported by Cavendish of 5.48 ± 0.38. Baily remarked that,

It cannot escape observation that the general mean result, obtained from these experiments is much greater (equal to $1/25^{th}$ part) than that deduced either by Cavendish or Reich,[22] who both agreed in the very same quantity, namely 5.44:[23] but he does not assign any probable cause for this discordance. (Baily 1842, 121)

Baily was quite confident in his own result:

It is evident from the detail which he [Baily] has given of his own experiments, that perceptible differences not only arose according to the mode in which the torsion rod was suspended but also depended on the materials of which the suspension-lines were formed: *but it is somewhat singular that none of these mean results, in any of the classifications, are so low as that obtained by the two experimentalists above mentioned.* (Baily 1842, 121; emphasis added)

Baily had, in fact, made measurements with small masses of different sizes, different materials, and with different modes of suspension (Fig. 4.7). The results were quite consistent. Baily did not, however, provide an experimental uncertainty, nor did it enter into his discussion of the difference between his value for the density of the Earth and those obtained by Cavendish and Reich. Baily offered no detailed results for any individual experiments and offered only averages for different experimental

---

[19] Modern American usage regards an outhouse as a privy, an outdoor toilet. The *Oxford English Dictionary* defines outhouse as a building such as a shed or barn that is built onto or in the grounds of a house.

[20] As discussed below, Baily was not always so careful with his own measurements.

[21] Baily remarked that he was indebted to Professor Forbes of Edinburgh for these helpful suggestions.

[22] Reich performed his measurements from 1837 to 1840 and from 1847 to 1850, and in 1852. This was after Cavendish and both before and after Baily's work.

[23] See discussion below.

Balls.	Double silk.		Double wire.		Single wire.	
	No.	Density.	No.	Density.	No.	Density.
2¼-inch lead	148	5·60	130	5·62	57	5·58
2-inch lead.........	218	5·65	145	5·66	162	5·59
1½-inch platina ...	89	5·66	...	...	86	5·56
2¼-inch brass	46	5·72	...	...	92	5·60
2-inch ⎰ zinc	162	5·73	20	5·68	40	5·61
⎱ glass	158	5·78	170	5·71	...	...
⎰ ivory	99	5·82	162	5·70	20	5·79
2¼-inch lead, with brass rod......			44	5·62		
2-inch lead, with brass rod			49	5·68		
Brass rod, alone			56	5.97		

Fig. 4.7 Baily's results for the density of the earth. (Source: Baily 1842)

runs with no uncertainty offered. I note here that, contrary to Baily's view, his result agreed with that of Cavendish within Cavendish's stated experimental uncertainty and within a reasonable estimate of the uncertainty in his own value.[24]

4.6 Criticism of the Critic

In 1878, however, Cornu and Baille pointed out a problem with Baily's method. They noted that using the fourth reading of the turning-point as the first one of the next experiment resulted in an error, "They showed that the rotation of the plank holding the masses could not be performed rapidly enough to get the masses into the new position before the arm had begun its return journey" (Mackenzie 1900, 119). They then calculated the results for the density of the earth using only the last three measurements in 10 of Baily's experiments. They found that for those 10 experiments the mean value for the density of the earth was reduced from 5.713 to 5.615. Applying the same percentage correction to Baily's mean value changed that result from 5.67 to 5.55. This argued that Baily's experimental technique provided

---

[24] Using the same method that Cavendish used, taking half the difference between the maximum and minimum values Baily reported (see Fig. 4.7), 5.58 and 5.97 we find an uncertainty of 0.20. We should increase it slightly to take into account the temperature effect found by Hicks, although it is difficult to place a numerical value on that uncertainty, but it should be smaller than 0.20.

Temperature F.	Number of series.	Number of daily means.	Number of observations.	Mean density.
36° (mean)	4	7	46	5 + ·7296
40 ± 2	12	22	128	·7341
45 ± 2	20	43	247	·6823
50 ± 2	18	38	302	·6799
55 ± 2	12	23	187	·6594
60 ± 2	13	31	333	·6495
65 ± 2	7	17	140	·5935
68 (mean)	4	11	96	·5828

Fig. 4.8 Hicks's analysis of Baily's results showing the dependence of the density of the earth on temperature. (Source: Hicks 1886)

results that had a small systematic uncertainty and that it resulted in a value for the density that was too high.[25]

W.M. Hicks in 1886 found yet another problem with Baily's measurements of the density of the earth. He found, by reanalyzing Baily's data, that the density varied with temperature.[26] Specifically, the density fell with a rise in temperature.

> I have recently been examining Baily's observations on the mean density of the earth in order to see if they showed any traces of a dependence of the attraction between two masses on their temperature. I was astonished to find in his numbers most decided signs of some temperature effect." (Hicks 1886, 156–57)

Hicks's results are shown in Fig. 4.8. He concluded, "The gradual fall of mean density with rise of temperature is most marked, the only exception being in the case of the lowest temperature (36°) which is slightly smaller than for the temperature of 40°" (Hicks 1886, 158). Mackenzie suggested that the most probable explanation of this effect was given by Poynting, who remarked that the experiments with light balls were performed in winter whereas those with heavy balls were done in summer. These results seem ironic given Baily's criticism of Cavendish's efforts to maintain a constant temperature during his measurements.

How then should we regard Baily's measurement? Baily was a reasonably careful observer. He did, as had Cavendish, worry about a possible effect caused by temperature gradients in the experimental apparatus. He found such effects and discarded his early results because of them. He also made modifications to his

---

[25] This reanalysis of other experimenter's results seems to have been quite common in the nineteenth century; cf. Mackenzie 1900, 143–44.

[26] Hicks used Baily's data included in his much longer account of his experiments (Baily 1843).

apparatus to minimize such effects. Nevertheless, as Hicks later showed, Baily was not completely successful. His results did show a temperature dependence. As Cornu and Baille showed there were also problems with Baily's experimental technique. Including a reasonable estimate of Baily's uncertainty (see note 23), we may conclude that Baily's result did agree, within uncertainty, with those of Cavendish and Reich. At the time, it is reasonable to consider it a contribution to measurements of the density of the Earth.[27]

4.7 An Oddity

There is an oddity in Cavendish's final result. He claimed that the average value he obtained from the first six measurements of the density, 5.48, those with the less stiff wire, was equal to that of the last 23 measurements, those found with the stiffer wire. He reported that both sets of measurements gave a density of 5.48 and that his final result was 5.48. This is not correct. The average of the first 6 measurements is 5.31 ± 0.22, whereas the average for the last 23 measurements is 5.48 ± 0.19. The average of all 29 measurements is 5.448 ± 0.22.[28] This discrepancy was first noted by Baily, and later by Poynting and others. They attributed this to an arithmetic error by Cavendish. Baily pointed out that if the third measurement (Fig. 4.6), published as 4.88, was in fact 5.88 the discrepancy disappears. Baily recalculated the density of the earth using Cavendish's original data for this experiment and found that the value is indeed 4.88.[29]

Baily's work is only a small part of the measurements of the density of the earth in that period. Mackenzie (1900) lists 20 measurements of the density between Cavendish's 1798 report and the end of the nineteenth century. These measurements used several different methods; the torsion pendulum, a balance, a simple pendulum, and the deflection of a plumb line by a hill or a mountain.[30] Several of these results were analyzed not only by the author, but also by other scientists. The results vary from 4.25 to 7.60. Given the recent discussions of the issue of replication in science examining this history might provide interesting insights.[31]

This uncertainty in the measurements of gravity in both the density of the earth and in the value of G has continued into the twenty-first century. The

---

[27] We may also conclude that Baily had excessive confidence in his result.

[28] Thus, the agreement with Reich referred to by Baily.

[29] Perhaps Cavendish made an error and used 5.88 rather than 4.88.

[30] Sir George Airy, the Astronomer Royal, remarked on the difficulty of gravity measurements. 'He measured the periods of two pendulums, one at the top of a mineshaft and the other deeper in the shaft. "We were raising the lower pendulum up the South Shaft for the purpose of interchanging the two pendulums, when (from causes of which we are yet ignorant) the straw in which the pendulum-box was packed took fire, lashings burnt away, and the pendulum with some other apparatus fell to the bottom. This terminated our operations for 1826 (Airy 1856, p. 299)".

[31] See Franklin 2018.

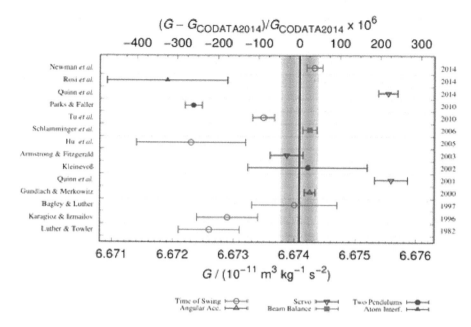

Fig. 4.9 Recent measurements of G. The vertical line shows the recommended value of G and the shading indicates the recommended uncertainty. (Source: CODATA 2014)

current value for G is $(6.67408 \pm 0.00031) \times 10^{-11}$ m^3/kg s^2, with a relative uncertainty of 47 parts per million. Although the precision has improved there is still considerable disagreement among the measurements (see Fig. 4.9). This is because the gravitational force is so weak when compared to other forces that it is difficult to eliminate confounding factors. I note that the ratio of the electrical force between the electron and the proton in the ground state of hydrogen is 2×10^{39} times as large as the gravitational force between them. Measurements of G and further analysis of the experiments are continuing.

References

Airy, G.B. 1856. Account of pendulum experiments undertaken in the Harton Colliery, for the purpose of determining the mean density of the earth. *Philosophical Transactions of the Royal Society of London* 146: 297–355.

Baily, F. 1842. An account of some experiments with the torsion-rod, for determining the mean density of the earth. *Philosophical Magazine* 21: 111–121.

———. 1843. Experiments with the torsion-rod for determining the mean density of the earth. *Memoirs of the Royal Astronomical Society* 14: 1–120, i–ccxlviii.

Boys, C.V. 1894. On the Newtonian constant of gravitation. *Nature* 50: 330–334.

————. 1895. On the Newtonian constant of gravitation. *Philosophical Transactions of the Royal Society* 186: 1–72.

Cavendish, H. 1798. Experiments to determine the density of the earth. *Philosophical Transactions of the Royal Society of London* 88: 469–526.

Cornu, A., and J.B. Baille. 1878. Sur la mesure de la densité moyenne de la terre. *Comptes Rendus des Séances de L'Académie des Sciences* 80: 699–702.

De Rujula, A. 1986. Are there more than four? *Nature* 323: 760–761.

Franklin, A. 2018. *Is It the same result? Replication in physics.* San Rafael: Morgan & Claypool.

Hicks, W.M. 1886. On some irregularities in the values of the mean density of the earth, as determined by Baily. *Proceedings of the Cambridge Philosophical Society* 5: 156–161.

Mackenzie, A.S. 1900. *The laws of gravitation; memoirs by Newton, Bouguer, and Cavendish.* New York: American Book Company.

Newton, I. 1999. *Mathematical principles of natural philosophy*, ed. and trans. I.B. Cohen and A. Whitman. Berkeley: University of California Press.

Poynting, J.H. 1892. On a determination of the mean density of the earth and the gravitation constant by means of the common balance. *Philosophical Transactions of the Royal Society* 182: 565–656.

————. 1913. *The earth: Its shape, size, weight and spin.* Cambridge University Press.

Chapter 5
Does the Present Overdetermine the Past?

Craig W. Fox

I am indebted to George Smith and Chris Smeenk for countless conversations related to this topic and for their thoughtful feedback on earlier drafts

5.1 Introduction

There has recently been a surge in philosophical interest in sciences that reconstruct the past (see, e.g. Turner 2005, 2007, 2016; Currie 2018, 2019; Anderl 2018). One recurring theme in this emerging literature is a response to skepticism about the possibility of such endeavors leading to genuine scientific knowledge. Among the considerations that might lead to such a skeptical view are that investigations of the past characterize unpredictable and unique particulars, as opposed to regularities of nature. Historical investigations are concerned with reconstructing token phenomena. Insofar as the scientific method is geared towards theorizing about types of phenomena, there is a difference in the kind and quality of evidence that these sciences can garner. The worry is that investigating the past requires scientists to infer past causes based on limited traces (distal effects) that they are lucky enough to find. Because the phenomena of interest are unique particulars, no new evidence can be generated,[1] and, in fact, the ravages of time tend to erase most of the traces that were made to begin with. Both the uniqueness of the target and the loss of traces amount to additional epistemic hurdles that investigators of history face. These additional hurdles seem to threaten the proper testing of theories and hypotheses

---

[1] Not that new evidence cannot be found, but that the phenomena of interest tends to be unrepeatable and so the evidence base is all old.

C. W. Fox (✉)
Hebrew University of Jerusalem, Edelstein Center for History and Philosophy of Science, Technology and Medicine, Jerusalem, Israel
e-mail: cfox49@uwo.ca

© The Author(s), under exclusive license to Springer Nature Switzerland AG 2023
M. Stan, C. Smeenk (eds.), *Theory, Evidence, Data: Themes from George E. Smith*, Boston Studies in the Philosophy and History of Science 343, https://doi.org/10.1007/978-3-031-41041-3_5

concerning the deep history of life, the Earth, the Universe, etc. But in a series of influential papers that have become a touchstone for those interested in the so-called historical sciences, Cleland (2001, 2002, 2011) has argued that we should not be as pessimistic as all of that. The skeptic would be right if not for the fact that historical scientists have at their disposal a distinctive methodology that takes advantage of a temporal asymmetry—the asymmetry of overdetermination—which negates the putative epistemic inferiority.

In what follows, I'll start with a careful exposition of Cleland's arguments for the thesis that the present overdetermines the past.[2] I'll then proceed to critical engagement with the overdetermination thesis, where I'll show that the argument for the overdetermination thesis is circular, and that the thesis is false.[3]

Before turning to the main business of the paper, I would like to put my cards on the table. Although I aim to undermine the case for optimism concerning scientific knowledge of the past, I'm not optimistic. Rather, I'm skeptical that optimism vs. pessimism is an enlightening question to be asked because I'm skeptical of the distinction between "experimental" and "historical" sciences. But reassessing this distinction is a much bigger journey, to which I hope the present paper contributes a small step. For more along these lines, see Fox (2021); Jeffares (2008, 2010).

5.2 Cleland's Case for Optimism

Cleland's account begins with an articulation of the method employed by experimental scientists. Her account of the experimental method, moreover, is a refinement of Popperian hypothetico-deductivism. Cleland stresses the importance of controlled experiments that are designed to safeguard against misleading results. Scientists manipulate the experimental apparatus as well as vary the test condition in order to make sure that the predicted outcome is actually the result of the test condition and not some other accidental feature of the setup. They also, in the face of negative results, conduct further controlled trials in order to test whether the negative result is due to faulty equipment or some other issue with the slew of auxiliary hypotheses. In so doing, scientists are "engaging in systematic, extended experimentation that sometimes resembles an attempt to falsify a hypothesis and sometimes resembles an attempt to protect a hypothesis from falsification, but is really aimed at something quite different, minimizing the very real possibility of misleading confirmations and disconfirmations in concrete laboratory settings"

---

[2] Currie (2018) has provided some important refinements to the overdetermination thesis. I think a case can be made that the argument here poses significant challenges for Currie's version of optimism, especially for the "ripple model of evidence." Sadly, I must leave engagement with Currie to another paper.

[3] I'm not the first to worry about the status of the overdetermination thesis. Indeed, Turner (2005, 2016); Forber and Griffith (2011) have argued that the thesis is a metaphysical conjecture that is compatible with local epistemic underdetermination. To my knowledge, this paper is the first to argue that the thesis is actually false.

(Cleland 2002, 478). Here Cleland stresses the now-familiar underdetermination of theory by evidence of Duhem (1991). Cleland's stressing of the need to protect against concluding that a hypothesis has been confirmed or disconfirmed too quickly—on the basis of too little or spurious evidence—fits well with what scientists often, in fact, endeavor to do.

But insofar as the target phenomena of historical science is very often unique, large-scale events that occurred long ago, scientists clearly cannot test their hypotheses in anything like the way that classical experimentalists can. Cleland characterizes the inferential method of prototypical historical science thus:

> [A]n investigator observes puzzling traces (effects) of long-past events. Hypotheses are formulated to explain them. The hypotheses explain the traces by postulating a common cause for them. Thus the hypotheses of prototypical historical science differ from those of classical experimental science insofar as they are concerned with event-tokens instead of regularities among event-types[...] [H]istorical scientists focus their attention on formulating mutually exclusive hypotheses and hunting for evidentiary traces to discriminate among them. The goal is to discover a "smoking gun." A smoking gun is a trace(s) that unambiguously discriminates one hypothesis from among a set of currently available hypotheses as providing "the best explanation" of the traces thus far observed. (Cleland 2002, 481)

The idea here is to formulate a set of alternative common cause explanations for known traces and then try to elaborate a list of the other traces that each scenario would or could have produced. Any potential trace that is on the list of possible traces from one cause but not the others is a potential smoking gun, for, if found, it would eliminate the others from consideration. Historical scientists, then, go out into the field in search of smoking guns.

Now let's see Cleland's argument for how eliminative smoking-gun reasoning constitutes testing that is on epistemic par with the method of testing in experimental work. What Cleland's arguments need to address are the following evidential challenges: (1) Some, indeed, many, processes destroy the traces left by other processes (Information Loss). (2) Traces overlap—they are not isolated— making traces ambiguous as to which past process they derive from (Ambiguity of traces). And (3) Given (1) and (2), why should we expect that there are traces persisting to today that are sufficient for us to reliably infer their cause (Insufficiency of traces)? It is in response to these evidential challenges that the principle of the common cause plays a crucial role. According to Cleland:

> Common cause explanation promises a solution to the problem of evidential warrant faced by narrative explanations in natural science. [...]Common cause accounts of explanation are traditionally justified by appealing to 'the principle of the common cause'[...that] (roughly speaking) assert[s] that seemingly improbable coincidences (correlations or similarities among events or states) are best explained by reference to a shared common cause. [...]The principle of the common cause presupposes an ostensibly metaphysical claim about the temporal structure of causal relations among events in our universe. (Cleland 2011, 17f.)

The principle that justifies eliminative smoking-gun reasoning is that, in the absence of a common cause, the puzzling traces found so far are highly improbable. Now let's see how Cleland grounds this ostensibly metaphysical claim. The hope is that the argument will address the three evidential challenges raised above. If that argument does not work, then common cause explanations will be in no better standing than narrative explanations, which, according to Cleland, depend for their epistemic standing on vague notions of coherence and explanatory power (see Cleland 2011).

5.3 Common Cause and the Asymmetry of Overdetermination

Appeal to common cause explanation for epistemic justification is not without its problems (see Sober 1988, 2001; Turner 2005). One pressing issue is that the principle of the common cause stands in need of its own grounding or justification. Whereas other attempts to do so simply appeal to further metaphysical claims that, themselves, stand in need of justification, Cleland argues that the

> use of the principle of the common cause by historical natural scientists rests upon a substantive thesis about the nature of the world for which there exists overwhelming empirical evidence, namely, the thesis of the asymmetry of overdetermination. (Cleland 2002, 570)

The asymmetry of overdetermination serves two purposes for Cleland. It gives empirical justification for common cause reasoning. And, despite the additional epistemic hurdles inherent in studying the past, we should be optimistic about the ability of historical scientists to produce genuine knowledge of the deep past by smoking-gun-style reasoning. As we'll see, the asymmetry of overdetermination is supposed to insure that not all traces from some past event are necessary for inferring what happened so that smoking-gun-style inferences are (or can be when nature is generous) on epistemic par with experimental scientific inferences.

The asymmetry of overdetermination is a claim about how causes and effects are asymmetrically connected in time. The present state of a system, say an experimental setup, does not determine the outcome, even if the setup is prepared in accord with known dynamical laws. The outcome of the experiment can be circumvented in any number of ways, and so the future state of a system is underdetermined by the present. The future, in other words, is open; it is underdetermined by the present.

The present, on the other hand, overdetermines the past. The idea is that causes generally have many and widespread effects or traces such that not all traces are necessary for inferring what happened. One doesn't need every shard of glass from a window smashed by a brick; one needs only the brick and a few shards. Information loss, then, is not grounds for pessimism or skepticism regarding historical inquiry.

> [The asymmetry of overdetermination] tells us that a strikingly small subcollection of traces is enough to substantially increase the probability that a past event occurred, and that there are likely to be many such subcollections. The existence of so many different reliable possibilities of identifying past events provides the rationale for the historical scientist's emphasis on finding a smoking gun[...] One can never rule out the possibility of finding a smoking gun, and this is a consequence of an objective fact about nature, namely, most

past events are massively overdetermined by localized present phenomena. (Cleland 2002, 491f.)

Such is the claim, now for the argument for it. The argument is actually due to Lewis (1979) and it turns on yet another asymmetry, the "asymmetry of miracles." Though Cleland thinks it works for her purposes, Lewis' concern was not with historical inquiry, but with an analysis of counterfactuals. As is well-known, Lewis preferred analyzing counterfactuals in terms of possible worlds. One of the difficulties for Lewis-style possible world semantics is in specifying the "nearness" of possible worlds. Consider the following scenario: At one point during his Presidency, Richard Nixon supposedly seriously considered launching a nuclear assault against the Soviet Union. The counterfactual to consider is, "If Nixon had pressed the button to launch the U.S.A.'s nuclear weapons at the Soviet Union, the world would have been very different." It seems intuitively obvious that this counterfactual is true. At a minimum, the result would have been complete annihilation of the Soviet Union. But given the doctrine of mutually assured destruction, the Soviet nuclear arsenal would have been immediately unleashed in retaliation, and so the world would have been different, indeed. For Lewis's analysis, truth is to be cashed out in terms of possible worlds that are near the actual world in terms of matters of fact and law-like relations and in which the antecedent is true. If the consequent is true in all of the nearby possible worlds in which the antecedent is true, then the counterfactual is true.

So let W_a be the actual world and W_1 be a possible world that is exactly the same as W_a except that a tiny miracle occurs just when Nixon considers pressing the button. Say a few different neurons are made to fire and so he presses the button. Since the counterfactual should come out true, and W_1 is very near W_a, it is evident that close possible worlds can diverge radically after the event in question. But what about a world, W_2, in which the same tiny miracle makes Nixon press the button, but there's an additional tiny miracle, one that induces a short circuit, and so nothing happens? In W_2, the counterfactual is false, even though W_2 matches W_a in matters fact not only in the time prior to the event, but thereafter as well, save for the tiny miracles. So Lewis' criteria for closeness of possible worlds must exclude W_2.

According to Lewis, there is actually a big difference between W_1 and W_2, such that W_2 and W_a aren't nearly so similar after the event. The problem is that W_2 would diverge radically from W_a—albeit the divergence would be very different from the way W_1 diverges. The tiny (short-circuit-inducing) miracle is insufficient for the subsequent history of W_2 to match W_a. Although the second miracle prevents the holocaust, there are a great many effects that radiate out from the life-saving miracle. The trigger wire has a short-circuit, Nixon is left wondering what happened, "Maybe Nixon's memoirs are more sanctimonious … they have a different impact on the character of a few hundred out of the millions who read them. A few of these few hundred make different decisions at crucial moments of their lives—and we're off!" (Lewis 1979, 45f.). So the tiny extra miracle in W_2 does lead to a world that is very different from the actual world.

Alternatively, for the worlds to remain so similar, the second miracle in W_2 wouldn't be tiny at all, for it would involve a massive trace erasing operation including miracles erasing other miracles, for miracles would leave traces, too. W_2, then, is similar to W_a in matters of particular fact but only insofar as it's infected with a massive outbreak of lawlessness.

The asymmetry of overdetermination now follows from the asymmetry of miracles. What we learn from the Nixon counterfactual analysis is that the asymmetry of miracles is really a fact about traces. In order for a possible world to radically diverge from the actual world in matters of particular fact, a tiny change prior to an event in question is all that is needed. For the past to be different but the future the same requires an enormous number of miraculous things must occur. There is, then, essentially only one lawful way for the world to arrive at its present state. And all relevant traces from a past event cannot get erased. In Lewis' words:

> Whatever goes on leaves widespread and varied traces at future times. [...] It is plausible that very many simultaneous disjoint combinations of traces of any present fact are determinants [a minimal set of conditions jointly sufficient, given the laws of nature, for the fact in question] thereof; there is no lawful way for the combination to have come about in the absence of the fact...If so, the abundance of future traces makes for a like abundance of future determinants. We may reasonably expect overdetermination toward the past on an altogether different scale from the occasional case of mild overdetermination toward the future. (Lewis 1979, 50)

The argument, for Cleland, amounts to this: The asymmetry of miracles entails that there will always be some traces that persist and we don't need all of them. Moreover, a world in which there are traces that indicate some past occurrence but where that past occurrence didn't occur is a world in which the links between determinants (traces) and determined (the past cause) are broken, and this requires miracles (in the form of violations of natural law) or postulating a great many improbable coincidences. This follows from the fact that, were the past different, the present would also have to be very different. So, it seems, common cause explanations are far more probable, and though we may never find the right subcollection of traces that suffices, we are assured that they are out there and so we can never rule out the possibility of finding a smoking gun.

Crucial to the defense of Cleland's view is an argument as to how exactly the overdetermination thesis empirically grounds the principle of the common cause. Although Cleland does not explicitly spell out this step in the argument, the following is an attempt to fill out how the overdetermination thesis grounds the principle of the common cause that is in the spirit of Cleland's position. The asymmetry of miracles assures us that cover-ups are extremely difficult. Consider W_2 from the Nixon case, the world in which a tiny miracle occurred after Nixon presses the button that prevents the coming nuclear holocaust. In that world, there would be abundant traces that suffice for Nixon's having not pressed the button. These traces would be, in some sense, false. But, as Lewis argued, these would not be the only traces in W_2. The asymmetry of miracles guarantees that future times sensitively depend upon earlier times, such that Nixon's having pressed the button would necessarily have many other effects that would in turn have other effects and so on, making the world very different. In addition to the abundant "false traces" that make it seem as though Nixon didn't press the button, there would be abundant traces that show that he did, and so his having pressed the button would be overdetermined, nevertheless. So, the asymmetry of miracles justifies common cause reasoning because in order for there to be traces that do suffice for common cause inference and no traces that indicate something funny happened, the only lawful way for this to occur is that the common cause event occurred.

The method of testing employed in experimental work is appropriate because the future is underdetermined, and so scientists take advantage of the fact that they can manipulate and control phenomena in order to rule out false positives and false negatives. By contrast, historical scientists cannot conduct controlled manipulation, but this in no way diminishes their epistemic credentials, for they do not face the problem of underdetermination. To be sure, sometimes nature is ungenerous, and so smoking guns might be incredibly hard to find. But we can be confident that they exist and, when found, they provide comparable evidential value to controlled experiment. Moreover, the asymmetry of overdetermination thesis is a substantive claim about a contingent feature of the world that supports both the claim that there are sufficient traces because not all traces can be erased and the claim that we don't need to worry too much about misleading traces.

So the line of argument must be something like this: no slice of the past left zero traces into the present. Indeed, every slice of the past has left many lines of evidence, so information loss is not so great a threat as one might have worried. Moreover, the asymmetry of overdetermination and the asymmetry of miracles provide, for Cleland, the necessary grounding for the principle of the common cause—safeguarding common cause reasoning from worries about spurious correlations—which justifies Cleland's optimism that historical scientists can reliably investigate the past.[4]

5.4 Against Overdetermination

So far, I've been concerned with presenting the most positive case for optimism by way of the overdetermination thesis as I can. In this penultimate section, the tone turns decidedly negative as I extend Elga (2001)'s argument to this context.

---

[4] For the purposes of this paper I'm leaving aside the question of how to assess, at such a level of generality, Cleland's claim about epistemic parity—that historical scientists can achieve knowledge on epistemic par with experimental knowledge. The epistemic standing of a scientific claim is a local matter, to be reckoned by detailed assessment of the local evidence as well as the background knowledge entering into its interpretation. I do not see how to meaningfully compare claims from radically different domains. To be sure, some knowledge claims regarding the past are likely permanent additions to the body of scientific knowledge about the world, while some experimental claims are not. But does this establish the epistemic parity between the putative historical method and the experimental method? I think not.

I've shown that Cleland's proposed historical scientific methodology appeals to the principle of the common cause, which she tries to empirically ground by way of the overdetermination thesis. Moreover, I showed how the asymmetry of overdetermination seems to follow from the asymmetry of miracles. From the asymmetry of miracles, we get that causes have many and widespread effects that radiate out, effects becoming causes in their own right, producing more effects and so on. And the hoped-for epistemic fruit derives from this evidential redundancy and what seems plausible in Lewis' analysis—that there's only one lawful way that the present could have come about.

The problem with this line of reasoning is that the asymmetry of overdetermination isn't actually grounded or justified in any way; not only is it smuggled into the analysis, it is false. The manner of the smuggling is by way of what seems to be a very sensible move that Lewis makes at the start. Lewis explicitly ruled out of the analysis so-called backtracking counterfactuals. He recognized that when we employ or consider counterfactuals, we sometimes consider not how the world would be had the past been different. Rather, sometimes we consider how the past might've been different and yet the present the same. Lewis thought that this context of use for counterfactuals represented a non-standard resolution that need not be handled by his semantics, which were aimed at the standard resolution of counterfactuals. Indeed, the backtracking context was to be handled by an alternative semantics. He then, by fiat, excluded from consideration whether there might be closest possible worlds that radically violate the asymmetry of overdetermination, worlds in which the present is lawfully compatible with indefinitely many pasts.

The claim, then, is that the asymmetries of overdetermination and of miracles, along with any epistemic goods one might try to draw from them, do not hold; they only seem to if one refuses to consider how the past might differ while the present doesn't. To demonstrate this claim, all that needs to be done is to show how to construct possible worlds that satisfy Lewis' closeness criteria but have alternative pasts that converge to the same present and future. Happily, Elga (2001) has done just this.

Elga considers the following counterfactual about Greta who cracked an egg into a hot frying pan at 8:00. The counterfactual to consider is "Had Greta not cracked the egg, at 8:05 there would not have been a cooked egg on the pan." Though the circumstances are going to get pretty weird, what Elga's argument shows is that, contrary to the asymmetry of miracles, intervening after the cracking of the egg does not require a huge miracle in order to make the counterfactual false.

What Elga does, but Lewis wouldn't have dreamed of because of the prohibition on backtracking, is construct possible worlds from the actual world by inserting a tiny miraculous intervention *after* the triggering event, i.e. at 8:05 and then evolving the laws *backwards* in time. Consider the actual world at 8:05. If we run the tape backwards, we will observe what looks like strange behavior. The egg will slowly grow cooler, uncooking until 8:00 when it jumps off the pan and into the shell. The shell heals itself and Greta puts it into the refrigerator. Now, because of the way in which statistical mechanics works,[5] it would take only a tiny intervention on a very small portion of the phase space in order to make the backwards evolution from 8:05 look the same as the forward evolution. This new possible world is gotten by making such a change, such that continuing to run the evolution backwards from 8:05, the egg just sits there in the pan. This is problematic for Lewis, and for attempts to make use of the asymmetry of overdetermination, for this world is a world in which the antecedent is true but the consequent false, and yet it satisfies the criteria for closeness of possible worlds—it contains only a tiny violation of natural law and matches the actual world in matters of particular fact *after* the tiny miracle at 8:05, which shouldn't be possible.

To bring this discussion back around to Cleland, consider this matter from the standpoint of traces:

1. At 8:05 the actual world contains traces of Greta's having cracked the egg.
2. So immediately after 8:05, W_3 (the possible world gotten by running the laws backward from 8:05) also contains those traces.

But according to the asymmetry of overdetermination:

3. Since W_3 is a world in which Greta doesn't crack the egg, immediately before 8:05 W_3 does not contain those traces.
4. Therefore a large trace-manufacturing miracle occurs in W_3 at 8:05.

But 3 and 4 are wrong. To be sure, W_3 before 8:05 is an incredibly strange world. There's a very large spatio-temporal region that behaves thermodynamically rather strangely because the history of W_3 is gotten by running the laws in reverse. There's a cooked egg in Greta's pan that is more rotten the further back in time you go, but gets progressively less rotten as you approach 8:05, at which point it is freshly cooked. By 8:05 it appears that the egg had just recently been cracked on the pan and cooked, when, in fact, it was never raw and never cracked on the pan. Moreover, this is all in accord with the relevant natural laws. To be sure, W_3, at 8:05 contains a small spatio-temporal region infected by the change induced by the tiny miracle, and the size of that region grows dramatically as you go to earlier times. But this strangely behaving region, incredibly unlikely as it is, is perfectly lawful. Elga put the point nicely: "In general, the existence of apparent traces of an event (together with the laws and with the absence of evidence that those traces have been faked) falls far short of entailing that the event occurred" (2001, 324).

Moreover, the oddities of statistical mechanics may not be essential to this argument. I've presented Elga's formulation of the argument, but a moderately complex system governed by time-symmetric dynamics might work. Perhaps all that is needed is sensitive dependence to initial conditions such that a small change results in an ever-larger difference in the system as the laws are evolved backwards in time.

---

[5] See Elga (2001) for more details. Cf. also Eckhardt (2006); and Albert (2000) for more.

5.5 Conclusion

The upshot of the argument in the previous section is twofold: First, it shows that the asymmetry of overdetermination is not a substantive claim about the way the world is. Rather, the overdetermination thesis was smuggled into the metaphysical analysis of causation from the beginning. And it is only by restricting attention to ways the world would be different in the future—as opposed to ways the world could be the same—that the asymmetry thesis seems plausible. At best, the justification for the asymmetry thesis, then, is question begging. Note, too, that this has been pointed out as well by leading theorists of causation. Indeed, Paul et al. (2013) reject Lewis's analysis of causation for just this reason. This leaves Cleland's account of historical method in a difficult position, for it presupposes a deeply problematic account of causation. The only way forward, then, for deriving epistemic goods from the asymmetry of overdetermination is to find an analysis of causation that justifies it without circularity. Unfortunately, no such metaphysical analysis of causation is available at the moment.

The second upshot of this argument is that it shows that the link between a system's present and its past, between traces and the causes that produce them, is much more tenuous than is supposed in either Cleland or Lewis. Elga's argument shows that we live in a world governed by laws that not only make the future open—alternative futures can result from the same past—and that the same present can arise from indefinitely many pasts. And so the problem in the historian's temporal direction is no less underdetermination than it is for the experimentalist. Recall that one of the main burdens that the asymmetry of overdetermination was to bear was grounding the principle of the common cause. The asymmetry of overdetermination was to do this by guaranteeing that spurious and misleading correlations of traces generally don't occur. But now that the asymmetry thesis has been shown to be question-begging, there's no longer any guarantee that improbable traces accounted for by a plausible common cause aren't highly misleading. If we then rely on the asymmetry of overdetermination to underwrite smoking-gun reasoning, it seems that we are then in the exact same problematic position of an experimentalist who declares victory too soon in the face of tough epistemic circumstances. Causes tend to have various and widespread effects, yes, but spurious traces can ripple out too, and so scientists must be admonished to mind the risks of common cause reasoning, of falling for false positives that result from underdetermination of the past by the present. It takes a lot more to establish a knowledge claim about the past than the finding of a few traces compatible with a common cause.

This paper has examined whether the asymmetry of overdetermination and its potential to ground common cause reasoning makes sense from the standpoint of metaphysics. I want to close with a few brief remarks on whether the asymmetry of overdetermination makes sense from the standpoint of physics. I think the elephant in the room, so to speak, is the problem of equilibrium. Once a system reaches some dynamical equilibrium point, its history is erased. The problem here is that a system at equilibrium is compatible with indefinitely many past trajectories.[6] Sober (1991) portrayed this issue in terms of a ball at the bottom of a bowl. So long as the ball is *not* at the bottom of the bowl, the system (ball + bowl) retains some information as to the starting point of the ball. But once the ball reaches equilibrium, at rest in the bottom of the bowl, the system has no information as to the ball's starting point. The reason is that when we take a system at equilibrium and try to evolve the system backwards in time, the laws tell us that, with overwhelming probability, the system was always at equilibrium. The past of a system at equilibrium is massively underdetermined. To be sure, some of these problems for the overdetermination thesis result from the way we, i.e. computationally feeble, humans have to resort to coarse-graining complex systems. I'll grant, for the sake of argument, that a Laplacian Demon might not have the same problems of underdetermination. But this is of little consequence given that the "smoking gun method" is supposed to give agents like us the epistemic boost needed to overcome the evidential challenges we face.

Moreover, this kind of underdetermination applies not just to thermodynamic equilibrium. Even simple cases in Newtonian mechanics, where forces are paradigmatic causes, can make problems for the asymmetry of overdetermination. The difficulty is due to the composition of forces. Forces add and subtract from each other and so all that we generally can determine is the net force acting on a body or system of bodies. And net forces can be decomposed into arbitrarily many different configurations of contributing forces. This feature of causation in physics not only applies to determining the future trajectory of a system, but also to determining its past trajectory. Indeed, even if its past (or future) trajectory is known, the decomposition of causes governing the trajectory is, generally, massively underdetermined. Physical theories do not tell us that traces are univocal or that causes must have many and widespread effects; indeed, they are compatible with ambiguous effects, which can be lost altogether.

The asymmetry of overdetermination, it seems, is a no-go. At best, it is a metaphysical conjecture justified by a question-begging argument. At worst, it is simply false, given the known dynamics of the world. Moreover, since the asymmetry thesis was to provide the long-needed grounding for the principle of the common cause, it seems that common cause reasoning remains problematic. All of this means that Cleland's smoking gun reasoning fails to underwrite the hoped-for optimism about historical science. But does this failure underwrite wholesale skepticism toward the knowability of the past? I think the answer is that the knowability of the past is subject to the same problems of underdetermination that the present and future are, plus additional problems of information loss. Knowing the past isn't hopeless, but it is extra hard—and so, our credences should reflect this.

---

[6] See, e.g. Sober (1991). For a more technical discussion of the physics of equilibrium, see Albert (2000); Eckhardt (2006).

References

Albert, D.Z. 2000. *Time and chance*. Harvard University Press.

Anderl, S. 2018. Simplicity and simplification in astrophysical modeling. *Philosophy of Science* 85: 819–831.

Cleland, C.E. 2001. Historical science, experimental science, and the scientific method. *Geology* 29: 987–990.

———. 2002. Methodological and epistemic differences between historical science and experimental science. *Philosophy of Science* 69: 447–451.

———. 2011. Prediction and explanation in historical natural science. *British Journal for the Philosophy of Science* 62: 551–582.

Currie, A. 2018. *Rock, bone, and ruin: An optimist's guide to the historical sciences*. MIT Press.

———. 2019. Simplicity, one-shot hypotheses and paleobiological explanation. *History and Philosophy of the Life Sciences* 41: 10.

Duhem, P.M. 1991. *The aim and structure of physical theory*. Princeton University Press.

Eckhardt, W. 2006. Causal time asymmetry. *Studies in History and Philosophy of Modern Physics* 37: 439–466.

Elga, A. 2001. Statistical mechanics and the asymmetry of counterfactual dependence. *Philosophy of Science* 68: 313–324.

Forber, P., and E. Griffith. 2011. Historical reconstruction: Gaining epistemic access to the deep past. *Philosophy and Theory in Biology* 3: 1–19.

Fox, C.W. 2021. *Back to the beginning: An empiricist defense of scientific stories about the past*. Electronic Thesis and Dissertation Repository. URL: https://ir.lib.uwo.ca/etd/7798.

Jeffares, B. 2008. Testing times: Regularities in the historical sciences. *Studies in History and Philosophy of Biological and Biomedical Sciences* 39: 469–475.

———. 2010. Guessing the future of the past. *Biology and Philosophy* 25: 125–142.

Lewis, D. 1979. Counterfactual dependence and time's arrow. *Nous* 13: 455–476.

Paul, L.A., N. Hall, and E.J. Hall. 2013. *Causation: A user's guide*. Oxford University Press.

Sober, E. 1988. The principle of the common cause. In *Probability and causality*, 211–228. Springer.

———. 1991. *Reconstructing the past: Parsimony, evolution, and inference*. MIT Press.

———. 2001. Venetian sea levels, British bread prices, and the principle of the common cause. *British Journal for the Philosophy of Science* 52: 331–346.

Turner, D. 2005. Local underdetermination in historical science. *Philosophy of Science* 72: 209–230.

———. 2007. *Making prehistory: Historical science and the scientific realism debate*. Cambridge University Press.

———. 2016. A second look at the colors of the dinosaurs. *Studies in History and Philosophy of Science* 55: 60–68.

Chapter 6
Newton's Example of the Two Globes

Monica Solomon

6.1 Introduction

At the end of the Scholium Newton includes a long paragraph about two globes revolving around their center of gravity and held together by a tensed cord. It has been interpreted as a thought experiment (Sect. 6.2) meant to show how the properties of true circular motion defined as absolute motion can be determined in a three-dimensional empty universe. I start by showing that this reading of Newton's example as a bona fide thought experiment is riddled with interpretation problems and that it is less straightforward than so far assumed (Sect. 6.3).

My alternative relies on understanding the contrast between considering the globes to be a fictional scenario and using it as a quantitative model of a dynamical interaction. I argue here against the former and for the latter: the scenario is an idealized model of a quasi-isolated system of two interacting bodies. I introduce and briefly motivate the reading in Sect. 6.4. Section 6.5 complements existing translations of the passage with a carefully analyzed manuscript source. Then I flesh out the picture: Newton's scenario shows us that it is possible to build a model of a two-body system using his definitions, with no reference to another body. Specifically, he provides a model for the analysis of true motion in terms of *quantities* (which I take to be the quantities invoked in the definitions) and their changes. We analyze changes in these quantities (and not their absolute values) by means of the three laws of motion, and in the process no other bodies are used as reference points. Section 6.6 briefly explains how my interpretation faces up to the problems which the standard reading faced in Sect. 6.3. Finally, in Sect. 6.7, I give other examples from Newton's works illustrating similar models. I conclude by

M. Solomon (✉)
Department of Philosophy, Faculty of Humanities and Letters, Bilkent University, Bilkent, Ankara, Turkey

© The Author(s), under exclusive license to Springer Nature Switzerland AG 2023
M. Stan, C. Smeenk (eds.), *Theory, Evidence, Data: Themes from George E. Smith*, Boston Studies in the Philosophy and History of Science 343, https://doi.org/10.1007/978-3-031-41041-3_6

95

pointing out that this example, far from being an obscure passage of Newton's, paves the way to understanding the theory of the solar system presented in the *Principia*.

6.2 The Reception of the Globes Scenario in the Literature

Newton begins the Scholium by saying that "although time, space, place, and motion are very familiar to everyone, it must be noted that these quantities are popularly conceived solely with reference to the objects of sense perception." Unlike the entrenched use of quantities up until that moment, the *Principia* of 1687 puts forward new kinds of quantities,[1] and now it becomes "useful to distinguish these quantities into absolute and relative, true and apparent, mathematical and common." (Newton 1999, 408).

Then he introduces distinctions pertaining to time, space, place, and motion, and definitions of absolute 'time,'[2] 'space,'[3] 'place,'[4] and 'motion'.[5] The true and absolute motion of a body is distinct from its apparent and relative motions. Newton says that, although the parts of absolute space are not seen and make no impression on the senses, we are able to determine the true motion of bodies by means of the properties, causes, and effects of this motion. To show that the determination of true motion of bodies is not utterly hopeless, he discusses at the end of the Scholium the example of two globes revolving around the common center of gravity, while being held together by a tensed cord.

In the first stage of the description, the endeavor to recede from the center is known from the tension in the string. Then, using impressed forces on the faces of the globes, we can determine the direction of revolution (whether clockwise or counterclockwise from the perspective of an observer at rest with a bird's eye view on the globes). Finally, in the second part of the same paragraph, if we assume that there are some fixed bodies which maintain the same positions among themselves,

---

[1] Some of the novel quantities, such as the quantity of matter and the various quantities of centripetal force, are described in the set of definitions at the beginning of the *Principia*. The space-time Scholium, as it is now called, is a commentary pertaining to the set of definitions. The definitions are of: quantity of matter, quantity of motion, inherent force of matter (*vis insita*), impressed force, centripetal force and three measures of it (absolute quantity of centripetal force, accelerative quantity and motive quantity). (See Newton 1999, 403–408)

[2] "Absolute, true, and mathematical time, in and of itself and of its own nature, without reference to anything external, flows uniformly and by another name is called duration." (Newton 1999, 408)

[3] "Absolute space, of its own nature and without reference to anything external, always remains homogenous and immovable. Relative space is any movable measure or dimension of this absolute space; such a measure or dimension is determined by our senses from the situation of the space with respect to bodies and is popularly used for immovable space." (Newton 1999, 408–9)

[4] "Place is the part of space that a body occupies, and it is, depending on the space, either absolute or relative."(Newton 1999, 409)

[5] "Absolute motion is the change of position from one absolute place to another; relative motion is change of position from one relative place to another." (Newton 1999, 409)

we could compare the relative motions of the globes among these bodies with the tension in the cord and determine whether the motion belongs to the globes or not.

This is the example, in a nutshell. On the one hand, it faced - what I will call - the classical interpretation: on this reading, the example of the globes is lumped together with the example of a rotating water bucket. Their joint role is to show the *existence* of absolute motion (and by inference to the best explanation, the existence of absolute space).[6] The tension in the cord shows the endeavor to recede from the center. The existence of the endeavor to recede from the center signifies in turn the existence of real motion. Such motion is not motion with respect to any body, since it is implicitly assumed that there are no other bodies in the universe and there is no change in relative distance between themselves. Therefore, *this is* absolute motion. Some authors would go further and clarify the implicit inference: because absolute motion exists and since absolute motion is motion with respect to absolute space, then absolute space also exists.[7]

Naturally, the classical reading very often refers directly to the globes as an instance of a thought experiment:[8] the globes are moving in absolute space, which this interpretation takes to be an imagined empty universe. For instance:

> And if we accept the thought experiment with the globes in an otherwise empty space, the relevant motion cannot be motion with respect to any material body. Newton concludes that the motion must be motion with respect to absolute space: the spinning bodies successively occupy different locations in space itself. In this way, absolute motions are connected to forces and hence to observable effects. (Maudlin 2012, 23)

Recently however, against this classical reading, Laymon (1978) and Rynasiewicz (1995a, b, 2014, 2019) point out that, since the example comes at the end of the inquiry in the Scholium, it does not follow the pattern of argumentation of previous examples, such as the rotating bucket. According to this recent reading, we *assume* that true motion and absolute motion coincide; we no longer seek to prove either the existence of absolute motion, or that a body's true motion should be defined as motion with respect to absolute space.[9] The role of the globes example now becomes a matter of epistemology: how is one to distinguish absolute motion

---

[6] See Maudlin (2012), Nagel (1961), Van Fraassen (1970).

[7] See, for instance Maudlin (2012, 15): "Newton produces powerful empirical evidence for the existence of absolute motion (and hence absolute space and time) using considerations of the causes of motion."

[8] Arthur (2018), Barbour (1989, 629–40), Berkeley (1721), DiSalle (2006, 33–4), Earman (1989, Ch 4), Laymon (1978), Mach (1919, 229) Maudlin (2012, 22–5), Westfall (1971, 443–5)

[9] Rynasiewicz (2019) understands the distinction between the true and absolute motion of a body on the one hand, and the apparent and relative motions, on the other hand, as one of a metaphysical kind. The former has an elevated ontological status, more reality or existence perhaps, than the latter. See also Huggett (2012) and DiSalle (2002, 2006) on the connection between true and absolute motion. The clearest presentation I found in Brading, *Philosophy and the Physics Within*, Ch 3 (ms). My own view departs from all of these, but this is not the place to develop it. I take it from the recent literature that, at least in the case of the globes, Newton builds the description such that there is a single quantity of true motion pertaining to each globe, and that the challenge is to capture the factors which change this quantity, and only those.

from apparent motions, since we lack direct access through our senses to parts of space? Rynasiewicz (2019) argues that the globes scenario is a thought experiment supposed to illustrate "how to recognize the true motion of individual bodies and in actuality to separate [*discriminare*] from the apparent." (p. 18) The idea seems to be that, although in reality we do not have direct access to absolute motion and its properties, we can coherently conceive in imagination of true motion being different from apparent motions.

All available interpretations share a basic assumption: absolute motion can be conceived and changed in absolute space, where no bodies exist, whereas relative and apparent motions can be said to exist only by referring to changes in position among other bodies. And they all take the example to be one which is imagined unfolding in absolute space simpliciter.

We start by imagining an empty universe in which only the globes and the cord are present. Should the bodies not move, then there would be no tension in the cord (from Newton's own previous comments). But since there is tension in the cord, that means we know they *really* are in motion. (They have true motion.) This motion can only be conceived as motion with respect to absolute space, since there are no other bodies in the universe. In other words, they have a circular motion relative to absolute space. But a circular motion has a direction, so Newton shows how we can establish the direction by impressed forces on the faces of the bodies and see how the tension in the rope changes.

Now, for the second part of the scenario, imagine a different setup: the same globes and the cord, except now there is the "sphere" of fixed stars in the background. We notice a change in the relative position of the globes among the stars. Do we know if the globes really are moving only by attending to the relative changes of position? We do not; unless we attend to the effects of true motion (the endeavor to recede from the center). The tension in the string tells us that the motion belongs to the globes and not to the stars.[10] Since the stars are at rest in absolute space, we could use them as a backdrop reference frame: we now can infer the direction of motion from the relative changes of position of the globes among the stars. The first case was an instance of true motion conceived with respect to absolute space, while the second one illustrates how apparent motion is insufficient for the determination of true motion. They are both stages in a single thought experiment . . . or are they?

6.3   Is the Globes Scenario a Thought Experiment?

To my knowledge, there hasn't been an explicit justification of why we should interpret Newton's example in the manner introduced above, as a thought experiment. While some might find this reading natural and intuitive, as an interpretation of

---

[10] In Sect. 6.3 we shall see that this inference does not hold, given Newton's own qualifications about relativity of motions. The most direct criticism of it I found in Barbour (1989, 643–4).

Newton's text it deserves some scrutiny. Let me mention explicitly some possible reasons one might invoke for this interpretative category and point out some problems for it along the way. In the next section I introduce an alternative.

The reasoning goes presumably like this: the scenario is naturally a fictional thought experiment because it assumes some things which we cannot possibly observe in actuality. For instance, in the first part of the passage we assumed there were no other bodies in the universe. We start with absolute space imagined to be the space of the universe emptied of all material bodies. Then we "add" in imagination two bodies shaped like globes and a tensed cord between them.[11] Obviously, two bodies connected by a cord moving in an empty universe can only happen in a thought experiment.

This interpretation which relies strongly on imagination from the get-go is faced with several difficulties. To simplify, I will select two main challenges derived from the first part of the example (when the globes and the cord are supposed to be alone in absolute space), and two additional problems for the second part of the example (when we consider the fixed stars). The first problem for the empty-universe-with-globes reading derives in fact from the strongest textual support for it. The strongest supporting reason seems to be in a sentence Newton includes halfway through the paragraph. There, by the use of the tension in the cord and impressed forces we can conclude:

> In this way both the quantity and the direction of this circular motion could be found in any *immense vacuum*, where nothing external and sensible existed with which the balls could be compared. (Newton 1999, 414; my emphasis)

The idea is to consider the 'immense vacuum' above to be another name for the all-encompassing absolute space. Whether we want to equate a vacuum and absolute space is a different point, which I believe requires more argumentation, if only because Newton uses different terms for them and because nowhere does he define one in terms of the other. They could be assimilated to each other in imagination only if we *imagine* absolute space as a vacuum.

In any case, not every use of 'vacuum' should be taken to refer to absolute space. Specifically, this particular instance of vacuum makes perfect sense as an instance of a pocket of vacuum. Newton himself seems to have struggled with how and where to qualify the introduction of vacuum. He deleted an initial mention of vacuum ("two globes revolving in a vacuum") and qualified the single mention of the term ("*any* immense vacuum"; my emphasis). Consequently the emphasis shifted: from where we imagine the globes to be to what is involved in the determination of the direction of their motion. That is, the important point is that we do not use other bodies for the determination of motion. Clearly, then, the vacuum mentioned here is not the *single* all-encompassing universe emptied of matter in our imagination. It could be *any* pocket of vacuum (which could be quite immense) in which the two revolving

---

[11] A side note: we immediately face the question of how to understand the gravity of those two globes in such an empty universe.

globes find themselves because in *any vacuum the method introduced by Newton for determining the properties of their motions is the same* and it does not use reference to other bodies.[12]

The second problem involves the mention of impressed forces in the first part of the experiment (before the fixed stars are added to the scenario). The presence of impressed forces is in tension with the assumption that we supposed there are no other bodies in existence. The concept of 'impressed force' is a notion introduced by Newton in the *Principia* and it covers physical forces (such as pressure, impulses, or several kinds of centripetal force)—forces which have their seat or exist because of other material bodies. Hence there should be an action of other external bodies on the two-body system. On the assumption of empty space, however, this is not possible. If there are impressed forces in the system, then the space cannot be empty. How could it be both an empty space and non-empty at the same time?

Additional difficulties arise when we continue reading this as a thought experiment. There are problems concerning how to understand the transition between the first part and the second part of the scenario. For instance, about how those fixed stars come into being (a move that can make sense only in imagination), or whether we need to devise another thought experiment in which we no longer start with an empty universe. I will leave these kinds of worries aside for the moment because they are not directly relevant to my argument.[13]

More importantly, the inferences presented in the second part, after the introduction of the fixed bodies in the background, become problematic. Consider that we have the two globes revolving as they did in the beginning, only now we also have a set of bodies with fixed positions among themselves in the background. If we attend to the tension in the cord when the globes are revolving *uniformly* around their center of gravity, we could *not* infer that motion belongs to the globes *or* to the stars (contrary to Newton's own claim). To see why, consider the case of the stars revolving uniformly, but doing so at a different constant rate of rotation than the globes. To put it differently, if the motion of the globes and the stars' rotation are both uniform, but with different rates of revolution, Newton's own conclusion concerning true motion versus apparent motion is not correct. We could *not* infer that rotation pertains *only* to the globes.

This could not have been a minor blunder. Newton could have spotted this 'mistake' while revising the *Principia* for the second and third editions. In fact, the opposite happens: in the second edition Newton adds the sentence "and that the bodies at rest," strengthening his conclusion about which set of bodies are moving and which are at rest—a rather bold move, if all the motions are presumed uniform (and not varying) Smith (in-press b).

---

[12] There is a great similarity between this strategy and current methodology of studying the properties of binary star systems. Most stars are in fact binary systems. (csiro.au)

[13] For instance, Mach (1919) faults Newton with the assumptions entering into this thought experiment because it looks like the universe assumed is very different in crucial aspects from the universe we know to observe and inhabit. As he puts it, "the universe is not given twice."

Finally, received views relegate the example to obscurity: it is unclear how the connection with the project of the *Principia* proceeds, if at all. Given that it is a thought experiment, it is unlikely that it will be helpful for other real situations. Newton, then, seems to just pay an insincere lip service when he concludes the paragraph by saying:

> But in what follows, a fuller explanation will be given of how to determine true motions from their causes, effects, and apparent differences, and, conversely, of how to determine from motions, whether true or apparent, their causes and effects. For this was the purpose for which I composed the following treatise. (Newton 1999, 415)

As far as I can tell, nobody took these last sentences seriously. The argument of the *Principia*, the thought presumably goes, is far too complicated to have this scant example give us an insight into it. This is a fair point *if* we read the example as primarily a thought experiment concerning two bodies in a situation which is unlike any real instance in our physical universe. The example was presumed to have no actual connection to the "fuller explanation" provided by the *Principia*, despite Newton devoting a very long paragraph on it and saying exactly the opposite at the end of it. In the next section, I put forward a new approach that is in line with Newton's claim and argue that the passage can indeed be a guide to the methodology of the *Principia*.

6.4 Taking the Example as an Idealized Model: The Very Idea

The minimal description of the scenario is seen as an artifact of imagining the globes to be the sole inhabitants of absolute space. But it isn't the only way to read the example. Admittedly, Newton's description *is* minimal: the bodies lack any qualitative description, and any element introduced in the argument, such as the impressed forces or the fixed stars, shows up at a particular time, followed by specific inferences. But minimalism is justified, I argue, not because we start with the 'barest' environment of all: absolute space. Instead, the minimal description is the natural feature of idealized models, the result of considering a set of bodies as an isolated system under consideration in which measurement is performed of some quantities by means of other quantities. This section introduces the core idea of this interpretation.

Alongside recent scholarship, I take the example to have an epistemic function as well: it illustrates how true and apparent motions are to be presented and distinguished. I disagree, however, with existing views on how this role is achieved. Instead of conceivability, I rely on methodology. Instead of taking the example to describe actions and scenarios impossible to realize in reality,[14] I consider that the

---

[14] On the contrary, Newton says that the example aims to "actually" (*actu*) distinguish apparent and true motions. (See Sect. 6.4)

scenario is an idealized model of a system of two interacting bodies. Instead of relying on an individual's ability to conceive such motions in imagination starting from nothing, I argue that acts of imagination are subordinate to the process of idealization, in which features not under investigation or considered irrelevant are abstracted away. The last paragraph may look like an imagined case taking place in an empty universe, but that is because it is the result of idealization of physical situations in which two bodies affect each other's motions in determinate ways. Whereas received views take the lack of descriptive or contextual details as a problem, I suggest we should take them as features of the scenario curated to fit the role of a specific model.[15]

Specifically, there are two methodological factors incorporated in the scenario which help the example fulfill its function of an idealization.

The quasi-isolation feature: According to it, some bodies are included in the description, and some are left out. Included bodies are interacting in an idealized manner: their actions are expressed by means of forces defined by Newton, and not by specifying a causal mechanism of sorts. Their interactions are fully captured by the three laws of motions, and there are no requirements on specifying a mechanical cause for how the actions are done. The isolation of a system of bodies is a deliberate act of investigation. We do not isolate some random set of bodies. Bodies not specifically included in the description are *not* annihilated: their existence and influence are simply bracketed for the time being. Actions of these other bodies are to be accounted for in terms of impressed forces only.

The quantitative feature: All the relevant actions of bodies in our system are tracked solely by the set of quantities or forces Newton introduced previously in the *Principia*. Effects of actions always quantified. The quantities involved in the description will depend on the satisfaction of the quasi-isolation criterion. If the quasi-isolation is done in an appropriate manner, then the relations among quantities of motions and properties of forces have the strong form of a double conditional: one such quantity changes iff another quantity changes accordingly.

Such is the view that I propose. Although separating these two factors is my own doing, I suggest here that they belong to Newton's methodology proper. Moreover, I would argue that they are constitutive of Newtonian idealizations in general.[16] The globes example may be the first of such idealizations, but certainly not the last in the *Principia*.

In many of his writings, Smith has emphasized in various ways the crucial role of Newtonian idealizations in testing the theory of Newtonian gravity throughout its long history. Here is one such example:

---

[15] The model I have in mind is akin the two-body problem in physics, and not, say, mechanical models for causal interaction of two bodies. Newton, of course, does not use the word "model." For recent work on this understanding of Newtonian models see Ducheyne (2005) and Ducheyne (2012, esp. chap. 2). Ducheyne focuses on planetary models. I share much with Ducheyne's arguments, especially the idea that the models in Book 1 are not restricted to mathematics. But I also think that the two features which I introduce here are to be more systematically applied and embedded into Newton's natural philosophy, going beyond models of planetary motions.

[16] This claim will be developed elsewhere. This paper restricts itself to arguing that the presentation of the globes scenario is best understood as an illustration of the result of applying these two features. Specifically, it demotes the understanding of the scenario as a thought experiment. (That is not say that thought experiments in natural science do not use models.) Briefly put, this scenario is closer to reasoning in physics proper than we have seen it so far represented in the literature.

The purpose is to shift the focus of ongoing research onto systematic discrepancies between the idealizations and observation, asking in a sequence of successive approximations, what further forces or density variations are affecting the actual situation? The theory of gravity and deductions from it become not so much explanations or representations of known phenomena, but instruments in ongoing research, revealing new discrepancies between, for example, true and idealized orbital motions. [...] The theory of gravity gets tested in this process through its requiring that every deviation from any Newtonian idealization be *physically significant* – that is, every deviation has to result from some unaccounted for density variation or force, gravitational or otherwise. So the test question is not whether calculation agrees with observation, but whether robust physical sources can be found for the discrepancies between calculation and observation. (Smith 2014, 277)

My interpretation builds on Smith's insight that Newton's methodology for natural philosophy in the *Principia* aims to form the laws of forces of nature by modeling ever more complex systems of bodies interacting through those forces. In this process, idealizations of a specific kind take center stage.

Let's see how they could briefly play out in Newton's example. In principle, the two globes could also be affected by other bodies acting by means of impressed forces. The idealization consists in bracketing the existence of these bodies at different times and taking the two globes as a (possibly) isolated system of interacting bodies. In such a system, any changes in motion are described only quantitatively. The changes in the quantity of true motion could be compared with changes in two other different quantities. One is the tension in the rope, the other is the relative rate of rotation among the fixed stars. Newton's goal in the scenario, I argue, is to separate these two types of correlations in the first stage, and then to connect them, in the latter part. The former set of quantities (the quantity of true circular motion and the tension in the cord) could indicate the source of physically significant changes (caused by other bodies affecting the quasi-isolated system). The latter set (the quantity of true circular motion and relative rate of rotation), by itself, could not. They could be combined under certain conditions. But making this distinction clear is achieved not by a thought experiment, but by a carefully constructed and worded scenario, to which we now turn.

6.5   The Original Text and the Model of Interaction

Let us now have a closer look at Newton's own words. Instead of using Cohen and Whitman's authoritative translation, I include an (unpublished) translation by George Smith and Anne Whitman with notes sending to the original Latin or to deleted passages:[17]

It is most difficult [*difficillimum est*] indeed to identify [*cognoscere*] the true motions of individual bodies, and ~~in practice~~[18] ?actually? to discriminate them from apparent ones,

---

[17] ?- shows insertions

[18] The adverb deleted and replaced by "*actu*" cannot quite be made out, and hence this translation is a guess.

because those parts of that immobile spaces in which bodies are truly moved do not strike against [*incurrunt*] the senses. ~~Xxxx xxxx xxxx which individual immobile xxxx xxxx true nevertheless are to be reckoned [verum tamen disputando], and that we can sometimes gather something partly from the forces that are the causes and effects of the true motions, partly~~ ?Nevertheless, the case [*causa*] is not absolutely hopeless [*prorsus desperata*]. For arguments are forthcoming [*suppetunt*] partly? from the apparent motions that are the differences of the true motions [and] ?partly from the forces that are the causes and effects of the true motions?. For example [*ut*], if two globes at a given distance from one another connected by an intervening cord were being revolved [*revolverentur*] ~~in a vacuum~~ around the common center of gravity, the endeavor of the globes to recede from the axis of motion ~~will~~ would become known [*innotesceret*] from the tension in the cord, and then the quantity of circular motion could be computed. Thereupon if ?no matter what? equal forces were to be impressed [*imprimerentur*] at the same time on alternate faces of the globes increasing or lessening the circular motion, the increase or decrease of the motion ~~could be still be learned~~ ?would become known? from the added or diminished tension in the cord, and therefrom ?finally? on which faces of the globes the forces would have to be [*deberent*] impressed for the motion to be increased maximally could be found, that is, the posterior faces, or those which follow [*sequuntur*] in the circular motion. Moreover, learning the face that follows and the opposite face that precedes, the direction of the motion would be identified [*cognosceretur*]. ~~This circular [motion] in an immense vacuum~~ In this manner both the quantity and the direction of this circular motion could be found in ?whatever [*quovis*]? immense vacuum where nothing external and sensible exists with which the globes could be compared. If now some distant bodies were set in that space maintaining given positions among themselves [*inter se*], as the fixed stars are in our regions, it could indeed not be known from the relative translation of the globes among the bodies [*inter corpora*] whether this motion was to be attributed to these [globes] or those [bodies]. But if the cord were examined and its tension were found to be that which the motion of the globes would require [*requireret*], it would be legitimate [*liceret*] to conclude the motion to be of the globes,[19] and finally from the translation of the globes among the bodies to gather the direction of this motion. Moreover, to gather the true motions from their causes, effects and apparent differences, and conversely their causes and effects from motions whether true or apparent, will be shown more extensively [*fusior*] in the following. For to this end I composed the Following Treatise.[20]

The text is fairly clean, with very few changes. One modification involves qualifications of the use of 'vacuum' and the insertion of a stronger conclusion which attributes motion to the globes and rest to the fixed stars. We also witness the use of the subjunctive throughout the passage directing the reader to hypothetical situations. My approach is to consider this hypothetical situation as a description of an idealized case, in which given certain assumptions, some consequences follow. In other words, the hypothetical situation is not *fabled* one. It has a hypothetical form where "ut" has the power of introducing an example. And it describes a carefully abstracted instance of interacting bodies, not the result of freely running creative

---

[19] In the second edition of the *Principia* Newton inserted at this point the further clause, "and that the bodies were at rest."

[20] These passages are quoted from a longer manuscript by George Smith and Anne Whitman. Appendix 5 consists of variorum translations of selections from the version of *Liber Primus* Newton submitted to Cambridge Library under the auspices of Lucasian Lectures, Dd. 9.46 (pp. 36–215 of Whiteside 1989). See Whiteside (1989, 45–6), translated by George E. Smith; and the folio numbers (11 and 12) in Dd. 9.46. Smith (in press-b).

imaginative powers. We start from reality and abstract away some features, whereas in fabled imaginings we start from scratch and add features as we please.

It is worth noting, for instance, that there is nothing in the language to direct our thoughts to a scenario *created* by imagination. There is no language of creation when it comes to the fixed stars either: the bodies are set (not imagined, not created). By comparison, in the earlier manuscript *De Gravitatione*, the famous passage that invites the reader to imagine some entities similar to the bodies of our experience starts by saying, "Fingamus itque spatia vacua per mundum disseminari quorum [....]. Sed si fingamus [...]." (Hall and Hall, 1963, 106) In the case of *De Gravitatione* the wording elicits the act of imagination, hypothesis, and fabrication, whereas the Scholium confines its language to that of putting forward the givens of the problem and describing hypothetical results (but not imagined): the verbs are such as "revolverentur," "innotesceret," "cognosceretur," etc.- all expressing the hypothetical of a problem setting.

Let me flesh out more my interpretation on offer here, by emphasizing the two methodological factors mentioned previously. First, the quasi-isolation factor: we start with the world of experience as we encounter it, a universe with (many) bodies in motion. Somewhere, in a part of this universe, in a pocket of vacuum, two globes interact with each other by means of the tensed rope between them. They revolve around their common center of gravity as Newton suggests. Our imagination functions as abstraction: in the first part of the scenario (before considering the fixed stars), it is required to bracket the existence of other bodies as points of reference, but not to annihilate them altogether. This, I argue, is how the role of imagination is in service of using Newton's powerful concept of impressed force in order to provide a fairly complete mathematical description of the motions (to a first approximation) without invoking other bodies.

Second, there is the quantitative dynamics. According to the view on offer here, the quantity of **true** motion will be a quantity ascribed to bodies in a carefully specified system of interaction (and not with respect to absolute space). The changes in such quantities are going to be considered **absolute** because they are changed by impressed forces *acting explicitly from outside the system*. Basically, for my reading, all that matters are variations in quantities, and the causes and the effects of such changes. To put it differently, we ask ourselves: What are the things that make a difference to the system of bodies under investigation, and what difference do they make?[21]

---

[21] This question has often showed up during George's course and it is a recurrent pattern in the history of testing Newtonian gravity. See Smith 2014.

Some quantities can be thought of in a purely mathematical manner, even though they are measured by appealing to sensible quantities.[22] To my mind, this means that quantities involving absolute space, for instance, cannot be evaluated by reference to parts of absolute space, but they can still be used to model a particular system of interacting bodies. This kind of measurement is done by reference to sensible things, and the parts of absolute space "do not impinge on the senses."

I suggest that in this case we know or measure absolute quantities pertaining to some true motions in a *specialized sense*: by evaluating whether the changes in such quantities characterizing motions are caused by impressed forces in harmony with expected effects of such motions (such as the endeavor to recede from the center as a physical, real effect on bodies).[23] This means for me that we will be using mainly inferences concerning *relations among quantities*: the quantity of true circular motion of the globes is not to be known as an absolute number or value on its own, in isolation from any other kinds of quantities.

Thus, I take it that the epistemic challenge of the last paragraph of the Scholium is a challenge that we encounter when in the physical world we try to *measure* certain quantities (not when we try to conceive them). I suggest that the challenge of measuring true and absolute quantities of a particular, concrete instance of real bodies interacting is answered *not* by providing a number or a specific value. But by modeling the situation as a specific quasi-isolated system of two interacting bodies using only the definitions and his laws of motion. Provided we built such an idealization, then we answer the challenge when we fully identify all the other quantities which determine uniquely the quantity of true motion and which are explicitly stated as interdependent by means of mathematical relations, within a particular system of bodies.

This is how this methodology is illustrated by the two globes scenario. Each of the bodies is assumed to have a quantity of matter (Definition 1). Their revolution is around the center of gravity which suggests that gravity is also part of this scenario. The bodies are in circular motion which means that their motion is the result of two forces – the inertial force and the centripetal force keeping them on the trajectory (Definitions 3, 5, and the first two laws of motion). The centripetal force is clearly determined: we have its center. Physically, each body attracts the other one; mathematically, this interaction is described by considering the center of gravity as the center of attraction. The tension in the cord is a measure of the strength of the action of each body on the other (Law 3). What the scenario idealizes is the interaction of two bodies in which all the three laws of motion concur.

When we have a tensed rope between two bodies, we find ourselves in a lucky epistemic position because the tension is the expression of an equilibrium between

---

[22] The preceding paragraph stressed again the distinction between relative and "actual" quantities. The former are sensible measures of the latter. When we refer to quantities involving time, space, place, motion in the absolute sense, we use a "manner of expression which is out of the ordinary and purely mathematical."

[23] According to the reading on offer here, the bucket experiment follows the same method.

two endeavors to recede from the center. Those endeavors are quantities directly correlated with the amount of quantity of circular motion of each body, because they both are known to vary directly with the mass of the body, the period of rotation, and the radius of circular motion (and only with those quantities.) So, on my account, a quantity of circular motion can be assumed to be "computed" by a different quantity when variations in all the quantities which feature into the mathematical understanding of the former determine variations in the other quantity as well; and vice versa.

Notably, true motion is not known as a specific number or a specific value simpliciter. The true motion of the two bodies can be known only insofar as it changes. And it varies if and only if impressed forces act on those two bodies. I take it that this is also the insight provided by Hoek (2022) concerning motion according to Law 1. Rather than conceiving the motion invoked in Law 1 as the motion of force-free bodies, we conceive of it as the motion that is changed only by impressed forces. Similarly, the quantity of circular motion has one value only insofar as it does not change by means of impressed forces.[24]

Once the bodies and their relevant quantities are set to a first approximation (which involves both an attempt at quasi-isolation and quantitative modeling), we can now use a geometrical space to represent their motions and changes in those quantities. Note how in Newton's description, all reference points are fully determined by the quantities internal to the system of bodies, and not the other way around. The center of motion is not a point picked at random, but chosen by the properties of the system. In this case, the center of gravity depends only on the quantity of matter of the two bodies and their spatial separation.

But when represented in a geometrical space, the motion of the two globes gains two other features: the plane and the direction of rotation. Why are these two geometrical features important? Because they complete the geometrical description of the motion of the globes and they can be determined by the pairs of impressed forces (Definition 3, and the first corollary to the laws of motion). Again, the idea of properties being determined by other entities means that there is a one-to-one mathematical correlation. Only one pair of (equal but of opposite directions) impressed forces corresponds to the plane and the direction of rotation: those that maximize the change in the quantity of circular motion.

However, the fact that we use impressed forces tells us that the system is only *quasi*-isolated: impressed forces act as causes for changes in the quantity of motion within the system, while their physical source is abstracted away. The crucial bit is that we do *not* need other bodies when describing those changes: neither to describe their influence on the two globes, nor to use them as reference points for the motion. The description relying only on Newton's definitions and laws is self-contained and

---

[24] Focusing on changes in the motions of bodies as quantities also shows how specifying the bodies the motion to take place in a vacuum is a significant detail. It points out to the lack of resistance for the motions and, therefore, it provides a separate reason for considering the bodies a system unto itself. That is, the isolation of the system is well supported from a dynamical point of view as well.

complete, provided we start with a quasi-isolated system. The idea is that the true motion is to be fully determined by interactions internal to the system and changed by external actions on the system which are captured by means of impressed forces.

Let us now move to the second part of the scenario. Consider the *same system* and the *same changes in the tension* produced by impressed forces, but with a sphere of fixed stars surrounding the system of two globes. Or to put it differently, for the second part of the paragraph, we use imagination for a different abstraction: we bracket the existence of any other bodies except the globes, the rope, and the fixed stars. If we focus only on the changes in the relative positions of the globes among the stars, we will still notice changes in the rate of revolution. Based on these relative motions, this (apparent) rate of revolution would seem to increase or decrease. However, if we use only these changes in spatial positions, we could not say whether the quantity of true motion changes or not. That is, *we could not say whether impressed forces do act on our system or not*. Accounting for external physical influences on our system can only be determined when we consult the measure of true motion internal to the system (the tension in the cord).

6.6   Answering the Previous Challenges

So far, I have explained how the system of revolving globes becomes an idealization in virtue of it being quasi-isolated and because the inferences mentioned in the scenario rely on relations between quantities. But does my interpretation fare any better with respect to the previous interpretative challenges? I believe it does and now I turn to address them directly.

First, on my account, the two parts of the scenario (before and after introducing the fixed stars) are integrated. Consistent with Newton's words, the bodies are in motion from the get-go. We do not face the problem of creating new fixed stars in the scenario, messing up the dynamics recreated in our imagination: my reading does not assume we talk about two different universes, with different matter distributions, at different times in the argument. Additionally, it was assumed, from the beginning, by the isolation factor, that the fixed bodies which are considered in the second part of the scenario do not act causally on the globes.[25]

Second, the inference on which Newton relied in the last part of the scenario is no longer problematic. Previous readings assumed that in the second part of the scenario, the globes are taken to be in *uniform* circular motion when they are compared with the fixed stars. In that case, we could not say whether the motion pertains to them or to the stars. It could be that both the globes and the fixed stars revolve, but at different rates of rotation. Simply put, the objection is the following:

---

[25] These kinds of assumptions are crucial in idealizations of dynamically interacting systems. As George Smith's works show, when we clearly spell out assumptions, we make it easier for ourselves to subject these idealizations to systematic revision Smith (2007, 2012, 2014).

unless the tension in the cord *changes in a manner non-correlated with changes in the relative positions*, we would not be able to conclude that the bodies are moving and the stars are at rest.

However, the situation is different if we witness a correlation between the change in the tension in the cord and the variations in relative rotation when systematic pairs of impressed forces act on the globes. We might be tempted to think that we are looking to determine whether the stars *truly* are at rest: staying in the same place in absolute space from infinity to infinity. But this would go against Newton's previous qualification about true rest.[26] Newton does not conclude that the fixed bodies are at true rest. We are not interested in finding the true rest of the stars, but whether they can be *taken to be at rest* (i.e., provide a workable standard of rest) for the dynamical interactions involving the two globes. This allows us to conclude that, for all intents and purposes of studying the system of the globes, the fixed bodies are at rest in a restricted sense, not in a true sense.

Therefore, in my reading, both the existence of an immense vacuum and the existence of the fixed stars are compatible in the example. The first stage of the scenario deliberately excluded the fixed stars as reference points. The later stage adds them back (as reference points) and explains the conditions under which they can be used as a standard of rest for the true motions of the system. Imagination, as I said, can still be used, but not for a thought experiment.

The scenario serves the role of an idealization or a model for other systems of interacting bodies. In the following section I show that this model shows up in two other examples in Newton's own writings and how they all share the focus on quantities of motion and forces of a quasi-isolated system.

Before proceeding further, however, let me address a possible problem for my interpretation. Newton mentions a very specific pair of impressed forces and how their direction can be changed to map the direction of the circular motion. One could say that this can only be done by design: that is, we do it as an intervention. Thought experiments thrive on deliberate, fictional interventions or on the participation of the agent in the imagined scenario, whereas the idealized model put forward here has less room for actions caused by intentions.

My answer is that, as far as I can tell, there is nothing in Newton's own language[27] to suggest that the reader is assumed to *do* something. He is positing a situation in which certain kinds of impressed forces have a particular kind of effect.

---

[26] See Newton's discussion of the distinction between absolute space and relative space. Newton (1999, pp. 409–10)

[27] "Thereupon if₂no matter what₂ equal forces were to be impressed [*imprimerentur*] at the same time on alternate faces of the globes increasing or lessening the circular motion, the increase or decrease of the motion ~~could be still be learned₂~~would become known₂ from the added or diminished tension in the cord, and therefrom₂finally₂ on which faces of the globes the forces would have to be [*deberent*] impressed for the motion to be increased maximally could be found, that is, the posterior faces, or those which follow [*sequuntur*] in the circular motion." (See Sect. 6.4)

Granted, the reasoning presented is of a hypothetical kind,[28] but the focus is on the inferences connecting the direction of rotation to hypothetical actions of impressed forces. That is, the hypothetical language is an answer to the following question: *"what could we infer* if we start with these givens about the situation?" In contrast, it does *not* address a wholly different kind of question, pertaining to fictional actions, such as *"what would happen* if we do this (*imagines herself giving them a push with her hands in imagination*) to the globes?"

6.7 The Idealization in Action

If we take the example of the globes to be an idealized model for interactions of bodies in general, then we are no longer surprised to find surprisingly similar instances in two unlikely places. This section will describe two such instances, one from the Queries to the *Opticks*, the other one from a draft of *Liber Secundus*. For lack of space, I will not discuss these examples in detail. My focus is limited to the conceptual connections which the globes example illustrates. Since they are deep and significant for the methodology of natural philosophy, I hope to rescue Newton's scenario from obscurity and to do justice to Newton's final sentences of the Scholium.

First, a strikingly similar example shows up in the famous Query 31, added to the second edition of the *Opticks*, where Newton discusses active and passive principles. I call it 'striking' because the Queries (and especially Query 31) have always been the starting point for understanding Newton's more speculative thoughts on the principles of nature and procedures of experimental philosophy, and not usually associated with methodological considerations. Here is the passage:

> And thus Nature will be very conformable to herself and very simple, performing all the great Motions and the heavenly Bodies by the Attraction of Gravity which intercedes those Bodies, and almost all the small ones of their Particles by some other attractive and repelling Powers which intercede the Particles. The *Vis inertiae* is a passive Principle by which Bodies persist in their Motion or rest, receive Motion in proportion to the Force impressing it, and resist as much as they are resisted. By this Principle alone there never could have been any Motion in the World. Some other Principle was necessary for putting Bodies into Motion; and now they are in Motion, some other Principle is necessary for conserving the Motion. For from the various Composition of two Motions, 'tis very certain that there is not always the same quantity of Motion in the World. For if two Globes joined by a slender Rod, revolve about their common Center of Gravity with an uniform Motion, while the Center moves on uniformly in a right Line drawn in the Plane of their circular Motion; the Sum of the Motions of the two Globes, as often as the Globes are in the right Line described by their common Center of Gravity, will be bigger than the Sum of their Motions, when they are in a Line perpendicular to that right Line. By this Instance it appears that Motion may be got or lost. (Newton 1718, 397)

---

[28] The general hypothetical form is basically an inference: if such-and-such effects are present, then such-and-such claims are true. See Sect. 6.5.

Above, the example of two globes joined by a rod is introduced as a test model for composing kinematically the motions of each body. The simple composition of motions, without regard to their generation or physical causes, has the unpalatable consequence that, in the motion of the globes just described, we get different evaluations for the same quantity of motion. Thus, just as in the globes scenario, understanding circular motion in light of Newton's laws and definitions is again important.

We take that the significant component motion, (the vis inertiae), of the globes is on the tangent (that is where they move should they be released). The motion of the center of gravity will be perpendicular on those motions when the globes are in the right line described by the center of gravity. And the motion will be in the same line when they are in a perpendicular line. If we compare the effect of the uniform motion on the direction of the tangential motion of each of the globes, we get smaller and larger values for each globe. And thus, we might (rashly) conclude, that each globe equally loses and gains motion.

Following this passage, Newton explains that, in fact, bodies *lose* more motion than they have. So, kinematics by itself is not a guide for analyzing systems of bodies. We need to consider forces from the start, to focus on quantities of true motion, and *afterwards* evaluate the composition of motions.

The quoted passage shows that forces (as active and passive principles) are necessary in accounting for the true quantities of motion, while apparent variation in such quantities derive from misapplied compositions of motions. Similarly, the key component for changes in the system of the two globes were impressed forces. By contrast, the kind of mathematical composition of motion that disregards forces as causes and effects ends up in conclusions contrary to experience. We need to take into account what motion is communicated to the bodies within the system and what motion is taken from them by other bodies (no matter how fluid or viscous). This is in part what Newton's laws of motion help us model: how motion is communicated among bodies.

The second example speaks directly to the role of idealization for the methodology of the *Principia*. In article 21 of *Liber Secundus*,[29] Newton struggles to describe the kind of interaction Jupiter and the Sun share in virtue of mutual gravitational attraction. He explains that the action of gravity between any two planets (say, Jupiter and the Sun) is a simple action, which can be treated twofold. But in order to exemplify this conceptualization, Newton uses the model of two globes hooked together by a rope revolving around their common center of gravity.

> The Sun attracts [*trahit*] Jupiter and the other Planets, Jupiter attracts [*trahit*] its Satellites and similarly the Satellites act on one another and on Jupiter, and all the Planets act on themselves mutually. And although, in a pair of Planets, the action of each on the other can be distinguished and can be considered as paired actions by which each attracts [*trahi*] the other, ~~they are not two but a simple operation between two termini. By the contraction of one rope insofar as between~~[30] yet inasmuch as these are actions between two bodies, they are not two but a simple operation between two termini. **Two bodies can be drawn [*trahi*]**

---

[29] Smith (in press-a), '*Liber secundus*.'
[30] The bolded emphasis is mine throughout.

to each other by the contraction of a single rope between them. The cause of the action is two-fold, namely the disposition of each of the two bodies; the action is likewise two-fold, insofar as it is upon two bodies: but ~~the operation by which the Sun~~ insofar as it is between two bodies it is simple and single. There is not, for example, one operation by which the Sun for example attracts [*trahit*] Jupiter and another operation by which Jupiter attracts the Sun, but a single operation by which the Sun and Jupiter endeavor to approach each other. By the action by which the Sun attracts [*trahit*] Jupiter, Jupiter and the Sun endeavor to approach each other (by Law 3), and by the action by which Jupiter attracts the Sun, Jupiter and the Sun also endeavor to approach each other. Moreover, the Sun is not attracted [*attrahitur*] by a twofold action towards Jupiter, nor Jupiter by a twofold action towards the Sun, but there is one action between them by which both approach each other. [...][31]

I will cut through the complexity of the actual argument to point out that the passage relies on a model of the interaction between Sun and Jupiter which is almost identical to the system of the two globes connected by a rope. Here Newton tries to offer a model for conceptualizing the interaction of two such bodies through gravity.[32] The operations of each globe on the other are exerted at the same time through the means of the rope. Not without its problems, this model becomes a paradigm for thinking about gravitational attraction more broadly.[33] The tension in the string now acts as a placeholder for the equality of two actions: of how much each body attracts the other.[34] But this makes it also a model for the equality of action and reaction. Similarly, in the globes example, the tension in the cord was a measure both of the endeavor to recede from the center and of the quantity of true circular motion (which mathematically is determined by the inherent force and the centripetal force).

6.8   Conclusions

Let us take stock of this journey. I started by introducing the current interpretations of the last paragraph of the Scholium (Sect. 6.2), and the problems we encounter if we read the example of the two globes as a thought experiment (Sect. 6.3). My approach is to consider the scenario to be an idealization of two quasi-isolated interacting bodies illustrating the inferences which such an idealization allows based on Newton's definitions of laws of motion (Sect. 6.4). I think that the original

---

[31] This is yet another one of the great contributions for which I am grateful to George Smith. On the one hand, there is the dedication to the analysis of the text. On the other hand, there is the generosity in sharing these materials with generations of researchers.

[32] Recall that the two globes were taken to revolve around their center of gravity (and not some arbitrary point), a well-defined mathematical point which assumes some understanding of gravity.

[33] Newton proposes here a view of analyzing the motion of an isolated system of two bodies acting through a central potential. Our current physics sensibilities recognizes this as a two-body problem and it is one the paradigmatic examples taught in celestial physics. Yet until Newton there was nobody who formulated the motion of two bodies under gravity in this manner.

[34] Law 3 is inevitably included in the model.

language in the manuscript source supports my reading (Sect. 6.5) and does not describe a fictional thought experiment. I flesh out more fully my approach (Sect. 6.5) and then explain how it faces up to the original interpretation challenges (Sect. 6.6). Section 6.7 provides the proof that Newton himself took the system of the globes to be a model by introducing two other examples from his writings: one from the Queries to the *Opticks* and the other from a manuscript about the System of the World, predating the *Principia*.

We are now, I believe, in a position to draw a final connecting line between the two-globes scenario, other examples which Newton used, and Newton's methodology more generally. I think this connection could be described using George Smith (2020)'s charming (adopted) phrase of "putting questions to nature." When summing up his discussion of experiments in Newton's *Principia*, Smith writes that the experiments

> [...] posed questions about the relationship between forces of some kind, which are not directly accessible, and various other quantities more accessible to observation. They all, of course, involved intricate design and hence contrived situations. But what enabled them to have at least the promise of extracting answers from nature about forces was their presupposing, in every case, Newton's first two laws of motion. Their design involved, first, one or more theoretical relationships derived from these two laws between forces and quantities that would be accessible to observation in certain circumstances and, then, the physical realization of those circumstances. To say this in a more customary manner, all five of them involved one or more relationships derived from the first two laws that licensed a theory-mediated measurement of something pertaining to forces in specific circumstances. (Smith 2020, 18)

My aim for this paper was to argue that the scenario of the globes in the *Principia* is a model for how to "cross-examine Nature herself" as well (Bacon 1620, 232). In my view the example shows how to describe a system of bodies and forces relying on Newton definitions and laws and what kind of inferences are licensed by its idealization factors. I suggest that measuring the true motion of the globes by two different kinds of quantities is a case of successful cross-examination. The later procedure of putting this model to work in ever more complex, actual physical circumstances is a fuller development of this strategy, just as Newton himself concluded. George Smith has sometimes described this methodological strategy "theory-mediated measurement" and proved that it is a staple of continuous success of physical inquiry into the nature of gravitation.

References

Arthur, R., 2018. Thought experiments in Newton and Leibniz. In *Routledge companion to thought experiments*, ed. M. Stuart, Y. Fehige and J. R. Brown, 111–127. Routledge.

Bacon, F. 1620/2000. *The new Organon*, ed. L. Jardine and M. Silverthorne. Cambridge University Press.

Barbour, J.B. 1989. *Absolute or relative motion? A study from a Machian point of view of the discovery and the structure of dynamical theories*, The discovery of dynamics. Vol. I. Cambridge University Press.

Berkeley, G. 1721/1992. *De Motu and the analyst*, ed. and trans. D. Jesseph, Dordrecht: Kluwer.

DiSalle, 2002. Newton's philosophical analysis of space and time. *Cambridge companion to Newton*, ed. I. B. Cohen & G. E. Smith, 33–56. Cambridge University Press.

DiSalle, R. 2006. *Understanding spacetime. The philosophical development of physics from Newton to Einstein*. Cambridge University Press.

Ducheyne, S. 2005. Mathematical models in Newton's *Principia*. A new view of the 'Newtonian Style.'. *International Studies in the Philosophy of Science* 19: 1–19.

———. 2012. *The main business of natural philosophy*. Springer.

Earman, J. 1989. *World enough and space-time: Absolute versus relational theories of space and time*. Cambridge, MA: MIT Press.

Hoek, D. 2022. Forced changes only: A New Take on the Law of Inertia. *Philosophy of Science*, 1–17. https://doi.org/10.1017/psa.2021.38. https://www.cambridge.org/core/journals/philosophy-of-science/firstview/firstview

Huggett, N. 2012. What did Newton mean by 'absolute motion'? *Interpreting Newton: Critical essays*, ed. E. Schliesser and A. Janiak, 196–218. Cambridge University Press.

Laymon, R. 1978. Newton's bucket experiment. *Journal of the History of Philosophy* 16: 399–413.

Mach, E. 1893/1902/1919. *The science of mechanics. A critical and historical account of its development*. Translated from German by Thomas J. McCormack, The Open Court Publishing Company.

Maudlin, T., 2012. *Philosophy of Physics: Space and time*, Princeton University Press.

Nagel, E. 1961. *The structure of science*. Harcourt. Brace & World.

Newton, I. 1718/1730. *Opticks*, based on fourth ed. London, 1730. Reprinted by Dover, 1952.

Newton, I. 1999. *The* principia*: Mathematical principles of natural philosophy*. Trans. I. Bernard Cohen and Anne Whitman. University of California Press.

Rynasiewicz, R. 1995a. 'By their properties, causes and effects.' Newton's Scholium on space, time, place and motion – I. The text. *Studies in History and Philosophy of Science* 16: 133–153.

———. 1995b. 'By their properties, causes and effects.' Newton's Scholium on space, time, place and motion – II. The context. *Studies in History and Philosophy of Science* 26: 295–321.

Rynasiewicz, R. 2014. Newton's views on space, time, motion. *Stanford encyclopedia of philosophy*, ed. E.N. Zalta, Summer 2014 Edition. http://plato.stanford.edu/entries/newton-stm/

———. 2019. Newton's Scholium on time, space, place and motion. *Oxford handbook of Newton*, ed. E. Schliesser and Chr. Smeenk.

Smith, G.E. 2007. Newton's *Philosophiae Naturalis Principia Mathematica. Stanford Encyclopedia of Philosophy*, ed. E.N. Zalta. https://plato.stanford.edu/entries/newton-principia/

———. 2012. How Newton's *Principia* changed physics. *Interpreting Newton. Critical essays*, ed. A. Janiak and E. Schliesser, 360–395. Cambridge University Press.

———. 2014. Closing the loop. *Newton and Empiricism*, ed. Z. Biener and E. Schliesser, 262–345. Oxford University Press.

———. 2020. Experiments in the *principia. Oxford handbook to Newton*, ed. E. Schliesser and Chr. Smeenk. https://doi.org/10.1093/oxfordhb/9780199930418.013.36.

Smith, G.E. (in press-a). '*Liber secundus*: A variorum translation,' with Anne Whitman.

Smith, G.E. (in press-b). A variorum translation of several pages from *The preliminary manuscripts for Isaac Newton's 1687* principia*, 1684–1686*, ed. D. T. Whiteside, Cambridge University Press, 1989, (p. 45–46), corresponding to folio numbers (11 and 12) in Dd. 9.46.

van Fraassen, B. 1970. *An introduction to the philosophy of time and space*. New York: Random House.

Westfall, R. 1971. *Force in Newton's physics*. New York: Wiley.

Whiteside, D.T. 1989. *The preliminary manuscripts for Isaac Newton's 1687 principia, 1684–1686*. New York: Cambridge University Press.

Chapter 7
On Metaphysics and Method in Newton

Howard Stein

Abbreviations

CSM	Descartes 1985
NC	Turnbull 1959
NPL	Cohen 1958
NUP	Hall & Boas Hall 1962
Opticks	Newton 1952
Principia	Newton 1934

When I was a student, reigning opinion held that Newton, although unquestionably in the foremost rank of the great among scientists, was a shallow and unoriginal philosopher. In a work whose reputation at that time was high, E. A. Burtt put it

---

Editor's note: This paper is based on a talk originally given at the University of North Carolina (Greensboro) in March 1989, at the conference "How Theories are Constructed: The Methodology of Scientific Creativity," organized by Jarrett Leplin. Some of the talks from that conference have been published as part of Leplin 1995, but this paper has not been published previously. This version is based on an unpublished manuscript, edited for length—mainly by removing three discursive footnotes, and an introductory paragraph peculiar to the presentation of the paper at the conference. Stein explores the key themes broached in this paper further, in response to comments from the conference, in his unpublished manuscript, "Further Considerations on Newton's Method," which can be found (along with a longer version of the current paper) at http://www.strangebeautiful.com/other-minds.html. The places of the three elided footnotes are indicated in the text of the current version of the paper.

H. Stein (✉)
Department of Philosophy, University of Chicago, Chicago, IL, USA
e-mail: hstein@uchicago.edu

© The Author(s), under exclusive license to Springer Nature Switzerland AG 2023
M. Stan, C. Smeenk (eds.), *Theory, Evidence, Data: Themes from George E. Smith*, Boston Studies in the Philosophy and History of Science 343, https://doi.org/10.1007/978-3-031-41041-3_7

thus: "In scientific discovery and formulation Newton was a marvelous genius; as a philosopher he was uncritical, sketchy, inconsistent, even second rate."[1]

Among Burtt's criticisms of Newton are the following two—one in the topic of method, the other in that of metaphysics. On the former head, Burtt says: "Would that in the pages of such a man we might find a clear statement of the method used by his powerful mind in the accomplishment of his dazzling performances, with perhaps specific and illuminating directions for those less gifted...!" (Note that he is asking for something like what Descartes proposed to develop in the *Regulæ:* as it were, an *algorithm* for scientific discovery—a *prescription* for "the accomplishment of dazzling performances.") "But," Burtt continues, "what a disappointment as we turn the leaves of his works! Only a handful of general and often vague statements about his method, which have to be laboriously interpreted and supplemented by a painstaking study of his scientific biography...."

Yet this, Burtt adds, is better than what Newton gives us on the second head, that of "an exact and consistent logical analysis of the ultimate bearings of the unprecedented intellectual revolution which he carried to such a decisive issue." For Newton, according to Burtt, exhibits, in his metaphysical views, the characteristic vices of "the positivist mind" that "decries metaphysics"; in particular, "[h]is general conception of the physical world and of man's relation to it ... was taken over without examination as an assured result of the victorious movement whose greatest champion he was destined to become."[2]

You will of course not expect here a defense of Burtt's opinions. On one point, however, he is right—or almost right: it is certainly true that for a proper understanding of Newton's general remarks about method, it is necessary to study their exemplification in his work—not, however (I think), as Burtt puts it, to study "his scientific biography," but his scientific *writings*.

The depreciation or patronizing of Newton by philosophical commentators is pretty limp in comparison to William Blake's blazing attack:

> The Atoms of Democritus
> And Newton's Particles of light
> Are sands upon the Red sea shore

Newton becomes a demon figure in Blake's poetic mythology, emblematic of that state of nightmare life-in-death-like sleep of the soul, of rigidly limited vision, that Blake called "Ulro." He writes:

> Now I a fourfold vision see
> And a fourfold vision is given to me
> Tis fourfold in my supreme delight
> And threefold in soft Beulah's night
> And twofold Always. May God us keep
> From Single vision and Newton's sleep.

---

[1] Burtt 1955, 208.

[2] Burtt 1955, 229–30.

I don't pretend to give evidence on Newton's behalf that might change Blake's mind; to *those* realms of the imagination that Blake was concerned with, Newton does indeed seem to have been quite blind (if one may judge from the few remarks attributed to him about poetry and music). But in another (and important) sense, Blake's characterization is very mistaken: "single vision" is just what does *not* characterize Newton. I shall argue that a key element of his scientific/philosophical mentality—contributing crucially, on the one hand, to "the accomplishment of his dazzling performances," and on the other hand making some of his accomplishments very hard for his contemporaries to accept or to grasp, and his philosophical position hard to appreciate at (what I believe to be) its true value—was what can be described as a remarkable capacity for multiple vision.

The conception of scientific inquiry held by the main community of natural philosophers in the period of Newton's first investigations—by those of the Royal Society of London, founded in 1660, where Hooke was Curator of Experiments; and of the Académie Royale des Sciences in Paris, founded in 1666, with Huygens the dominant figure—embraced three principal modes of inquiry: experiment, to discover phenomena; the search for (to use a phrase of Huygens') "hypotheses by motion" to explain them; and the deduction, or mathematical derivation, of the consequences of established principles. The first two modes, pursued together, are abundantly illustrated in Hooke's *Micrographia,* published in 1665; the third, in its purest form, in the very beautiful work of Huygens on the problems of impact and of centrifugal force (both achieved in the 1650s, although for long unpublished) and on the compound pendulum (a subject that occupied him from the late '50 s through the '60 s; the results of these studies form one of the principal parts of Huygens' great treatise of 1673 on the pendulum clock).[3] And all three modes are exemplified, in mutual interaction, in Huygens' *Treatise on Light,* written (as he tells us) in 1677 (although not published until 1690).

That Newton, in those first investigations of the mid-to-late 1660s—in particular, his investigations of light—had discovered a new way of inquiry, is (I think) quite dramatically shown by the controversies they gave rise to. The most distinguished of Newton's critics were the two men I have mentioned: Hooke and Huygens. Hooke, as curator of experiments, had the duty of repeating experiments reported to the Royal Society, with a view to checking the accuracy of the reports. Newton's letter of February 6, 1672, to Oldenburg (Secretary of the Royal Society and publisher of its *Philosophical Transactions*), containing his "New Theory about Light and Colors,"[4] was read to the assembled membership on February 8.[5] One week later,

---

[3] An equally brilliant accomplishment in the same mode, also embodied in that work, is the discovery of the curve of isochronous vibration for constrained gravitational motion (the so-called "tautochrone" curve).

[4] See *NC*, 92–102. A facsimile reproduction of the letter as it was published in the Royal Society's *Philosophical Transactions* (No. 80, February 19 1671/2, pp. 3075–87), is given in *NPL*, 47–59. (It should be noted that Oldenburg, in publishing the letter, omitted one brief but rather interesting passage that will be of some concern to us below.).

[5] *NC*, 103, n. 1 (continued from p. 102).

Hooke sent his critical discussion to Oldenburg. The keynote is struck in its opening sentences:[6]

> I have perused the Excellent Discourse of Mr. Newton about colours and Refractions, and I was not a little pleased with the niceness and curiosity of his observations. But though I wholy agree with him as to the truth of those he hath alledged, as having by many hundreds of tryalls found them soe, yet as to his Hypothesis of salving the phænomena of Colours thereby I confesse I cannot yet see any undeniable argument to convince me of the certainty thereof.

So the experiments are applauded, the explanatory hypothesis not. Hooke is at pains to make clear, however, that his objection is not to Newton's hypothesis as such, but to his claim to have *established* that hypothesis beyond reasonable doubt:[7]

> For all the expts & obss: I have hitherto made, nay and even those very expts which he alledged, doe seem to me to prove that light is nothing but a pulse or motion propagated through an homogeneous, uniform and transparent medium: And that Colour is nothing but the Disturbance of yt light by the communication of that pulse to other transparent mediums, that is, by the refraction thereof: that whiteness and blackness are nothing but the plenty or scarcity of the undisturbed Rayes of light; and that the two colours (then which there are noe more uncompounded in Nature) are nothing but the effects of a compounded pulse or disturbed propagation of motion caused by Refraction. But how certaine soever I think myself of my hypothesis, wch I did not take up without first trying some hundreds of expts; yet I should be very glad to meet wth one Experimentum crucis from Mr. Newton, that should Divorce me from it. But it is not that, which he soe calls, will doe the turne; for the same phænomenon will be salved by my hypothesis as well as by his without any manner of difficulty or straining; nay I will undertake to shew an other hypothesis differing from both his & mine, yt shall do the same thing.

And toward the end of his discussion, Hooke remarks:[8]

> Nor would I be understood to have said all this against his theory as it is an hypothesis, for I doe most Readily agree with him in every part thereof, and esteem it very subtill and ingenious, and capable of salving all the phænomena of coulours; but I cannot think it to be the only hypothesis; not soe certain as mathematicall Demonstrations.

Newton's reply to "ye Theoretique part" of Hooke's critique begins with the issue of "hypotheses":[9]

---

[6] *NC,* 110–4; *NPL,* 110–5. (One minute point: The text in *NPL* is a facsimile of the transcription given in volume III of Thomas Birch's *History of the Royal Society* [Oldenburg did not publish Hooke's critique in the *Philosophical Transactions*—although he did publish Newton's reply to Hooke]. In that version, some rectifications of Hooke's spelling are made [e.g., "wholly" for Hooke's "wholy" in the first passage here quoted]; but one alteration is curiously for the worse: in that same passage, "phenomæna" where Hooke had "phænomena.").

[7] *NC,* 110–1; *NPL,* 111.—There is in this passage a more substantial editorial alteration in the text given by Birch (from the Register Book of the Royal Society): where, in the first sentence quoted, Hooke writes that the experiments and observations seem to him to prove that "light is ... a pulse or motion [etc.]," the text in Birch—hence that in *NPL*—reads, "white is ... a pulse or motion [etc.]" It appears that the text in Birch corresponds to the copy Newton saw (cf. *NC,* 115, n. 3).

[8] *NC,* 113; *NPL,* 8.

[9] *NC,* 173; *NPL,* 118.

I shall now take a view of Mr Hooks Considerations on my Theories. And those consist in ascribing an Hypothesis to me wch is not mine; in asserting an Hypothesis wch as to the principall parts of it is not against me; in granting the greatest part of my discourse if explicated by that Hypothesis; & in denying some things the truth of wch would have appeared by an experimentall examination.

After taking up in turn the topics of the first three of those clauses, Newton comments on the third of them—"granting the greatest part of [his] discourse if explicated by [Hooke's] Hypothesis"—that "I do not think it needful to explicate my Doctrine by any Hypothesis at all." And proceeding to the fourth clause— "things the truth of wch would have appeared by an experimentall examination"—he repeats: "You see therefore how much it is besides the businesse in hand to dispute about *Hypotheses.* For wch reason I shall now in the last place proceed to abstract the difficulties involved in Mr Hooks discourse, & without having regard to any Hypothesis consider them in generall termes."[10]

In a draft rebuttal, Hooke says of this, a little petulantly:[11]

I see noe reason why Mr. N. should make soe confident a conclusion that he to whome he writ did see how much it was besides the busness in hand to Dispute about hypotheses. for I judge there is noething conduces soe much to the advancement of Philosophy as the examining of hypotheses by experiments & the inquiry into Experiments by hypotheses. and I have tha Authority of the Incomparable Verulam to warrant me.

Here we see an assertion of the two modes of empirical inquiry I have mentioned, with an appeal to the authority of Bacon.

Huygens' point of view I should describe as significantly different from Hooke's, although the two agree both in general that hypotheses are not beside the point, and in special that Newton has not established that there are more than two fundamental colors. Thus Huygens writes:[12]

It seems to me that the most important objection, put to him in the form of a Quære, is this: Whether there are more than two sorts of colors? For I, for my own part, believe that a Hypothesis that should explain mechanically and by the nature of motion the colors *Yellow* and *Blue* would suffice for all the others .... Moreover I do not see why Mr. Newton does not content himself with the two colors, Yellow and Blue. For it will be far easier to find a

---

[10] *NC,* 177; *NPL,* 123.

[11] *NC,* 202; cf. Bacon, *Novum Organum:* "Now my directions for the interpretation of nature embrace two generic divisions: the one how to educe and form axioms from experience; the other how to deduce and derive new experiments from axioms." (Bacon 1960, 130).

[12] *NC,* 255–6. The translation from Huygens' French is mine; Oldenburg published Huygens' remarks (without identifying him by name), in his own translation, in the *Philosophical Transactions,* No. 96 (July 21 1673), pp. 6086–7—see *NPL,* 136–7.—In consulting the latter source, care should be taken to note that what Oldenburg prints immediately following (pp. 6087–92; *NPL,*137–42) as Newton's reply is in fact his reply to Huygens's *second* letter (printed by Oldenburg in the *next* number of the *Transactions* [p. 6112; *NPL,* 147] after Newton's actual reply to the first letter [pp. 6108–11; *NPL,* 143–146]). (There is a half-apology for this mix-up in Oldenburg's heading to Newton's reply, p. 6108.).

hypothesis by motion that should explain these two differences, than for so many diversities as there are of other colors. And until he has found that hypothesis, he will not have taught us in what the nature and difference of colors consists, but only this accident (which certainly is very considerable) of their different refrangibility.

Whereas Hooke's grounds of contention with Newton can be characterized as essentially methodological (he rejects Newton's claim to have avoided hypotheses and to have established his results securely), those of Huygens can be characterized—in the sense in which I am using the word—as essentially metaphysical: he denies that Newton has instructed us as to the *nature* of light, and he objects to Newton's conclusion as hard to reconcile with any acceptable account of that nature.

The conception of "being as such," or of the *rerum natura,* that Huygens appeals to is that of the "mechanical philosophy," which in Huygens (who early threw off the shackles of Cartesianism) took the special form of the *atomistic* mechanical philosophy. Matter and space are distinct; there are void spaces, and matter itself consists of ultimate indivisible, impenetrable, rigid—or, in the terminology of Huygens and Newton, "hard"—particles. All natural process consists in the motions of these particles. The motion of any one particle is uniform and rectilinear unless and until it is constrained to change by its impingement upon another particle (or particles). Therefore, the fundamental laws of nature, namely the laws of the motion of the fundamental particles, are *the laws of impact of hard bodies.* These laws of impact Huygens himself had discovered (his recognition of the absurdity of the rules of impact proposed by Descartes was the crucial moment in his break with the Cartesian philosophy and the beginning of his own great independent work in physics). The task that remained, then, for fundamental physics was the discovery of suitable "mechanical models," as we should now say, for the representation of natural phenomena. But, it should be emphasized, mechanical models of a quite special kind. Comparison with a much later view may be instructive: Toward the end of the last century, a number of physicists were dissatisfied with the received view of the structure of mechanics, and one kind of improvement that was sought was the elimination of the concept of *force.* Independently, J. J. Thomson and (later) Heinrich Hertz proposed to eliminate "forces" in favor of what, in analytical mechanics, are called "constraints."[13] A way of putting this is to say that *there is no potential energy:* all energy is kinetic, what appears macroscopically as potential energy is really kinetic energy of "hidden" motions. Now, a Huygensian mechanical model, or "hypothesis by motion," is a system of this type—with two qualifications: First, Hertz requires his "constraints" to satisfy a rather strong condition of *continuity,* and (as we shall see in a moment) this condition can *never* be satisfied in a Huygensian system. Second, and on the other hand, Hertzian constraints are of a very general kind, whereas for Huygens the only constraints are those implied by the strict impenetrability and unalterable shape of the fundamental particles. An immediate consequence is that the changes of motion of the fundamental particles

---

[13] See, respectively, Thomson 1885; and Hertz 1956.

are not smooth—not continuous—but instantaneous. (This is a point raised by Leibniz as one of the fatal objections against atomism.)[14]

There is one point in which the sketch I have given does not quite do justice to Huygens' basic position: I have ignored here his very interesting reservations about the "absolute" conception of motion. But waiving this matter, what I have called the Huygensian metaphysical principles are also, by all the evidence known to me, the principles accepted by Newton in the period we are now considering. In particular, when he writes of "the lawes of motion" he means the laws of impact, and what *we* call the laws of "perfectly elastic" impact *he* ascribes to "bodyes which are absolutely hard."[15]

How does all this bear on Newton's optical investigations and the controversy over them? So far as the metaphysical issue—the "nature" of light and colors—is concerned, Newton has in fact set it almost entirely aside. This is Huygens' complaint; and just as in his answer to Hooke Newton had written, "I do not think it needful to explicate my Doctrine by any Hypothesis at all," so to Huygens' remark that until he has found a mechanical explanation "he will not have taught us in what the nature and difference of colors consists," Newton replies:[16]

> [T]o examin how colours may be thus explained Hypothetically is besides my purpose. I never intended to show wherein consists the nature and difference of colours, but onely to show that *de facto* they are originall & immutable qualities of the rays which exhibit them, & to leave it to others to explicate by Mechanicall Hypotheses the nature & difference of those qualities.

Three questions arise: What is the actual content of Newton's allegedly non-hypothetical "Doctrine"? How in fact does he claim to have established this doctrine? And is his claim to have established it with "certainty," with no hypothetical or "conjectural" element, a defensible one?[17]

Let us begin from Hooke's challenge to what Newton called the *Experimentum Crucis*. That experiment, Hooke says, cannot "divorce him" from his own hypothesis. What was the experiment, and what did Newton claim to have established by it?

Having observed—as no one had before him[18]—that the shape of the colored image produced by a prism was incompatible with the presumed constancy of the ratio of the sines of the angles of incidence and refraction, Newton isolated, successively, different portions of the dispersed spectral light, and found that each portion by itself, when refracted through a second prism, yielded an image

---

[14] See, e.g., Part II of *Specimen Dynamicum*, in Leibniz 1970, 447.

[15] See the early manuscript "The Lawes of Motion," now in *NUP*, 157–64; and *ibid.*, 162.

[16] *NC*, 264; *NPL*, 144.

[17] See *NC*, 100; *NPL*, 57 (and cf. infra).

[18] [Editor's note: Here we omit a lengthy footnote regarding Thomas Harriot's optical contributions, obtained roughly 60 years before Newton's, but not published. Stein notes that "although [Harriot's] results would indeed *predict* an oblong shape for the dispersed solar image, the evidence does not suggest that he made observations of that sort at all."].

compatible with that law; but that these several portions were refracted differently from one another: the light at the end of the spectrum that was refracted least in the production of that spectrum continued to be refracted less, when segregated, than that at the other end; and correspondingly for the intermediate places in the spectrum. You will notice that I have—perhaps rather clumsily—avoided in this statement any mention of colors. So does Newton: his description of the *Experimentum Crucis* itself contains not a word about color. And the conclusion he draws is very precisely stated:[19]

> And so the true cause of the length of that Image was detected to be no other, then that *Light* consists of *Rays differently refrangible,* which, without any respect to a difference in their incidence, were, according to their degrees of refrangibility, transmitted towards divers parts of the wall.

In a terminology that Newton proceeds immediately to introduce, he claims to have established a distinction—that is, an actually existing one: a distinction *in rerum natura*—between "uniform" or "homogeneous" light, and "difform" or "heterogeneous" light; to have established, moreover, that there is an indefinite variety of kinds of the former; and that ordinary sunlight is a "heterogeneous mixture" of rays of a continuous range of refrangibilities.[20] Since the *Experimentum Crucis* actually *exhibits* this distinction—that is, *produces* (approximately) homogeneous lights whose different refrangibilities it shows, and shows moreover that the diverse refrangibilities of the several homogeneous lights correspond exactly with the different degrees of refraction of the refracted parts of sunlight from which they were obtained, it is very hard to fault Newton's assertion that his experiment had "detected" these facts.

It is in the second part of his paper that Newton turns to the theory of colors. Here he contents himself with a statement of the "doctrine" itself, with some brief indications of a few of the many experiments on which it rests.[21] The doctrine of colors can be summed up more briefly than in Newton's original account. It consists essentially in two points: (1) The visible appearance, the *"Species,"* of any light—that is, the appearance it gives rise to when it falls on the retina of a seeing eye—is entirely determined by its physical constitution out of the rays of the various homogeneous sorts in definite quantities. (2) On the other hand, lights that differ in their physical constitution can produce the same visual appearance of color; thus, Newton remarks, "The same colours *in Specie* [that is, *in visual appearance*] with [those of homogeneous light] may also be produced by composition."[22] Or in other words, combining the two points: there is a *many-to-one mapping* from the physical constitutions of lights, in the sense of "physical constitution" established by the *Experimentum Crucis,* to their perceptual effects or "Species."

---

[19] *NC*, 95; *NPL*, 51.

[20] *NC*, 95, 96; *NPL*, 51, 53; cf. also Stein (unpublished).

[21] *NC*, 97; *NPL*, 53.

[22] *NP*, 98; *NPL*, 54.

There are many points of detail worth considering, with respect to Newton's investigation itself, to his account of it, and to the controversy;[23] but like Newton in his letter to Oldenburg, I have to limit myself here to a few instances. I have said earlier (and not yet explained the point) that Newton's capacity for multiple vision has been a source of confusion—or misconception—to his readers from his time to our own. I have now to add that another such source lies in the fact that Newton is a quite exceptionally precise writer; making allowances for human fallibility, a remarkable proportion of his statements say exactly what he means to say. Perhaps it seems paradoxical that precision of statement should be a cause of misapprehension; but in fact most readers do not *read* with the precision Newton expects of them. To take a rather small point: it has been instanced, in favor of the view that "there is not ... a single term of which it is true to say that it *could not* (without changing or extending its meaning) be used to refer to unobservables," that "'Red,' for example, was so used by Newton when he posited that red light consists of *red corpuscles.*"[24] But what Newton actually says, in his letter to Oldenburg, is that some rays— the least refrangible ones—are "disposed to exhibit a Red colour"; not that they "are red."[25] If, therefore, he does occasionally apply the adjective directly to the word "ray," that is patently a derivative use—an "extension of meaning." Indeed, so sensitive is Newton to this that later, in the *Opticks,* he gives us the following as a formal "Definition":[26]

> The homogeneal Light and Rays which appear red, or rather make Objects appear so, I call Rubrifick or Red-making; those which make Objects appear yellow, green, blue, and violet, I call Yellow-making, Green-making, Blue-making, Violet-making, and so of the rest. And if at any time I speak of Light and Rays as coloured or endued with Colours, I would be understood to speak not philosophically and properly, but grossly, and accordingly to such Conceptions as vulgar People in seeing all these Experiments would be apt to frame. For the Rays to speak properly are not coloured. In them there is nothing else than a certain Power and Disposition to stir up a sensation of this or that Colour. For as Sound in a Bell or musical String, or other sounding Body, is nothing but a trembling Motion, and in the Air nothing but that Motion propagated from the Object, and in the Sensorium 'tis a sense of that Motion under the Form of Sound; so Colours in the Object are nothing but a Disposition to reflect this or that sort of Rays more copiously than the rest; in the Rays they are nothing but their Dispositions to propagate this or that Motion into the Sensorium, and in the Sensorium they are Sensations of those Motions under the Forms of Colours.

Another recent commentator—Professor Shapiro, whose positive service in clarifying Newton's optical work (and in making his Cambridge lectures available in a splendid edition) is surely well known—has claimed that the theory presented in the letter to Oldenburg is flawed, and is so because of a confusion of Newton's: "The source of this flaw is that he did not initially define the entities to which his theory applied. For instance, in the third proposition [of the section on color]

---

[23] Cf. Stein (unpublished).

[24] Pitcher 1977, 261, n. 7; with a reference there to Putnam 1962, 243.

[25] *NC*, 97; *NPL*, 53.

[26] Newton, *Opticks*, 124.

he says that 'the species of colour, and degree of Refrangibility proper to *any particular sort of Rays,* is not mutable by Refraction, nor by Reflection....' He does not define 'particular sorts of rays'; thus both greens, for example—that exhibited by a homogeneous light, and that, "the same *in Specie,*" exhibited by a mixture of yellow and blue—could be called 'particular sorts of rays,'[27] since, as he has just told us, their color is the same."[28]

But Newton has not told us that "their color is the same"; he has told us that their colors are "the same *in Specie*"—in appearance. And he *has* told us what he means by "a particular sort"; for this is implied both by his use of the words "uniform" or "homogeneous" *versus* "difform" or "heterogeneous" ("one particular sort" = "one particular *form* or *kind,*" therefore a particular sort is uniform or homogeneous), and by the very sentence quoted from Proposition 3, where a "proper degree of refrangibility" is ascribed to a particular sort. But when Newton, in his reply to Hooke, refines his formulation to make the point clearer, Shapiro says that Newton recognized the inadequacy of his original distinction but "in his typically sly manner" wrote Hooke that he was merely restating that distinction. This seems to me an all too typical instance of the strange fashion for Newton-bashing: examples abound of cases in which attempts by Newton to clarify his meaning, even by citing his own words and pointing out their content, have subjected him to charges of evasion, deception, dishonesty, and malice.

Let us consider an example of greater moment. Recall that in the passage I have quoted above on "unobservables," it was said that Newton "postulated that red light consists of *red corpuscles.*" I have commented on the adjective; but what of the noun: "corpuscles"?

The "hypothesis" that Hooke (in contrast to Huygens) thinks that Newton's paper was chiefly concerned to defend is described by him in the following passage:[29]

> But grant his first proposition that light is a body, and that as many colours or degrees thereof as there may be, soe many severall sorts of bodys there may be, all wch compounded together would make white, and grant further, that all luminous bodys are compounded of

---

[27] Evidently a slip of the pen for "could be called *rays of the same particular sort.*" It should perhaps be noted that Newton was mistaken in thinking that a green could be made by composing homogeneous blue and yellow lights. At best, one can obtain in this way a very unsaturated green—a "greenish white." It is possible that Newton eventually recognized this fact, although he never (so far as I know) announced it; for in the *Opticks,* in discussing the composition of colors (Book I, Part II, Proposition IV), whereas he tells us that "a Mixture of homogeneal red and yellow compounds an Orange, like in appearance of Colour to that orange which in the series of unmixed prismatick Colours lies between them," he says something rather different about green in relation to yellow and blue: "And after the same manner other neighboring homogeneal Colours may compound new Colours, like the intermediate homogeneal ones, as yellow and green, the Colour between them both, and afterwards, if blue be added, there will be made a green the middle Colour of the three which enter the Composition. For the yellow and blue on either hand, if they are equal in quantity they draw the intermediate green equally towards themselves in Composition, and so keep it as it were in Æquilibrium, that it verge not more to the yellow on the one hand, and to the blue on the other, but by their mix'd Actions remain still a middle Colour" (*Opticks,* 132–3).

[28] Shapiro 1980, 222.

[29] See Hooke's paper in *NC,* 113–4; *NPL,* 114.

such substances condensd, and that, whilst they shine, they doe continually send out an indefinite quantity thereof every way *in orbem,* which in a moment of time doth disperse itself to the outmost and most indefinite bounds of ye universe; granting these, I say, I doe suppose, there will be noe difficulty to demonstrate all the rest of his curious Theory: Though yet, methinks, all the colourd bodys in the world compounded together should not make a white body; and I should be glad to see an expt of the kind.

Apart from his serious misunderstanding of the relation of light to the composition of bodies and the way in which the various colors together make white, and his intimation of something like instantaneous propagation of light, the view Hooke here attributes to Newton is indeed one that Newton inclined to—as probable. On the other hand, the only reference to such a view in Newton's paper occurs in a paragraph that follows the presentation of the entire "Doctrine," and suggests *consequences* of that doctrine. What Newton wrote is: "These things being so, it can be no longer disputed, whether there be colours in the dark, nor whether they be the qualities of the objects we see, no nor perhaps, whether Light be a body." Then, after a brief argument (surprisingly scholastic and unconvincing) in favor of this last claim, he continues: "But, to determine more absolutely, what Light is, after what manner refracted, and by what modes or actions it produceth in our minds the Phantasms of Colours, is not so easy. And I shall not mingle conjectures with certainties."[30]

In his reply to Hooke, Newton acknowledges that "from my Theory I argue the corporeity of light," but denies—quite accurately—that this is a "Hypothesis" or "fundamentall supposition" upon which the theory rests, or, indeed, "any part of [the theory itself], wch was wholly comprehended in the precedent Propositions." (For this he is condemned as "dishonest" by another eminent critic.)[31] And he points out further that he has taken great care, in the exposition of his theory proper, to avoid language that in any way implies commitment to any explanatory mechanical hypothesis: "But I knew that the Properties wch I declared of light were in some measure capable of being explicated not onely by that, but by many other Mechanicall Hypotheses. And therefore I chose to decline them all, & speake of light in generall termes, considering it abstractedly as something or other propagated every way in streight lines from luminous bodies, without determining what that thing is ...."[32] Newton, in fact, *never* speaks of "red corpuscles"; whether using the "vulgar" or the "philosophic" mode, he always says "red—or rubrific—*rays.*"

What one sees here is characteristic of Newton's public utterances: he has exercised a very remarkable discipline to separate—as he puts it—"conjectures" from "certainties"; that is to say, from results supported sufficiently strongly by experiments to merit confident belief. Three aspects of this performance were quite novel for the time, and—as (I hope) we have in part seen—were, and to some extent still are, widely misunderstood: first, the claim of such confidence for theoretical

---

[30] *NC,* 100; *NPL,* 57.

[31] Kuhn 1958, 40.

[32] *NC,* 173–4; *NPL,* 118–9.

conclusions about the non-apparent *physical constitution* of something; second, the claim of "theoretical" status—that is, the status of principles that genuinely concern physical constitution—for anything other than "mechanical" explanations; and third, that attitude of mind that can simultaneously incline towards one sort of mechanical explanation, be willing to consider possibilities alternative to that one, and separate out those results of investigation that are secure irrespective of such explanation. Newton's latter-day critics have been too prone to find in this capacity of his for holding both several levels of theory, and several alternative hypothetical explanations, simultaneously in view, evidence of vacillation, inconsistency, and even hypocrisy.[33]

When Newton says that hypotheses "are not to be regarded in," or "have no place in," experimental philosophy, he in no way intends to imply that they have no *use* in such philosophy.[34] In 1672, in his reply to the second letter on his theory of light by I. G. Pardies, Newton wrote that "hypotheses should be subservient only in explaining the properties of things, but not assumed in determining them; *unless so far as they may furnish experiments.* For if the possibility of hypotheses is to be the test of the truth and reality of things, I see not how certainty can be obtained in any science; since numerous hypotheses may be devised, which shall seem to overcome new difficulties."[35] And in Book III of the *Opticks,* whose first edition was published in 1704, he introduces the famous series of conjectures that constitutes the major part of that Book as follows:[36]

---

[33] Thus, for instance, Kuhn speaks of Newton's "retreat from the defense of metaphysical hypotheses which [he] believed and employed creatively," as "attested by the inconsistencies in his discussions and use of hypotheses throughout the optical papers printed below [*sc.,* in *NPL*]"; and details these "inconsistencies" as follows: "In the first paper light was a substance. In the letters to Pardies light was either a substance or a quality, but the definition of light rays in terms of 'indefinitely small ... independent' parts made light again corporeal. In the same letter Newton proclaimed that his observations and *theories* could be reconciled with the pressure *hypotheses* of either Hooke or Descartes, but in the letter to Hooke he forcefully demonstrated the inadequacy of all pressure hypotheses to explain the phenomena of light and colors.... In 1672 he denied the utility of hypotheses when presenting a theory which he believed could be made independent of them, but in dealing with the colors of thin films in the important letters of 1675/6 he employed explicit hypotheses, presumably because the new subject matter of these letters could not otherwise be elaborated" (Kuhn 1958, 43–4). Newton's critics, as I have remarked, not infrequently attach moral culpability to what they see as his errors. Kuhn in particular attributes to Newton (*ibid.,* 39) a "fear of exposure and the correlated compulsion to be invariably and entirely immune to criticism," accuses him (p. 40) of dishonesty in his response to Hooke (this has already been noted—cf. n. 31 above), and asks whether Newton "is not ... convicted of an irrationally motivated lie" in his reply to Huygens. [Editorial note: we have removed an extended discussion in which Stein critically assesses the evidence offered by Kuhn for these attributions.]

[34] Newton, *Opticks,* 404; and *Principia,* 547.

[35] *NPL,* 106 (emphasis added).

[36] Newton, *Opticks,* 338–9 (emphasis added).

When I made the foregoing Observations, I design'd to repeat them with more care and exactness, and to make some new ones .... But I was then interrupted, and cannot now think of taking these things into farther Consideration. And since I have not finish'd this part of my Design, I shall conclude with proposing only some Queries, *in order to a farther search to be made by others.*

—It is precisely the use he had assigned as proper to hypotheses some 32 years earlier.

It is worth remarking that one chief source of the view of hypotheses prevalent in the period we are considering was the strange history of Descartes' program. For although Descartes, in the *Regulæ*, deprecated "conjecture" in stronger terms than Newton ever did, going so far as to say that what we "conjecture" is not even worth *investigating,*[37] yet in his work in natural philosophy he did find it either necessary or useful to introduce hypotheses. And, although I myself think (in opposition to some recent opinion) that Descartes never abandoned his belief both that he had derived his own most basic physical views—e.g., his theory of the nature of light—from stringently established principles, and that eventually such a derivation of *all* natural knowledge was to be attained, it is clear that the hypothetical mode of constructing mechanical explanation was what continued to influence such natural philosophers as Huygens, who had lost all confidence in Descartes' program of *demonstration* while continuing to embrace the program of *mechanical explanation.*

In *Le Monde,* Descartes introduces what he characterizes as a "fable" of the creation by God of a "new world," governed by principles which (in the sequel) we are intended to conclude are those that in fact govern our own world. In developing the structure of this world, he remarks, "nous prenons la liberté de *feindre* cette matière"—the "new matter" created for the new world—"à nostre fantaisie": "we are taking the liberty to *feign* this matter to our fancy."[38] In his *Principia,* the place where the method of hypothesis is set out, and which corresponds in content to the creation fable of *Le Monde,* is Part III, §§42–47. The headings of §§43, 44, 45, and 47 make a rather extraordinary sequence: "§43. That it is hardly possible but that causes from which all the phenomena are clearly deduced are true. ... §44. That I nevertheless wish those I set forth here to be considered only as hypotheses. ... §45. That I shall even assume here some which are indisputably false. ... §47. That the falsity of these suppositions does not prevent what will be deduced from them from being true and certain." Part IV opens with a section headed: "That the false hypothesis we have already used is to be retained here, in order to explain the true nature of things." And the next section begins: "And so let us feign that this Earth we inhabit was formerly composed [like the Sun], ... and that it was situated in the center of a vast vortex." The phrase "and so let us feign" is, in Latin, "fingamus itaque": the verb, *fingere,* to which the French *feindre* is cognate, is the same as used by Newton in his celebrated declaration, *"Hypotheses non fingo"*—

---

[37] See Rule Three—in *CSM*, 13; and cf. also the optical example in Rule Eight—*ibid.*, 28–9.

[38] Descartes 1979, 50 (French)/51 (English)—emphasis added; my own translation of *feindre* (Mahoney there has "the liberty of *imagining* this matter").

"I do not feign hypotheses." To "feign" hypotheses is, for Newton, to put forth conjectures—indeed, even confessed falsehoods—as a basis on which to establish truths. Newton knew his Descartes very well, and there seems to me little doubt that his choice of words made a deliberate reference to these passages.[39] It seems likely, too, that in his first Rule of Philosophizing in Book III of his own *Principia* Newton's demand that we admit only "*causas rerum naturalium ... quæ et veræ sint et earum phænomenis explicandis sufficiant*"—"causes of natural things that are true and suffice for the explanation of their appearances"—contains an equally deliberate reference to Descartes' introduction of *false* causes.[40]

I must still comment on the sense of "certainty" in Newton's refusal to "mingle conjectures with certainties." Hooke, it will be remembered, said of Newton's theory, "I cannot think it to be the only hypothesis; not soe certain as mathematicall Demonstrations." This last clause alludes to the passage in Newton's paper in which he makes the transition from the discussion of refrangibility to that of color. Newton had written:[41]

> A naturalist would scearce expect to see ye science of those [that is, of colors] become mathematicall, & yet I dare affirm that there is as much certainty in it as in any other part of Opticks. For what I shall tell concerning them is not an Hypothesis but most rigid consequence, not conjectured by barely inferring 'tis thus because not otherwise or because it satisfies all phænomena (the Philosophers universall Topick), but evinced by the mediation of experiments concluding directly & without any suspicion of doubt.

Hooke's reading of the passage as implying "the certainty of mathematical demonstrations" is certainly understandable; but careful reading shows that that is not what Newton in fact implied: rather, he said that he had discovered a *mathematical theory* (or science) of colors, and that this theory—based upon experiment—was as certain as any part of optics. In his reply to Hooke, he points this out very clearly, and goes on to explain (a) that *physical* principles, which are always based on *experiment,* can never rise to the certainty that attaches to mathematical demonstrations, and (b) that a science is mathematical if its principles are such as to allow the determination of phenomena by mathematical argument *from* those principles. (One should remember Newton's statement about geometry, 14 years later, in his preface to the *Principia:* that "geometry is founded in mechanical practice," and that "it is the glory of Geometry that from those few principles, fetched from without, it is able to produce so many things.") But the passage is worth

---

[39] I am not sure that Newton knew *Le Monde;* but it is certain that he had read the *Principia* very closely.

[40] In the first edition, this rule is "Hypothesis I"; and there are small verbal differences—in particular, there is some grammatical confusion of number and mood, with *vera* and *sufficiunt* for *veræ* and *sufficiant.*

[41] *NC,* 96–7. The passage was omitted by Oldenburg when he published Newton's paper, and therefore does not appear in the Cohen edition; likewise for the corresponding passage in Newton's reply to Hooke.

quoting, in part, verbatim, for a little further instruction it contains on Newton's conception of method—both in 1672, and for the rest of his life:[42]

> [T]he Propositions themselves [of the theory of colors] can be esteemed no more then *Physicall Principles* of a Science. And if those Principles be such that on them a Mathematician may determin all the Phænomena of colours that can be caused by refractions .... I suppose the *Science of Colours* will be granted *Mathematicall* & as certain as any part of *Optiques.* And that this may be done I have good reason to beleive, because ever since I became first acquainted with these Principles, I have with constant successe in the events made use of them for this purpose.

What this last sentence intimates is that, in Newton's view, what he calls "deduction from experiments" is not the last word in establishing the "certainty" of physical principles. Indeed, there *is* no last word—that is the difference between "mathematical" and "physical" certainty. Rather, physical principles once established, the process of their subsequent "proof" by experiments in principle never ends[43] (Newton, I think, has this point in a far more cogent form than does Karl Popper). And a satisfactory level of "certainty" is achieved, not, in general, by an "*Experimentum Crucis,*" but by a sufficiency of such experimental "proof."

Thus far, under the head of "multiple vision," I have touched on the distinction of metaphysical principles, or ultimate explanatory aims, from a level of theory securely established by evidence, although short of such aims; on the distinction of the latter—the "certainties"—from the conjectures or hypotheses that may serve to adumbrate possible explanations and to guide research; and, so far as concerns these conjectures, on Newton's willingness to entertain alternative hypotheses—a clear corollary of the distinction itself between "certainties" and "conjectures." Another (related) aspect of this *multiply nuanced* visionary capacity is Newton's vision of the future progress of science—which contrasts rather strikingly with the almost desperate claim of Descartes in his *Discours de la Méthode* to be on the verge of establishing, himself alone, essentially all possible human scientific knowledge.[44] It is, to me, rather poignant to read in the inaugural set of lectures delivered by the 27 year old Newton (to what audience?!) at Cambridge this passage, occurring in a

---

[42] *NC*, 187–8.

[43] This is discussed in considerable detail in Stein (unpublished), as is the claim made in the following sentence, and the related question of just what Newton meant by a "deduction from phenomena."

[44] See, e.g., *CSM*, 145 (in Part VI of the *Discourse*): "For my part, if I have already discovered a few truths in the sciences ..., I can say that these discoveries merely result from and depend upon my surmounting of five or six principal difficulties .... I even venture to say that I think I need to win only two or three other such battles in order to achieve my aims completely, and that my age is not so far advanced that I may not in the normal course of nature still have the time to do this." Cf. his remarks in Part II (*ibid.*, 116) that "there is not usually so much perfection in works composed of several parts and produced by various different craftsmen as in the works of one man," and in Part VI (*ibid.*, 146–8) that, in effect, no cooperative effort of thought has been or is likely to be of any use to him; that "if there was ever a task that could not be accomplished so well by anyone other than its initiator, it is the one on which I am working"; and that the one kind of aid he needs is that of hired hands to carry out experiments under his supervision.

digression that excuses the introduction of the subject of colors into the lectures of a professor of mathematics:[45]

> [S]ince an exact science of [colors] seems to be one of the most difficult things that Philosophy is in need of, I hope to show—as it were, by my example—how valuable mathematics is in natural Philosophy. I therefore urge geometers to investigate Nature more rigorously, and those devoted to natural science to learn geometry first. Hence the former shall not entirely spend their time in speculations of no use to human life, nor shall the latter, while working assiduously with a preposterous method, perpetually fail to reach their goal. But truly with the help of philosophizing Geometers and Philosophers who practice Geometry, instead of the conjectures and probabilities that are being marketed everywhere, we shall finally achieve a natural science secured by the highest evidence.

But also in metaphysics proper we find examples of this disciplined ability of Newton's to entertain conceptions at more than one level, and to exercise critical discrimination among these levels. Here the most astonishing document is the unfinished paper *De gravitatione et æquipondio fluidorum,* first published in 1962 and dated by most experts to the mid-to-late 1660s, roughly three-fourths of which—some 23 pages in Latin, 26 in the English translation—is occupied with a metaphysical digression concerning the nature of space and body, discussed in the light of the exigencies of physics on the one hand, and the teachings of theology on the other.[46] In my opinion, this fragment deserves to be considered one of the most interesting metaphysical disquisitions of the seventeenth century; and if it does indeed derive from Newton's early years, it is remarkable testimony not only to the depth of his thought in that period, but to the extraordinary coherence of his scientific and philosophical development. I should like, in the rest of this paper, to outline the doctrine there expounded, and to put it in relation to what I believe to have been a great transformation of Newton's vision of physics that occurred during the composition of the *Principia.*

The metaphysical digression occurs almost at the very beginning of the essay, immediately following four Definitions: of place, body, rest, and motion. Place is defined as "a part of space which a thing fills adequately":[47] that is, the basic notion is "the place of a thing"—$p(A)$; and its definition identifies the place of A as that precise part of space which is occupied, point for point, by A, to the exclusion of anything else of the same kind as A. For Newton explains that in using the verb *implere*—"to fill"—it is just this exclusion of like things that he intends. Body is defined, correspondingly, as that which fills a place; and Newton remarks that if he were not considering only bodies—*impenetrable* things—he would have defined place more generally as a part of space in which a thing *is* adequately. Rest then is defined as remanence in the same place, motion as change of place.

---

[45] Newton 1984, 87, 89 (Latin original on 86, 88); I have departed slightly from Shapiro's translation.

[46] *NUP,* 90–121 (Latin), 121–56 (English). The metaphysical digression occupies pp. 91–114, 123–48.—Unfortunately, the English translation given by the Halls is very seriously defective; some instances will be of concern to us below.

[47] Not, as the Halls have it, "evenly."

Remarking, next, that these definitions are at variance with the notions of Descartes, both in taking space to be distinct from body and in referring motion to the parts of space rather than to the positions of contiguous bodies, Newton undertakes to defend his view and (as he puts it) "to dispose of [Descartes'] fictions"—or "feignings": *Figmenta* (derived in fact from *fingere*).

The first part of the ensuing discussion is a summary of Descartes' doctrine about the nature of motion, carefully supplied with references to the *Principia Philosophiæ,* followed, first, by a citation of further passages in which Descartes himself contradicts his own position, and then by a series of arguments demonstrating the utter incoherence of Descartes' conceptions as a foundation for the physical theory of motion. (Some commentators persist in finding in this text an essentially theological basis for Newton's rejection of Descartes.[48] For my part, I can only say that, although it is true that theology figures significantly in the later portions of Newton's discussion, which also touch on the defense against atheism, a reading of this argument as a whole that finds its *basis* to be theological seems to me as absurd a *misreading* as Hooke's of the paper on light and colors, which found its basis to be the corpuscular theory of light.)

After this negative critique of Descartes on place and motion, Newton turns to his own positive view, first of space, and then—more tentatively—of body. He begins his account of space with a strange-sounding declaration: "Perhaps," he says, "it may now be expected that I shall define extension to be substance or accident, or else nothing at all"; and at once denies all of these. I take it that the possibility "nothing at all" is a reference to the Greek atomists, who called the void τὸ μή ὄν, "non-being"; Newton is thus dissociating himself from the atomist, Aristotelian, and Cartesian traditions, all three. His principal reason for denying that space is substance is that "it does not support [or "stand under"] characteristic affections of the sort that denominate substance,[49] namely actions, such as are thoughts in a mind and motions in a body." This reference to *action* as characteristic of substance is, he says, not the usual definition of the philosophers, but it is understood tacitly (he claims) by them all. On the other hand, in an important respect, space—so far from being "nothing"—is much more "something" than is any accident, and approaches rather to the nature of substance: for we can clearly conceive it as existing without any subject, as when we imagine extramundane spaces or places empty of bodies.

Space has, according to Newton, "its own mode of existing which fits neither substances nor accidents"; it is (and here some *explication de texte* will be required—and some discussion of the translation) "as it were an emanative effect of God, and an affection of every being [or "thing"]." In the published translation,

---

[48] E.g., Westfall 1980, 302: "The gravamen of [Newton's] charge [against Descartes] was atheism"(!).

[49] The Hall's translation here (*NUP*, 132) is very bad: "it is not among the proper dispositions that denote substance." The Latin reads (*ibid.*, 99): "non *substat* ejusmodi proprijs *affectionibus* [etc.]" (emphasis added).

the last phrase is rendered: "a disposition of all being."[50] The Latin is *omnis entis affectio*. Now, the standard rendering of "affectio" into philosophical English is certainly "affection"—not "disposition," whose connotation is quite inappropriate. As for *omnis entis*, it can bear either the abstract construction, "of all being," or the concrete one I have preferred; evidence of Newton's intention is afforded by comparison with a later passage, which we shall come to presently.

The statement that space is *tanquam Dei effectus emanativus*, "as it were an emanative effect of God," has unquestionable affiliations with Neoplatonic philosophy; but as we shall see in a moment, Newton gives this notion a very unusual twist. First, we must understand that the term "emanation" implies a source of existence that does not involve an act of creation, but rather a derivation or "flowing" from the essence of something. It is of some interest that the *Oxford English Dictionary* quotes a poem of Henry More to exhibit this usage; and also cites a use in a treatise on logic of 1628 with the simple meaning "logical consequence."[51] Now, the crucial passage in Newton's ontological discussion of space (and also of time, or duration) is rendered as follows in the published translation:[52]

> Space is a disposition of being *qua* being. No being exists or can exist which is not related to space in some way. God is everywhere, created minds are somewhere, and body is in the space that it occupies; and whatever is neither everywhere nor anywhere does not exist. And hence it follows that space is an effect arising from the first existence of being, because when any being is postulated, space is postulated. And the same may be asserted of duration.

Once again we find "disposition" instead of "affection" for *affectio*. But two other defects are more serious, both concerning the phrase "an effect arising from the first existence of being." One would not guess from the translation that "effect arising from" is in the Latin *effectus emanativus*—"emanative effect," the same words as before. And "the first existence of being"—which is a monstrous phrase, impossible for Neoplatonism or any other philosophy I know of—is in violent discord with the very syntax of the original: *"et hinc sequitur quod spatium sit entis primario existentis effectus emanativus,"* "and hence it follows that space is an emanative effect of the first-existent being" (*existentis* is an adjective, modifying *entis*, whose genitive case it shares, and *primario* is an adverb, not an adjective): "first-existent," not "first existence"; and not "existence of being," but "of first-existent being." That the concrete construction (with an article supplied in English) rather than the abstract construction is here intended seems to me clear beyond a doubt from Newton's own explication: *"quia posito quodlibet ente ponitur spatium"*—"for if I posit any being whatever, space is posited." So I translate the whole passage thus:

> Space is an affection of a being just as a being. No being exists or can exist that is not in some way related to space. God is everywhere, created minds are somewhere, and a body [is] in the space it fills, and whatever is neither everywhere nor anywhere is not. And hence

---

[50] *Ibid.*, 132; Latin, 99.

[51] The latter citation reads: "1628. T. Spencer *Logick* 199: This truth is necessary by emanation, and consecution."

[52] *NUP*, 136; Latin, 103.

it follows that space is an emanative effect of the first-existent being, for if I posit any being whatever space is posited. And the like may be affirmed of Duration.

Note particularly here that although Newton has said that space is "as it were an emanative effect of God," this passage explicitly *does not* derive space from theology. Space is "an emanative effect of *the first-existent thing*," and that, according to Newton's theology, is indeed God; but "if I posit *any* thing, space is posited."

The remaining part of the metaphysical digression discusses the nature of body. Here, Newton says, the explication will be more uncertain, because body is a divine *creation*, i.e., exists by an act of the divine *will*, and it is not given us to know all possible ways in which God could have produced the effects we discern. But he proceeds to describe one possible manner in which God could have created beings similar, in all ways known to us, to the bodies we are acquainted with. He quite clearly mimics, in this, the creation story of Descartes; and mimics, also, Descartes's phrasing in the *Principia Philosophiæ*: "And so let us feign"— "*fingamus itaque*"—says Newton, that God causes some previously empty region of space to become impervious to penetration by bodies; that, as it were (this is my own gloss), he creates a *field of impenetrability*. "Feign" in the second place that this impenetrability is not always preserved in the same part of space, but rather is allowed "to be transferred hither and thither according to certain laws," but in such a way that the shape and size of the impenetrable region are conserved. If there are several such *mobile impenetrability fields,* then since *ex hypothesi* they cannot penetrate one another, the laws governing their migrations must be such as to determine their behavior in the event of impingement upon one either of another, or—if we make such a distinction—of a body of the "ordinary" kind; in other words, those laws must include laws of impact.

Finally, according to Newton, a third supposition is necessary, if these newly created regions of impenetrability are to have all the essential attributes of the matter we know: they must be able to interact with minds—able, that is, to excite perceptions in the latter, and susceptible in turn of being moved by them. If all three conditions are met, Newton says, the mobile impenetrable spaces would be, to us, entirely indistinguishable from what we call "bodies."[53]

In his further discussion of this analysis of the nature of body, Newton remarks that it replaces the obscure—he says "unintelligible"—notion of a substrate or "substance" endowed with a "substantial form" with the clear one of extension (of this, he has said, we have "an Idea the clearest of all")[54] to which there have been imparted clearly specified "forms" (impenetrability, laws of motion, laws of interaction with minds).[55] He adds:[56]

---

[53] For all of this, see *NUP*, 138–40; Latin, 105–6.

[54] *Ibid.*, 99: "extensionis Ideam habemus omnium clarissimam"; translation, 132: "we have an exceptionally clear idea of extension."

[55] *Ibid.*, 140; Latin, 106.

[56] *Ibid.*, 141; Latin, 107.

If there is any difficulty in this conception it is not with the form God imparts to space, but with the manner by which he imparts it. But that is not to be taken for a difficulty since the same [point] occurs with regard to the manner by which we move our limbs, and nonetheless we believe ourselves able to move them.

This one obscurity—our deficiency of knowledge of the *laws* of minds and of their interaction with bodies—is touched on by Newton in one more place, after he has passed on to argue the "usefulness of the Idea of body" he has described in illuminating the principal truths concerning God and his relation to the world. "Substantial reality," he says,[57]

is rather to be ascribed to Attributes of this kind which are of themselves real and intelligible and do not need a subject in which they inhere, than to some subject ... of which we can in no way form any Idea.... In the same way if we had an Idea of that Attribute or power by which God through the sole action of his will can create beings, we might perhaps conceive that Attribute subsisting as it were of itself and involving his other attributes.

In other words, this "substance-free" conception of substances—as regularly connected systems of attributes—might be extended, not only to created minds as well as bodies, but even to God himself. But, with his characteristically firm discrimination between (a) what is clear and not clear and (b) what we know and do not know, Newton ends this section of his account with what seems to me still the wisest statement on record concerning the mind-body problem: "But so long as we remain unable just to form an Idea of that Attribute [of God], or of our own power by which we move our bodies, it would be rash to say what may be the substantial foundation of minds."

Substantiality, then, is to be divorced from the notion of a "substrate," but associated with that of *powers of acting;* and such powers are to be conceived, to be made intelligible, to be understood, through *laws* of motion.—Whitehead, in Chap. IX of *Science and the Modern World,* cites William James' essay "Does Consciousness Exist?" as inaugurating a new stage in philosophy, and compares that essay with Descartes' *Discourse on Method* and *Meditations.*[58] Descartes, Whitehead tells us, sounded the keynote of modern philosophy: the notions of "ultimate mind" and "ultimate matter," and the problem of their relationship. James "clears the stage of the old paraphernalia; or rather he entirely alters its lighting": he denies that consciousness is a "stuff," asserting that it is a "function." In these terms, Newton, in the text we have been considering, suggested not only that minds, but— and more emphatically—that *bodies* are best conceived, not through the notion of "stuff," but through that of "functions."

In its application to bodies, this Newtonian metaphysic readily accommodates the metaphysical view of fundamental explanation I have earlier associated with Huygens (which, as I have also said before, Newton seems in his early years to have shared). One need only stipulate that the laws of migration, or motion, of the regions of impenetrability are just the laws of impact that Huygens had discovered.

---

[57] *Ibid.,* 144–5; Latin, 110–11.

[58] Whitehead 1967, 141–4.

These are—again, as already noted—exactly what Newton did refer to in his early manuscripts as "the laws of motion," and as the laws of impact of "hard bodies."

In Newton's terminology, a body is called "solid" (in contrast to "fluid") if it resists division; and called "hard" (in contrast both to "soft" and to "elastic") if it is rigid, or non-deformable. The hypothesized regions of impenetrability created by God, having constant shape and size, are, therefore, *solid* and *hard.* They are of course—*ex hypothesi*—impenetrable. And if they satisfy suitable laws of motion— e.g. the known laws of impact—they must have *mass,* since mass is a parameter entering those laws.

Now, it is very interesting to note that in Newton's *Principia* a certain reservation appears concerning the laws of impact of hard bodies. In the scholium to the Laws of Motion, Newton mentions "the theory of *Wren* and *Huygens,*" by which "bodies absolutely hard return one from another with the same velocity with which they meet"; and then adds: "But this may be affirm'd with more certainty of bodies perfectly elastic."[59] It is still more interesting to see that in his last pronouncement on this subject—in Query 31 of Book III of the *Opticks*—Newton *definitively rejects* the Huygensian theory of ultimate interaction: he writes, "Bodies which are either absolutely hard, or so soft as to be void of Elasticity, will not rebound from one another. Impenetrability makes them only stop."[60]

It is quite clear that this revision in Newton's view about the impact of ultimate particles is no mere technical change: the new position is utterly incompatible with the principle that all interaction is "mechanical" by contact.[61] In other words, although Newton never explicitly commits himself to action at a distance as a fundamental principle, that is what is unequivocally implied by this statement— introduced in the guise of a passing remark.

The *Principia* itself, in its opening sections—the Definitions and Laws— develops the conceptual structure for the new form of explanation, precisely to the extent required for physics itself. The central notion is that of a *vis naturæ*—a "force of nature"—or, as Newton also calls it in his Preface (here once more echoing Cartesian terminology),[62] a *potentia naturalis*—a "natural power." A significant body of comment on Newton has been led astray by failure to recognize just what Newton's concept of "force" is. One such force of nature is what Newton calls the *vis insita* of bodies: their "innate (or inherent) force" or "force of inertia" *(vis inertiae).* All other forces of nature are forces of *interaction;* such a force, when it acts on a body, is said to be "impressed" upon it; and the relationships among the

---

[59] Newton, *Principia*, 25 (Scholium to the Laws of Motion and their Corollaries).

[60] Newton, *Opticks*, 398.

[61] Indeed, on this view, if all interactions occurred through impacts of fundamental particles every interaction would entail an agglomeration of matter, and elasticity of any sort would be impossible: there would be nothing that could cause a rebound, or a repulsive force of any description. In particular, the ether Newton suggests in the celebrated twenty-first Query of the *Opticks* (350–2), whose "exceeding great elastick force" may be the cause of gravity, is itself inexplicable on "mechanical" principles, given Newton's cited position about the impact of "hard" bodies.

[62] See, e.g., Descartes' Rule Eight; *CSM*, 29.

innate and impressed forces and the motions of bodies are governed by the three Laws of Motion. These, although Newton says with some—not complete—justice that they are implicit in the work of his predecessors, are, in the *Principia,* put to the service of a radically new plan of inquiry.

Of course what precipitated the new conceptual scheme was Newton's discovery of the first known interactive force of nature. Notice, by the bye, that what may seem (to put it anachronistically) Newton's heterodox use of the word "force" is in fact essentially the same usage that is employed today, when physicists speak of the "four fundamental forces"; the one Newton discovered—gravitation—remains still among the fundamental ones.

But it is in the *Opticks* rather than the *Principia* that Newton makes most clear and explicit just what this general notion of a force of nature *is.* A force of nature is a *law* of nature—of a suitable type; thus, for instance, we can speak of "the same force" when the same law governs—a point of critical importance for Propositions IV–VII of Book III of the *Principia.* Among forces there is one major dichotomy: between the force of inertia, which Newton calls a "passive Principle," and all others, which he calls "active Principles" (what I have referred to as "forces of interaction"). Here are the central passages. First: "The *Vis inertiæ* is a passive Principle by which Bodies persist in their Motion or Rest, receive Motion in proportion to the Force impressing it, and resist as much as they are resisted."[63] Thus it is the *conjunction* of *all three* "Laws of Motion" that characterizes the "force of inertia." And then:[64]

> It seems to me farther, that these [fundamental] Particles have not only a *Vis inertiæ,* accompanied with such passive Laws of Motion as naturally result from that Force, but also that they are moved by certain active Principles, such as that of Gravity, and that which causes Fermentation, and the Cohesion of Bodies. These Principles I consider, not as occult Qualities, supposed to result from the specifick Forms of things, but as general Laws of Nature, by which the things themselves are form'd; their Truth appearing to us by Phænomena, though their Causes be not yet discover'd.

It should be clear how naturally this new scheme fits into the deeper metaphysics of *De gravitatione.* The "general Laws of Nature" associated with each *vis naturæ*— or rather, those associated with the truly *fundamental* or *ultimate* forces—are simply to be included among the laws governing the migrations of the ultimate bodies (the fields of impenetrability). Newton suggests, both in the preface to the *Principia* and in the *Opticks,*[65] that all the active forces may be what we should call "central force fields" associated with the fundamental particles. Among his speculative suggestions as to what "active principles" of this kind there may be to be found out in later research, one in particular deserves to be mentioned. Near the beginning of the lengthy last Query in the *Opticks* he writes: "The Attractions of Gravity, Magnetism, and Electricity, reach to sensible distances, and so have been observ'd by vulgar

---

[63] Newton, *Opticks,* 397.

[64] *Ibid.,* 401.

[65] *Ibid.,* 397.

Eyes, and there may be others which reach to so small distances as hitherto escape Observation; *and perhaps electrical Attraction may reach to such small distances, even without being excited by Friction.*"[66] I cannot resist once more quoting Blake— one of his "Proverbs of Hell": "What is now prov'd was once, only imagin'd."

7.1 Concluding Remarks

Descartes begins his philosophy with metaphysics; immediately after the *cogito*, with God, upon whom, he maintains, all of his physics rests (and by whose guarantee it is true beyond a doubt). Newton introduces into the beginning of his natural philosophy only just that part of what I have called his metaphysics that he regards as (a) adequately supported by prior evidence, and (b) necessary for the development of physics. The rest, in so far as it appears at all in his scientific work, does so at the *end:* in the *Principia,* in the General Scholium, where after some theological discussion he concludes:[67] "And thus much concerning God; to discourse of whom, *from the appearances of things,* does certainly belong to Natural Philosophy."; in the *Opticks,* at the end of the last Query, in a section that begins:[68]

> All these things being consider'd, *it seems probable to me* that God in the Beginning form'd Matter in solid, massy, hard, impenetrable, moveable Particles, of such Sizes and Figures, and with such other Properties, and in such Proportion to Space, as most conduced to the End for which he form'd them....

Observe that this is the same creation story as in *De gravitatione*—with its deeper but more speculative ontology suppressed. The "other Properties," besides those detailed in the list "solid, massy, hard, impenetrable, moveable," with "Size" and "Figure," all of which we encountered in *De gravitatione,* are those involved in the active forces of nature. Note, too, Newton's striking phrase about these (quoted earlier): "general Laws of Nature, *by which the Things themselves are form'd,*" echoing *De gravitatione*'s remark about the clear "forms" (contrasted with the obscure—occult—scholastic "substantial forms") imparted, not to obscure "prime matter," but to clear "parts of extension": the clear forms are fields of force, and Newton is suggesting that some among these—the most fundamental of them—may be what constitutes the ultimate *natura rerum.*

But always only "may." In the *Opticks,* Newton says of the active principles that their truth appears by the phenomena, "though their Causes be not yet discover'd"— and a little later, "I scruple not to propose the Principles of Motion above-mention'd, they being of very general extent, *and leave their Causes to be found out.*" In the

---

[66] *Ibid.,* 376 (emphasis added).

[67] That is, concludes the theological discussion—not the scholium as a whole. In the Cajori edition of the *Principia,* the passage occurs on p. 546. (Emphasis added.).

[68] Newton, *Opticks,* 400.

preface to the *Principia* he says, "I hope the principles here laid down will afford some light either to that, *or some truer,* method of Philosophy."

Newton was neither a modest nor a moderate man; but these expressions are not a pose. They express a deep conviction about the nature of scientific inquiry itself, and are one in spirit with the appeal for a disciplined search for mathematical principles of natural science that we have seen at the beginning of Newton's career, in the *Optical Lectures.* That a man of such imperious disposition never—into his most advanced age—made disproportionate claims about his contributions to science is one aspect of a tremendous discipline which he exerted—on the whole with remarkable success—in keeping the "imagin'd" parts of his vision distinct from the "prov'd"; the "conjectures" from the "certainties."

References

Bacon, F. 1960. *The New Organon and related writings,* ed. F.H. Anderson. New York: Bobbs Merrill.

Burtt, E.A. 1955. *The metaphysical foundations of modern physical science.* 2nd ed. Garden City: Doubleday.

Cohen, I.B., ed. 1958. *Isaac Newton's papers and letters on natural philosophy.* Harvard University Press.

Descartes, R. 1979. *Le monde,* ed. and trans. M.S. Mahoney. New York: Abaris Books.

———. 1985. *The philosophical writings.* Trans. J. Cottingham, R. Stoothoff and D. Murdoch. Cambridge University Press.

Hall, A.R. and M. Boas Hall, (eds). 1962. *Unpublished papers of Isaac Newton.* Cambridge: Cambridge University Press.

Hertz, H. 1956. *The principles of mechanics presented in a new form* [1899]. Trans. D.E. Jones and J.T. Waley. New York: Dover Publications.

Kuhn, Th. 1958. Newton's optical papers. *NPL,* 27–45.

Leibniz, G.W. 1970. *Philosophical papers and letters,* 2nd edition, ed. and trans. L. Loemker. Dordrecht: D. Reidel.

Leplin, J. ed. 1995. *The creation of ideas in physics: Studies for a methodology of theory construction.* University of Western Ontario series in philosophy of science, vol. 55. Springer.

Newton, I. 1934. *Mathematical principles of natural philosophy and his system of the world.* Trans. F. Cajori. University of California Press.

———. 1952. *Opticks,* 4th edition [1730]. New York: Dover Publications.

———. 1984. *The optical papers,* ed. and trans. A.E. Shapiro. Cambridge University Press.

Pitcher, G. 1977. *Berkeley.* London: Routledge & Kegan Paul.

Putnam, H. 1962. What theories are not. In *Logic, methodology, and philosophy of science,* ed. E. Nagel, P. Suppes, and A. Tarski, 240–251. Stanford University Press.

Shapiro, A.E. 1980. The evolving structure of Newton's theory of white light and color. *Isis* 71: 211–235.

Stein, H. unpublished. *Further Considerations on Newton's Methods.* http://www.strangebeautiful.com/other-minds.html.

Thomson, J.J. 1885. On some applications of dynamical principles to physical phenomena. *Philosophical Transactions* 176: 307–342.

Turnbull, H.W., ed. 1959. *The correspondence of Isaac Newton.* Vol. I. Cambridge University Press.

Westfall, R.S. 1980. *Never at rest: A biography of Isaac Newton.* Cambridge University Press.

Whitehead, A.N. 1967. *Science and the modern world [1925].* New York: The Free Press.

Chapter 8
Working Hypotheses, Mathematical Representation, and the Logic of Theory-Mediation

Zvi Biener and Mary Domski

8.1 Introduction

In the General Scholium, Newton famously remarks that in the program of "experimental philosophy" that he pursues in the *Principia mathematica* (1687), "hypotheses, whether metaphysical or physical, or based on occult qualities, or mechanical, have no place" (Newton 1999, 943).[1] Newton targets Descartes' vortical explanation of planetary motion as one such hypothesis, and in Section 9 of Book 2, he presents a series of arguments intended to show that the Cartesian hypothesis is incompatible with observed Keplerian planetary motion and cometary motion. Based on these arguments, Newton asserts in the scholium that concludes Section 9 that "the hypothesis of vortices can in no way be reconciled with astronomical phenomena and serves less to clarify the celestial motions than to obscure them" (*ibid*, 790, emphasis added).[2]

---

[1] The General Scholium was added to the second edition *Principia* (1713) and retained in the third (1726). A similar though less forceful dismissal of "hypotheses" can be found in Newton's correspondence. See Shapiro 2004 for discussion of Newton's use of the term "experimental philosophy" to separate his program of natural philosophy from the "hypothetical" natural philosophy that he associates with Descartes and Leibniz.

[2] The calculations that Newton uses to establish the incompatibility between Descartes' Vortex Hypothesis and Kepler's Area Rule rest on incorrect assumptions, as first pointed out by George

---

Z. Biener (✉)
University of Cincinnati, Cincinnati, OH, USA
e-mail: bienerzi@ucmail.uc.edu

M. Domski
University of New Mexico, Albuquerque, NM, USA

© The Author(s), under exclusive license to Springer Nature Switzerland AG 2023
M. Stan, C. Smeenk (eds.), *Theory, Evidence, Data: Themes from George E. Smith*, Boston Studies in the Philosophy and History of Science 343, https://doi.org/10.1007/978-3-031-41041-3_8

While Newton was eager to distance his program of natural philosophy from Descartes', questions arise about just how different their methods were when Descartes' "hypothetical" natural philosophy is put into conversation with the portions of Book 2 that bear directly on the argument against the vortex hypothesis, specifically the portions dedicated to the problem of fluid resistance. These have been central to George E. Smith's path-breaking work on the *Principia* and our understanding of Descartes' and Newton's differing methodologies.

The goal of this paper is to articulate Smith's insights and to explore what we believe are some of their implications. First, we examine a few of the apparently 'Cartesian' moves that Newton makes in Book 2 and highlight how Smith's notion of 'working hypotheses' can be used to draw a clear line between Newton's method and Descartes' (§8.2, §8.3). The notion of 'working hypotheses' gives us further occasion to elaborate on the more general logic of theory mediation that Smith attributes to Newton (§8.4). While much work on theory mediation focuses on how theory-mediated measurements fix theoretical parameters, we focus on how they allow phenomena to be constituted and individuated. This lets us locate in Smith's framework two types of evidence to which working hypotheses can appeal: what we call 'conditional and 'independent' evidence. Considering their relative merits allows us to extend Smith's account of theory-mediation (in §8.5). Ultimately, we argue that Smith's nuanced portrait of Newton's non-Cartesian methodology opens up a way of appreciating how choices of mathematical formalism can themselves be considered working hypotheses.

8.2   Fluid Resistance and Working Hypotheses in the First Edition of the *Principia*

In all editions of Book 2 of the *Principia* (1687, 1713, 1726), Newton supposes that the overall resistance encountered by a body moving in a fluid arises from different features of the fluid, such as the fluid's inertia and what today we call viscosity. He considers each of these to be due to *independent* physical mechanisms and represents the contribution of each mechanism to the overall resistance encountered by a body as a function of the body's velocity relative to the fluid.[3] Put differently, in all editions of Book 2, Newton represents the resistance due to each physical mechanism as one of several additive components $(a_n v^n)$ that contribute to the total

---

G. Stokes in a paper of 1845. See Cohen 1999, Chapter 7 and Smith 2005, Notes 4 and 5 for further discussion of Newton's errors.

[3] In the third edition *Principia*, for instance, Newton holds a three-mechanism view such that the overall resistance acting on a body moving through a fluid is produced by the fluid's density, internal friction, and tenacity (or absence of lubricity, or slipperiness), and in this case, the tenacity is taken to be independent of the moving body's velocity, whereas the internal friction is taken to be proportional to the body's velocity. See the Scholium to Book 2, Section 3 (Newton 1999, 678–678) and Smith's contribution to Cohen 1999 (pp. 188–194).

resistance encountered by a body ($F_{\text{Resistance}}$). We call this Newton's 'fundamental assumption' of fluid resistance.[4] In its most general form, it can be represented as follows, allowing that n may be fractional:

$$F_{\text{Resistance}} = a_0 + a_1 v + a_2 v^2 + \cdots + a_n v^n \tag{8.1}$$

In the first edition of the *Principia*, v^2 is presented as the dominant term and taken as the effect of the moving body pushing the fluid medium out of its way (i.e., the effect of the body overcoming the inertia of the medium). The other two additive components—the a_0 and $a_1 v$ terms of the fundamental assumption—are both taken to arise from what Newton refers to as a "defect of lubricity," or slipperiness, in the fluid medium (Smith 2005, 157).[5] In other words, with these considered to be independent physical factors, Newton had to establish "separate force laws for [these] different mechanisms of resistance" (Smith 2005, 134). This meant he had to disaggregate the contribution of the fluid's inertia from the fluid's lubricity in such a way that he could identify the appropriate coefficients for each of the additive components above.

In the first edition, Newton tried to accomplish this disaggregation by means of pendulum experiments.[6] Assuming air resistance is the primary cause of pendulum decay, he expected the arc loss per swing to be a function of the various powers of v, according to the fundamental assumption. The arc loss could then be expressed as follows, with A_i being constants for a given pendulum:

$$\delta_{\text{arc}} = A_0 + A_1 V_{\text{max}}^1 + A_2 V_{\text{max}}^2 + \ldots A_n V_{\text{max}}^n \tag{8.2}$$

By starting the pendulum at different heights, Newton was able to generate different sets of results, which allowed him to infer the values of A_i by solving simultaneous equations.[7]

---

[4] Although Smith's account of Newton's reasoning in Book 2 isn't cast in exactly the same terms, our rendering of Newton's 'fundamental assumption' of fluid resistance captures the additive view of fluid resistance that Smith attributes to all editions of the *Principia* (See Smith 2000, 2001a, 2005). Our rendering is also sufficiently general to accommodate the fact that in different editions of Book 2 Newton treated the overall resistance encountered by a body as arising from different physical mechanisms (Smith 2005, 157).

[5] In other texts, including the later editions of the *Principia*, Newton used the term "tenacity" (*tenacitas*) to refer to the absence of slipperiness, or lubricity, in the fluid. See Smith's contribution to Cohen 1999 (pp. 188–194).

[6] Pendulum experiments are useful for studying projectile motion since their properties are easier to measure than the properties of free fall (although Newton ended up using free-fall measurements in the second edition of the *Principia*). Moreover, Newton lacked a fully general solution for projectile motion in resisting media, so constrained motion proved important.

[7] To calculate a single swing arc loss, Newton actually measured the loss of $^1/_8$ or ¼ of the overall arc, and divided by the number of swings. He worked out how to express arc loss per swing in terms of pendulum length, and total resistive force as a ratio to bob weight. He used a cycloidal

However, the overall data that he collected presented Newton with some problems. He tried several linear combinations:[8]

$$A_0 + A_1 V + A_2 V^2$$

$$A_1 V + A_2 V^2$$

$$A_{\frac{1}{2}} V^{\frac{1}{2}} + A_1 V + A_2 V^2$$

$$A_1 V + A_{\frac{3}{2}} V^{\frac{3}{2}} + A_2 V^2$$

But the result was always the same: No matter which combination he chose, he could not establish stable values for all of the A_i, or even establish a stable range of values that accommodated all of the measurements that he collected from the pendulum experiments. Ultimately, in the published edition of the first edition of the *Principia*, Newton used the final linear combination listed above, and he presented values for the A_i based only on a subset of the experimental data that he had collected. He offered no explanation for either of these choices.

The arbitrary, even speculative, character of Newton's reasoning in this case may well seem "hypothetical". As Smith notes, the failure to fix the A_i and determine the relevant powers of v (i.e., the failure to disaggregate the various contributions to resistance) indicates that Newton's pendulum experiments provided no real test of his fundamental assumption. In other words, the experiments did not entail—although they were compatible with—Newton's physical understanding of the causes of resistance.[9] This is why the charge of Cartesianism arises. Descartes' physical explanations often posited microphysical mechanisms that were compatible with the phenomena they were meant to explain, but were by no means uniquely entailed by them. Moreover, those microphysical mechanisms lacked independent evidence. They seemed plausible from a mechanical point of view, but there was no reason to believe in them except for their (purported) success in saving the very phenomena they were meant to save. This is what we mean when we call them "hypothetical." For example, in Part III of the *Principles*, Descartes asks us to imagine that the insensible particles that make up fire have both a rectilinear and a circular motion. When they collide with other bodies, we are also to imagine that

---

pendulum, where maximum velocity is proportional to overall arc length. For additional details see Propositions 30 and 31 of Book 2 of the first edition *Principia*, and Smith 2001a, 259ff.

[8] The v^2 term is common because at high velocities the total resistance varied as nearly v^2. See Smith (2000, 130, footnote 19).

[9] Smith notes that "[r]egardless of why Newton presented the findings in the way he did … [the] experiments do not begin to allow reliable conclusion to be drawn about contributions to resistance involving an exponent of v less than 2." Moreover, despite Newton's efforts to focus on this term in subsequent experiments (some in water), here too "his results were disappointing" (2001a, 262).

these particles adapt their shape in such a way that they can fill the narrow spaces between larger pieces of matter (Descartes 1984, 110ff; III.52ff). Descartes supplies this imagery to explain, among other things, why light is emitted from every part of the Sun and why it travels in straight lines (Descartes 1984, 115–118; III.60–64).[10] Newton's fundamental assumption, as articulated above, seems methodologically similar. His treatment of resistance and Descartes' treatment of fire both rely on claims about the behavior of unobservable micro-mechanisms that are supposedly responsible for the phenomena they are trying to explain, but which are not entailed by these phenomena, and for which there is no other evidence. It is in this respect that it becomes fair to ask: Was Newton guilty of pursuing the Cartesian brand of natural philosophy that he publicly rejected?[11]

Smith has given us good reasons to reject this reading. He has long argued that Newton's model of theorizing departed substantially from the Cartesian hypothetico-deductive model (CHD), even in the context of fluid resistance.[12] A full discussion of Smith's rich account of the "Newtonian Style" is beyond our scope, but a quick contrast with CHD will bring out the features with which we will be primarily concerned in this essay. On the CHD, the theoretician's role is to construct theories whose predictions agree as nearly as possible with observed phenomena.[13] When predictions and phenomena agree, the theoretician has accomplished her task. When they don't, it's back to the drawing board—the theoretician must modify her existing theories, perhaps even replace them. *How* she modifies or replaces them, however, is entirely up to her. Because the only empirical evaluative criterion for the success of a theory is that the theory, *taken as a whole,* saves the phenomena,

---

[10] For a critical discussion of the "hypothetical" explanations that Descartes presents in Part III of the *Principles of Philosophy*, see McMullin 2008.

[11] Other examples from Book 2 could also be used to motivate the same question. Perhaps the most famous among them is Newton's treatment of the speed of sound in the second edition *Principia*. In Book 2, Section 8, Newton compares his theoretical value for the speed of sound with the experimental data that had been compiled by William Derham, and he reports a perfect fit between the two (Newton 1999, 778). But to establish this fit, Newton relies on the alleged existence of "vapors lying hidden in the air," which have a "crassitude" that increases the speed of sound. Newton has no direct evidence of these hidden vapors, just as Descartes has no direct evidence that the insensible parts of fire adapt their shapes to fill the spaces between the parts of matter. Both are posited as generally intelligible but empirically untestable ways to make sense of what has been empirically verified. For the debate over whether Newton's introduction of the "crassitude" of air is best viewed as a "fudge factor" or as a good faith attempt to account for Durham's results, compare Truesdell 1970 and Westfall 1973 with Cohen 1999, 361–362.

[12] See Smith 2001a, 2002b, and 2005, 2016. In Smith 2002a and 2016, Smith specifically contrasts Newton's methodology with the Cartesian hypothetico-deductive method as it is described by Christiaan Huygens in the Preface to his *Treatise on Light* (1690) (Smith 2002a, 139–140; 2016, 189). For discussion of how Newton's method for establishing universal gravitation departs from a more generally construed hypothetico-deductive model of scientific reasoning, see Ducheyne 2012 and Harper 2011.

[13] We are not using "prediction" in a technical sense. The point is simply that however Cartesian theories (or models, qualitative descriptions, etc.) make claims about the world, the Cartesian theoretician seeks agreement between those claims and the world.

a mismatch between theory and observation provides no information about which theoretical changes are necessary. The phenomena only indicate whether a theory fits better or worse than its competitors. This is why competing but empirically equivalent theories are a real prospect for the CHD. In the CHD, at least generally, information flow is unidirectional. Theories tell us what to think about the world, but the world doesn't tell us what to think about the content of our theories.[14]

In contrast, information flow in the Newtonian framework is bi-directional. Smith stresses that Newton tried to construct theories by means of "if-and-only-if" propositions that mutually bind some theoretical feature and some feature of the phenomena.[15] These propositions allow both for seeing what observations would follow given certain values of relevant theoretical parameters, and for seeing what values these parameters would take given certain observations. The latter allows phenomena to measure theoretical parameters in light of previously established "if-and-only-if" propositions.[16] Moreover, because theory-mediated measurements presuppose some specific theoretical propositions (as opposed to the theory as a whole), when measurements go wrong—i.e., when they cannot be made to agree with one another or with other relevant measurements—they implicate the parts of the theory that made them possible, and thus point to possible revision. As Smith puts the contrast with CHD, the aim of Newtonian theory is to "find ways in which the world [c]ould provide conclusive answers to theoretical questions," not just ways to "conjectur[e] answers and then [test] the implications of these conjectures" (Smith 2002a, 147; 2016, 197–8). In other words, a primary aim of the Newtonian theoretician is to construct a theory sufficiently powerful to allow the world to force her hand in future theory development. This theoretical coercion, if you will, is the centerpiece of the "Newtonian style." Newton's claim that that the law of gravity was "deduced" from the phenomena takes on new significance in this light (Newton 1999, 790). The claim is not merely that there is some empirical support for the law. It is that within the theoretical framework of the laws of motion, and in light of the then-current best measurements of planetary and terrestrial motions, the world

---

[14] Not only can the world not tell the Cartesian how to revise her theory, arguably, the world cannot even tell her if the theory is generally true. This is because, at least as Descartes presents it in Part III of the *Principles*, the hypotheses that are posited to explain observed phenomena are not to be accepted as true, no matter how much empirical evidence might be amassed in their favor. Hypotheses are more or less acceptable depending on their consistency with Descartes' metaphysically derived laws of nature, and depending on their consistency with the phenomena. But in general, every theory that Descartes posits to explain the visible world, including his vortex theory of the heavens, is "to be taken only as an hypothesis {which is perhaps very far from the truth}" (Descartes 1984, 105, III.44). For further discussion, see Domski 2019.

[15] Smith also notes that when "[Newton] is unable to establish a strict converse [of an if-then statement], he typically looks for a result that falls as little short of it as he can find" (Smith 2002a, 146; 2016, 197).

[16] Ideally, for these inferences to facilitate good measurements, additional constraints are necessary. For example, they should submit to *quam proxime* reasoning, as emphasized in Smith 2002a and 2016.

constrains further theory choice to such an extent that accepting the law of gravity becomes the most reasonable course of action.[17]

Smith has also argued that Newton's treatment of motion in resisting media is methodologically consonant with his deduction of gravity (Smith 2001a). In particular, Smith has provided a means of re-describing Newton's seemingly arbitrary hypotheses in a way that sidesteps the Cartesian paradigm. For Smith, Newton's hypotheses—like the fundamental assumption—are best viewed as "working hypotheses" (Smith 2000, 2005). Although these are not directly entailed by the phenomena, they are not final pronouncements either, as Descartes' hypotheses seem to be.[18] Rather, they are theoretical posits—perhaps first approximations—that play a critical role in theory development. Specifically, they facilitate theory-mediated measurements that otherwise would have been impossible. By so doing, they let new evidence be constructed and deployed for/against the theory in question. In essence, they enable new questions to be put to the world and empower the world to answer. And because these new measurements implicate the parts of the theory that made them possible, the evidence they generate is not for/against the theory *simpliciter*, but for/against those very working hypotheses. When things go wrong— when negative evidence accrues—they point the way to theoretical revision.

Smith grants that the fundamental assumption that guides Newton's treatment of motion in resistive media in Book 2 is not as robust a working hypothesis as we find in other parts of the *Principia*. In accepting it, the theoretician's hand is not forced to the extent that it is forced in regard to the law of gravitation, for example. Nonetheless, the fundamental assumption, Smith argues, retains the primary virtue of Newton's other working hypotheses. It serves as an enabling posit that opens a fruitful and well-defined pathway for further inquiry, and as such, it gives us a fruitful way of distinguishing Newton's method from Descartes'.

Smith applies a similar framework to the revised sections of Book 2 that were published in the later editions of the *Principia*. In these sections, Newton pursues

---

[17] We say "the *most* reasonable" because theoretical choice is not *entirely* constrained. Questions regarding the inductive scope of accepted claims—how they are "rendered general by induction"—remain open. A more guarded reading of Newton's claim that gravity is "deduced from phenomena" is that within the theoretical framework of the laws of motion, given the then-current best measurements of planetary and terrestrial motions, and *within the range of then-current measurements*, the world constrains further theoretical choice to such an extent that accepting the law of gravitation *within that range* is the most reasonable choice. As Newton notes in the General Scholium, his evidence shows that gravity extends "as far out as the orbit of Saturn, as is manifest from the fact that the aphelia of the planets are at rest, and even as far as the farthest aphelia of the comets, provided that those aphelia are at rest" (589). See also Harper 2011 and Biener 2016.

[18] We say "*seem* to be" because we believe the traditional reading of Descartes is not entirely correct. As we read Part III of the *Principles*, and as noted in Note 14 above, Descartes is not offering his hypotheses as true and final pronouncements about the workings of nature. They are explanatory devices that, he says, could be "very far from the truth," and consequently, they are revisable and essentially different in kind from the three laws of nature of Part II, which are final and firm pronouncements about the true workings of nature. For further discussion of the "truth" that Descartes associates with his laws of nature, see Domski 2018.

a different strategy for establishing overall fluid resistance than he did in the first edition. But again, the fundamental assumption plays a central role in his reasoning. And again, he seems to employ the "Cartesian" style hypotheses we described above.

8.3 Fluid Resistance and Working Hypotheses in the Second and Third Editions of the *Principia*

In Book 2, Section 7 of the second and third editions of the *Principia*, Newton examines the resistance forces on bodies that move through two general types of fluid: rare and continuous.[19] A rare fluid consists of particles that do not interact. Specifically, on Newton's account, a rare fluid consists of non-interacting particles "that are equal and arranged freely at equal distances from one another" (Newton 1999, 728).[20] When a body moves through this type of fluid, the fluid's particles impact the body and thus impede the body's motion. However, since the particles do not interact, there is no change in the fluid constitution ahead, behind, or around the moving body. Consequently, in this type of fluid, resistance is treated by Newton as a function only of the front surface area of the body, the density of the fluid, and their relative velocity. Determining the resistance offered by a continuous fluid, in contrast, is more complicated. Real-world examples of this type of fluid include water, hot oil, and quicksilver, and they are continuous insofar as they consist of "solid particles effectively in contact with their immediate neighbors" (Smith 2005, 129). When a body moves through this sort of fluid, it "does not impinge directly upon all the particles of fluid which generate resistance"—as is the case in a rare fluid—"but presses only the nearest particles, and these press others and these still others" (Newton 1999, 735).

To get some purchase on how much resistance the fluid offers in this scenario, Newton first limits his treatment to continuous fluids that are inelastic and lack (in contemporary terms) internal friction and viscosity. He also initially considers only the resistance encountered by spherical bodies that move through the idealized continuous fluid he has defined.[21] He reasons that the inertial resistance encountered

---

[19] For more general discussions of the problematic and sometimes mistaken reasoning that can be found in these editions of Book 2, Section 7, see Truesdell (1970), Westfall (1973) and the Smith essays cited herein, especially Smith 2005. Rouse Ball's *An Essay on Newton's "Principia"* (1893) is also a noteworthy commentary on Book 2. Ball makes no attempt to hide the mistakes of Section 7, but also finds there "much that is interesting in studying the way in which Newton attacked questions which seemed to be beyond the analysis at his command" (Ball 1893, 99; cited in Cohen 1999, 181).

[20] Newton also briefly discusses motion in an "elastic" fluid, which is a rare fluid in which there are repulsive forces between the particles.

[21] Newton discusses other shapes in the scholia to Propositions 37 and 38. In what follows, we limit our discussion to the case of moving spheres since it is most relevant to Newton's purported

by a body moving in this type of fluid is due mainly to the body's needing to push forward the column of fluid ahead of it. By symmetry considerations, this force is equivalent to the force exerted in a gravitational field on a stationary body by the column of fluid above it. The contiguous parts of the fluid press downward on one another to exert some total force (their weight), which, in the non-gravitational case, is the total (inertial) force a moving body needs to overcome when it "presses . . . the nearest particles, and these press others and these still others". To determine this total weight and thereby generate a model of resistance for spherical bodies, Newton expands on his treatment of the efflux problem—the problem of determining the velocity of water escaping from a hole in a large vessel.

The efflux problem is treated in Book 2, Proposition 36, and its extension to the case of resistance begins in Corollary 7 (see Fig. 8.1; Newton 1999, 740).[22] The specific task is to determine the weight of the column of water supported by the circular disk PGQ, which is at rest with respect to the cylinder ABCD but in relative motion with respect to the descending water flowing out through EP and QF. Earlier in the proposition, Newton asks us to imagine that "the interior of the vessel around the falling water ABNFEM is filled with ice, so that the water passes through the ice as if through a funnel" (ibid, 734). In Corollary 7, he further adds that PHQ—the column of water above the resting disk PGQ—should be considered frozen. This is because the fluidity of these regions "is not required for the very ready and very swift descent of the water" (ibid, 740). To determine the weight of PHQ—which, as noted above, is equivalent to the force PHQ exerts on the disk PGQ—Newton compares its weight to the weight of the "cylinder of water whose base is that little circle and whose height is GH." He determines that the weight of the column PHQ will be greater than 1/3 of the weight of the cylinder (Corollary 7) and less than 2/3 of the weight of the cylinder (Corollary 8). Then, in Corollary 9, he provides a precise value: "The weight of the water sustained by the little circle PQ, when it is extremely small, is very nearly equal to the weight of a cylinder of water whose base is that little circle and whose height is $^1/_2$GH" (ibid, 741). In other words, Newton establishes that the weight of the column of water supported by the disk PGQ is very nearly half the weight of the cylinder of water that has PGQ as its base and a height of GH.[23]

---

refutation of Descartes' Vortex Hypothesis—a hypothesis that concerns the motion of (roughly) spherical bodies through the heavens.

[22] For discussion of Newton's general (and sometimes mistaken) approach to the efflux problem in the second edition, see Maffioli 1994 and Westfall 1980.

[23] In the first edition of the Principia, Newton claimed that a stationary body suspended in the flowing efflux stream of water would support the full weight of the cylinder of water above it. He also claimed in the first edition that the water escaping the vessel in the efflux problem would have a velocity equal to a body falling from the full height of the tank. In the second edition, both of these values are cut in half, and as a result, in the second edition, Newton lowered the value for the inertial resistance on the front face of the body by a factor of four (Smith 2000, 119–121). See Smith 2001a for a detailed account of Newton's first edition treatment of the weight supported by a resting body in a moving fluid, and also for an English translation of the relevant portions of Book 2, Section 7 from the first edition.

Fig. 8.1 The efflux problem
and fluid resistance in the
second edition *Principia*

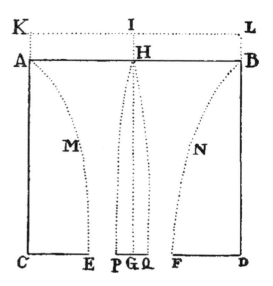

Fig. 8.1 The efflux problem and fluid resistance in the second edition *Principia*

How exactly does Newton justify this value? In the first instance, he reasons that although the specific shape of the column PHQ is not well defined, its shape (whatever it is) can be set in proportion to the weight of the cylinder of water that has PGQ as its base and a height of GH. Specifically, he asks us to imagine that the cataract of water falling around PHQ "falls with its whole weight and does not rest or press on PHQ but slides past freely and without friction, except perhaps at the very vertex of ice, where at the beginning of falling the cataract begins to be concave" (*ibid*, 740). In this scenario, the angles PH and HQ, which are formed at the vertex of the column, will be convex towards the cataract just as AME and BNF are convex. Consequently, PHQ "will be greater than [that is, it will inscribe] a cone whose base is the little circle PQ and whose height is GH, that is, greater than $^1/_3$ of a cylinder described with the same base and height." (*ibid*, 740). This provides the lower bound. To establish the upper bound, Newton compares the frozen column with a half spheroid "whose base is the little circle and whose semiaxis or height is HG," which meets P and Q at right angles (*ibid*, 741). Because the angles of PHQ are all acute, the entire frozen column above PGQ will "lie within the half-spheroid," and is therefore less than "$^2/_3$ of a cylinder whose base is that little circle and whose height is GH" (*ibid*, 741). Immediately thereafter, in Corollary 9, Newton concludes:

> The weight of the water sustained by the little circle PQ, when it is extremely small, is very nearly equal to the weight of a cylinder of water whose base is that little circle and whose height is ½ GH. For this weight is an arithmetical mean between the weights of the cone and the said half-spheroid. (*ibid*, 741)

Newton's use of the arithmetical mean is elegant. However, Newton does not explain why it is physically compelling. Instead, he provides some experimental results concerning the general relation between the velocity of the efflux stream and the width of the efflux hole (*ibid*, 735–736). But none of these results provide any support for his use of an arithmetical mean, which raises questions about his claim that the weight supported by the disk PQ is equal to the weight of the cylinder that has a height of ½ GH.

Indeed, looking ahead in the text to the experimental results that Newton presents both to support his general theory of resistance and to disconfirm Descartes' Vortex Hypothesis, one might think that Newton's choice of ½ was reverse-engineered.[24] What's at stake in particular is Newton's theoretical assumption that the force acting on the rear of a body moving through a fluid is negligible. To see the significance of the assumption, consider that when a body moves through a fluid, the volume of fluid that it displaces comes rushing in behind it. It's not unreasonable to think that the force exerted by the rushing fluid on the rear of the body could offset, to some degree, the resistance force on the front. It is also not in principle unreasonable to think that the force acting on the rear of the body could be large enough to entirely counteract the force on the front. On this line of thinking, it would be possible for, say, planets to move in an aetherial fluid without encountering a net resistance, and thus without any observable retardation of their motion.[25]

Newton attempts to block this line of reasoning. He argues on qualitative grounds that, in fact, the force acting on the rear of the body is negligible. For example, he reasoned that in a continuous fluid pressure is propagated instantaneously, and consequently "generates no motion . . . , [and] thus neither increases nor decreases the resistance" (*ibid*, 744).[26] He offered no direct experimental evidence to support these particular claims; instead, after he's established that the weight of the water sustained by the disk in the flowing efflux stream is one half the weight of the cylinder that is above the disk, what we get, starting in Proposition 40, are the results of a series of free-fall experiments that are based on this theoretical model. The charge of reverse-engineering could arise at this juncture because Newton's choice of ½ is uniquely suited to allow these experiments to confirm the idea that the force on the rear of the moving body is negligible and thus to preclude the possibility of a non-resistive Cartesian aether. In other words, Newton chose a value that left unchallenged his account of the physical mechanisms responsible for resistance. This account includes not only his speculations about the propagation of pressure, but the division of fluids into physical kinds and the use of weight as a proxy for inertial resistance. Most importantly, it includes the additive, fundamental assumption that was discussed earlier. Without that assumption, Newton could not subtract away internal friction and viscosity from real fluids and thereby isolate inertial resistance.

---

[24] Smith explicitly considers this possibility in Smith 2005.

[25] In fact, in 1752, Jean d'Alembert showed that the drag force on a body moving with constant velocity in an incompressible and *inviscid* fluid is zero.

[26] The quoted inference may seem unbelievably terse, but Newton's justification of it and its premises is, in fact, unbelievably terse, no more than a few sentences.

Taking a closer look at the experimental results that Newton presented in Proposition 40 of Section 7 raises further questions about the testability of the principles that are foundational to Newton's treatment of resistance, and to his purported challenge to Descartes' Vortex Hypothesis. Prior to presenting these results, Newton had used his solution to the efflux problem and its extension to calculate a normalized value for the inertial resistance acting on a spherical body moving through an idealized continuous fluid. That is, in Proposition 38, he presented a way of quantifying the "resistance that arises from the inertia of matter of the fluid" in terms of the fluid's density, the sphere's weight and diameter, and their relative velocity. In Proposition 40, he compares this theoretical value with real-world measurements of resistance, which, he says, "can be investigated" by measuring a body's velocity and time of descent through a specified distance of fall. And thus, the announced task of the experiments that he presents is "[t]o find from phenomena the resistance of a sphere moving forward in a compressed, very fluid medium" (*ibid*, 749).

Newton reports that "I got a square wooden vessel, with an internal length and width of 9 inches (of a London foot), and a depth of 9 ½ feet, and I filled it with rainwater; and making balls of wax with lead inside, I noted the times of descent of the balls, the space of the descent being 112 inches" (*ibid*, 750–751). He provides the measurements he collected when conducting twelve separate experiments using the setup he describes, each of which involved balls of varying weight falling through water. The final two experiments that he details, Experiments 13 and 14, are different than the first twelve; in these cases, Newton reports what was observed when he dropped balls from the top of St. Paul's Cathedral in London and measured the rate of their fall through the air.

In many of these fourteen cases, Newton provides experimental data of the times of fall that very nearly matches the times that are predicted by his theory of resistance. Measuring the time of descent by the oscillations of a pendulum, he reports in Experiment 5, for instance, that the times of fall he observed for four balls were 28 ½, 29, 29 ½, and 30, and "By the theory they ought to have fallen in the time of very nearly 29 oscillations" (*ibid*, 754). However, in other cases there is a notable discrepancy between Newton's theoretical prediction and experimental results. For instance, when completing Experiment 3, "Three equal balls, each of which weighed 121 grains in air and 1 grain in water, were dropped successively in water" and allowed to fall 112 inches (*ibid*, 752). Newton notes that, "[a]ccording to the theory, these balls should have fallen in a time of roughly 40 seconds" (*ibid*, 752). But they didn't. They instead fell "in times of 46 seconds, 47 seconds, and 50 seconds" (*ibid*, 752). Newton lists several possible reasons for this discrepancy:

I am uncertain whether their falling more slowly is to be attributed to the smaller proportion of the resistance that arises from the force of inertia in slow motions to the resistance that arises from other causes, or rather to some little bubbles adhering to the ball, or to the rarefaction of the wax from the heat either of the weather or of the hand dropping the ball, or even to imperceptible errors in weighing the balls in water. (*ibid*, 752)

In light of these possibilities, the conclusion that Newton reaches is not that his theory should be adjusted but that the experimental setup should be refined to account for the interfering factors that he's listed. He claims, in particular, that "the weight of the ball in water ought to be more than 1 grain, so that the experiment may be made certain and trustworthy" (*ibid*, 752).

There are other cases in the Scholium to Section 7 where Newton reports significant discrepancies between theory and data but about which he makes no comment at all. And in some of these cases, the reported discrepancies appear to challenge the reliability of his theoretical account of inertial resistance. Notice, in particular, that the vertical-fall experiments were conducted using the real continuous fluids of water and air. In principle, the balls falling through these fluids will always encounter greater resistance, and move more slowly, than bodies moving through ideal continuous fluids, which have no internal or surface friction. Consequently, since Newton's theory of inertial resistance is a theory of motion in *ideal* continuous fluids, it should always be the case that the time of a ball's fall through water and air is greater than what the theory predicts. And yet, Newton provides experimental results indicating that the times of fall through real continuous fluids is *lower* than a body's fall through an ideal continuous fluid. In his report of Experiment 11, for instance, Newton states that three balls of equal weight "fell in the times of 43 ½, 44, 44 ½, 45, and 46 oscillations," and he simply ends the description of this experiment by remarking that "[b]y the theory they ought to have fallen in the time of roughly 46 5/9 oscillations" (*ibid*, 755). The lack of an explanation for the discrepancy here is noteworthy, because this is a case where the observed times of fall contradict the theoretical model. It should never be the case that the observed fall of bodies through real continuous fluids is swifter than the theoretical prediction for fall through ideal continuous fluids. Yet this is exactly what Newton reports—and he reports it without any attempt to account for the discrepancy by locating possible problems with his experimental setup, and thus, without any reason why this data should not count as an exception to his theory, and thus, should not count as adequate grounds for modifying or even rejecting it.

So, while Newton concludes the presentation of his experiments with the qualification that "*almost all* the resistance encountered by balls moving in air as well as in water is correctly shown by our theory" (*ibid*, 759; emphasis added), there are particular cases where the failure to establish a fit between theory and experiment raises questions about Newton's methodology. He presents experimental evidence that seems to undermine his theoretical approach to inertial resistance, and with no explanation for why the data in these cases might be unreliable, his theory is left vulnerable to being challenged by his own experimental results. Smith puts it this way:

> While most of the vertical-fall data lie close to his theory, some of the experimentally determined resistances were non-trivially larger than the theory implied, and some were *less*. Newton noted the cases where the experimental resistances were significantly larger, calling attention to the high velocities in these cases and arguing that when the velocity becomes high enough, there is a loss of fluid pressure on the rear face of the moving body, contrary to the assumption made in deriving his theory. He says nothing, however,

about the cases in which the experimental resistances fell non-trivially below his theoretical value ... In putting forward his theory, Newton remarked that it is supposed to give the least resistance that can occur insofar as the viscous and surface-friction effects will augment the resistance from fluid inertia. Unless the experimental resistances falling below his theory are attributed to experimental error, therefore, they cannot help but raise questions about whether the theory is correct. (Smith 2005, 144–145; emphasis in the original)

The experimental results also cannot help but raise questions about the Cartesian character of Newton's methodology. It is the fundamental, additive assumption with which we began that entails that real resistance will always be greater than ideal, merely inertial resistance. Therefore, it is this assumption that is directly challenged by the experimental results that Newton presents. That Newton does not reconsider its merits, even after producing legitimate empirical grounds for doing so, appears to leave the fundamental assumption on the same footing as a vortex model of the heavens, and leaves Newton appearing as guilty as Descartes of postulating a general explanatory device that is (somewhat) consonant with the phenomena but is by no means tested by them.[27]

8.4   Working Hypotheses and Newton's Logic of Theory-Mediation

Smith applies the same strategy for defusing the charge of Cartesianism made about the first edition of Book 2 to the revisions found in the later editions of the *Principia*. In brief, he urges a reading according to which the questionable proposals that Newton makes in Book 2, Section 7 are not proposals that are meant to be directly verified by experimental evidence, and also are not proposals that are forwarded as final pronouncements that gain support from the evidence available. According to Smith, these proposals should be understood as serving a different function. Broadly speaking, they convince us of the acceptability of the theory insofar as they demonstrate the fruitfulness of the theory. Furthermore, they reveal how the theory can be extended and they serve as the basis for further evaluation of the theory itself.

To better understand the nuance of Smith's suggestion, we can return to Newton's choice of the precise ½ value in the case of the stationary disk of Corollaries 7 through 10 of Problem 36. On Smith's account, Newton did not choose this specific value simply because his theoretical assumption about the action on the rear of a body dictated it. Relatedly, he did not choose ½ in order to block a Cartesian counter-argument to his claims about the detectable effects of inertial resistance on the

---

[27] Or, as Smith puts the point: "Newton's vertical-fall data for water and air provide no real evidence for his theory of resistance in inviscid fluids. To whatever extent Newton's presentation gives an impression that these data do constitute evidence for the theory, that presentation is misleading. The only element of good science here appears to be the vertical-fall data themselves" (Smith 2005, 145).

motion of spheres through continuous fluids.[28] Rather, there are other, interrelated considerations at play here. Newton is showing that there is a value for the weight of the column of water above the stationary disk that: [1] is consistent with his theoretical considerations (it is a value that falls between the upper and lower bounds he's calculated); [2] is consistent with the general theoretical assumption that the action on the rear of the body is negligible; but moreover, [3] allows him to extend his theory in such a way that he can isolate and measure the inertial resistance acting on the front face of the body. And in addition, the value he proposes [4] gives him the means of investigating why there might be discrepancies between what the theory predicts and what obtains in the real world. These factors, taken together, make Newton's chosen value of ½ a "working hypothesis," a proposal radically different in kind from a Cartesian hypothesis. As Smith puts it:

> Newton's choice of ½ is best regarded as a *working hypothesis*, a first approximation on which further research can be predicated. The choice of ½ need not then have been mere wishful thinking. Evidence would have accrued to it to the extent that the further research predicated on it would have succeeded in yielding stable results on the magnitude and variation of non-inertial mechanisms of resistance. Failure of such research to yield stable results would have given grounds for modifying or abandoning the ½ number. This strategy might well have seemed the best hope not only for investigating the non-inertial mechanisms, but also for providing a basis for refining the ½ number. (Smith 2000, 124)

The ½ choice is thus not "deduced from the phenomena" insofar as there was insufficient empirical evidence to infer that the weight of the column of water is half the weight of the cylinder. Newton's hand is in no way forced into endorsing the ½ value. But the choice is justified nevertheless for two types of reasons. First, it is consistent in various ways with Newton's other theoretical commitments. Second, it allows Newton to obtain a precise value for the theoretically expected inertial resistance of a fluid, and thus provides a baseline against which to compare experimental results. In other words, specifying a precise value for the column of water allows for the construction of further 'down-stream' evidence. The theoretical backing of Newton's theory of inertial resistance does not stand as the sole reason for accepting it as a possibility worth testing. It is *a* reason, of course. But on Smith's reading, working hypotheses are possibilities worth accepting because they have both theoretical backing and also because of the fruits they bear for further, more sustained research.

The choice of ½ also show that the vertical-fall experiments cannot be fully understood apart from this promise of further research. The experiments are not offered to confirm a theoretical model in a straightforward hypothetico-deductive way. That is, the results Newton reports aren't meant to convince us that the theory should be accepted because what the theory predicts obtains in real-world situations. Instead, the results are presented in the context of the theory—or better, in conversation with the theory—such that they create resources for further research. More precisely, Newton is showing us that when the theory is applied to the real-

---

[28] Smith explicitly considers these options in Smith 2000, 122–123.

world situations he describes, the theory can give rise to the sort of evidence that allows for further extension of the theory. It does so by making deviations from the baseline the subject of further research. On this point, Smith emphasizes Newton's claim (made after concluding the presentation of the theory and turning to the vertical-fall data) that "This is the resistance that arises from the inertia of matter of the fluid. And that which arises from the elasticity, tenacity, and friction of its parts can be investigated as follows" (Newton 1999, 749). On Smith's reading, this remark "does not imply that the data are to be taken as evidence for the theory" (Smith 2005, 145). It implies that the data have provided Newton adequate support to begin the project of extending his theory of resistance.

Two intertwined features of this account are particularly important to us. First is the constitutive role of working hypotheses in individuating new phenomena and, by so doing, rendering them theoretically meaningful. The ability of working hypotheses to initiate novel research lies in this power to individuate. This is a point worth putting in its full generality: Departures from baseline theoretical values cannot be understood as distinct phenomena outside the context of a relevant theory. By 'phenomena', we mean a description of a real or idealized physical system or process that individuates it from the systems or processes in which it is embedded. Once properly individuated, such phenomena can be used as evidence, or explananda, or be invoked in other explanations. Take, for example, the retardation effects due to the inertial component of fluid resistance in fall. These 'phenomena' are parasitic on, first of all, our understanding of free fall in a vacuum (call the latter the 'primary' phenomena). Putting a value on this retardation (the secondary phenomena) also depends on the fundamental additive assumption and a few lower-level parameters (like the ½ value). These entail expected times of fall, which we can compare both to the expected times of fall in a vacuum and to real-world observations. These comparisons allow us to gauge if our assumption about the retardation are correct. If real-world observations are close to our expectations but depart from them in a systematic way, we can conceive of those departures as tertiary phenomenon and seek to quantify and explain them in their own right. And if these explanations yield relatively good predictions, but still depart from real-world observations in a systematic way, we can continue to forth-order, fifth-order, and higher-order phenomena. But nothing in the real-world observations of falling bodies corresponds to how much a body slows down due to inertial fluid resistance or how much its slow-down departs from our expectations. Carving real observations into theoretically meaningful components (i.e., constituting new phenomena) is only possible when some enabling theoretical assumptions—the working hypotheses—are made. The power of this iterated approach is this: a successful explanation of some n^{th} phenomena relies on and must conform to all the explanations that came before it. Consequently, explanations of high-order phenomena provide repeated checks on explanations of low-order phenomena. They also provide checks on the way in which all lower-level explananda—the phenomena in question—were constituted. The last point is worth emphasizing: the success of a chain of explanations provides compelling evidence that we have decomposed phenomena in more-or-less the right way, at least for the purposes of the investigation at hand. We might say that it provides evidence that we have carved nature at more-or-less the right joints. And because constituting the series of phenomena relies on the working hypotheses, the success of a chain of explanations thus provides compelling evidence for accepting the working hypotheses as fundamental inferential posits.

We will call the sort of evidence that higher-order phenomena provide for the working hypotheses 'conditional' evidence. This sort of evidence is based on phenomena that would not be constituted as phenomena without assuming the hypothesis that the evidence is meant to support. The value of this sort of evidence should be clear from our discussion thus far. However, we would also like to call attention to the prospect of a complement that Smith only briefly considers. We will call it *independent* or non-conditional evidence. It is independent in the following sense: it serves as evidence for some hypothesis, but on the basis of phenomena that can be constituted independently of the hypothesis in question. Smith notes that evidential reasoning before Newton was based almost entirely on evidence of this kind. He writes that "the quantities central to the mathematical theories of motion under uniform gravity laid out by Galileo and Huygens were all open to measurement without having to presuppose any propositions of the theories themselves (Smith 2016, 193). Consequently, any mismatch between actual measurements and theoretical predictions provided no specific information about what theoretical propositions required revision (as discussed in §8.2). Evidence of this sort plays no role in the portions of Book 2 we've been considering. However, things look different in the case of the laws of motion, a set of claims that Smith also identifies as working hypotheses (Smith 2001b, 335). Considering these will allow us to discuss non-conditional evidence in more detail.

8.5   Mathematical Representation in Newton's Logic of Theory-Mediation

Support for the laws of motion exemplifies the two notions of evidence we have sketched above. The case for conditional evidence is familiar to students of Smith's work. The laws allow deviations from straight-line motion to be identified *as phenomena*, and thus to become evidentially meaningful. The deviations provide evidence concerning forces and their sources, and this evidence is used in conjunction with the laws to generate further, higher-order deviations and higher-order evidence. When the evidence yields a coherent account of forces and sources, the laws gain empirical support and the process can continue. Yet the laws are merely "working hypotheses" in the sense that the world can be such that this process ultimately fails: the process can yield incoherent information, or we may be unable to find appropriate forces and sources to explain some phenomena. When this happens, the laws themselves need to be reconsidered. For our purposes, what's important in this picture is that the laws themselves allow us to constitute the phenomena that serve as evidence for them.

As a matter of historical development, the laws gained empirical support of this sort through a process that extended far beyond the *Principia* (Smith 2014). However, Newton also drew attention to the conditional nature of the evidence for the first and second laws as the initial, prima facie reason for accepting them. In the scholium that follows the laws, he writes that "By means of the first two laws and the first two corollaries Galileo found that the descent of heavy bodies is in the squared ratio of the time" (*ibid*, 424). That is, he suggests that by *assuming* the laws, Galileo was able to generate results that were recognized in Newton's time to be true. Newton suggests the same for the laws of collision of the famous Royal Society competition of 1666-8: "*From the same laws* and corollaries and law 3, Sir Christopher Wren, Dr. John Wallis, and Mr. Christiaan Huygens … found the rules of the collision and reflections of hard bodies" (*ibid*, 424; emphasis added). According to Newton, the three mathematicians did not reason *to* the laws of motion. Rather, they started with the laws of motion and derived from them generally accepted results. In other words, they generated conditional evidence in the laws' favor. Newton presents this evidence as the presumptive reason we should accept the laws.[29]

But Newton also suggests there was independent (non-conditional) evidence for the third law. He writes that by means of a 10-foot pendulum he found "within an error of less than three inches in the measurements—that when the bodies met each other directly, the changes of motions made in the bodies in opposite directions were equal, and consequently that the actions and reaction were always equal" (*ibid*, 426). He provides a calculational method for conducting these experiments with bodies of varying elasticity (*ibid*, 425) and reports studying the collisions of bodies of wool, cork, and steel (*ibid*, 427). What's notable here is that Newton doesn't assume the law in order to describe the phenomena under study, and then use that phenomena to generate evidence in favor of the law. That is, the phenomena he uses as evidence are not constituted by the third law itself. Of course, many other assumptions are needed to make pendular collisions serve as evidence for the third law (assumptions about elasticity, the center of mass of the pendulum bobs, the first law, etc.). But the third law is not required. The situation is the same in Newton's defense of the third law for attractions. Newton argues that in a body whose parts do not attract one another

---

[29] Newton offers a reconstructed history to suit his needs, not one we should take as fact. First, he was in no position to comment on Galileo's processes of discovery. He knew little of Galileo (Cohen 1992). Second, Newton would have known that the laws of Wren, Wallis, and Huygens were based on a variety of principles. Those principles can be recovered within the framework of Newtonian laws, but they are not identical to them. Why Newton didn't make this explicit is a complicated question (see Biener and Schliesser 2017). For our purposes, it is only important to notice that he portrayed the evidence generated by Galileo, Wren, Wallis, and Huygens as conditional. For more on the Royal Society competition, see Jalobeanu 2011. For further discussion of mathematical certainty in the reasoning that Newton attributes to Galileo, Wren, Wallis, and Huygens, see Domski 2018.

equally and oppositely (i.e., where one part exerts an unbalanced force on the other), motion would be generated, and thus the system would move "indefinitely with a motion that is always accelerated" (*ibid*, 427–428). This, he claims, "is absurd." The absurdity is derived from unstated notions about infinitely accelerated motion, but does not derive from a provisional acceptance of the third law itself.[30] We do not need the third law to constitute the (hypothetical) phenomena that Newton judges absurd. Newton further adds that indefinitely accelerated motion is "contrary to the first law." In that sense, collisions and attraction both provide *non-conditional* evidence for the third law, but that evidence *is* conditional on the first law. In general, evidence may be non-conditional in relation to some propositions, but conditional in relation to others. Newton provides further examples from the study of simple machines, much to the same effect.

At least in this highly delimited context, it might seem that the non-conditional evidence for the third law puts us in a Huygensian/Galilean evidential predicament. To reiterate, in this predicament empirical evidence is collected without assuming the theoretical propositions being tested. Thus, the evidence can confirm/disconfirm those propositions, but cannot provide more specific information about how they might be revised in the face of disconfirming evidence. Is the preliminary evidence for the third law like this? We don't think so, but in order to see this we must look beyond Newton's laws to more fundamental mediating theoretical assumptions. To this end, Newton's defense of the third law—specifically, his comparison of the rebound motions in two-pendulum collisions—is instructive. The core of Newton's claim is that, in pendular collisions, "when the bodies [the pendulum bobs] met each other directly, the changes of motions made in the bodies in opposite directions were equal, and consequently that the actions and reaction were always equal" (*ibid*, 426). Newton establishes this by showing that "the quantity of motion— determined by adding the motions in the same direction and subtracting the motions in the opposite directions—was never changed." The process requires a host of assumptions. Some assumptions are quite specific: e.g., that the distance between a bob's center of oscillation and its center of mass is negligible. Other assumptions are more general: e.g., that air resistance retards motion but small corrections can be introduced. We'd like to draw attention to further assumptions that are more general still and that bear on the evidence for the third law from other experimental contexts (say, evidence regarding simple machines): for example, that velocities, directions, and some measure of 'bulk' are the relevant mathematical properties for the study of collision, that they can be assigned coherently to bodies, and that quantities of motion calculated by means of them can be combined vectorially according to the parallelogram rule—the latter in order to combine and compare motions along one direction with motions along another.

Of course, these assumptions were broadly accepted by Newton's time. Yet this does not make them trivial. Identifying the relevant mathematical properties for

---

[30] Newton's thought experiment is actually more complicated, but the gist is the same. See Newton 1999, 427–428.

the study of motion was a hard-fought achievement of pre-Newtonian mechanics. Similarly, measuring these properties raised a host of conceptual issues that natural philosophers grappled with. For instance, scholars have long recognized that Newton, as part of his argument against the Cartesian definition of motion, explicitly confronted the problem of identifying locations in space and durations in time in a mathematically consistent way.[31] Debates also surrounded the parallelogram rule.[32] First and foremost was the issue of which motive quantities or features of motion the rule governs. Second, using the rule requires that the relevant quantities are represented as proportionately sized line segments in a diagrammatic space. Whether the choice of representation makes physical sense—viz, whether quantities and the features of motions they define combine in real space as they do in this space—is an empirical matter. But investigating the matter requires some choice of mathematical representation, which partially or completely determines what constitutes the phenomena of motion.[33] Newton's enunciation of the parallelogram law makes it clear that, for him, "magnitude and direction are jointly constitutive of a 'motion'" (Miller 2017, 188).[34] This conception is also implicit in the Laws of Motion and it dictates what "motion" means in the *Principia*. We'd like to suggest that such choices about mathematical representation should also be understood as working hypotheses.

We can see this more clearly by considering a working hypothesis of this kind, but in an ultimately failed context, where a philosopher's understanding of motion and its mathematical representation come apart. The later Descartes' stated position is that the magnitude of a body's motion (which is proportional to its speed) and the direction of its motion are essentially separate: a change in the magnitude of motion ought to have no effect on its direction, and conversely.[35] Descartes treats the measure of each as an independent quantity (anachronistically, a scalar), most famously in his rules of collision. However, in the *Optics*, Descartes treats motion differently. There, he derives the sine law of refraction by analogizing the behavior of light to the motion of a tennis ball. His diagram is (Descartes 2001, 92). Descartes imagines a moving ball changing its speed as it reaches a barrier line (akin to light passing from, say, air to a vat of water). The motion of the ball is represented as a line inclined to, and terminating in, the barrier line. Descartes' task is to determine how to continue the line on the other side of the barrier (the 'refracted' side).

---

[31] See DiSalle 2002 for Newton's position and Reid 2012 for the larger context.

[32] The following is indebted to Miller 2017.

[33] We will shortly discuss a case where the definition implicit in the choice of mathematical representation does not match a philosopher's stated definition.

[34] The parallelogram rule is in the first corollary to the Laws (Newton 1999, 417).

[35] We are simplifying a bit. For Descartes, the direction of motion is called a 'determination', and there is considerable debate about how it relates to the force of motion (see McLaughlin 2000). Nevertheless, he writes that "The first part of this law is proved by the fact that there is a difference between motion considered in itself, and its determination in some direction; this difference makes it possible for the determination to be changed while the quantity of motion remains intact" (*Principles*, Book II, Art. 41; Descartes 1984, p. 62).

Descartes' procedure relies on decomposing the line representing motion by means of the parallelogram rule into two orthogonal components, one parallel and one perpendicular to the barrier. He claims that motion along the parallel component will be unaffected by the barrier, and so halving the speed (as the problem stipulated) will only alter the vertical component. But because motion is changed along the vertical component without any corresponding change in the horizontal component, the line representing motion on the 'refracted' side of the barrier must have a different inclination to the barrier than the original line. This inclination is the resulting angle of refraction.

The details of this construction are interesting, but ultimately inessential for us. This is because by Descartes' own lights, a change in a body's speed should have no effect on its direction. Whatever its details, the construction is clearly in conflict with his stated position. Why? It seems that Descartes' choice of mathematical representation leads him down a garden path. Instead of treating the magnitude and direction as two independent quantities (as is his stated position), he represents them jointly as the direction and length of a line segment. This representation allows him to apply a mathematical operation—the vectorial decomposition of the line segment by the parallelogram rule—that in effect ties together the magnitude of motion with its direction. If we follow Descartes' stated position, the magnitude of a body's motion and its direction should be distinct phenomena, requiring distinct explanations and able to be invoked separately in explanations. Much of the time, this is how Descartes approaches them: he has two conservation laws (one to explain change in magnitude and one to explain changes in direction)[36] and he uses speed and direction separately to explain features of his laws of collision.[37] We can think of the independence of magnitude and direction as itself a "working hypothesis": it provides fundamental inferential posits regarding motion and tells us how to individuate the phenomena under study. But when Descartes uses the parallelogram rule to decompose a line segment that represents a directed motion, he puts in place an implicit, but conflicting working hypothesis—one that treats motion vectorially. This way of treating motion provides a different set of fundamental inferential posits, and so constitutes and individuates phenomena differently. For example, two independent conservation laws should no longer be needed to explain motion, because motion is one thing: a magnitude-with-direction.

For Newton, there is no schism between stated position and mathematical representation. The Laws treat motion vectorially and motion is represented vectorially in the work's first diagram—the parallelogram rule—and in the remainder of the work.[38] But like Descartes, Newton could have chosen a wrong representational vehicle, one that would have entailed a different description of the phenomena of

---

[36] *Principles*, Book II, Art. 37, 39 (Descartes 1984, 59–60).

[37] See letter to Clerselier, 17 February 1945 (Adam and Tannery 1974–1986, IV, 183–188).

[38] To be more precise, Newton's purpose in his exposition of the parallelogram rule is to show how to compose forces. But the construction does so by treating the effects of forces, i.e., motions (Newton 1999, 417).

motion. That his choice of representation fits with his stated commitments does not change its logical status. It is a working hypothesis in itself, one that could be accepted without the first two laws (as it was accepted by Newton's predecessors).[39]

And so, finally, we can point to the theoretical commitments that do mediate what we earlier presented as the non-conditional evidence for the third law. What that evidence relies on, and provides a check on, is the idea that motions can be treated vectorially—that "the quantity of motion [can be] determined by adding the motions in the same direction and subtracting the motions in the opposite direction" (426). This might seem far from eye-opening in the Newtonian context, but it is also far from trivial—it is one of the most basic assumptions of the *Principia*. That it was widely accepted also does not change its logical status. It is a basic posit about how the phenomena of motion were to be described and individuated, and subsequent successful treatments of motion provided accruing evidence that it was, indeed, correct. Importantly, like other working hypotheses, it also ultimately turned out to be of limited scope: finite rotations, although they can be represented by a magnitude (the angle of rotation) and a direction (the axis of rotation) combine vectorially, but with a sharp limitation: they add, but not commutatively.[40]

We can clarify the point by using another of Smith's contrasts between the *Principia*'s methodology and the idea that the *Principia* as a whole ought to be evaluated on the CHD model:

> Newton's laws of motion made generic claims about the relation between unbalanced static forces and motions. In reaching so far beyond the available evidence, I submit, they acquired the status of working hypotheses. The role they play in the *Principia* is one of enabling phenomena of planetary motion (and, in Book 2, phenomena of motion in resisting media) to become evidence pertaining to physical forces . . . Taking the laws of motion to have the status of working hypotheses [can suggest] that Newton's theory can be viewed as "a single extraordinarily complex hypothesis." But this surely is not the best way of viewing the theory, for it radically obscures the logic of the evidential reasoning . . . The laws of motion have a very different logical status . . . than the law of gravity. (Smith 2001b, 335–336)

We might say that, in a Quinean image of knowledge, the laws occupy a central position. What we are trying to suggest is that perhaps even more centrally are choices about what to represent in mathematical formalism and how to represent it.

8.6 Conclusion

We began this essay by contrasting the "Newtonian Style" with the Cartesian, hypothetico-deductive method (CHD). Following Smith, we stressed (in Sect. 8.2) that the Newtonian style relies on theory-mediation to turn otherwise inchoate phenomena into well-behaved *evidence*, that is, to render phenomena theoreti-

---

[39] See Miller 2017.

[40] We thank the editors of this volume for this point.

cally meaningful. We also emphasized (in Sect. 8.3) that theory-mediation—often through the introduction of working hypotheses—is required to turn real-world observations into well-defined *phenomena*. Following Smith again, we used the fundamental assumption of Newton's theory of fluid resistance as our primary example. But we've also extended his framework (in Sect. 8.4) by introducing the complementary notions of conditional and independent evidence, and we did so to distinguish between evidence for a working hypothesis that is based on phenomena that are constituted by the working hypothesis itself, and evidence for a working hypothesis that is based on phenomena that are independently constituted. We argued that Newton provides preliminary evidence for the third law that is independent of that law. Because that evidence is not mediated in the same way as conditional evidence, we asked whether it must play the restricted role evidence plays on the CHD method. We concluded (in Sect. 8.5) by arguing that it need not, but that in order to see how it is theoretically mediated we must reach beyond the laws of motion to fundamental choices about how to represent motion mathematically.

References

Adam, Ch., and P. Tannery. 1974–1986. *Oeuvres de Descartes*. 2nd ed. Paris: Vrin.

Biener, Z. 2016. Newton and the ideal of exegetical success. *Studies in History and Philosophy of Science, Part A* 60: 82–87.

Biener, Z., and E. Schliesser. 2017. The certainty, modality, and grounding of Newton's laws. *The Monist* 100: 311–325.

Cohen, I.B. 1992. Galileo and Newton. In *Atti delle celebrazioni Galileiane* (1592–1991), 181–208. Trieste: Lint.

———. 1999. *A guide to Newton's* principia. In *Newton, the principia*, 1–370. Berkeley: University of California Press.

Descartes, R. 1984. *Principles of philosophy*. Trans., with explanatory notes by V.R. Miller and R.P. Miller. Dordrecht: D. Reidel.

———. 2001. *Discourse on method, optics, geometry, and meteorology*. Trans. P. Olscamp. Indianapolis: Hackett.

DiSalle, R. 2002. Newton's philosophical analysis of space and time. In *The Cambridge companion to Newton*, ed. I.B. Cohen and G.E. Smith, 1st ed., 33–56. Cambridge University Press.

Domski, M. 2018. Laws of nature and the divine order of things: Descartes and Newton on truth in natural philosophy. In *Laws of nature*, ed. W. Ott and L. Patton, 42–61. Oxford University Press.

———. 2019. Imagination, metaphysics, mathematics: Descartes's argument for the vortex hypothesis. *Synthese* 196: 3505–3526.

Ducheyne, S. 2012. *"The main business of natural philosophy": Isaac Newton's natural-philosophical methodology*. Dordrecht: Springer.

Harper, W. 2011. *Isaac Newton's scientific method: Turning data into evidence about gravity and cosmology*. Oxford University Press.

Jalobeanu, D. 2011. The Cartesians of the Royal Society: The debate over collisions and the nature of body (1668–1670). In *Vanishing matter and the laws of motion: Descartes and beyond*, ed. D. Jalobeanu and P. Anstey, 103–129. New York: Routledge.

Maffioli, C. 1994. *Out of Galileo: The science of waters, 1628–1718*. Rotterdam: Erasmus.

McLaughlin, P. 2000. Force, determination and impact. In *Descartes' natural philosophy*, ed. S. Gaukroger, J. Schuster, and J. Sutton, 81–112. New York: Routledge.

McMullin, E. 2008. Explanation as confirmation in Descartes's natural philosophy. In *A companion to Descartes*, ed. J. Broughton and J. Carriero, 84–102. Malden: Blackwell.

Miller, D.M. 2017. The parallelogram rule from pseudo-Aristotle to Newton. *Archive for History of Exact Sciences* 71: 157–191.

Newton, I. 1999. *The mathematical principles of natural philosophy*. Trans. and ed. I.B. Cohen, and A. Whitman. Berkeley: University of California Press.

Reid, J. 2012. *The metaphysics of Henry More*. Springer.

Shapiro, A. 2004. Newton's experimental philosophy. *Early Science and Medicine* 9: 185–217.

Smeenk, Ch., and G.E. Smith. ms. *Newton on constrained motion*.

Smith, G.E. 2000. Fluid resistance: Why did Newton change his mind? In *The foundations of Newtonian scholarship*, ed. R. Dalitz and M. Nauenberg, 105–136. Singapore: World Scientific.

———. 2001a. The Newtonian style in Book II of the *Principia*. In *Isaac Newton's natural philosophy*, ed. J. Buchwald and I.B. Cohen, 249–313. Cambridge: MIT Press.

———. 2001b. Comments on Ernan McMullin's 'The impact of Newton's *Principia* on the philosophy of science'. *Philosophy of Science* 68: 327–338.

———. 2002a. The methodology of the *Principia*. In *The Cambridge companion to Newton*, ed. I.B. Cohen and G.E. Smith, 1st ed., 138–173. Cambridge University Press.

———. 2002b. From the phenomenon of the ellipse to an inverse-square force: Why not? In *Reading natural philosophy: Essays in the history and philosophy of science and mathematics*, ed. D. Malament, 31–70. Peru: Open Court.

———. 2005. Was wrong Newton bad Newton? In *Wrong for the right reasons*, ed. J. Buchwald and A. Franklin, 127–160. Springer.

———. 2014. Closing the loop: Testing Newtonian gravity, then and now. In *Newton and empiricism*, ed. Z. Biener and E. Schliesser, 262–351. Oxford University Press.

———. 2016. The methodology of the *Principia*. In *The Cambridge companion to Newton*, ed. R. Iliffe and G.E. Smith, 2nd ed., 187–228. Cambridge University Press.

Truesdell, C. 1970. Reactions of Late-Baroque mechanics to success, conjecture, error, and failure in Newton's *Principia*. In *The Annus Mirabilis of Sir Isaac Newton, 1666–1966*, ed. R. Palter, 192–232. Cambridge: MIT Press.

Westfall, R.S. 1973. Newton and the fudge factor. *Science* 179: 751–758.

———. 1980. *Never at rest: A biography of Isaac Newton*. Cambridge University Press.

Chapter 9
Newton's *Principia* and Philosophical Mechanics

Katherine Brading

9.1 Introduction

Newton's *Principia* reconceptualizes rational mechanics and physics, and offers a novel unification of these heretofore distinct disciplines. In this paper, I argue for a reading of the *Principia* that insists on a strict distinction between the rational mechanics (in Books 1 and 2) and the physics (in Book 3), in which the Definitions and the Axioms/Laws play a surprising dual role that both distinguishes the rational mechanics from the physics and unifies them into a single project: a philosophical mechanics.

"Philosophical mechanics," as Marius Stan and I use that term, applies to projects that seek to combine rational mechanics with physics, as those enterprises were then understood. Looking back from our present-day vantage point, it can be difficult to appreciate the gulf that existed between these two, and the difficulties that were involved in bringing them together into a single project. My goal in this paper is to make vivid just how radical a reconceptualization of mechanics and physics—and of the relationship between them—Newton's *Principia* achieves.

The value of this reading depends upon its utility in resolving interpretational puzzles and in bringing to light features of the *Principia* that otherwise lie hidden. These include the dual status of the Definitions and Axioms/Laws, the relationship of the mathematical to the physical, and the innovative reconceptualizations of mechanics and physics that make the *Principia* possible.

More generally, the value lies in the framework offered for re-thinking physics, mechanics, and natural philosophy in the eighteenth century. How we read Newton's *Principia* affects how we understand its place in the wider philosophical landscape

K. Brading (✉)
Department of Philosophy, Duke University, Durham, NC, USA
e-mail: katherine.brading@duke.edu

© The Author(s), under exclusive license to Springer Nature Switzerland AG 2023
M. Stan, C. Smeenk (eds.), *Theory, Evidence, Data: Themes from George E. Smith*, Boston Studies in the Philosophy and History of Science 343, https://doi.org/10.1007/978-3-031-41041-3_9

163

of the eighteenth century. At the beginning of the century, physics was a dependent subfield of philosophy, practiced by philosophers, whereas theoretical or rational mechanics fell under the authority of mathematicians (see Sect. 9.2). By the end of the century, these disciplinary boundaries and domains of authority had been redrawn.[1] Stan and I argue that these transformations reflect an extended effort throughout the eighteenth century to provide a satisfactory philosophical mechanics of the material world, the failures and successes of which embody the central metaphysical, epistemological, and methodological lessons of natural philosophy in the age of reason.[2] By showing that Newton's *Principia* is powerfully understood within the framework of philosophical mechanics, this paper offers a first step towards re-evaluating physics and mechanics in the eighteenth century.

The main body of this paper is taken up in explicating and arguing for my interpretation of Newton's *Principia* as a text in philosophical mechanics. In Sect. 9.2, I set out my view. I describe the rational mechanics (9.2.1), the physics (9.2.2), and then the philosophical mechanics (9.2.3) of the *Principia*. In Sect. 9.3, I remove oversimplifications, discuss further considerations arising from the secondary literature, and address some puzzles present in the existing literature that my view solves.

9.2 Newton's *Principia*: A Project in Philosophical Mechanics

My view is this. Newton's *Principia* combines rational mechanics (Books 1 and 2) with physics (Book 3). The *Definitions*, and the *Axioms, or Laws of Motion*, come prior to Books 1–3, and therefore pertain to both the rational mechanics and the physics. The result is a novel integration of rational mechanics and physics into a single project, a project in philosophical mechanics.

It can be hard to see why there should be anything of interest here, especially for any present-day reader who, turning to the *Principia*, sees simply a book in physics; even more so, for anyone who uses the terms "classical mechanics" and "classical physics" interchangeably. But this reaction rests on a conceptualization of the relationship between mechanics and physics that happened after Newton. So we first need to familiarize ourselves with the terms as Newton understood them, and as they were understood more widely at the time.

---

[1] Why, and how, did all this happen? And with what philosophical significance? One thing we know for sure is it is not, as older narratives suggest, that the scientific revolution happened in the seventeenth century, culminating in Newton's *Principia*, after which physics was independent of philosophy so that the eighteenth century was a period of working out the details and solving problems within "Newtonian mechanics" or "Newtonian physics." See also Brading and Stan, 2021.

[2] We make our case in Brading and Stan, 2023.

9.2.1 *Rational Mechanics*

What is rational mechanics, according to Newton? In the Preface to the first edition of his *Principia*, published in 1687, Newton offered a taxonomy of mechanics in which he divided "universal mechanics" into three: practical mechanics, rational mechanics, and geometry. In so doing, he changed the domain of geometry and set out a bold new agenda for rational mechanics: the *exact* treatment of *any motions* under *any forces* whatsoever. To understand the epistemological status of rational mechanics, and its relationship to physics, we first need clarity about Newton's taxonomy of mechanics, which is as follows.[3]

Newton begins by noting that practical mechanics lacks the exactness pertaining to geometry. But this lack of exactness, he states, arises from imperfections in the mechanic rather than in the subject-matter of mechanics. A perfect mechanic, able to "work with the greatest exactness,"[4] would be able to produce perfect circles and straight lines. Moreover, the successful completion of this task is presupposed in geometry:

> To describe straight lines and to describe circles are problems, but not problems in geometry. Geometry postulates the solution of these problems from mechanics, and teaches the use of the problems thus solved. (Newton 1999, 382)

In other words, we presume the exact production[5] of the geometrical figures that we then use as the basis for developing the problems of geometry.[6] However, for Newton the means of production are not part of geometry; rather, the exact production of geometrical figures is the domain of what Newton calls "rational mechanics."

Rational mechanics is *exact* in the same sense that geometry is exact, but whereas geometry includes only *magnitude*, rational mechanics concerns also *motion*.[7] This is because rational mechanics concerns the production or generation through motion of the figures (curves, shapes, solids) whose properties are the subject-matter of geometry. Crucially, these motions arise from the application of forces, and there is no restriction to curves that are constructible by traditional methods. Rational mechanics, Newton says, is

> the science, expressed in exact propositions and demonstrations, of the motions that result from any forces whatever and of the forces that are required for any motions whatever. (Newton 1999, 382)

---

[3] See also Garrison 1987; Domski 2003; Guicciardini 2009; Smeenk 2016; and references therein.

[4] Newton 1999, 381.

[5] I use the terms "production" and "generation" here, rather than "construction," because there is no requirement that the curves by constructible by traditional means. See Domski 2003 and 9.3.1.

[6] See the similar discussion Newton offers in his treatise on geometry, as discussed and quoted in Guicciardini 2009, 300. Mechanics is the means by which the objects of geometry are produced, and geometry presupposes that these objects have been so exactly.

[7] Guicciardini 2009, 298.

We can sum up this conception as follows. On the one hand, like practical mechanics, but unlike geometry, rational mechanics considers motions and forces, and it considers any motions and forces whatsoever. On the other hand, like geometry and unlike practical mechanics, rational mechanics is exact.[8]

This taxonomy of universal mechanics differs from that of Newton's contemporaries in the relationships it describes between geometry and mechanics.[9] Moreover, this conception of universal mechanics alters the domains of both geometry and rational mechanics. Newton *extended* the subject-matter of geometry beyond that admitted by Descartes and others, by including curves produced by any motions resulting from any forces.[10] He *restricted* the domain of geometry by placing the problems of generation in the domain of mechanics. And he *extended* the domain of mechanics beyond the mechanical powers associated with the five machines (the lever, pulley, winch, wedge and screw of Archimedes) to incorporate *natural* forces:

> But since we are concerned with natural philosophy rather than the manual arts, and we are writing about natural rather than manual powers, we concentrate on aspects of gravity, levity, elastic forces, resistance of fluids, and forces of this sort... (Newton 1999, 382)

This incorporation of natural forces and motions within the domain of rational mechanics is, for our purposes, the most important philosophical move that Newton makes in setting out his new taxonomy. For it is this that connects rational mechanics to physics, in a specific way that we discuss below.

Books 1 and 2 of the *Principia* are books in rational mechanics in the above sense: they provide an exact treatment of the generation of curves by means of forces. In them, Newton is attempting unprecedented generality: no matter what motions we consider when we turn our attention to the physical world, the mathematics to treat those motions in terms of forces is to be found in rational mechanics.

However, as books in rational mechanics they are incomplete: they lack any definitions and axioms from which the demonstrations in Books 1 and 2 are to proceed. For these, we turn outside Books 1 and 2, for it is prior to these books that Newton gives us his *Definitions* and *Axioms, or Laws of Motion*. In my view,

---

[8] In Cohen and Whitman's translation of the *Principia*, Cohen writes in a footnote (Newton 1999, 381) that "Newton's comparison and contrast between the subject of theoretical or rational mechanics and practical mechanics was a common one at the time of the *Principia*", citing a later reference (from 1704) that in turn appealed to the authority of John Wallis. However, there is a great deal more going on in this Preface than that distinction. For example, Guicciardini (2009, 294) *contrasts* Newton's view of the relationship between geometry and mechanics with that of Wallis, writing that Wallis defined mechanics as an *application* of geometry to the science of motion. As we have seen, this was not Newton's conception.

[9] See Domski 2003 and Guicciardini 2009.

[10] Guicciardini (2009) emphasizes that Newton's philosophy of geometry extends geometry beyond Descartes's account. Crucially, Descartes excluded "mechanical curves from the realm of the exactitude and certainty of geometry" (*ibid*, 299). For Newton, what matters for the purposes of geometry is that the resulting objects are exact, not the means by which they are constructed (*ibid*, 301).

with respect to Books 1 and 2, these definitions and axioms/laws are to be read as principles of rational mechanics. The reason that they are placed outside Books 1 and 2 is that they pertain to all three books of the *Principia*, as we will see in more detail below.

By opening his work with a set of definitions followed by a set of axioms, Newton structured his rational mechanics analogously to standard presentations of Euclidean geometry from the period. In so doing, he indicated the epistemic status to be accorded to the definitions and to the axioms, the standards that demonstrations in rational mechanics are expected to meet, and the appropriate criteria of justification for the claims made by rational mechanics. I will return to the relationship between the definitions and the axioms, and to their epistemic status, below. For now, I wish to highlight a different issue: the *character* of the mathematical challenge faced by Newton, to which his rational mechanics is a solution. De Gandt (1995) puts the issue beautifully in his discussion of Wren, Hooke, Halley, Newton and the search for a proof that the celestial motions are derivable from an inverse square force law. He writes:

> Wren required truly deductive demonstrations, and what Hooke provided did not convince him. But what would a demonstration in these matters be like? What examples could be used? ... In geometry, the criteria of proof were well established by virtue of a culture nourished by the books of the ancients and cultivated in discussions, courses, challenge contests, and discoveries. But what would constitute proof when it was a question of forces and motions? What indubitable principles could be adopted as a foundation? What mathematical tools should be used? (De Gandt 1995, 6)

Geometry, with its long history of development, had achieved a stability of method: a wide range of problems were solved and soluble by means of the same techniques, meeting the same standards of demonstration.[11] As yet, there was no science that connected forces and motions such that a variety of different problems were soluble under the umbrella of a single set of principles, methods, and standards.[12] In manuscripts from the decades before the *Principia*, we see Newton wrestling with this challenge as he sought to solve problems of increasing generality,[13] and in the rational mechanics of the book he offered his developed response.

We now have on the table the main elements of Newton's conception of rational mechanics. Thus far, I have used the terms "mechanics" and "rational mechanics"

---

[11] For example, Newton goes to great lengths to use (as far as possible) long-standing geometrical methods for his demonstrations in the *Principia*, notwithstanding being at the forefront of developments in mathematics. See also Landry 2023, who argues that, since the time of Plato, it is precision of definitions and stability of method that underwrites the practice of mathematics.

[12] As a simple illustration of this point, note that geometry considers magnitudes, but not the *directed* magnitudes required by problems involving forces and motions. So new rules of how to combine (i.e. to add, subtract, and so forth) directed magnitudes had to be introduced and justified.

[13] This account of Newton's work prior to the *Principia*, in which we see him striving to solve problems of increasing generality and through this process developing the conceptual innovations and resources that we find in the *Principia*, is due to Solomon 2017.

as if they are unproblematic terms. However, Newton's *Principia* played a crucial role in the reconceptualization of mechanics that began prior to the *Principia* and achieved an explicitly articulated form in Lagrange's *Mecanique Analitique* at the end of the eighteenth century. In addressing whether or not Newton's *Principia* is a treatise in mechanics, Gabbey (1992) emphasizes that the term was used in different ways at the time Newton was writing, and that we cannot assume that our contemporary usage reflects a category applicable to Newton's treatise.[14] He says:

> Mechanics as a discipline underwent radical changes in nature during the period covered by the prehistory, publication and early reception of the *Principia*. Correspondingly, the term "mechanics" as used by writers of the time often carried equivocal senses. (Gabbey 1992, 308)

In light of these complexities, it is important to be careful how we handle the term "mechanics." Moreover, we also have the seventeenth century rise to prominence of the "mechanical philosophy:" an approach to natural philosophy that was largely qualitative, at least in Descartes's hands, and distinct from the mathematical discipline of mechanics (more on this below, when we turn our attention to physics). So there is ample opportunity for confusion and misunderstanding surrounding the term "mechanics." Nevertheless, I think that Newton is clear in his conception of rational mechanics as he presents it in the *Principia*, and that Books 1 and 2 realize this conception. It is a discipline of mathematics, and it concerns the exact treatment of any motions under any forces whatsoever.

Rational mechanics provides us with one strand of Newton's project in philosophical mechanics. The second comes in Book 3, which is a book in physics. Once we have both strands in place, we will be in a position to see what is at stake and why all of this matters.

9.2.2 Physics

When the *Principia* came out, Newton faced accusations that he had produced a book in mechanics, but not in physics.[15] As noted above, this is puzzling to any reader today who, turning to the *Principia*, sees a book in physics, and even more so to one who uses the terms "classical mechanics" and "classical physics" interchangeably. But that very reconceptualization of mechanics in relation to physics happened after Newton. It was not in place at the time Newton wrote his *Principia*, nor for some time afterwards.

We can begin with Rohault's 1671 *Traité de physique* as an exemplar of how the term 'physics' was understood then. Rohault opens with a chapter entitled "The Meaning of the Word Physics, and the Manner of treating such a Subject,"

---

[14] For more on Gabbey's discussion of the *Principia* as a treatise in mechanics, see 9.3.3.

[15] See the review published anonymously in the *Journal des Sçavans* (2 August 1688) 153ff, and the English translation in Koyré 1965, 115.

in which he says that "we here use it [the word physics] to signify Knowledge of natural Things, that is, that Knowledge which leads us to the Reasons and Causes of every Effect which Nature produces."[16] Physics, thus understood, encompasses all "natural Things," including all of their causes and effects.

This scope for physics persists into the eighteenth century. Musschenbroek's early eighteenth century characterization of physics is, I think, particularly helpful. Physics, he says, is that part of philosophy which "considers the space of the whole universe, and all bodies contained in it; enquires into their nature, attributes, properties, actions, passions, situation, order, powers, causes, effects, modes, magnitudes, origins."[17] Excluded from physics are the study of spiritual beings, teleological investigations of things, metaphysics (which studies "such general things as are in common to all created beings," both physical and spiritual), moral philosophy and logic. For our purposes, the most important things to note are the inclusion of physics within philosophy, and the characterization of the domain of physics. According to Musschenbroek, the primary subject-matter of physics is bodies, and many aspects of bodies that we might think of today as being the subject-matter of metaphysics—such as the nature, powers, causes, effects and origins of bodies—fall within physics. This conception of physics was widely held at the time, both before and after the publication of Newton's *Principia*.[18]

The term "physics" was often used interchangeably with "natural philosophy," and the pursuit of physics fell within the remit of philosophers. Mechanics, on the other hand, was a distinct discipline, practiced by mathematicians. Gabbey (1992, 310) emphasizes that during the Renaissance and early seventeenth century, works in mechanics and in physics were written by distinct groups of authors between which there was little overlap. Even where there is overlap, the two subjects were treated largely separately.[19]

---

[16] My quotations are from the English translation: Rohault 1723, 1.

[17] Musschenbroek 1744, 1–2.

[18] See Brading and Stan 2021 and 2023.

[19] Though mechanics and physics were largely separate there were active attempts to bring them together, and there is a long history of "mixed mathematics." Famously, Galileo expanded the range of problems treated in mechanics and sought a mathematical theory of natural phenomena. In his case, he eschewed talk of causes, and therefore of physics as it was then understood. Kepler, on the other hand, transformed positional astronomy into physical astronomy via his commitment to a causal explanation of the motions of the planets and his demand that the mathematical theory and the causal explanation align exactly. And Huygens, with greater success than anyone else, took Descartes's qualitative "unification" of matter theory and mechanics seriously and used it to attempt a philosophical mechanics. There is more to be said about each of these cases, but the simplified picture is as follows. In mechanics, physical bodies were taken as given, abstracting only those properties (such as size, shape, and weight) relevant to the demonstrations. Mechanics could assume the existence of such bodies because another discipline – physics – was tasked with the general theory of bodies (of their nature, properties, and so forth). See also 9.2.3.

Rohault is an example of this: he presented his mechanics in a separate treatise from his physics.[20] The subject-matter of Rohault's mechanics is machines (artefactual devices), whereas that of his physics is the natural world. The two treatises proceed from different principles, by means of different methods, and with different goals. Moreover, in the Preface to his English translation of Rohault's mechanics, Watts describes mechanics as an application of geometry remarking that these applications are *unaffected by the Cartesian philosophy* that seems at times to be presupposed. He writes:

> If our Author in his Definition of Gravity, or in an Expression or two besides, seems to refer to the *Cartesian* Philosophy, now deservedly exploded, yet it is done in such a Manner, as does not in the least affect the Demonstrations, which will be equally true, whatsoever Hypothesis we follow in those Points. (Rohault 1717)

In other words, Rohault's mechanics is independent of his physics.

In short, physics, or natural philosophy, studied physical bodies including their natures, properties, causes and so forth. This was the work of philosophers. Mechanics, on the other hand, studied the behavior of machines, and belonged to the domain of the mathematicians. In the mid to late seventeenth century, notwithstanding Descartes's "mechanical philosophy," the integration of mechanics and natural philosophy into a single discipline did not yet exist.

Books 1 and 2 of Newton's *Principia* explicitly eschew consideration of the physical causes and effects of the motions of bodies, considering these issues only "mathematically." Newton makes this clear multiple times (for example, in the opening paragraph of Book 1, Section 11) and re-emphasizes the point at the beginning of Book 3 as he writes, "In the preceding books I have presented principles of philosophy that are not, however, philosophical but strictly mathematical [21]" While Books 1 and 2 offer a radical expansion of the domain of mechanics to include natural forces and motions, they are mathematical, and they do not consider causes. As I argued above, they are rightly considered books in mechanics. My point here is that they are not books in physics, as physics was then conceived.[22]

Nevertheless, Newton did intend the *Principia* to include physics. In the scholium to Proposition 69 of Book 1 of the *Principia*, Newton explains how he understands the relationship between rational mechanics and physics. Whereas rational mechanics is the exact mathematical treatment of any forces and motions whatever,

---

[20] Rohault's *Mechanics* (*Les mécaniques*) was published posthumously in 1682. An English translation by Thomas Watts was published, the second edition of which (Rohault 1717) is my source. The full title reads: *A Treatise of Mechanics: Or, The Science of the Effects of Powers, or Moving Forces, as apply'd to Machines, demonstrated from its first Principles.*

[21] (Newton 1999, 793).

[22] I simplify in presenting my thesis, and in two important respects. First, Newton is unfolding his rational mechanics with an eye to the physics he hopes to treat. This is no accident, of course, given the genesis of the *Principia*. Second, in various scholia in Books 1 and 2 he illustrates the applicability of his mechanics with physical examples. Neither of these caveats undermines the central thesis that Books 1 and 2 are books in rational mechanics. See 9.3.5 for detailed discussion of this point.

physics considers the actual motions of bodies and the actual forces responsible for those motions. Moreover, physics is also concerned with the causes of those motions. In proceeding from rational mechanics to physics, we first determine which forces are actual (i.e. which of the force laws explored by rational mechanics pertains in the behaviors of actual bodies), and then we seek the causes of these forces.[23] So the method has three steps: develop the general theory of forces and motions (rational mechanics); determine which forces are actual (physics); identify the causes (physics). This is crucial for understanding the successes and limitations of the *Principia*, as we shall see.

If we look at the scholium in more detail, we see that Newton begins by re-emphasizing the point that (at least in Book 1 of the *Principia*) he is giving a "mathematical" treatment of "force," "attraction" and "impulse;" that is to say, he is treating "not the species of force and their physical qualities," but the "quantities" of force and their "mathematical proportions." He then writes:

> Mathematics requires an investigation of those quantities of force and their proportions that follow from any conditions that may be supposed. Then, coming down to physics, these proportions must be compared with the phenomena, so that it may be found out which conditions of forces apply to each kind of attracting bodies. And then, finally, it will be possible to argue more securely concerning the physical species, physical causes, and physical proportions of these forces. ... (Newton 1999, 588–9)

Cushing says of this passage:

> That is, there are three different levels at which we must work: the mathematical (or deductive), where we analyze the implications of certain assumptions or axioms; the physical, where we use comparison with data to decide which of the many possible axioms or laws actually do correspond to nature; and, finally, the philosophical, where we seek the causes of these laws. In the *Principia*, he attempted to do the first two as a preparation for the third that he also felt to be important. ... In his life Newton never did succeed in constructing an explanation of the causes behind his laws of mechanics and of gravitation. (Cushing 1998, 95)

I agree with the overall message here about the three steps. The first lies in rational mechanics.[24] The third concerns causes, and—as we have seen—this places it within the domain of physics, as physics was conceived at the time. Cushing's label of "philosophical" for the third step is appropriate in the sense that physics, as it was then understood, was that part of philosophy that included the search for the causes and effects of the properties and behaviors of bodies. It is this third step in Newton's three-step methodology that his contemporaries would have recognized as physics.

For Newton, however, physics involves a prior step. His second step, in which we treat physical phenomena using the results developed in our rational mechanics, is also explicitly located within physics. This is what Newton does in Book 3. Specifically, he treats the force of gravity as a physical force. He uses the

---

[23] See 9.3.4 for Newton on causes and the relationship between forces and causes.

[24] I disagree with Cushing that Newton's axioms can be understood as assumptions, freely chosen. I discuss their epistemic status in 2.3.

mathematics of Books 1 and 2 to theorize gravitational phenomena, both terrestrial and celestial, with spectacular success.[25] While Books 1 and 2 of the *Principia* are books in rational mechanics, for Newton Book 3 is a book in physics.

Famously, however, Newton does not uncover the causes of the force of gravity. As he himself states in perhaps the most notorious passage of the *Principia* (added in the second edition):

> Thus far I have explained the phenomena of the heavens and of our sea by the force of gravity, but I have not yet assigned a cause to gravity. ... I have not as yet been able to deduce from phenomena the reason for these properties of gravity, and I do not feign hypotheses. (Newton 1999, 843)

This claim, and others like it, were part of a widespread dispute at the time over whether in the *Principia* Newton had indeed provided a physics. The difficulties were all but inevitable because of the prevailing conception of physics: it was that part of philosophy charged with providing *causal* knowledge of the natural world, and in particular of the behaviors of bodies. By Newton's own admission, the *Principia* begins but does not complete this task, and this left evaluation of what the *Principia* achieves unclear to many of his contemporaries. As the first reviewer of the *Principia* complained (see above), Newton seemed to him to have provided a perfect mechanics, but to have failed to provide a physics, for he had failed to provide a complete account of the *causes* of the gravitational behavior of bodies.

However, read in the context of the scholium to Proposition 69 of Book 1 (quoted above), Newton's meaning is clear. The achievements of Book 3 lie in step 2: he had completed step 2 but not step 3 of his three-step methodology. This is an achievement in physics: Newton had determined that there *is* a gravitational force acting among bodies, and what conditions that force satisfies. Step 2 contributes to our causal account of the world, and properly belongs to physics.[26]

Nevertheless, it is easy to see why Newton's three-step methodology was a source of great confusion among his contemporaries and eighteenth century critics.[27] Step 3, had Newton completed it, would have been recognizable to his contemporaries as physics. Placing Step 2 within physics was novel. Newton offered it as a bridge linking rational mechanics to traditional physics, and then made this bridge a part of physics: physics, in Newton's reconceptualization, begins not with the qualities of bodies, but with the mathematical treatment of the motions of physical bodies in terms of forces. With or without Step 3, this is a radical transformation of the goals and methods of physics.

---

[25] My account of Newton's three-step methodology differs from that offered by Cohen in his "Newtonian style." See 9.3.4 and 9.3.5.

[26] For Newton on forces and causes and the complexities in the relationship between them, see 9.3.4.

[27] On the incompleteness of Newton's causal account of gravitation see Janiak 2008, ch. 3, especially p. 64; Biener 2018, 3; and Janiak's discussion therein of Leibniz's and Clarke's attempts to grapple with the issue.

From our vantage point, we tend to use the terms "classical physics" and "classical mechanics" interchangeably, and to think of mechanics as a part of physics. I have emphasized that this is not how things stood during Newton's lifetime. Instead, it is appropriate to view the *Principia* as offering a conjunction of two books in mechanics with one book in physics, where that book in physics is, by the author's own criteria, incomplete. In so doing, we are able to see the highly revisionary conceptions of each that Newton offered. Moreover, it is not just that Newton transformed mechanics and physics individually, he also transformed the relationship between them. All three books of the *Principia* are gathered together into a single text, and the significance of this is our next concern.

9.2.3 Philosophical Mechanics

We have seen that, as of the late seventeenth and early eighteenth centuries, rational mechanics belonged to mathematics whereas physics was a subfield of philosophy. In our book, Stan and I argue that, from the late seventeenth through the eighteenth century, the guiding research program for much of natural philosophy was what we call "philosophical mechanics."[28] Projects in philosophical mechanics seek to integrate rational mechanics and physics into a unified treatment of bodies and their motions. In what follows, I argue that Newton's *Principia* is fruitfully understood as offering a philosophical mechanics.

Newton uses the term "natural philosophy" for his project, as the full title of the *Principia* makes clear and as he states in the Preface. However, he is also explicit that he intends this to include both rational mechanics and physics. Since the terms "physics" and "natural philosophy" were often used interchangeably at the time, the term "natural philosophy" risks masking the very distinction I have highlighted between rational mechanics and physics.[29] This is one reason why Stan and I chose to adopt the term "philosophical mechanics." It is a term of art, not found in the literature until the late eighteenth century, and even then with a meaning not quite

---

[28] Brading and Stan 2023.

[29] From Newton's correspondence with Halley, it is clear that his choice of title deliberately reflects the expansion in scope from Books 1 and 2 (developed in the "De Motu" manuscripts), of most interest to mathematicians, to include Book 3, of most interest to natural philosophers. See also Newton's introduction to Book 3 of the *Principia*, where he describes the first two books as not "philosophical" but instead "strictly mathematical," noting also that in these two books he illustrates the applicability of his mathematical results to actual bodies and motions in several scholia. For Book 3, it "remains for us to exhibit the system of the world from these same principles." Here, we think that the framework of philosophical mechanics is helpful in clarifying Newton's intended distinction between the "mathematical" work of Books 1 and 2 (rational mechanics) and the "philosophical" work of Book 3 (physics, as that term was understood at the time).

as we use it.[30] We introduce it for the conceptual work that it enables us to do as we work through developments in seventeenth and eighteenth century natural philosophy, physics, and mechanics.

We have seen that Books 1 and 2 of Newton's *Principia* are books in rational mechanics, whereas Book 3 is a book in physics. In addition to Books 1-3, the *Principia* contains a Preface followed by a set of *Definitions* and then three *Axioms, or Laws of Motion*. These precede Books 1-3. These are the main elements that Newton assembles into a single text under the title *Mathematical Principles of Natural Philosophy*. The structure of how they are put together is significant. First, it indicates that Newton conceived of his *Principia* as a unified project. Second, it shows that the Preface, the *Definitions*, and the *Axioms or Laws of Motion*, pertain to *all three books*: they are the hinge that joins Books 1 and 2 to Book 3, creating a unified whole.

Conceiving of the project in this way provides insight into the status of the *Definitions* and of the *Axioms, or Laws of Motion*. Newton's placement of these elements prior to and outside the three books of the *Principia* is surely deliberate.[31] In the context of rational mechanics, the *Axioms or Laws of Motion* have the status of axioms, whereas in the context of physics, they have the status of laws of motion.[32] This gives them a highly interesting character. On the one hand, in their different roles they differ in their justification. On the other, as a single set of principles they unify rational mechanics and physics in a very special way. I discuss both these aspects below.

First, however, notice how surprising this unification is. Descartes, in his physics, offered us his laws of nature. Nowhere in this physics does he offer us a mechanics, or provide axioms of mechanics.[33] Rohault, as we noted above, wrote two separate treatises, one on physics and one on mechanics. Both texts contain axioms, one set for physics and another for mechanics, but the axioms are utterly different. Rohault's axioms of physics begin:

> The first is, that Nothing, or that which has no Existence, has no Properties ... Secondly, It is impossible that something should be made of absolute Nothing; or that mere Nothing can become any Thing ... (Rohault 1723, 18–9)

---

[30] While the term was first used (to our knowledge) in a French text on mechanics by Prony (1800), Stan and I adopt it for our own purposes. See Brading and Stan, 2023 for further discussion.

[31] Biener 2018 opens by saying that the *Axioms, or Laws of Motion* are in Book 1, but I think this is not right. He then says that the placement of the Rules of Reasoning at the beginning of Book 3 mirrors this positioning, but he doesn't say very much about why we should think this, and I do not agree. In my opinion, the *Axioms, or Laws of Motion* are very deliberately placed prior to Books 1–3, and the *Rules of Reasoning* are similarly deliberately placed within Book 3. See Sect. 9.3 for further discussion of the status of the *Axioms, or Laws of Motion*.

[32] Adopting such a clean terminological distinction is conceptually helpful but oversimplifies the situation. For a more nuanced treatment, see 9.3.2.

[33] This notwithstanding the "mechanical philosophy" offered in his *Principles*, see above. Descartes does not use his laws of nature to develop a mechanics in the sense intended here: his laws are not principles from which he mathematically demonstrates the behavior of machines.

Whereas his first axioms of mechanics are:

> Ax. 1. In heavy Bodies, which are Regular and Homogeneous (that is, which have all their Parts equally heavy) and plac'd Horizontally, the Center of Magnitude is also the Center of Gravity.
> Ax. 2. The different Gravities of Homogeneous Bodies are one to another in Proportion to their Bulks. (Rohault 1717, 5)

Newton's use of a common set of axioms for both mechanics and physics is striking.

In the seventeenth century, works in mechanics and physics were most often written by different people, and the two subjects were treated largely separately, as Gabbey has discussed (see above). Nevertheless, as Gabbey goes on, discussion over the appropriate taxonomy for the materials treated under the umbrellas of mechanics and physics, and how these materials related to one another, was live and explicit. Newton was exposed to this not least through the lectures and writings of Isaac Barrow, John Wallis, and Robert Boyle, in which the mathematical principles of mechanics find a new place in the foundations of natural philosophy.[34] Thus Newton's unification does not come out of nowhere, and should be viewed in the context of this wider community project. It is the creativity and depth of Newton's contribution that stands apart.

Newton's use of the label "Axioms, or Laws of Motion," rather than simply "Axioms" or "Laws" is significant, for it provides information about the dual role that these principles are required to play. In labeling his principles "Axioms," Newton followed a standard practice in mechanics. Moreover, he structured his rational mechanics analogously to geometry, thereby providing information about the epistemic status of the *Axioms*. We will return to this important point below, but first a few words about *Laws*.

It is well-known that in Descartes' philosophy laws of nature play an important role, situated as they are between his metaphysics and his physics, providing a bridge between an immutable God and the changing world. However, while laws of nature are an important theme of seventeenth century natural philosophy, whether or not to include such laws, and if so their formulation, their appropriate placement within a philosophical system, and their metaphysical and epistemic status, remained open questions. For example, Descartes' laws of nature have no special place in Rohault's *Traité de physique*, and indeed only the first law receives this label in Rohault's exposition.[35] Nevertheless, it is clear from Newton's manuscripts that Newton's own laws of motion have their origins in Descartes' laws of nature, and it is also evident that Descartes' *Principles of Philosophy* was both inspiration and target for Newton's *Mathematical Principles of Natural Philosophy*. In retaining the terminology of laws, Newton is signaling two things. First, that his three principles pertain not only to mechanics, but also to physics. Second, that

---

[34] Gabbey 1992, 311–14.

[35] Rohault does not present Descartes's three laws of nature as laws, but he does write (1723, 47) that "it is one of the Laws of Nature, *that all Things will continue in the State they once are* unless any external Cause interposes," which is the beginning of Descartes's first law of nature.

they play a role in his physics similar to that which Descartes's laws play in his: specifically, just as Descartes aspires to a deductive structure for his physics, at least in principle, so too Newton's laws lie at the basis of deductive arguments in his.

The dual role of Newton's *Axioms, or Laws of Motion* in his mechanics and physics affects their epistemic status and justification. As axioms, they have the status of mathematical hypotheses in the specific sense that our demonstrations assume them as given and proceed from there. That is, taking them as axioms allows us to proceed as if they are true.[36] The wider the class of problems falling under the scope of the axioms (that is, soluble via mathematical reasoning from the axioms), the greater our justification for adopting them as axioms for rational mechanics. In short, their justification lies in their *generality*.[37]

As laws of motion, on the other hand, they have the status of empirical claims: they are claims about the physical world that may or may not be true. The wider the range of phenomena to which they may be successfully applied, the greater our justification for taking them to be true. In short, their justification lies in their *universality*. Universality of applicability justifies their status as laws of motion. As laws of motion, they rest on an inductive basis: as Newton is at great pains to emphasize in his scholia to the *Axioms, or laws of motion*, the laws are empirically well-supported.[38] And so, we take them to be true "until yet other phenomena make such propositions either more exact or liable to exceptions."[39] The standards and criteria for successful application to the phenomena are set by the community of physicists, just as the standards and criteria for successful solutions of problems in rational mechanics are set by the community of mathematicians.

This dual justificatory status for a single set of principles makes clear that Newton's *Axioms, or Laws of Motion* unify rational mechanics and physics in a very special way. They are justified twice over and independently, first within rational mechanics and second within physics, and they thereby form part of the hinge

---

[36] See De Risi 2016 for discussion of different epistemological approaches to the axioms of geometry in the early modern period. Here, I gloss over those differences: the important points are that axioms are (i) taken as true and (ii) required to meet appropriate conditions of justification. See Landry 2023. for elaboration of an "as if" interpretation of mathematical hypotheses. In early drafts Newton used the term "hypotheses" and then changed this for "laws". See 9.3.1 for more on the epistemic status of the axioms, and below for the relationship between the definitions and the axioms.

[37] Cohen (1980, 101) in his discussion of Newton's methodology, the "Newtonian style," highlights the importance for Newton, as a mathematician, of generalization.

[38] Though on this point see Sects. 9.3.2 and 9.3.6. Friedman 2001 takes this claim by Newton to be disingenuous and his constitutive interpretation of the laws offers an alternative account of their universality (see 9.3.6). For Descartes, the laws are similarly universal in their applicability, but their justification is *a priori*, depending upon prior accounts of the nature of matter and motion, and on the nature of God. For more on applicability, see 9.3.5.

[39] Newton 1999, 796. Newton allows that other regions of our universe may be subject to different laws. Nevertheless, Newton's method encourages us to take the laws as true throughout the universe until the phenomena show us that we must restrict their scope. See Smith 2014. See also Biener and Schliesser 2017, 317.

that joins rational mechanics to physics. They are a conduit through which rational mechanics may speak about physics, and through which physics may pose problems for rational mechanics.[40]

The *Axioms, or Laws of Motion* do not, of course, stand alone. They are preceded by the *Definitions*. These too, fall outside Books 1-3, and pertain to them all. I have often puzzled over the relationship between the definitions and the laws of motion. There is so much overlap between Definition 3 and the first law of motion, for example, so why did Newton need both? And why is one a definition and the other an axiom or law of motion? I think we can solve this by viewing the *Definitions* and *Axioms, or Laws of Motion* through the lens of philosophical mechanics.

Consider first rational mechanics, understood as an exact science analogous to geometry. De Risi (2016, 602) notes that in the early modern period many, though not all, mathematicians believed that the axioms of geometry were provable from the definitions, even when no such proof was explicitly given. It was the definitions that were considered the basic principles of mathematics. The postulates and axioms were statements, derivable from the definitions, to be used in the ensuing proofs. If we follow this approach then Newton's axioms should, in principle, be derivable from his definitions. However, even with this derivation in place, a problem remains: the definitions are neither self-evident nor do they rest on convincing empirical evidence. As such, they lack the epistemic status from which to derive a system of rational mechanics. The axioms, on the other hand, enable the solution of an unprecedentedly wide-ranging and general set of problems, as Book 1 demonstrates. This justifies their status as axioms, and confers justification on the definitions themselves insofar as they are necessary for the statement of the axioms. The status of the definitions as basic principles is legitimate insofar as the axioms are derivable from them. Thus, the axioms play an intermediary role between the definitions, whose status they justify, and the ensuing system of rational mechanics, whose mathematical derivation they facilitate.

The relationship between the *Definitions* and the *Axioms, or Laws of Motion* is rather different in the context of physics. To see this, we first need to consider how the *Definitions*, like the *Axioms, or Laws of Motion*, have a dual face. This is perhaps obvious, since the *Definitions* are clearly to be read in either a mathematical or a physical key, as appropriate. For example, the terms 'body' and 'force' can be taken to apply to physical bodies (e.g. "snow" in the explanation of Definition 1) and physical forces (e.g. "magnetic force" in the explanation of Definition 5), as well as to bodies and forces considered not physically but mathematically (e.g. in the discussion of Definition 8, where Newton is explicit about this). I understand every definition to have a dual reading, as either mathematical or physical, depending on whether the problem to be solved is a problem in rational mechanics (as in Books 1 and 2) or in physics (as in Book 3).

---

[40] This dual status ensures that the applicability of rational mechanics to problems in physics is not piecemeal, one problem, system, or scenario at a time; the shared axioms guarantee a (partially) shared modal structure for the problem space of rational mechanics and physics (see 9.3.5).

Particularly important, I think, is the dual face of the term 'measure.' Mathematically—in rational mechanics—when a quantity measures (an aspect of) something, it allows for its treatment as a magnitude: that is, for its treatment within the science of geometry. The role of the definitions is to set up those magnitudes that are then to be treated using the tools of geometry. (This point, that traditionally geometry was the science of magnitude, will be important in a different context later on, in Sect. 9.3, when we return to the issue of the status of Newton's axioms, and need to recall that geometry was not, at the time, always assumed to have *space* as its subject-matter.) For example, in introducing "quantity of motion" as a measure of motion in Definition 2, Newton tells us how to treat motion as a magnitude. In so doing, he sets up the conditions for motion to be treated using the tools of geometry.[41]

In the context of physics, the role of the *Definitions* is to articulate the relationships among the relevant quantities and parameters to which we have empirical access through measurement. For example, Newton relates 'quantity of matter' to two quantities whose empirical measurement is already well established (density and volume), and he relates 'quantity of centripetal force' to motion and time. In so doing, he sets up the conditions for mass and centripetal force to be treated through quantitative empirical measurements.

In this way, the dual face of the term 'measure' allows the *Definitions* to function successfully in both rational mechanics and empirical physics. Though a simple point, I think this is enormously important for understanding what Newton achieves in the *Principia*, and how he achieves it. In his lectures on Newton, George Smith has repeatedly emphasized the importance for Newton of quantities and their empirical measure, and Newton's remarkable ability to transform seemingly intractable puzzles about the natural world into problems that he could handle mathematically. Viewed through the lens of philosophical mechanics, we gain important insight into the philosophical moves that make this possible. Newton unifies rational mechanics with physics, not by collapsing them into a single discipline, but via a dual role for the *Definitions* and the *Axioms, or Laws of Motion*. Rational mechanics and physics each retains its own subject-matter, but through the *Definitions* and *Axioms, or Laws of Motion* the results of the former are applicable in the domain of the latter, and the problems of the latter fall within the domain of the former. It is in this sense that they are *mathematical* principles of *natural philosophy*.

For those who have emphasized Newton's claims that Book 1 is a book in mathematics, the problem of applicability has loomed large: what justifies the move from mathematics to physics? A variety of responses is available in the literature.[42]

---

[41] There is a sleight of hand here. Newton's geometry requires the treatment of (to use anachronistic terminology) vector quantities as well as scalar, and so new rules for the addition of such quantities must be introduced. The debate over the status of the parallelogram rule for the addition of forces is a prominent example of where this issue arises (see, for example, Miller 2017 and references therein).

[42] See, for example Cohen 1980; Ducheyne 2012, 79–80; Smeenk 2016.

My own reaction was, for a very long time, enormous puzzlement over what the problem was supposed to be, since (to my mind) Book 1 was clearly a book of physics—recognizably so from any modern training in physics—in which physical subject-matter is discussed in the language of mathematics, with no special problem of applicability. But this is not right. Neither geometry nor rational mechanics have physical entities as their subject-matter. Rather, it is through the dual aspect of the principles—the *Definitions* and the *Axioms, or Laws of Motion*—that we are able to use the language of rational mechanics to talk about the subject-matter of physics. These principles sit outside Books 1, 2 and 3, serving both the rational mechanics and the physics; it is this dual aspect that provides the bridge from the rational mechanics to the physics, and that ultimately unifies the two.[43]

There is, however, an important contrast between the scope of Books 1 and 2 as compared to that of Book 3. Newton intended generality for his rational mechanics. He sought a rational mechanics with resources appropriate for *whatever* forces and motions are found in the world, whether in manmade machines or natural bodies. This scope for Newton's rational mechanics contrasts sharply with that for his physics. The focus of Book 3 of the *Principia* is one force: gravitation. As a result, what Newton provides is an attempted philosophical mechanics of gravitation.

To sum up. The framework of philosophical mechanics is helpful in clarifying Newton's intended distinction between the 'mathematical' work of Books 1 and 2 and the 'philosophical' work of Book 3, and the relation between them. Books 1 and 2 are books in rational mechanics; Book 3 is a book in physics; and the three books are powerfully unified into a single system via the shared *Definitions* and *Axioms, or Laws of Motion*. This unification allows resources from Books 1 and 2 to be deployed in Book 3. More generally, the unification allows rational mechanics to be applicable within physics, and it allows physics to present rational mechanics with problems it might hope to tackle. Through his unification, Newton transformed rational mechanics (see 9.2.1), physics (see 9.2.2), and the relationship between them.

9.3   Elaborations and Refinements

In this section of the paper, I remove some oversimplifications and add detail to the account of Newton's *Principia* offered in Sect. 9.2, taking into account some of the secondary literature and the objections that arise therefrom.

---

[43] For more on applicability, see 9.3.5.

9.3.1 On Rational Mechanics: Its Relation to Geometry, and the Epistemic Status of the Axioms

I have claimed that Newton's rational mechanics concerns the production of geometrical objects (especially curves) through motion, where those motions arise from any forces whatsoever. This is the sense in which rational mechanics is prior to geometry.

Garrison draws a rather different conclusion in his reading of Newton's Preface. He moves from the claim that mechanics is prior to geometry to the conclusion that:

> For Newton, geometry had a very definite and particular content; geometry was about empirical (physical) objects extended in empirical (physical) space and constructed by God, nature, or man. ... Thus for Newton, physical-empirical objects were not an interpretation of geometry but rather the interpretation of geometry; no other was imaginable. (Garrison 1987, 612)

I sympathize with his urging that we not view Newton through post-twentieth century formalist glasses, but nevertheless I disagree with his conclusion, in two related respects.

First, his view collapses the mathematical into the physical, making physical objects the subject-matter of geometry. However, Newton was careful to distinguish the mathematical from the physical, as I emphasized at the end of Sect. 9.2, and as I discuss below (see 9.3.4 and 9.3.5).

My second disagreement with Garrison is related to the first, and concerns constructibility. Garrison suggests that the constructions of lines and circles, and of the subject-matter of geometry in general, must be carried out concretely, as a construction of physical-empirical objects. I see nothing in Newton to support this. While God is, indeed, the perfect artificer, and we may on those grounds suppose that the geometrical features of His created world are constructed exactly, rational mechanics does not require that such exact constructions have been implemented concretely rather than merely ideally in order to proceed. The very lack of concrete implementation is, after all, one element of what makes Newton's mechanics a *rational* mechanics. I take Domski's 2003 discussion of Garrison, as well as of Dear 1995 and Molland 1991, to be decisive on the issue of constructibility.

I endorse Domski's view that for Newton, "no considerations of constructibility enter into rational mechanics" and that Newton thereby frees geometry from the constraints of practical mechanics.[44] She argues that Newton rejected Descartes's distinction between "geometrical" and "mechanical" curves, and writes:

> In particular, although the origins of geometry rest on mechanical practice, for Newton, the domain of geometry is not restricted to those curves actually constructible by straightedges and compasses. The geometer does not simply start from those curves constructible by rulers and compasses and then idealize such constructions by substituting instruments with straight lines and circles. To do so would be to distinguish geometry and mechanics solely on the basis of the exactness of instruments employed in these fields, and for Newton, "this common belief is a stupid one." (Domski 2003, 1121–2)

---

[44] Domski 2003, 1123.

If this is right, then the question immediately arises as to what constraints there are on the curves of rational mechanics, the curves that are to serve as the subject-matter of geometry. In my view, Newton's criterion on admissible curves is simply that they are produced by forces and motions. This criterion becomes contentful via the definitions and the axioms, which tell us about forces, motions, and the relationships between them.

I have argued (9.2.3) that the justification for taking the axioms as true rests on their generality: we show their utility in solving a wide range of problems, to a high level of generality, and in so doing we justify their status as axioms. An alternative epistemology, in which a clear and evident mathematical proposition may be taken as an axiom, was present among seventeenth century French mathematicians, especially those in the Cartesian tradition,[45] and it is perhaps this that Domski 2018 has in mind when suggesting that Newton took his axioms to be rationally certain. But this was not the only epistemology of axioms available at the time. For example, De Risi notes that in Germany the idea that "all axioms should ultimately be proven from the definitions" persisted, and he lists Christian Wolff's axiomless textbook as an important example.[46] Even where the axioms and postulates were judged sufficiently evident as not to require proof, "many mathematicians in the Early Modern Age believed that axioms and postulates were in fact provable from definitions."[47] For this approach, it was the definitions that were considered the basic principles of mathematics, and the postulates and axioms were statements, derivable from the definitions, to be used in the ensuing demonstrations. It is this latter view of axioms that I think we see in Newton's *Principia*. My grounds for this claim are that I find it consistent with Newton's problem-oriented approach to mathematics and mechanics[48] and a fruitful way of interpreting the structure of the *Principia*.

9.3.2   On the Distinction Between Axioms and Laws of Motion

My claim is that Newton's *Axioms, or laws of motion* serve two distinct roles, one as principles of rational mechanics and the other as principles of physics. So far I have adopted a clean terminological distinction between "axioms" and "laws" to mark these two distinct roles. This oversimplifies the situation.

First, it is easy to find counterexamples. In Book 1 of the *Principia*, Newton says things like "it is required to find the law of centripetal force;" so there are laws in rational mechanics.

---

[45] De Risi 2016, 599.

[46] De Risi 2016, 601.

[47] De Risi 2016, 602.

[48] Solomon 2017.

More importantly, however, insisting on a clean terminological distinction risks masking the sense in which Newton's laws of motion are rightly thought of as axioms for physics, as first principles to be used in solving problems in physics. When Newton says,[49]

> as in Geometry the word Hypothesis is not taken in so large a sense as to include the Axiomes & Postulates, so in Experimental Philosophy it is not to be taken in so large a sense as to include the first Principles or Axiomes w[ch] I call the laws of motion[50]

I think he is clearly telling us that the laws of motion have axiomatic status in experimental philosophy.

Nevertheless, the laws of motion—taken as axioms of physics—differ from axioms of rational mechanics in their epistemological status for they are to be justified inductively. Newton himself goes on to state this clearly

> These Principles are deduced from Phænomena & made general by Induction: wch is the highest evidence that a Proposition can have in this philosophy (*ibid*).

These quotations are from an exchange with Cotes which led to the addition of Rule 4 in the second edition of the *Principia*. According to Rule 4, the laws of motion—like the proportions of universal gravitation—are to be treated as "either exactly or [*quam proxime*[51]] true notwithstanding any contrary hypotheses until yet other phenomena make such propositions either more exact or liable to exceptions."[52]

Biener and Schliesser 2017 take a different view of the significance of the term "axiom" in *Axioms, or Laws of Motion*. They argue that Newton introduced it to indicate his increasing confidence in the truth of his hypotheses/laws, and in their applicability to a wide range of phenomena. I think that Biener and Schliesser are right that Newton became increasingly confident about the universality of his laws, as laws of physics, and I return to this below (9.3.6). However, I think this is not the whole story. Domski 2018 distinguishes the truth of Newton's laws from their rational certainty: the former concerns their applicability to the phenomena, whereas the latter arises from their being mathematical. In her view, Newton's laws are empirical when taken physically, but when taken mathematically they have an axiomatic status and are to be taken as rationally certain. I think Domski is right to distinguish the physical from the mathematical, and that Newton's choice of the term "axiom" tells us something about the type of justification required for the axioms/laws when considered mathematically. While Domski appeals to rational

---

[49] Letter to Cotes, March 28, 1713.

[50] Edleston (1850, 154–5).

[51] George Smith has emphasized Newton's use of this peculiar phrase.

[52] Newton 1999, 796. One might expect Newton to provide significant inductive justification for the laws of motion, and find it puzzling that he did not. But in fact he did not take the onus to be on himself to provide such justification: in the scholia to the laws he indicated where he took such evidence to lie, and he appealed to authority as a way of evading having to make the case in detail. The point stands that in the context of physics Newton's laws are to be inductively justified.

certainty, I think that it is the generality of the problems solved by means of these principles that justifies their status as axioms (see 9.2.3).

With all of this in mind, we arrive at a more nuanced view than that presented in Sect. 9.2. The use of "Axioms, or laws of motion" as a single label for the three principles that ensue, and the placing of them outside of Books 1–3, emphasizes that these very same three principles serve as axioms for both rational mechanics and physics, and that these shared axioms turn out to be laws of motion. However, the epistemic status of axioms in rational mechanics differs from that in physics, and this distinction is crucial for understanding the philosophical structure of the *Principia*. Moreover, if we fail to recognize this difference then we may not feel the *surprise* we should on seeing that the same principles can serve as axioms for both. And so I stand by my use of "axioms" for rational mechanics and "laws of motion" for physics as a way of marking this philosophically important distinction.

9.3.3   On the Principia as a Text in Mechanics

In his paper "*Newton's Mathematical Principles of Natural Philosophy*: a treatise on 'mechanics',?" Gabbey presents us with a puzzle. On the one hand, the *Principia* seems to be a text in mechanics. Newton's own remarks in the Preface indicate this, and when we look back with hindsight, in the wake of later figures such as Euler and Lagrange, this seems an appropriate label.[53] On the other hand, Newton chose not to include the term "mechanics" anywhere in the title of the *Principia*, so perhaps it is not a text in mechanics afterall. Indeed, Gabbey interprets some of Newton's remarks in the *Principia* as casting doubt on the view that he thought of himself as writing a text in mechanics.

Gabbey's solution is to suggest that Newton equivocates in his use of the term, and he writes that this is understandable because the *Principia* is itself a revolutionary and transitional text.

I think my insistence that Books 1 and 2 concern rational mechanics and Book 3 physics, as Newton explicitly understood those fields of enquiry, allows us to resolve the puzzle without committing Newton to any equivocation. Newton's title is deliberate, for while his text incorporates mechanics (in Books 1 and 2), it is a mechanics that is targeted towards solving problems in physics, and Book 3 is a book in physics. Moreover, though I will not go through this explicitly here, the "problematic" remarks pointed to by Gabbey are also straightforwardly resolved in light of this understanding of the *Principia*.

So Newton's *Principia* is not a mechanics, but nor is it a physics: it is a *philosophical mechanics* combining a mechanics with a physics. Or so I have argued.

---

[53] See Gabbey 1992, 322.

9.3.4   On Newton's Three-Step Methodology and the "Newtonian Style"

A challenge to my reading comes from Cohen's "Newtonian Style." At first sight I seem to be following Cohen's interpretation unproblematically, for Cohen says Books 1 and 2 are "mathematical," whereas Book 3 is a book in physics.[54] However, our interpretations of what this means diverge in significant ways.

Newton's methodology, Cohen suggests, consists of three phases, and of an iteration between the first and second phases. In the first phase, Newton treats an idealized physical system mathematically without further attention to what is "physically realistic." In the second phase, the mathematically derived results are "compared and contrasted with the data of experiment and observation,"[55] and this allows a new phase one in which further deductions are made mathematically, and so on. This takes place in Books 1 and 2. The final stage of this process, phase three, takes place only when the iterative process of phases one and two reaches a level of development such that comparison with "realities of the external world" can take place,[56] and this is what Newton does in Book 3:

> In bk. three there is a transition from mathematical systems to the realities of the system of the world. . . . Then, and only then, does the question arise as to what can possibly "cause" such an "attraction." (Cohen 1980, 52ff.)

Let's begin with Books 1 and 2. For Cohen, the subject-matter of Books 1 and 2 is idealized physical systems. That is to say, though idealized, the quantities appearing in Books 1 and 2 have a physical interpretation: the bodies are physical bodies, idealized and treated mathematically. I agree with Cohen that Books 1 and 2, though mathematical, are developed with attention to the future applicability of the results thereby obtained. However, I do not think that we should read them as treating idealized physical systems. This obscures the character of Books 1 and 2, in which Newton seeks to give an axiomatized system of rational mechanics as a branch of mathematics treating motions and forces quite generally. As a consequence, it risks inviting a conflation of the mathematical with the physical and leads to a mistaken account of applicability, for more on which see 9.3.5, below.

Turning to Book 3, I agree with Cohen this is a book in physics, but I disagree with him about what this means. For Cohen, the third step in Newton's three-step methodology (see 9.2.2) is completed by Book 3 of the *Principia*, and does not include the consideration of causes. Cohen claims that phases 1–3 of his "Newtonian style" correspond to the three steps, with consideration of causes coming as a "sequel":

---

[54] Cohen 1980, 52ff.

[55] *Ibid*, 63.

[56] *Ibid*, 64.

> Each of the sentences in this paragraph corresponds to one of the three successive phases
> of the Newtonian method in the *Principia*. ... And it is only in a sequel to phase three,
> after the mathematical principles (established in phases one and two) have been applied to
> natural philosophy, that such questions as physical cause or the nature of a force need arise.
> (Cohen 1980, 85)

Later on, when explaining in more detail, Cohen says that phase three is "the use of the principles, laws, and rules found in phases one and two in the elaboration of the system of the world,"[57] and that its "sequel" is "the process of finding a cause of gravity and of understanding how gravity may operate."[58] He says that the move from phase two to phase three is as follows, with causes falling outside the three phases:[59]

> In phase two he found that certain forms of the basic construct (or system) led to an
> agreement with the phenomena to an extent that gave him confidence that the construct
> was not fictive ... Phase three consisted of the elaboration of the system of the world, the
> application of the mathematical principles to natural philosophy. ... In his private world,
> and not in the public world of the *Principia*, he then devoted himself to an exploration of
> the cause of the gravitating force. (Cohen 1980, 110)

In short, Cohen makes the third step—the discussion of physical causes—a "sequel" to his phase three.

The problem is that Cohen's reading does not agree with Newton's own words. According to Newton, Cohen's "sequel" *just is* the third step (see 9.2.2). Conceptually, Newton's three-step method seems clear. First, we have the mathematical treatment (Books 1 and 2). Second, we "come down to physics" and apply that mathematical system to the physical world, in order to identify the physical forces present in the world. Third, we seek "physical species, physical causes, and physical proportions" of those forces. This is the interpretation I advocate.

I have said that the search for causes belongs to the third step of Newton's three-step methodology, which might seem to imply a clean separation between forces (Step 2) and causes (Step 3). This oversimplifies in two ways. First, Newton himself sometimes writes that forces *are* causes, even though in other places he seems to distinguish them.[60] Treating forces as causes makes Step 2 pertain to causes, though it challenges the dominant conceptions at the time of what is involved in giving a causal account in physics, as Janiak (2008) discusses. Second, the issue is tricky because the term 'cause' remains untheorized in the *Principia*, and 'force' receives no general definition. I agree with Janiak (2008, 58) that Newton takes the term "cause"—as commonly understood at the time—to be adequate for his purposes, needing no further elaboration or clarification. In contrast to this, though we get no definition of force in general, in the *Definitions* Newton introduces technical terms

---

[57] *Ibid*, 102.

[58] *Ibid*, 111.

[59] *Ibid*, 110.

[60] See, for example, Newton 1999, 382, which seems to distinguish forces from causes, and Newton 1999, 412, which seems to say that forces are causes.

for the several different force concepts that he uses in developing his arguments. So the relationship between cause and force is complicated. Nevertheless, even if forces are not causes *per se*, knowledge of physical forces is causal knowledge and so, no matter how these complexities are resolved, the important point is that Step 2 as well as Step 3 contributes to our causal knowledge. Both are needed for a complete causal account; both are properly a part of physics (see 9.2.2).[61]

9.3.5 On the Mathematical, the Physical, and the Problem of Applicability

According to Cohen's 'Newtonian style,' the subject-matter of Books 1 and 2 is idealized physical systems, and the applicability of the results of Books 1 and 2 to actual physical systems is achieved by the progressive removal of the idealizations from which we began. These removals takes place in phase 2 (in Books 1 and 2) and phase 3 (in Book 3). This risks a conflation of the mathematical with the physical in a way that the *Principia* is explicitly structured to avoid. Moreover, it leads to a mistaken approach concerning the applicability of Newton's mathematical results to physical systems, or so I argue.

In Books 1 and 2, Newton makes heuristic use of concrete physical systems in developing his mathematical results, and he frequently discusses such systems in scholia. Moreover, the iterative process of his method involves a complex interplay between Book 3 and the resources developed in Books 1 and 2. Newton's method, highlighted by Cohen and demonstrated in detail by George Smith and Bill Harper lies at the heart of what makes Newton's methodology so powerful.[62] All of this makes the temptation to interpret Books 1 and 2 has having physical systems—albeit idealized ones—as their subject-matter very strong. Nevertheless, it would be a mistake.

In the introduction to Book 3, Newton describes the first two books of the *Principia* as being not "philosophical" but rather "strictly mathematical," while at the same time noting that in these two books he *illustrates* the applicability of his mathematical results to actual bodies and motions in several scholia. The *Principia* developed from Newton's attempts to devise mathematical tools for solving problems concerning the motions of physical bodies, and throughout Books 1 and 2 he provides scholia addressing the relevance of his mathematical results to the treatment of physical systems. What this shows is that Newton developed his mechanics with an eye to application. In this, I agree with Cohen.[63] Nevertheless,

---

[61] Newton begins this task for gravitation in the *Principia*, but does not complete it. On the incompleteness of Newton's causal account of gravitation, see Janiak 2008, ch. 3, especially p. 64, and Biener 2018, 3.

[62] See especially Smith 2002, 2014 and Harper 2011.

[63] Cohen 1980, 101.

Books 1 and 2 do not conflate mathematics with physics, nor do they presume a physical subject-matter for a mathematical theory. Newton says that with Books 1 and 2 in place, what remains is "for us to exhibit the system of the world *from these same principles*" (emphasis added). In moving from Books 1 and 2 to Book 3, we move from mathematics (specifically, rational mechanics) to physics. It is only in Book 3 that physical systems become our subject-matter.

The position I have developed here enables me to clarify a puzzle found in Janiak's treatment of the mathematical/physical distinction in Newton's *Principia*.[64] Janiak writes that Newton's mathematical treatment of force "is not mathematical in the sense that it deals solely with mathematical entities;" rather, it deals also with physical quantities.[65] But how does this come about? One answer, to which we might be tempted, is that physical quantities are the subject-matter of Books 1 and 2. I agree with Janiak that this is not the right way to go. According to Janiak, the treatment in Book 1 is "merely mathematical" and "it is only with the physical treatment of book III in the *Principia* that we have something beyond a merely mathematical treatise on motion in general."[66] Janiak suggests that we are able to move from the mathematical treatment to the physical treatment because "the mathematical treatment in Book I enables us to measure any centripetal force through a number of means, thereby enabling us to think of that force as a physical quantity."[67] Understood mathematically, however, Book 1 sets out dependencies among mathematical quantities: it is in this sense that one quantity "measures" another. So here is the puzzle left to us by Janiak. How is it that this mathematical treatment enables us to "*think of* that force as a physical quantity" (emphasis added)? A little more is needed. The right answer, I suggest, lies in the dual nature of the *Definitions* and *Axioms, or Laws of Motion*, and in particular the dual face of the term "measure" (see 9.2.3, above).

I emphasize the importance of separating the mathematical from the physical because failure to do so is widespread in the literature, and leads to a variety of interpretative problems. We have already seen that Garrison 1987 collapses the distinction between the mathematical and the physical, and with what consequences (see 9.3.1). Guicciardini rightly draws attention to Newton's paraphrasing of Pappus in the opening lines of the Preface, in which Pappus distinguishes between the part of mechanics that is "rational" and the part that requires "manual work," but then suggests that the exactness of rational mechanics is a consequence of the subject-matter being natural and governed by mathematical laws, in contrast to the inexactness associated with imperfections of artificially constructed machines.[68] I think this further elaboration is not supported by the text of the Preface, in two crucial respects. First, while Newton's focus is indeed on natural forces, he makes no

---

[64] Janiak 2008, ch. 3.

[65] Janiak 2008, 60.

[66] Janiak 2008, 64.

[67] Janiak 2008, 64.

[68] Guicciardini 2009, 296–7.

suggestion that the principles associated with manmade machines cannot be given an exact treatment in rational mechanics. Indeed, in a text dating from after the first edition of the *Principia*, Newton explicitly incorporates machines in his treatment.[69] Second, Newton makes no mention of mathematical laws in the Preface. This is important for my view because of the dual status that I attribute to the *Axioms, or Laws of Motion* as axioms for rational mechanics and laws of motion for physics. In my view, the exactness of rational mechanics arises directly from its formulation as an axiomatic discipline of mathematics; it does not arise from the physics, which is the domain of the laws.

Modern readers struggle to separate the rational mechanics of the *Principia* from the physics because today we think of mechanics as a branch of physics, and so when reading Book 1 of the *Principia* we see what seems to us to be physics. We read Halley's phrase "the Mathematico-Physical Treatise of the Eminent Isaac Newton" as offering a single label ("Mathematico-Physical") applied to each book of the *Principia*, rather than as a compound label applying to the treatise as a whole. Ducheyne's terminology of "physico-mathematical" for the content of Book 1 is an example of this tendency,[70] and this is how I saw things myself until recently (9.2.3).

However, conflation of the physical with the mathematical disguises the question of how Newton's mathematical results in Books 1 and 2 are applicable to physical systems. If we follow Cohen, we read Books 1 and 2 as treating idealizations of physical systems, and by a successive removal of these idealizations we arrive at mathematical treatments of actual physical systems. Insofar as this solves the applicability problem, it does so by assuming physical content for the mathematics throughout Books 1 and 2. In his highly interesting paper, Smeenk interprets Newton this way, situating him alongside Hobbes and Barrow, whom Smeenk says "collapsed the distinction between 'pure' geometry and physics."[71] According to Smeenk, Newton took geometry—and indeed the mathematics of the *Principia* quite generally—"to apply directly to physical objects, whose geometrical properties are immanent in sensation rather than directly apparent."[72] If this is right, then no problem of applicability can arise: "Since the properties are not ascribed to abstract mathematical entities with a distinctive ontological status, there is no place for a worry to arise regarding how mathematical entities can stand in relation to, or represent, physical objects."[73] However, the idealized physical systems treated in Book 1 are not themselves present in sensation either directly or immanently (think, for example, of the cases where the center of force is at rest, and where the force law differs from inverse-square), and so more must be said: the idealizations involved are theoretical, not material, and the question of their relationship to the

---

[69] See Gabbey 1992, 318.

[70] Ducheyne 2012, 80.

[71] Smeenk 2016, 310–11.

[72] *Ibid*, 314.

[73] *Ibid*, 314.

target physical system remains. Here is one possibility: the actual trajectories of systems are exact geometrical curves (immanent in sensation) and through Newton's method we seek to ever more closely approximate them using exact theoretical trajectories. The problem with this is that the mere instantiation of an *actual* curve by a physical system is not sufficient to underwrite the iterative *counterfactual* reasoning at the heart of Newton's method (reasoning that Smeenk deftly describes), with the unwelcome consequence that Newton's method collapses into curve-fitting. Smeenk would reject this outcome, of course, but I struggle to see how he avoids it.[74]

I want to suggest a different approach. We can turn to Domski 2018 for help. As we have seen, she distinguishes between the rational certainty of Newton's laws and the truth of those laws. The former arises from their being mathematical, and the latter from their applicability to the phenomena. In her view, Newton's laws have an axiomatic status insofar as they are considered mathematical, and therefore rationally certain.[75] Insofar as they are considered physically, their justification is empirical. According to Domski, experimental evidence confirms that the mathematically certain propositions are applicable to physical objects (idealized or otherwise): it shows their truth (2018, 58). This is her account of applicability in the *Principia*.

I think that this is right, so far as it goes, but that it leaves us with a puzzle. It seems that every mathematically certain proposition will have to be independently verified empirically. That a given proposition is *derived* from axioms that are rationally certain and empirically confirmed tells us nothing about the empirical status of that proposition: the connections are mathematical, not physical. The generality of the axioms considered mathematically seems only accidentally related to their universality considered physically. So the applicability of the overall mathematical structure of the *Principia* remains a puzzle. One answer might be to invoke formal causation: if formal causation is both a property of mathematical demonstrations from axioms and the means by which laws govern the behaviors or bodies and forces (Biener and Schliesser 2017), then we see how it is that the deductive structure of the rational mechanics is truth-preserving in the physics. I offer a different answer, one which lies in the dual role of the definitions and axioms/laws as principles of both rational mechanics and physics.

---

[74] What is the ontological status of the idealized physical systems that are the subject-matter of much of Books 1 and 2? Smeenk's view is that for Newton "'rational mechanics' offers as exact a description of material objects as the description of abstract [objects] allegedly provided by mathematics" (310). But idealized physical systems are not material objects, and often the idealization and the target material system have conflicting mathematical properties; so the question of the relationship between them remains. Smeenk also suggests that the object of geometry is space, and I disagree on this point too. I think that, for Newton, geometry remains the science of magnitude. Since space is rich in geometrical properties, geometry is a powerful tool for the investigation of those properties.

[75] As we saw in 3.1, in my view it is the generality of the problems that are solved by means of these principles that justifies our taking them as axiomatic.

I insist on a rigorous separation of the mathematical from the physical: the subject-matter of Books 1 and 2 is *not* physical systems. By appreciating that Book 1 is a book in rational mechanics and Book 3 a book in physics, as I have explained the meanings of those terms, we can remove the above difficulties. While Book 1 is a *mathematical* treatise, its *interest* lies in the fact that it is derived from definitions and axioms that have a dual face, as both mathematical and physical. This is why, in so many of the scholia in Book 1, Newton discusses physical evidence and applications for his results: he is demonstrating the interest of his results for those readers whose concerns lie in physics. I have argued that Newton's *Definitions* and his *Axioms, or Laws of Motion* are independently justified within rational mechanics and within physics, and that this dual status unifies rational mechanics and physics in a very special way. In particular, these principles form a conduit through which rational mechanics may speak about physics, and through which physics may pose problems for rational mechanics. This is because rational mechanics, as a mathematical theory, provides us with a language by which we can speak about anything that satisfies those axioms. In using the very same axioms for rational mechanics and for our laws of physics, we entitle ourselves to use the language of rational mechanics to speak about the physical world. The shared axioms yield a shared logical space for truth-preserving inferences and modal dependencies, Newton's innovative use of the latter being crucial to his method, as George Smith has emphasized. In my view, the best way to understand how the axioms are applicable is not that the laws of motion have axiomatic status in a physical theory of idealized physical bodies. Rather, these principles have two different roles, one as axioms in a rational mechanics (in which their subject-matter may be idealizations of physical systems), and the other as laws of motion in a physics (in which their subject-matter may be physical systems), and this dual character is the route through which applicability flows.[76]

9.3.6 On the Epistemological Status of the Laws, as Laws of Physics

I have said that the epistemological status of the laws, as laws of physics, lies in their universality (see 9.2.3): the wider the range of phenomena to which they may be successfully applied, the greater our justification for taking them as true, within the context of physics.

Biener and Schliesser (2017) argue that Newton introduced the term "axiom" as an indicator of this universality (see 9.3.2). They follow the evolution of Newton's

---

[76] As an interpretative issue concerning Newton's own position, the situation is more complex and I do not pretend to be able to argue conclusively for my approach. My claim is that my approach is consistent with Newton's statements and with his practices, and that it readily handles issues that otherwise remain unsolved.

terminology in the "De motu" manuscripts from his initial use of "hypotheses" through the introduction of "laws," to the introduction of "Axioms" in *Liber Primus*.[77] They point out that Newton replaced "hypotheses" with "laws" at the moment that he: (a) moved from the pessimism of the "Copernican Scholium" (Smith 2007, § 2), in which he despaired of ever determining the true motions of the planets, to optimism; and (b) recognized that his solution would use principles found Galileo's study of terrestrial gravitational motion. Looking at the second version of *Liber Primus*, they point out that the terminology of "axioms" seems to have been introduced at the same time that Newton connected his work to (c) papers on "Laws of motion" by Wren, Huygens and Wallis of the 1660s, in which they offered rules of collision, and (d) ancient mechanics (that is, the study of the five simple machines).

Biener and Schliesser suggest that Newton made these changes because he became increasingly confident that they are true of actual bodies, that they are applicable across a wide range of phenomena celestial and terrestrial, including the ancient machines and collisions, and that they are at least implicit in the mechanics of his predecessors. They claim that his concern, in making these terminological choices, was to signal "the truth of and broad agreement regarding his principles."

The documenting of the changing terminology over time is helpful, and I think Biener and Schliesser are right that the epistemic justification for the laws of motion, as laws in physics, lies in their universality. However, I disagree that this universality is marked by the term "axiom." Notice that (c) and (d) are topics in rational mechanics, so the introduction of the term "axiom" at this time might equally well indicate Newton's decision to attempt a unification of his results on forces and motions within a rational mechanics, as my interpretation would suggest. So I don't take the historical evolution to be decisive either way. Moreover, if I am right then the axioms require justification not just as laws of physics, but also as axioms of rational mechanics, and this demands that we show their generality (as distinct from their universality, see 9.3.2), which is what (c) and (d) allow. Finally, on my interpretation we gain insight from Newton's retention of both "law" and "axiom" in the label for his three principles: the term "law" indicates the universality demanded by physics and the term "axiom" indicates the generality demanded by rational mechanics.

Suppose we agree that it is the universality of the laws of motion that justifies our taking them as laws. One way to support claims of universality is through inductive evidence, and Newton's Rule 4 supports such an approach.[78] This is consistent with the view found in, for example, Harper 2011, according to which Newton's laws of motion are "empirical propositions that have already been sufficiently established to be accepted as guides to research."[79]

---

[77] Biener and Schliesser 2017, 314.

[78] For Rule 4 see Newton 1999, 796. On the inductive justification for the laws see 9.2.3 and 9.3.2.

[79] Harper 2011, 45. In this, they differ greatly from the axioms of natural philosophy found in Rohault's physics (Chapter V). For Rohault, the axioms are "important *Truths*, which are self-evident, and which being the Foundation of all Philosophical Truths, are consequently the principal *Axioms* of Philosophy."

An alternative account of universality is offered by Friedman, who does not accept the view that Newton's laws of motion are justified inductively.[80] He emphasizes that although Newton *presented* them as familiar and accepted, this both disguises the radical innovations in Newton's deployment of these laws, and brushes under the carpet Newton's use of the third law as holding between bodies at-a-distance in the derivation of universal gravitation. In Friedman's view,

> What characterizes the distinguished elements of our theories is rather their special *constitutive function*: the function of making the precise mathematical formulation and empirical application of the theories in question first possible. (Friedman 2001, 80)

For Friedman, Newtonian theory is best understood as having a tripartite structure: a mathematical part that sets up the spatiotemporal framework of the theory (Euclidean geometry); a "mechanical" part that sets up a correspondence between the mathematical part and concrete empirical phenomena (Newton's laws of motion); and a "physical or empirical" part, which attempts to formulate "precise empirical laws describing some concrete empirical phenomena" (the law of universal gravitation).[81] Of most interest for our purposes is the claim that the universality of the laws of motion arises from their constitutive function.

Friedman's overall partition of the *Principia* differs significantly from that presented here. Geometry, as I read it in the context of the *Principia*, is a science of magnitude primarily, and thereby of space (among other things), and the "mathematical part" is the rational mechanics of Books 1 and 2, not just the spatiotemporal framework of the theory. I agree with Friedman that Newton's laws of motion are critical for the applicability of the mathematical to the physical, but he and I differ in how this is achieved. We agree that Newton's treatment of universal gravitation in Book 3 is distinct from his treatment of the laws of motion, and for me this is because it takes place in the physics. Friedman's account partitions and conceptualizes the *Principia* very differently from my approach, but I agree with Friedman that there is a constitutive role for the axioms/laws. I think this holds for both the rational mechanics and the physics, and plays out differently in each. As a result, I would like to return to Friedman's account of the constitutive role of the axioms/laws in light of the dual role of the axioms/laws, but that is a task for another paper.

My point here has been that the justification for taking Newton's laws of motion as axioms for physics lies in their universality, and that this may be justified either inductively or through appeal to their constitutive role.

---

[80] Friedman 2001, 39.

[81] Friedman 2001, 80.

9.4   Conclusions

Stan and I use the label 'philosophical mechanics' for projects that seek to combine rational mechanics with physics into a single account of a given range of phenomena. I have argued that Newton's *Principia* is powerfully understood as just such a project. Approached in this way, we can unite into a coherent whole Newton's description of rational mechanics in the Preface, his claims that Books 1 and 2 are mathematical, his claim that in Book 3 we move from the mathematical to the physical, and his three-step methodology in which we begin with mathematics and then, moving on to physics, we first identify which forces obtain in the world and then consider the causes of those forces. The unification is achieved via the *Definitions* and the *Axioms, or Laws of Motion*, which serve as principles in both the rational mechanics and the physics. The interpretation and justification of these principles has a correspondingly dual aspect.

This interpretation of Newton's *Principia* offers a new angle on existing questions in the secondary literature, including the sense in which Books 1 and 2 are to be understood as "mathematical;" whether or not the *Principia* is a text in mechanics; why Newton came to adopt the dual label "Axioms, or laws of motion;" the epistemic status of the axioms; the relationship between the axioms and the Definitions; in what sense Book 3 is incomplete as a physics; and the problem of applicability (how it is that the mathematics of Newton's *Principia* is applicable to the physical world).

Equally important, for my purposes, is that it provides us with an example of a philosophical mechanics that was highly visible, powerful, and influential throughout the eighteenth century. On the one hand, it set standards for what a successful integration of rational mechanics with physics should look like. On the other hand, Newton's *Principia* offered an *incomplete* physics, and eighteenth century attempts and failures to complete the physics provide us with insights into the limitations of philosophical mechanics as a project for understanding the natural world. In our book, Stan and I follow parallel attempts to provide a philosophical mechanics of collisions. The rules of collision, dating back to Wren, Huygens and Wallis, lie at the heart of any rational mechanics of collisions. A successful philosophical mechanics of collisions would unite such a rational mechanics with a causal story of the collision process by which bodies undergo changes in their state of motion due to contact action.

The lens of philosophical mechanics allows for a comparison of the two options for body-body action that dominated natural philosophy at the time: contact action, as exemplified by collisions, and action-at-a-distance, as exemplified by Newtonian gravitation. The former led to a progressive research program, as the work of George Smith has done so much to demonstrate.[82] The latter did not. For philosophers of the early eighteenth century, this outcome was not in sight. Indeed, it was

---

[82] See especially Smith 2014.

Newtonian attraction, or action-at-a-distance, rather than action-by-contact, that was controversial and in doubt. By the end of the century, the positions were reversed. The successes and failures of each turned out to have long-lasting consequences for philosophy and physics. But that is a story for another day.[83]

To conclude, then. This paper concerns a rather abstract conception of the *Principia*. As such, it is far removed from the details of Newton's work, and the painstaking interplay between empirical data and mathematical reasoning that Smith emphasizes and so masterfully conveys in his research and teaching. My original plan had been to compare the evidential situation in the eighteenth century for gravitational physics/mechanics with that for non-gravitational physics/mechanics. This is more in keeping with Smith's approach, but I never got that far. My comparison was to be carried out within the framework of philosophical mechanics, and the "introductory" section on Newton's *Principia* as a text in philosophical mechanics grew to become the entire paper. Nevertheless, there is a common theme between this paper and Smith's work. In order to appreciate Newton's achievements, we have to go back to the areas of research in which he worked, as they stood prior to the *Principia*, to see the problem-space as it presented itself to people at the time. When we do this, we see in detail the extraordinary ways in which Newton went beyond what anyone else was able to do, transforming entire fields of research in the process. No-one has done more to demonstrate the value of this methodology for Newton scholarship than George Smith.

Acknowledgements I am grateful to Mary Domski, Elaine Landry, Chris Smeenk, and Monica Solomon for their feedback on an earlier draft, and especially to my collaborator in philosophical mechanics, Marius Stan. Their prompting makes me think harder and deeper, the agreements rewarding and the disagreements fruitful.

References

Biener, Z. 2018. Newton's *Regulae Philosophandi*. In *The Oxford handbook of Newton*, ed. E. Schliesser and C. Smeenk. Oxford University Press.

Biener, Z., and E. Schliesser. 2017. The certainty, modality, and grounding of Newton's laws. *The Monist* 100: 311–325.

Brading, K., and M. Stan. 2021. How physics flew the philosophers' nest. *Studies in History and Philosophy of Science* 88: 312–320.

Brading, K., and M. Stan. 2023. *Philosophical mechanics in the age of reason*. Oxford University Press.

Cohen, I.B. 1980. *The Newtonian revolution*. Cambridge University Press.

Cushing, J.T. 1998. *Philosophical concepts in physics*. Cambridge University Press.

De Gandt, F. 1995. *Force and geometry in Newton's* Principia. Trans. C. Wilson. Princeton University Press.

De Risi, V. 2016. The development of Euclidean axiomatics. *Archive for History of Exact Sciences* 70: 591–676.

---

[83] See Brading and Stan 2023.

Dear, P. 1995. *Discipline and experience*. University of Chicago Press.

Domski, M. 2003. The constructible and the intelligible in Newton's philosophy of geometry. *Philosophy of Science* 70: 1114–1124.

———. 2018. Laws of nature and the divine order of things: Descartes and Newton on truth in natural philosophy. In *Laws of nature*, ed. W. Ott and L. Patton, 42–61. Oxford University Press.

Ducheyne, S. 2012. *The main business of natural philosophy*. Springer.

Edleston, J., ed. 1850. *Correspondence of Sir Isaac Newton and Professor Cotes*. London.

Friedman, M. 2001. *Dynamics of reason: The 1999 Kant lectures at Stanford University*. CSLI Publications.

Gabbey, A. 1992. Newton's *mathematical principles of natural philosophy*: A treatise on 'mechanics'? In *The investigation of difficult things*, ed. P.M. Harman and A.E. Shapiro, 305–322. Cambridge University Press.

Garrison, J.W. 1987. Newton and the relation of mathematics to natural philosophy. *Journal of the History of Ideas* 48: 609–627.

Guicciardini, N. 2009. *Isaac Newton on mathematical method and certainty*. MIT Press.

Harper, W.L. 2011. *Isaac Newton's scientific method*. Oxford University Press.

Janiak, A. 2008. *Newton as philosopher*. Cambridge University Press.

Koyré, A. 1965. *Newtonian studies*. Harvard University Press.

Landry, E. 2023. *Mathematics: Method without metaphysics*. Philosophia Mathematica 31:56–80, https://doi.org/10.1093/philmat/nkac027

Miller, D.M. 2017. The parallelogram rule from pseudo-Aristotle to Newton. *Archive for History of Exact Sciences* 71: 157–191.

Molland, A. 1991. Implicit versus explicit geometrical methodologies: The case of construction. In *Mathématiques et philosophie de l'antiquité à l'age classique: Hommage à Jules Vuillemin*, ed. R. Rashed, 181–196. Paris: Éditions du CNRS.

Musschenbroek, P. van. 1744. *The elements of natural philosophy*. Trans. J. Colson. London: J. Nourse.

Newton, I. 1999. *The principia: Mathematical principles of natural philosophy*. Trans. I.B. Cohen and A. Whitman. University of California Press.

Prony, G. 1800. *Mecanique philosophique, ou, Analyse raisonnee des diverses parties de la science de l'equilbre et du mouvement*. Paris.

Rohault, J. 1717. *A treatise of mechanics: Or, the science of the effects of powers, or Moving forces, as apply'd to machines, demonstrated from its first principles*, 2nd ed. Trans. Thomas Watts. London.

———. 1723. *Rohault's system of natural philosophy, illustrated with Dr. Samuel Clarke's Notes*, Vol. I. Trans. John Clarke. London: J. Knapton.

Smeenk, C. 2016. Philosophical geometers and geometrical philosophers. In *The language of nature*, Minnesota studies in the philosophy of science 20, ed. G. Gorham, B. Hill, E. Slowik, and C.K. Waters, 308–338. University of Minnesota Press.

Smith, G.E. 2002. The methodology of the *Principia*. In *Cambridge companion to Newton*, ed. I.B. Cohen and G.E. Smith, 1st ed., 138–173. Cambridge University Press.

———. 2007. Newton's *Philosophiae Naturalis Principia Mathematica*. In *Stanford encyclopedia of philosophy*, Winter 2008 Edition, ed. E.N. Zalta. https://plato.stanford.edu/archives/win2008/entries/newton-principia/.

———. 2014. Closing the loop: Testing Newtonian gravity, then and now. In *Newton and empiricism*, ed. Z. Biener and E. Schliesser, 262–351. Oxford University Press.

Solomon, M. 2017. *On Isaac Newton's concept of mathematical force*. Ph.D. thesis. Notre Dame: University of Notre Dame. https://onesearch.library.nd.edu/permalink/f/1phik6l/ndu_aleph004620022.

Chapter 10
Newton on Quadratures: A Brief Outline

Niccolò Guicciardini

10.1 The Problem of Quadratures

10.1.1 Basic Definitions

When Newton began his mathematical career in the mid-1660s, the problem of determining the tangent to a plane curve was basically solved, either by kinematic methods *à la* Roberval (and this was often the case with "mechanical lines," what we call transcendental curves) or by pre-calculus methods *à la* Hudde (and this was the case with "geometrical lines", what we call algebraic curves) (Malet 1996). True: the formulation of an effective notation and the identification of a basic set of rules was yet to come. In the *anni mirabiles*, Newton expressed the rules as the foundation of a "new analysis" he was to dub "the method of fluxions," but did not publish. The notation and rules of the differential calculus were revealed to the Republic of Letters in 1684 by Leibniz, the result of his independent path to discovery.

In Newton's times, the great open problem was what we call, following Johann Bernoulli's terminology, the problem of "integration." In Newton's terminology this was a more geometrically understood problem, often called the problem of "quadratures," or of "squaring of curves." What did Newton, and his contemporaries, mean by that?

"Squaring a curve" meant calculating the area bounded by the curve. Most notably, "squaring the circle" meant calculating the area bounded by the circumference. "Squaring the hyperbola" meant calculating the area bounded by the hyperbola and, typically, by one asymptote (Jesseph 1999). So far so good. But, in order to

N. Guicciardini (✉)
Università degli Studi di Milano, Milano, Italy
e-mail: niccolo.guicciardini@unimi.it

© The Author(s), under exclusive license to Springer Nature Switzerland AG 2023
M. Stan, C. Smeenk (eds.), *Theory, Evidence, Data: Themes from George E. Smith*, Boston Studies in the Philosophy and History of Science 343, https://doi.org/10.1007/978-3-031-41041-3_10

197

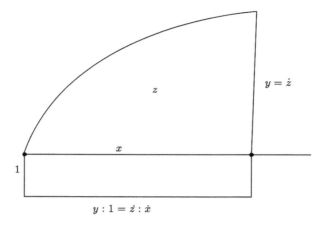

Fig. 10.1 Relations between area z, abscissa x, and ordinate y. ©Niccolò Guicciardini

understand how Newton framed the "problem of quadratures," we have to devote some attention to the basics of the method of fluents and fluxions.

Following in the steps of Isaac Barrow, Newton conceived plane curves as generated by the motion of a point P in the plane (see Fig. 10.1).[1] As the point moves (or "flows," as Newton would say) it traces a curve C. The Cartesian coordinates of the point, x and y, flow in time too, they are "flowing quantities." The "invention of the equation of the curve"[2] consists in finding an equation in x and y, so that C is the locus of points satisfying the equation. Since the coordinates "flow in time," we would say that the curve is expressed by an equation in parametric form. The symbol z denotes the "value of the area" subtended by the curve.[3]

Most often, Newton assumes that $\dot{x} = 1$. This choice simplifies the calculations since all higher order fluxions of x cancel. This assumption deserves our attention. Newton's choice $\dot{x} = 1$ is "equivalent" (what a difficult concept!) to Leibniz's choice $dx = constant$, a typical Leibnizian strategy about which Henk Bos has written magisterial pages (Bos 1974).

Further, Newton distinguished the role of the two "flowing quantities" x and y. He called x the "relate" and y the "correlate" quantity. It is clear that he had some primitive notion of the fact that the equation of the curve expresses a functional relationship between the two coordinates. Let us quote from the so-called *De methodis serierum et fluxionum* (1670ca):

---

[1] On Barrow's influence on Newton, see Feingold's chapter "Isaac Barrow: Divine, Scholar, Mathematician" in Feingold (1990, 1–104).

[2] See the term "inventio æquationis" in Newton (1967, VII, 306).

[3] "areae valor" see, e.g., Fig. 10.3.

Moreover, that the flowing Quantities may the more easily be distinguish'd from one another, the Fluxion that is put in the Numerator of the ratio $[\dot{y}/\dot{x}]$, or the Antecedent of the Ratio, may not improperly be call'd the Relate Quantity, and the other in the Denominator, to which it is compared, the Correlate: Also the flowing Quantities may be distinguish'd by the same Names respectively. And for the better understanding of what follows, you may conceive, that the Correlate Quantity $[x]$ is Time, or rather any other Quantity that flows equally, by which Time is expounded and measured. And that the other, or the Relate Quantity $[y]$, is Space, which the moving Thing, or Point, any how accelerated or retarded, describes in that Time. (Newton 1736, 28–9).[4]

We have to refer to this typical Newtonian construction in order to understand the "problem of quadratures." As the point P traces the curve C, the ordinate y sweeps the surface, whose area is z, subtended by the curve (see Fig. 10.1). Newton proved that the ratio of the fluxion of the area to the fluxion of the abscissa, $\dot{z}/\dot{x}$, is equal to the ratio of the ordinate to a unit segment $y/1$: a statement that we consider equivalent to the fundamental theorem of the calculus, and that Newton often expressed worrying little about geometrical homogeneity as $\dot{z} = y$.

How did Newton prove the above theorem? Newton considered a small increment of the abscissa $B\beta = \dot{x}o$, which he variously termed an "infinitely," "indefinitely" small increment, or a "moment" (see Fig. 10.2). The correlate infinitely small increment of the area z is the rectangle $B\beta HK = \dot{z}o$. It was "easy" to prove that the ratio $\dot{z}o/\dot{x}o$ is proportional to the ordinate $\beta H = y$, since $B\beta \times \beta H = B\beta HK$.[5] As Newton would write, "the fluxion of the areas will be as the ordinates."[6] How Newton came to justify this procedure in terms of limits is a fascinating and well-researched topic, which I will ignore in this chapter.[7]

All this premised. In Newton's mathematical works, the problem of "quadrature" is the following. Given the "equation of a curve," that is an equation in x and y, determine an equation in x and z. The first equation tells how to find the ordinate y for any given value of x (belonging to the interval in which the equation is well-defined). The second equation tells how to find the area z of the subtended surface for any given value of x (belonging to the interval in which the equation is

---

[4] "Præterea quo fluentes quantitates a se invicem clarius distinguantur, Fluxionem quæ in Numeratore Rationis disponitur, sive Antecedentem Rationis haud impropriè Relatam Quantitatem nominare possum, et alteram ad quam referetur, Correlatam; ut et fluentes Quantitates ijsdem respectivè nominibus insignire. Et quo sequentia promptiùs intelligantur, possis imaginari Correlatam Quantitatem esse Tempus vel potiùs aliam quamvis æquabiliter fluentem quantitatem qua Tempus exponitur et mensuratur, et alteram sive Relatam Quantitatem esse spatium quod mobile utcunque acceleratum vel retardatum in illo tempore transigit." Cambridge University Library, MS Add. 3960.14: pag. 26. In Newton 1967, III, p. 88 and 91.

[5] I follow here (with some liberty) the proof in Newton's *De analysi* (Royal Society, MS LXXXI, no. 2). For a detailed commentary, wholly respectful of Newton's notation, see (Guicciardini 2009, 105).

[6] "arearum fluxiones erunt ut ordinatae." See the quotations and discussion of *De quadratura* in (Guicciardini 2009, 204–6).

[7] It might help the reader to note that the rectangle $B\beta HK = \dot{z}o$ is equivalent to Leibniz'a dA. These "equivalences" raise a whole series of issues concerning the comparison between the Leibnizian and the Newtonian algorithms I cannot broach here. See, e.g. Bertoloni Meli (1993).

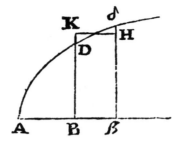

Fig. 10.2 Newton on the relations between area-problems and tangent-problems, from *De Analysi* (1669). Source: Newton (1711), p. 19. Isaac Newton, *Analysis per Quantitatum Series, Fluxiones, ac Differentias: Cum Enumeratione Linearum Tertii Ordinis*. Edited by William Jones. London: Pearson, 1711, p. 19. Source: Author's copy

well-defined). Thus, the problem of quadratures meant associating an equation to a given equation. As we shall see, Newton tabulated these correspondences between equation and equation in pages-long "catalogues" (see Figs. 10.3 and 10.6).

Newton had no problems in thinking about the equation in x and z as the equation of a new curve K with abscissa x and ordinate z, and in defining $\dot{z}/\dot{x}$ as the slope of this new curve. Also in this case, he was following Barrow's lesson. That is why the problem of quadratures was often called the "inverse problem of tangents."[8] The equation in x and y, since $y = \dot{z}/\dot{x}$, measures the slope of the tangent of K. Solving the problem of "quadratures" means therefore: given the equation in x and y of the slope of the tangent of K, determine the equation in x and z of the curve K. We are given the equation of the slope, we have to find the curve. This is exactly the inverse of what we solve with the direct method (in Leibniz's terms the differential calculus) in which we are given the equation of a curve and we have to determine the equation of the slope of the tangent.

Two last observations are in order.

First. What we have just seen implemented in the curve K depends upon a general feature of how Newton understood plane "curves." in Newton's times, curves were of intrinsic interest for the geometer (for example, determining the properties of the cycloid elicited great interest in the middle of the Seventeenth Century). But curves were not only studied for the sake of geometry: they were used as representation tools. They allowed representations of functional relationships between continuous magnitudes, as in the medieval "latitude of forms." This was of enormous importance in natural philosophy. Today, we would say that the mathematicised natural philosophers of the scientific revolution employed "graphs." Galileo, with his triangle representing the variation of speed in relation to time in free fall, had made these graphs famous. Mathematicians such as Evangelista Torricelli, Christiaan Huygens, Newton, and Pierre Varignon also represented the

---

[8] On the origins of the inverse method of tangents in Descartes, see Scriba (1961).

functional dependence of continuous magnitudes, such as times, spaces, speeds, and forces, in terms of graphs—plane curves whereby, for example, the abscissa represents a distance and the ordinate the intensity of force, or the abscissa the time and the ordinate the speed (Roche 1998, 87). Varignon was particularly influential in promoting the systematic use of these graphs, but Newton employed them extensively, most notably in his *Principia* where some curves represent the variation in time or space of speed, force, resistance, pressure, weight, and so on. The curve K represents the variation of the "correlate" quantity (the area z) as dependent upon the variation of the "relate" quantity x (we would say, of z in function of x).[9] One of Newton's examples in the *De Methodis* is

$$z^2 + axz - y^4 = 0, \tag{10.1}$$

where z is the area of the segment ABD of a circle whose diameter is a, abscissa $AB = x$, and ordinate $BD = \sqrt{ax - x^2}$.

Second. Newton was aware that a whole class of interesting problems could be reduced to the "problem of quadratures," namely to the squaring of a plane curve. These problems were, for example, the rectification of curves, the cubature of solids, the calculation of barycenters, the finding of trajectories, the path of light in the atmosphere, etc. Many advanced problems in geometry, statics, mechanics, optics could be solved "given the quadrature of curvilinear figures." Thus one of Newton's tasks consisted in reducing these problems to the quadrature of a plane curve, to the calculation of the area of the surface subtended to a curve.[10] This is an important research project that Newton pursued his whole life. It was clear to a few cognoscenti that the "inverse problem of tangents," or the problem of "quadratures," was central both for the progress of geometry and of mathematical natural philosophy. In proleptic and anachronistic terms (if I may be allowed), this is in essence the problem of integrating ordinary differential equations.

---

[9] "Quinimò si in æquatione quantitates involvantur quæ nullâ ratione geometricâ determinari et exprimi possunt, quales sunt areæ vel longitudines curvarum: tamen relationes fluxionum haud secus investigantur" (Cambridge University Library, MS Add. MS. 3960.14, pag. 20) = "To be sure, even if quantities be involved in an equation which cannot be determined and expressed by any geometrical technique, such as the areas and lengths of curves, the relations of the fluxions are still to be investigated the same way" (Newton 1967, III, 79).

[10] Many propositions in the *Principia* indeed open with the clause "granting a method for squaring curvilinear figures." In these propositions the problem at hand is reduced to the squaring of a plane curve that represents the "correlation" (let us use this term in order to avoid "functional dependance") between continuous magnitudes, such as the distance and the speed of a falling body. For a well-known example, see (Newton 1999, 530).

10.1.2 Evaluations of Newton's Method

As I claimed at the outset, the problem of quadratures was deemed of the utmost importance. As a young Newton wrote in the middle of the 1660s:

> If two Bodys A & B, by their velocitys p & q describe the lines x & y & an Equation bee given expressing the relation twixt one of the lines x, & y [and] ye ratio of their motions q & p; To find the other line y. Could this bee ever done all problems whatever might bee resolved. (Newton 1967, I, 403)

This statement shows how the importance of this problem was clear in Newton's mind since his youth. A few years later, in the so-called *De methodis serierum ed fluxionum* (1671ca.), Newton underlined the importance of quadrature problems as follows[11]

> Observing that the majority of geometers, with an almost complete neglect of the ancients' synthetic method, now for the most part apply themselves to the cultivation of analysis and with its aid have overcome so many formidable difficulties that they seem to have exhausted virtually everything apart from the squaring of curves and certain topics of like nature not yet fully elucidated. (Newton 1967, III, 33)

From the 1670s to the mid-1690s, Newton tried very hard to find a simple rule for the "squaring of curves" and "similar problems," such as rectification and cubature. This was the great open issue for the mathematicians working at the forefront of the research on calculus. In their correspondences and manuscripts, the historian often finds hopes, failed projects, and false announcements about the finding of the golden rule allowing one to integrate as easily as one can differentiate. Just to make an example, in the *Acta Eruditorum* for 1694 one could read, as a premise to a paper by Guillaume de L'Hospital:[12]

> Leibniz proposed a new kind of calculus, which he calls differential ... with the help of which one can easily determine tangents to curves, both geometrical and mechanical, the maxima and minima, the points of inflexion, the evolutes of Huygens, and the caustics of Tschirnhaus ... but what remains to be done is very difficult, that is to find the inverse of that calculus, that is a general method for describing curves if the property of their tangent is given. Several very curious problems depend on this calculus, such as the quadrature of indefinite solids, the dimensions and the surfaces of these solids, the rectification of curved lines, the centre of gravity and oscillation, and many other physical questions.(L'Hospital 1694, 193–4)

Yes, the direct method (differentiation) had been solved: mathematicians, especially thanks to Leibniz's wonderful notation and algorithm, now had a set of simple rules. But what about the inverse (integration)? The extant manuscript evidence suggests that Newton, no matter how incredibly wonderful his early discoveries on fluxions might appear to us, thought to have achieved only its simplest part.

---

[11] Translation by D. T. Whiteside.

[12] The translation from Latin is mine.

I will now surmise a thesis which is difficult to prove. By reading Newton's mathematical work, correspondence, and considering his policy of publication, one gets the impression that his methods designed to attack the problem of quadratures struck him as a confused set of recipes, each valid for a class of curves. The best policy—if I may pursue this tentative historical reading—was to delay publication, hoping to attain the big result: a simple rule to integrate all differential equations! Needles to say, the integral calculus has remained, until our days, to some extent a matter of guesswork.

How did Newton fared in tackling the inverse problem? In general, histories of mathematics devote little attention to his work, and for very good reasons. In the 1690s, Leibniz, the brother Bernoullis, and later Italians such as Jacopo Riccati and Giulio Carlo Fagnano, initiated a program in the integration of differential equations that was to enjoy great success. The results of Newton and his acolytes appear a subchapter that can be ignored.[13]

I should add, in passing, that I dislike the bracketing of dead-ends in the history of mathematics. Failed programs can be as interesting, historically, as the most successful ones. Further, the narrative of a crisis in British mathematics ignores the highly influential works of Roger Cotes (on integration of irrational functions) and Colin Maclaurin (on elliptic integrals). Be that as it may, Jean-Étienne Montucla's (and Joseph de Lalande's) philo-Continental evaluation has more than a few grains of truth.[14]

> The reader should not conclude that Newton resolved the problem [of the integration of differential equation] completely; this would not fit well with what we have said above. Newton's method only delivers the sought relationship [between the independent and the dependent variables] as an infinite series [...] This is why geometers, reserving Newton's method for the most desperate cases, have sought means, both for integrating in finite terms, when this is possible, and for separating the indeterminates. (Montucla 1799–1802, III, 165)

Were Newton's methods of quadrature so desperate, as Montucla and Lalande claim? The purpose of this chapter is to give a brief outline of what Newton achieved in the field. We will encounter methods and results to which Newton gave the utmost importance. These methods are rarely considered by scholars, even by Newtonian scholars, with the exception of course of those, who like George—the dedicatee of this volume—are familiar with the "technical" Newton. My purpose here is not to address the specialists in the history of seventeenth-century mathematics, but rather to offer a reader-friendly primer in Newton's quadrature techniques. I will not help the reader by adapting the notation to our standards: I will strictly adhere to Newton's notation. But I will try to make things as simple as possible by choosing the simplest examples, so simple that the machinery of Newton's quadrature techniques will not show its breadth and strength. Very often we will crack a nut with a caterpillar! I am sure that many of my readers will be surprised by the fact that we will encounter a Newton that is an algebraist rather than a geometer. Therefore, it is in order to round off this introductory section with some cautionary remarks.

---

[13] A recent paper devoted to Newton's quadratures is Malet (2017).

[14] The translation from the French is mine.

10.1.3 Caveat

Newton did not have the concept of function and consequently did not understand the concept of integration as an operator acting on functions. The primary objects to which his mathematics is applied are geometrical: in this chapter we will encounter plane curves. Newton had learnt, however, how to associate an equation to most plane curves. These equations could be either "finite equations" in two variables (let's say, algebraic equations) or "infinite equations" (infinite series). He manipulated these equations, as we shall see, algebraically, showing an extraordinary dexterity in symbolic manipulations. We have however to pay attention not to project on Newton's algebraic methods a modernity that does not belong to their author's mindset.

First, the equations express how geometrical magnitudes embedded in a diagram are related one to another: they should not be understood as functional relationships between real numbers, as it would be the case nowadays. Second, the geometrical and kinematic properties of curves enter into the picture in ways that are no longer into place and that are typical of the mathematical culture of Newton's times. Most notably, the continuity of the motion generating the curve is essential for justifying the limiting procedures implied in Newton's method ($\dot{z}/\dot{x} = y$ is defined as the slope of the tangent of K by appealing to a limit procedure that Newton termed the "method of first and last ratios"). It is often the case that Newton deploys other geometrical properties of the curves in his demonstrations (for example, that the curve is monotonic, or that the curve has a finite curvature, and so on), where we would rather use numerical reasonings. Before commenting the pages of the rather "algebraic" Newton we will encounter in this chapter, we have to pay attention to the geometrical interpretation, rather than numerical, of the symbols occurring in the equations.

But we have to pay attention not to fall into the opposite mistake consisting in labelling Newton as a "geometer," without further qualification. It is often stated that Newton's method was "geometrical," which is true, as I just said. But once Newton associated an equation to a curve, he could proceed by relying mostly upon symbolic manipulation. As I said before, this symbolic aspect of Newton's mathematics is not very well appreciated nowadays. Yet, it was appreciated in his times so much so that Newton's quadrature techniques were developed and extended in the Eighteenth Century by British and Continental mathematicians, most notably by Roger Cotes, Abraham De Moivre, Colin Maclaurin, and Leonhard Euler. I hope that the readers non familiar with this algebraic side of Newtonian mathematics will derive some instruction, and even some intellectual pleasure, by scratching the surface of the treasure trove of algebraic techniques Newton achieved when with a "complete neglect of the ancients' synthetic method," he devoted himself "to the cultivation of analysis."

10.2 Quadratures by Means of "Finite Equations"

As we have seen in the previous Sect. 10.1, in some cases Newton knew how to associate to a curve an equation relating the abscissa to the ordinate of the curve. Squaring the curve meant finding an equation relating the abscissa to the area of the subtended surface.

A first method for squaring curves was based on the so-called "fundamental theorem" of the calculus. Once Newton realized (in 1665) that the algorithm for the squaring of curves is the inverse of the algorithm for determining the tangent, he soon set himself the task of writing down "tables of curves which can be squared by means of finite equations."[15] He applied the direct method of fluxions (differentiation) to increasingly complex equations (with particular interest for equations in which a trinomial is bound by a square root $\sqrt{e + fz^\eta + gz^{2\eta}}$, where η is a rational number and e, f, g are constants) (see Fig. 10.3). The inverse relationship between the algorithm for the squaring of curves and the algorithm for determining the tangent implied that the equations generated by the directed method can be squared in finite terms. I make a simple example.

The direct method applied to the equation of a higher parabola $t = z^n$ leads to the equation of its tangent $y = nz^{n-1}$. Thus, we know that the area subtended to $y = nz^{n-1}$ is $t = z^n$.

It should be noted that the inverse relationship between the two algorithms (what we call the differential and the integral algorithms) was "in the air." Most notably, Barrow had included a proof in his lectures, which Newton certainly knew. Newton, however, was the first to apply the inverse relationship to the construction of what we would call integral tables. This method delivers quadratures in finite, exact form. However, in most cases, Newton realized, one has to resort to approximation techniques.

The so-called *De methodis serierum et fluxionum*, composed in 1670–1, contains two "catalogues of curves" in which one finds tabulated the equations of several curves, divided into different "orders," and the "values of their areas." These catalogues were reprinted, with some variants, in *De quadratura curvarum*, which Newton published as an appendix to the *Opticks* in 1704.[16]

We will briefly consider these two catalogues, following Newton's notation. It is an awkward notation, and the fact that Newton uses δ to represent an arbitrary constant does not help! I have to ask the reader an effort here to adapt to Newton's notation, since I will refer to images of Newton's manuscripts. I aim at adhering

---

[15] Newton's terminology varied. Curves "may sometime be squared by means of finite equations also, or at least compared with other curves (such as conics) whose area may after a fashion, be accepted as known" (Newton 1967, III, 237). Translation by D. T. Whiteside.

[16] Newton might have continued to add and tweak the *De methodis* until the middle of the 1670s. The manuscript (missing the first folio) is Cambridge University Library, MS Add. 3960.14. It is edited and translated in Newton (1967–81, III, 32–328).

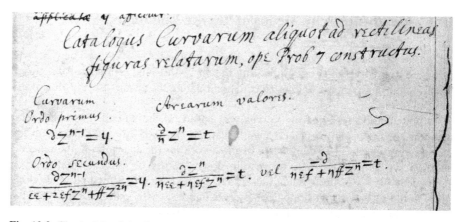

Fig. 10.3 The incipit of the first catalogue. The first catalogue of the *De methodis* summarizes some of Newton's results on quadratures. According to Newton's conventions, δ, e, f, g are constants, η can be a "positive or negative integer or fraction." The curves to be squared have abscissa z and ordinate y, while the areas of the surfaces they subtend are denoted by t or τ. Most of the equations of the curves in the first catalogue involve radicals of the form $R = \sqrt{e + fz^{\eta}}$ or $R = \sqrt{e + fz^{\eta} + gz^{2\eta}}$. Source: Cambridge University Library, MS Add. 3960.14: pag. 77. Reproduced by kind permission of the the Syndics of the Cambridge University Library

strictly to Newton's text, since I want to convince my readers that I am not "algebraizing" the great geometer: in this works Newton was an algebraist.

Let us consider the first tabulated curve of the first catalogue (see Fig. 10.3). We see two columns in which several equations are tabulated. Let us look at the simplest equations: the ones that open the catalogue. Newton uses z and y for the variable abscissa and ordinate of a plane curve. This curve is conceived as generated by the motion of a point P in the plane. As the point moves (or "flows," as Newton would say) it traces a curve. The Cartesian coordinates of the point, z and y flow in time too. We would say that the curve is expressed by an equation in parametric form. Newton further assumes that $\dot{z} = 1$. This choice simplifies the calculations since all higher order fluxions of z cancel.

Further, δ and η are constants: from the context it is clear that Newton assumes that they can be any fractions (in our terms they are rational numbers).

The symbol t or τ denote the "value of the area" subtended by the curve. Since we know that the fluxion of $\tau = (\delta/\eta)z^{\eta}$ is $\dot{\tau} = \delta z^{\eta-1}$, Newton concluded that the "value of the area" of the curve $y = \delta z^{\eta-1}$ is $\tau = (\delta/\eta)z^{\eta}$. A simple consequence of the theorem we have just seen above.

This first catalogue thus provides a list of classes of curves that have equations such that they can be squared by "anti-differentiation," as we would say nowadays.

10.3 Quadratures by Means of the Binomial Series

The binomial theorem was Newton's main result for approximating quadratures, when an exact quadrature "by means of finite equations" is not at hand. In the middle of the 1660s, Newton deployed the binomial for expanding into power series the equation of the hyperbola $y = (1 + x)^{-1}$ and the circle $y = \sqrt{1 - x^2}$. He squared term-wise obtaining series expansions for the logarithm and the elementary trigonometric and hyperbolic functions (see Fig. 10.4).

So, for example, for the hyperbola, Newton wrote

$$y = (1 + x)^{-1} = 1 - x + x^2 - x^3 + x^4 - \cdots , \qquad (10.2)$$

a result that he considered valid when x is small. Newton knew that the area under the hyperbola and over the interval $[0, x]$ for $x > 0$ (and the negative of this area when $-1 < x < 0$) is a measure for $\ln(1 + x)$. Thus, by (to speak anachronistically) "term-wise integration," Newton could express $\ln(1 + x)$ as a power series:

$$x - \frac{x^2}{2} + \frac{x^3}{3} - \frac{x^4}{4} + \frac{x^5}{5} - \cdots . \qquad (10.3)$$

He did not, however, write that the above series is equal to "$\ln(1 + x)$." Until the mid Eighteenth Century, European mathematicians rarely used a notation for transcendental functions. They represented them via geometrical constructions, such as the hyperbolic surface in the case of the logarithm.

One should stress that Newton did not deploy a notation for the logarithm and he deployed a notation for the sine, cosine and tangent only in his manuscript lectures on trigonometry for Newton and his contemporaries these "functions" were actually visualized by geometrical relationships. The logarithm was visualized by the area subtended by a hyperbola and the trigonometric magnitudes were actually "exhibited to the eye," as Newton would say (Newton 1967, VIII, 113, 115), by geometrical relationships between lengths and areas in a circle or in a hyperbola.

10.4 Quadratures by "Comparison to the Areas of Conic Sections"

The above ideas were implemented, to express ourselves anachronistically, in the integration of an ample class of irrational functions in terms of logarithms, trigonometric and hyperbolic functions. I mean that Newton achieved results, just to pick an example, that one would translate into modern Leibnizian notation, such as the following.

$$\int \frac{dw}{\sqrt{w^2 - b^2}} = \text{arccosh} \frac{w}{b} \qquad (10.4)$$

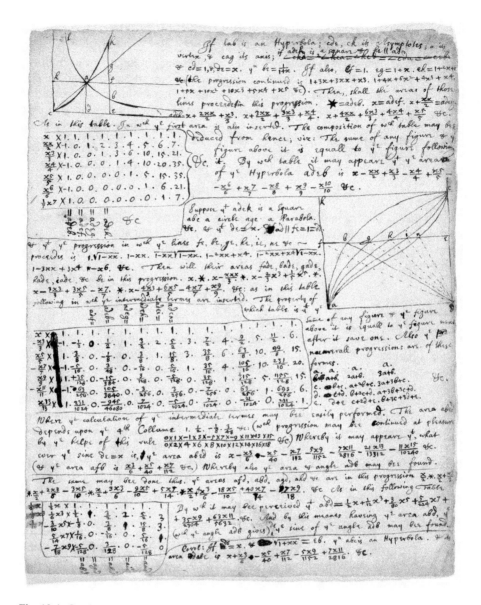

Fig. 10.4 Quadratures of the hyperbola and circle via the binomial theorem Autumn 1665? Source: Cambridge University Library, MS Add.3958.3, fol. 72r. Reproduced by kind permission of the Syndics of Cambridge University Library

What we just wrote is totally anachronistic, of course ... but it helps to understand what is going on here. We just follow the example of Tom Whiteside who had no qualms in translating Newton's mathematics into Leibnzian modernized symbols. This practice helps ... and it is dangerous. As historians of mathematics, we are always trying to strike a balance between domesticating and foregnizing translations, to use Lawrence Venuti's terminology (Venuti 1995).

In more historically accurate terms, Newton reduced the quadratures of classes of rather complicated curves to the squaring of conic sections. What we mean here, of course, is that Newton considered curves which have complicated equations in two variables (which he knew algebraically expressed plane curves). If one could not square a curve "by means of finite equations", one could try to reduce its equation to the equation of a conic section (such as an hyperbola as in the just considered example). The (approximate) quadrature was thus guaranteed by means of "infinite equations" (that is, by means of infinte series). Indeed, conic sections could be (approximately) squared via the binomial theorem.

Let us be clearer here. What Newton did is to transform a "difficult" equation in two variables of a plane curve (such as (10.5) below) into a well-known equation of a conic section (such as (10.6)) by a substitution of variables. The fact that Newton talks about "curves" rather than "equations" is indicative of the fact that the former, not the latter, are the primary objects of his "mathematical ontology" (to use metaphysical heavy-artillery). I mean to repeat here what I wrote in the introductory Sect. 10.1. The objects of Newton's method of fluxions are geometrical: primarily plane curves. In some cases, Newton knew how to "invent" the equation of the curve. In the case at hand, Newton considers curves whose equations are rather complicated for his times, since roots of trinomials occur, as we shall immediately see.

In Newton's words (which we comment in square brackets):

> Curves ... may sometime be squared by means of finite equations [by the first catalogue in Fig. 10.3] also, or at least compared with other curves (such as conics) [by the second cataloogue in Fig. 10.6] whose area may after a fashion, be accepted as known [by the binomial theorem]. (Newton 1967, III, 237)

Newton considered a series of equations (in which in general square roots of trinomials occur) and reduced them to the form of a conic section via a substitution of variable. Some of these equations deserves our attention, since they proved essential for the solution of several problems in the *Principia*. He deviced this technique, tabulating long list of equations (what we would call as integral tables) very soon (see the page from the College Notebook in Fig. 10.8). Let us take a closer look at this method of quadrature.

It is featured in a second catalogue of "curves related to conic sections" occuring in the *De methodis* a few pages after the first catalogue we just considered above (Newton 1967, III, 241–55). This second catalogue too was republished, with some variants, in appendix to the *Opticks* (1704) as part of the *Tractatus de quadratura curvarum*. Basically, by appropriate substitutions of variables, Newton reduces the

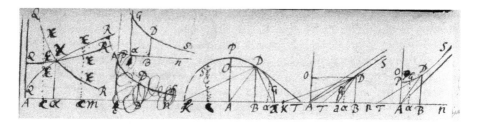

Fig. 10.5 Auxiliary conics for the second catalogue of curves. The quadrature of the curves listed in this catalogue is reduced, via substitution of variable, to the quadrature of central hyperbolas (that is, logarithms), ellipses (that is trigonometric functions), and focal sectors of hyperbolas (that is, hyperbolic functions). Source: Cambridge University Library, MS Add. 3960.14: pag. 75. Reproduced by kind permission of the Syndics of the Cambridge University Library

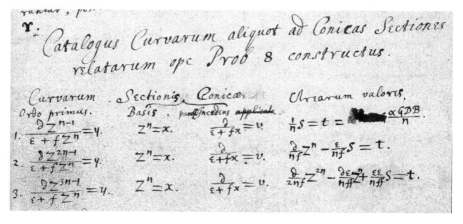

Fig. 10.6 The incipit of the second catalogue. These quadratures were relevant in propositions of the *Principia*, such as Props. 41 and 92, Book 1. Source: Cambridge University Library, MS Add. 3960.14: pag. 79. Reproduced by kind permission of the Syndics of the Cambridge University Library

quadrature of a series of equations divided into several "orders" to the quadrature of conic sections. The conics appear at the top of the table and can be seen in Fig. 10.5.

We see (Fig. 10.6) that in this second catalogue there are four columns. The first column lists, as in the first catalogue, the equations of the curves to be squared divided into orders. As before, their abscissa is z, ordinate y, and area τ. Then we have a second column where the reader learns how to change variable (today we would write $x = f(z)$). The next column lists the equations of conic sections, with abscissa x, ordinate v, and area s. The sought area τ is expressed in terms of the conic area s in the fourth column (in Leibniz's notation, one would write $\tau = \int y\,dz$ and $s = \int v\,dx$). The conic areas, to repeat for clarity, are represented at the top of the table (see Fig. 10.5). As we know, Newton could square conic sections by means of infinite series that we would understand as the Taylor series of log, trigonometric and hyperbolic functions.

All this is best explained by a couple of examples. The simplest is the first case of the first order see (Fig. 10.6). We have to square (and remember, δ is fastidiously a constant!)

$$y = \frac{\delta z^{\eta-1}}{e + f z^{\eta}}. \tag{10.5}$$

With a change of variable $z^{\eta} = x$, the quadrature of the curve is reduced to the quadrature of a hyperbola

$$v = \frac{\delta}{e + fx}. \tag{10.6}$$

Newton would write:

$$y\dot{z} = \frac{\delta}{x(e+fx)}\frac{x\dot{x}}{\eta} = \frac{1}{\eta}\frac{\delta}{e+fx}\dot{x} = \frac{1}{\eta}v\dot{x}, \tag{10.7}$$

and from this conclude that

$$\tau = \frac{1}{\eta}s = \frac{\alpha GDB}{\eta}, \tag{10.8}$$

where $s = \alpha GDB$ is the area subtended to the hyperbola.[17] Newton, of course, knew that s is the logarithm $(\delta/f)\log(fx + e)$, but he would express the area as an infinite series:

$$\tau = \frac{\delta}{\eta f}(1 + \frac{fx}{e} - \frac{f^2 x^2}{2e^2} + \frac{f^3 x^3}{3e^3} - \dots), \tag{10.9}$$

and rather than write a symbol for the logarithm he would represent this transcendental magnitude geometrically in terms of the area subtended under an hyperbola αGDB (see the hyperbola, second from left, in Fig. 10.5).

We have explained in Sect. 10.3 that in Newton's times transcendental functions (as we would say) were represented by a geometrical construction rather than by a symbolic expression. Another clear indication that geometrical interpretation is presupposed in the algebraic reasonings of Newton and his contemporaries. Algebra was not independent from geometry, it was a language used to express geometrical relationships. Just remind what I wrote in the opening Sect. 10.1: the objects of the method of fluxions are curves, the equations (when found) express geometrical relationships embedded in the curve.

---

[17] In Leibnizian notation: $\tau = \int y \, dz = (1/\eta) \int v \, dx = (1/\eta)s$.

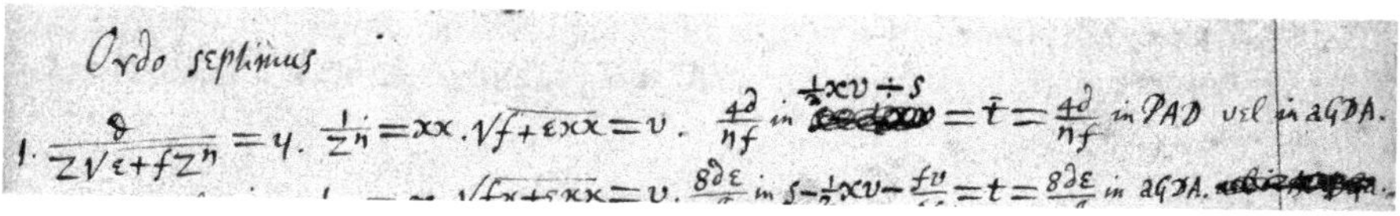

Fig. 10.7 First case of *ordo septimus* in the second catalogue. Source: Cambridge University Library, MS Add. 3960.14: pag. 81. Reproduced by kind permission of the Syndics of the Cambridge University Library

For the purpose of the reader of the *Principia*, a relevant quadrature occurs as the first in *ordo septimus* (see Fig. 10.7).[18] For $\eta = 2$, this translates into the statement that, if a curve has equation

$$y = \frac{\delta}{z\sqrt{fz^2 + e}},$$ (10.10)

then its area is

$$\tau = \frac{2\delta}{f} PAD \text{ or } \frac{2\delta}{f} \alpha GDA,$$ (10.11)

where PAD and αGDA, are the conic sectors in Fig. 10.5. The confusing symbol δ occurs again! Newton was writing all this in 1671, and it seems as if a devil had inspired him to use δ for a constant, a symbol that nowadays we would rather associate to a differential operator.

Note that the change of variable in the second column is very much the one we would employ in a modern treatise on Newtonian mechanics (see Fig. 10.7). Indeed, Newton sets $x = 1/z$. The differential equations for central force motion which depends on this integral is much easier to integrate after this substitution, as Newton explained to Gregory in 1694, and as later authors would discover (e.g. Johann Bernoulli in 1710 and Jacques Binet in 1810) (Newton 1961, 348–54).

In the third column, we find the equation of the auxiliary conics $v = \sqrt{ex^2 + f}$ ($e < 0$ for the ellipse, $e > 0$ for the hyperbola).

Via the change of variable indicated in second column (read $1/z^2 = xx$) the required quadrature is reduced to a simpler form as follows:

$$y\dot{z} = \frac{\delta\dot{z}}{z\sqrt{fz^2 + e}} = \frac{-\delta x}{\sqrt{f/x^2 + e}} \frac{\dot{x}}{x^2} = -\frac{\delta\dot{x}}{\sqrt{ex^2 + f}}.$$ (10.12)

---

[18] In the second *Tabula* of *De quadratura*, this corresponds to the first case of the fourth *Forma*. This quadrature is often employed by Newton in the *Principia*. Newton discussed the use of this and similar integrations in the *Principia* with David Gregory and Roger Cotes. At several points, he even planned to add a treatise on quadratures as an appendix to the *Principia*, in order to allow expert readers to see how the quadratures used in the main text could be achieved. See, Guicciardini (2016).

In modern notation, this corresponds to reducing the sought integral, via substitution of variable, to a easier integral

$$\delta \int \frac{dz}{z\sqrt{fz^2 + e}} = -\delta \int \frac{dx}{\sqrt{ex^2 + f}}. \tag{10.13}$$

Nowadays, we would integrate this in terms of the trigonometric and hyperbolic functions. Thus, for example, we would write (for $e > 0$):

$$\delta \int \frac{dx}{\sqrt{ex^2 + f}} = \frac{\delta}{\sqrt{e}} \text{arccosh} \left(\sqrt{\frac{e}{f}} x \right) + C. \tag{10.14}$$

But Newton, his contemporaries and immediate successors, as we have been repeating, did not express this quadrature in symbolic terms. Rather, they made recourse to a construction in terms of the auxiliary conics. Therefore, the quadrature is provided as a geometric construction in the fourth column on Newton's second catalogue as:

$$\tau = \frac{2\delta}{f} \left| \frac{1}{2} xv - s \right| = \frac{2\delta}{f} PAD, \tag{10.15}$$

where s is the area of the surface subtended under the conic whose abscissa is x and ordinate is v (see Fig. 10.7).[19] Indeed, the absolute value of the difference between the area of the triangle with sides x and v and the area of the region subtended under the conic is equal to the area of the conic sector PAD, in terms of which Newton performs the required quadrature. Or, in modern symbols:[20]

$$\tau = \frac{2\delta}{f} \left| \frac{1}{2} x \sqrt{ex^2 + f} - \int \sqrt{ex^2 + f} dx \right|. \tag{10.16}$$

---

[19] Newton did not use the modern symbol for the absolute value $|\frac{1}{2} xv - s|$ but rather one that he found in Barrow's works. Newton wrote $\div$ for "the Difference of two Quantities, when it is uncertain whether the latter should be subtracted from the former, or the former from the latter."

[20] To recapitulate. The first case of the seventh order translated into Leibnizian notation is as follows. For $\eta = 2$, Newton evaluates the integral $\int \delta/(z\sqrt{e + fz^2}) dz$ (δ, e, f constants). By substitution of variables $z = x^{-1}$, he reduces it to the conic area $s = \int v dx = \int \sqrt{f + ex^2} dx$. Namely,

$$\tau = \int \frac{\delta}{z\sqrt{fz^2 + e}} dz = \frac{2\delta}{f} \left| \frac{1}{2} xv - s \right| + C = \frac{2\delta}{f} \left| \frac{1}{2} x \sqrt{f + ex^2} - \int \sqrt{f + ex^2} dx \right| + C.$$

We note that for the conic $v^2 = ex^2 + f$ (when again we take the hyperbola as our example, $e > 0$) the sector PAD has area equal to

$$PAD = \frac{1}{2}\sqrt{f}\sqrt{\frac{f}{e}}\text{arccosh}\left(\sqrt{\frac{e}{f}}x\right),$$ (10.17)

which makes Newton's geometric quadrature (10.15) "equivalent" to the modern analytic solution (10.14).

Of course, by bracketing the term "equivalent," I hint at a historiographic issue of translation from Newton's to modern terminology and notation that any serious historian of mathematics should broach with great care. I have no space, however, to deal with this issue in this chapter.

10.5   Fluxional Equations

Newton dealt also with what he called "fluxional equations." In Leibnizian terms, he integrated differential equations expanding the integrand into a power series. Newton deployed his algorithmic techniques of series expansion by long division, root extraction, and what he called "resolution of affected equations," a technique similar to that developed much later by Victor Puiseux.

In a simpler case, Newton considered an equation in which "two fluxions together with one only of their fluent quantities are involved" (Newton 1967, III, 91).[21] From Newton's many examples, I propose[22]

$$(\dot{y}/\dot{x})^3 + ax(\dot{y}/\dot{x}) + a^2(\dot{y}/\dot{x}) - x^3 - 2a^3 = 0.$$ (10.18)

Applying his technique for the resolution of affected equations, Newton obtained

$$\frac{\dot{y}}{\dot{x}} = a - \frac{x}{4} + \frac{x^2}{64a} + \frac{131x^3}{512a^2} + \frac{509x^4}{16384a^3} + \cdots,$$ (10.19)

which can be squared term-wise, thus obtaining the relation between the fluents:

$$y = ax - \frac{x^2}{8} + \frac{x^3}{192a} + \frac{131x^4}{2048a^2} + \cdots.$$ (10.20)

This is one of the basic techniques of series expansion employed in the 1669 treatise *De analysi per aequationes numero terminorum infinitas*. It should be noted that the approximation is valid for $x \approx 0$: Newton, that is, obtained a local approximation of the fluent (the integral, in Leibnizian terms).

---

[21] To help the reader's understanding, Newton considers equations of the following form: $f(x, \dot{x}, \dot{y}) = 0$. In the equation (10.18), indeed, two fluxions $\dot{x}$ and $\dot{y}$ as well as a fluent x occur.

[22] Newton (1967), III, 89–91.

In slightly more complex cases, an equation is given in which either two fluxions $\dot{x}$ and $\dot{y}$ together with both the fluent quantities x and y occur (Case 2), or more than two fluxions are present (Case 3). Newton here implemented an algorithm of successive approximations, where again the aim was to express $\dot{y}/\dot{x}$ as an infinite series such as (10.19).

Unfortunately, Newton was not particularly interested (or successful) in the integration of fluxional (differential) equations in closed form. His solutions were always approximate and, as a matter of fact, provided a solution only locally (in the interval of convergence of the series solution). This makes a notable difference by comparison with the techniques that were to be deployed, when Newton was not in the "prime of his age," by the Bernoulli since the 1690s. The Bernoulli sought solutions of differential equations in closed form and in this context developed techniques (of separation of variable, of integrating factors, etc.) that still occupy center stage in our treatises on integration. Montucla and Lalande's judgement was not that groundless! I will expand on this theme in the closing Sect. 10.7.

10.6   The Last Theorems in De Quadratura

Since the mid 1670s Newton sought to generalize the results obtained in Table 2 (which I briefly discussed in Sect. 10.4). For example, he considered curves to which very complex equations were associated. As before, he names the abscissa z and the ordinate y. Newton's first (the simplest!) equation is breathtaking:

$$y = z^{\theta-1} R^{\lambda-1}(a + bz^\eta + cz^{2\eta} + \ldots) \tag{10.21}$$

where

$$R = e + fz^\eta + gz^{2\eta} + hz^{3\eta} + \cdots \tag{10.22}$$

Newton sought to express the area t subtended to the curve (10.21) in the form

$$t = z^\theta R^\lambda S = z^\theta R^\lambda (A + Bz^\eta + Cz^{2\eta} + \ldots) \tag{10.23}$$

The result he achieved is the following. As always I strictly adhere to Newton's notation. He set $r = \theta/\eta,\ s = r + \lambda,\ t = s + \lambda,\ v = t + \lambda,\ \ldots$. Then the area t under the curve (10.21) is

$$t = z^\theta R^\lambda \left(\frac{a/\eta}{re} + \frac{b/\eta - sfA}{(r+1)e} z^\eta + \right.$$
$$\frac{c/\eta - (s+1)fB - tgA}{(r+2)e} z^{2\eta} +$$
$$\left. \frac{d/\eta - (s+2)fC - (t+1)gB - vhA}{(r+3)e} z^{3\eta} + \cdots \right),$$

where each A, B, C ... is the coefficient of the preceding power of z, namely, $A = (a/\eta)/(re)$, $B = (b/\eta - sfA)/((r + 1)e)$, etc. (Newton 1964–67, I, 145). Nowadays, we would say that the coefficients are obtained by a recursive algorithm.

We can just check that the algorithm works as it should with the very basic $y = z^\alpha$. We set $\theta = \alpha + 1$, $a = e = 1$, and $b, c, \ldots, f, g, \cdots = 0$, so that formula (10.21) is $y = z^\alpha \cdot 1^{\lambda-1} \cdot 1$. We substitute in (10.23), so that the area is

$$t = z^{\alpha+1} \cdot 1^\lambda \cdot \left(\frac{1/\eta}{((\alpha + 1)/\eta) \cdot 1} \right) = \frac{z^{\alpha+1}}{\alpha + 1}. \tag{10.24}$$

This is reassuring!

The formula (10.23) is proved in Theorem 3 of *Tractatus de quadratura curvarum*, a work that Newton composed in the early 1690s and published in appendix to the *Opticks* (1704). The Newtonians referred proudly to these theorems on quadratures. I will not deal here with these theorems in any detail.[23] I will just note the following.

First, even though Newton writes S and R as infinite series, the examples he proposes are polynomials. Second, Newton is aware that a constant of integration must be added, but he skillfully chooses what we would call the limits of integration so that this constant is zero. Third, the solutions (such as formula (10.23)) are most often infinite series, even though Newton proposes examples in which the algorithm zeros coefficients after a few steps.

The last point is definitely the most important. Newton obtains power series solutions that work only in the interval of convergence. There is no hint about how to evaluate this interval. One can say that Newton knows "by hand" that the series converges when z is "close" to a certain value. Newtonians, such as James Stirling, got very interested in evaluating this interval and improving the speed of convergence. This was one of the most important research fields promoted by Newton in the eighteenth-century "fluxional" school. Do you still beleive that the Newtonian heritage caused a decline in British mathematics because it was "geometrical" (see Figs. 10.8 and 10.9)?

10.7 Concluding Remarks

Newton's approach to quadrature cannot but appear inferior to the Leibnizian methods, especially those developed in the Basel school by Jacob Bernoulli, Johann Bernoulli and Jacob Hermann ... not to speak of Leonhard Euler. It should be noted, however, that Newton developed his quadratures in the decade from the middle 1660s to the middle 1670s: he continued perfecting them until the early

---

[23] See the commentary by Gerhard Kowalewski in Newton (1908) and the recent beautiful paper Malet (2017).

Fig. 10.8 Early quadratures by substitution of variable in the College Notebook (mid 1660s) Source: Cambridge University Library, MS Add. 4000: fol. 156v. "The area *abc* of the line" (first column) is "equall to the area of the line" (second column), "supposing that" (third column). The third column tabulates the pertinent variable substitution. Reproduced by kind permission of the Syndics of the Cambridge University Library

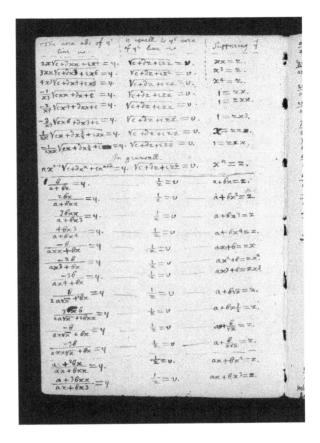

1690s, when the Basel program in integration got off the ground. The differences between Newton's methods of quadratures (pursued, to name a few, by Newton, Roger Cotes, Brook Taylor, Colin Maclaurin, Thomas Simpson) and the methods of integration pursued in the Leibnizian network (most notably in the Basel school of the Bernoullis) are many. The topic is very complex and would deserve a whole paper, if not a book. Here I note just the following.

The Newtonians preferred integration via infinite series to integration in closed form.[24] Most of Newton's methods of quadrature were in terms of series expansions. His methods of quadrature expounded in *De quadratura* (1704), celebrated by Nicolas Fatio de Duillier and David Gregory (see Sect. 10.6), provide only a local solution (they can be calculated in the interval of convergence of the series solution). The Leibnizians, instead systematically searched solutions in closed form. That

---

[24] This should be read as a tendency, not as a rule. It is true that Cotes and Maclaurin, for example, contributed to techniques of integration in closed form. Of course, series solutions were studied also in the Bernoulli's school.

$$[183]$$

Et fi rectangulorum illorum coefficientes nume-
rales fint refpective

$$\tfrac{1}{n}\theta = r. \qquad r\text{-}|\text{-}_\lambda = s. \qquad s\text{+}|\text{-}_\lambda = t. \qquad t\text{-}|\text{-}_\lambda = v. \ \&c.$$

$$r\text{+}|\text{-}_\mu = \acute{s}. \quad s\text{+}|\text{-}_\mu = \acute{t}. \quad t\text{-}|\text{-}_\mu = \acute{v}. \quad v\text{-}|\text{-}_\mu = \acute{w}. \&c.$$

$$\acute{s}\text{-}|\text{-}_\mu = \ddot{t}. \quad \acute{t}\text{+}|\text{-}^\mu = \ddot{v}. \quad \acute{v}\text{-}|\text{-}^\mu = \ddot{w}. \quad \acute{w}\text{+}|\text{-}_\mu = \ddot{x}.\&c.$$

$$\ddot{t}\text{-}|\text{-}^\mu = \dddot{v}. \quad \ddot{v}\text{+}|\text{-}^\mu = \dddot{w}. \quad \ddot{w}\text{-}|\text{-}^\mu = \dddot{x}. \quad \ddot{x}\text{-}|\text{-}_\mu = \dddot{y}. \ \&c.$$

area Curvæ erit hæc

$$z^\theta R^\lambda S^\mu \ in \ \frac{\tfrac{1}{n}a}{rek} + \frac{\tfrac{1}{n}b \ {-sfk \atop -s'el}A}{\overline{r\text{-}|\text{-}1},ek}z^n + \frac{\tfrac{1}{n}c \ {-s\overline{+1},fk \atop -s'+1,el}B \ {-tgk \atop {-t'fl \atop -t'em}}A}{\overline{r\text{+}2},ek}z^{2n} +$$

$$+ \frac{\tfrac{1}{n}d \ {-s\overline{+2},fk \atop -s'+2,el}C \ {-t\overline{+1},gk \atop {-t'+1,fl \atop -t''\text{-}|\text{-}1,em}}B \ {-v \ hk \atop {-v' \ gl \atop {-v''fm \atop -v'''en}}}A}{\overline{r\text{+}3},ek}z^{3n} + \&c.$$

Ubi A denotat termini primi coefficientem datam
$\frac{\tfrac{1}{n}a}{rek}$ cum figno fuo $+$ vel $-$, B coefficientem datam
fecundi, C coefficientem datam tertii, & fic deinceps.
Terminorum vero, a, b, c, &c. k, l, m, &c. unus
vel plures deeffe poffunt. Demonftratur Propofitio
ad modum præcedentis, & quæ ibi notantur hic ob-
tinent. Pergit autem feries talium Propofitionum in
infinitum, & Progreffio feriei manifefta eft.

PROP.

is: solutions to differential equations expressed in terms of elementary algebraic operations, the logarithm, or trigonometric functions.[25]

An example of a Newtonian solution of a fluxional (i.e. differential) equation can help explain this feature.[26] Let us consider the fluxional equation

$$\dot{y} = 1 - 3x + y + x^2 + xy, \tag{10.25}$$

with initial condition $y(0) = 0$. An approximation in the vicinity of the origin is $\dot{y} = 1$ (we retain only terms of zero-degree on the right-hand side). Squaring (i.e. integrating), we obtain $y = x$. Substituting this value in the original equation and retaining only terms of first-degree on the right-hand side, we get an approximation in the vicinity of the origin: $\dot{y} = 1 - 2x$. Squaring, we obtain $y = x - x^2$. Substituting this value in the original equation and retaining only terms of second-degree on the right-hand side, we get an approximation in the vicinity of the origin: $\dot{y} = 1 - 2x + x^2$. Squaring, we obtain $y = x - x^2 + x^3/3$. This process can be reiterated so that one stepwise obtains the higher-order terms of the series. Such a method allows a numerical approximation of the solution in the interval of convergence of the series, in this case in the vicinity of the origin, but does not deliver any information on the global features of the solution (in modern terms, we can approximate the graph in the vicinity of the origin).

Leibniz and the Basel mathematicians instead searched for integrations of ordinary differential equations in closed form, that is, in terms of algebraic or transcendental functions. The example of the solution of the catenaria problem can illustrate the advantage of closed solutions over series expansions. The solution obtained, $y(x) = K + c\cosh((x - x_0)/c)$, provides information on the global character of the solution (in modern terms: we can plot a graph over the R-axis). Another example might be the solution of the differential equation for damped oscillations. Getting a solution in the form $y(t) = A\exp^{-\zeta\omega_0 t}\sin\left(\sqrt{1 - \zeta^2}\omega_0 t + \phi\right)$, rather than as a power series, provides information on the global characteristics of the solution (its amplitude A, phase ϕ, the undamped angular frequency ω_0, the damping ratio ζ, the decay time $\tau = 1/(\zeta\omega_0)$).

The difference between the Newtonian and the Leibnizian programs in integration is correlated with the difference in the programs in mechanics they investigated. Newton was exploring the mathematization of mechanical problems of unthinkable difficulty for his age (from the study of planetary perturbations to that of tidal motions). The mathematical models he required were non-integrable in closed form: the best he could do was to devise numerical approximations, and his integration techniques via infinite series provided just that. One might note, for example, that the important problem of central force motion can be integrated in closed form in terms of elementary (circular and hyperbolic) functions only for

---

[25] The classic paper on the different approach to the inverse method in Newton's and Leibniz's works is Scriba (1964).

[26] Discussed in Giusti (2007, 45–6).

a handful of force laws ($F = kr^n$, for $n = 1, -2, -3$), exactly those covered by Newton in the *Principia*.[27] The Newtonians were interested in integration via infinite series because in most cases this was the only viable choice, given the imperfect knowledge of integration techniques at the time. Leibniz and the Basel mathematicians, instead, devoted their attention to problems (such as the brachistochrone, the catenaria, and orthogonal trajectories) that—as we have just seen—were considered rather dull questions to ask in Newton's milieu; yet they led to interesting problems in integration in closed form. This difference between the Newtonian and the Leibnizian integration methods was emphasized by one of Johann Bernoulli's most gifted pupils, Pierre-Louis de Maupertuis, who in 1731 wrote:

> It is true that this method of series that we owe to M. Newton is general, and the only absolutely general method that the integral calculus has; but it is also true that the solutions gotten by using it are very far from the elegance of the solutions found by integration or quadratures; one should only consider it as the last resort in the hopeless cases.[28]

This evaluation is echoed by the quote from Montucla and Lalande's *Histoire* with which I opened this chapter.

After this very cursory comparison between the Leibnzian integration techniques and Newtonian quadratures, it is apposite to conclude with a few brief observations. Newton's methods are notable for (at least!) four reasons. For the sake of clarity, I will express these in anachronistic terms. To be purists we should retranslate back into the historical actors' language and conception (as explained is Sect. 10.1.1), which we have followed in this chapter: most notably, the algebraic expression "integrating a function" should be understood geometrically as the "squaring of a curve."

1. Newton deployed the (so-called) fundamental theorem in order to integrate by anti-differentiation. In the 1660s, the inverse relationship between differentiation and integration was understood by a number of mathematicians (including Barrow), but Newton (and Leibniz), it seems, were the first to understand that the fundamental theorem could be used in the making of integral tables.
2. Newton used the binomial theorem to integrate the equations of conic sections via power series expansion. He obtained the series for the logarithm, the exponential, the trigonometric and hyperbolic (direct and inverse) functions.
3. Newton was interested in integrating irrational functions in which typically a trinomial is bounded under a radical. By substitution of variable, Newton reduced these equations to equations of conic sections that he could integrate by the binomial theorem. Nowadays we express his results by integrating irrational functions in terms of elementary transcendental functions (ln, sin, sinh, etc.).

---

[27] In propositions I.10, I.11–13, I.41 (Cor. 3). For a limited number of other exponents n, closed solutions in terms of elliptic functions (not available to Newton's contemporaries) are possible.

[28] Translated in (Greenberg 1995, 253). Maupertuis's statement remained unpublished (it can be found in the minutes of the Academy of Sciences), yet it must have reached Montucla or Lalande who echoed it in the *Histoire*, see Sect. 10.1.

These Newtonian integrations are essential in order to understand some results in the *Principia*.
4. Most of Newton's methods provide a solution, but only locally. Newton integrated rather complex irrational functions and ordinary differential equations via power (and even fractional power) series expansions. The solution is given only locally in the interval of convergence. Leibniz and his followers, instead, sought solutions in closed form and developed techniques that proved to be more successful.

As Montucla and Lalande were to state at the end of the Eighteenth Century, Newton's methods were used only in "the most desperate cases." The progress of the theory of integration during the Eighteenth Century made Newton's methods appear just archaic. Indeed, most of Newton's methods of quadrature are quite general, but they have a drawback: they provide only local solutions. The mathematician, however, will appreciate Newton's foresight in understanding the potential of the fundamental theorem and his algebraic dexterity in integrating, by variable substitutions, irrational functions in terms of "elementary functions." The historian, especially when inspired by George Smith's lectures on the *Principia*, will appreciate that behind some geometrical constructions featured in the great classic there are the substitution of variables and quadrature by reduction to conic areas that appear in Newton's *De methodis* of the early 1670s. Newton the discoverer of the fluxional calculus and Newton the author of the *Principia* are, after all, the same man, as his close correspondents knew very well. Or better, he was the same mathematician who left in all his mathematical works an unmistakable signature: *ex ungue leonem.*

Acknowledgments I would like to thank Marius Stan for his correspondence and competent editing. Toni Malet for sharing many insights on Newtonian mathematics with me. This research was funded by the Department of Philosophy "Piero Martinetti" of the University of Milan under the Project "Departments of Excellence 2018–2022" awarded by the Ministry of Education, University and Research (MIUR).

References

Bertoloni Meli, Domenico. 1993. *Equivalence and Priority: Newton versus Leibniz*. Oxford: Clarendon Press.

Bos, Henk. 1974. Differentials, Higher Order Differentials and the Derivative in the Leibnizian Calculus. *Archive for History of Exact Sciences* 14(1): 1–90.

Feingold, Mordechai (ed.). 1990. *Before Newton: The Life and Times of Isaac Barrow*. Cambridge: Cambridge University Press.

Giusti, Enrico. 2007. *Piccola Storia del Calcolo Infinitesimale dall'Antichità al Novecento*. Roma: Istituti Editoriali e Poligrafici Internazionali.

Greenberg, John L. 1995. *The Problem of the Earth's Shape from Newton to Clairaut: The Rise of Mathematical Science in Eighteenth-Century Paris and the Fall of "Normal Science"*. New York: Cambridge University Press.

Guicciardini, Niccolò. 2009. *Isaac Newton on Mathematical Certainty and Method*. Cambridge (Mass): MIT Press.

Guicciardini, Niccolò. 2016. Lost in translation? Reading Newton on inverse-cube trajectories. *Archive for History of Exact Sciences* 70(2): 205–241.

Jean-Étienne Montucla. 1799–1802. *Histoire des Mathématiques*. Revised by Joseph Jérôme L. de Lalande, 4 vols. Paris: Henri Agasse.

Jesseph, Douglas M. 1999. *Squaring the Circle: The War Between Hobbes and Wallis*. Chicago: University of Chicago Press.

L'Hospital, Guillaume F. A. 1694. Solutio problematis geometrici. Acta Eruditorum (May 1694): 193–194.

Malet, Antoni. 1996. *From Indivisibles to Infinitesimals: Studies on Seventeenth-Century Mathematizations of Infinitely Small Quantities*. Bellaterra: Universitat Autònoma de Barcelona.

Malet, Antoni. 2017. Newton's Mathematics. In *The Oxford Handbook of Newton*, ed. Eric Schliesser and Chris Smeenk (online edn, Oxford Academic, 6 Feb. 2017).

Newton, Isaac. 1704. *Opticks: Or, A Treatise of the Reflexions, Refractions, Inflexions and Colours of Light, Also Two Treatises of the Species and Magnitude of Curvilinear Figures*. London: for S. Smith and B. Walford, p. 183.

Newton, Isaac. 1711. *Analysis per Quantitatum Series, Fluxiones, ac Differentias: Cum Enumeratione Linearum Tertii Ordinis*. Edited by William Jones. London: Pearson.

Newton, Isaac. 1736. *The Method of Fluxions and Infinite Series*. London: H. Woodfall for J. Nourse.

Newton, Isaac. 1908. *Newtons Abhandlung über die Quadratur der Kurven. (1704). Aus dem lateinischen übers. und hrsg. von Dr. Gerhard Kowalewski*, Leipzig, Ostwalds Klassiker der exakten Wissenschaften: W. Engelmann.

Newton, Isaac. 1961. *The Correspondence of Isaac Newton*, Volume III (1688–1694). Edited by Herbert W. Turnbull, Cambridge: Cambridge University Press.

Newton, Isaac. 1964–1967. *The Mathematical Works of Isaac Newton*. Edited by Derek T. Whiteside. 2 vols. New York: Johnson Reprint Corp.

Newton, Isaac. 1967–1981. *The Mathematical Papers of Isaac Newton*. Edited by Derek T. Whiteside. 8 vols. Cambridge: Cambridge University Press.

Newton, Isaac. 1999. *The Principia: Mathematical Principles of Natural Philosophy ... Preceded by a Guide to Newton's Principia by I. Bernard Cohen*. Translated by I. Bernard Cohen and Anne Whitman, assisted by Julia Budenz. Berkeley: University of California Press.

Roche, John. 1998. *The Mathematics of Measurement: A Critical History*. London: The Athlone Press.

Scriba, Christoph J. 1961. Zur Lösung des 2. Debeauneschen Problems durch Descartes: Ein Abschnitt aus der Frühgeschichte der inversen Tangentenaufgaben. *Archive for History of Exact Sciences* 1(4): 406–419.

Scriba, Christoph J. 1964. The Inverse Method of Tangents: A Dialogue between Leibniz and Newton (1675–1677). *Archive for History of Exact Sciences* 2(2): 113–137.

Venuti, Lawrence. 1995. *The Translator's Invisibility: A History of Translation*. London and New York: Routledge.

Chapter 11
A Tale of Two Forces: Metaphysics and its Avoidance in Newton's *Principia*

Andrew Janiak

Isaac Newton did more than any other early modern figure to revolutionize natural philosophy, but he was often wary of other aspects of philosophy. He had an especially vexed relationship with metaphysics. As recent scholarship has highlighted, he often denounced metaphysical discussions, especially those in the Scholastic tradition (Levitin 2016). He insisted that he himself was not engaging with the aspect of philosophy that played such a prominent role in the work of his predecessors, especially Descartes, and his critics, especially Leibniz. In the twentieth century, the most influential interpretation of Newton proclaimed that he successfully avoided metaphysics, and indeed, transformed natural philosophy into something akin to modern physics precisely by excising metaphysical disputes from the enterprise. From this point of view, Descartes' natural philosophy was unsuccessful because it was hampered by metaphysical baggage and Leibniz resisted the most profound conclusions of Newton's natural philosophy due to his continued allegiance to outmoded metaphysical ideas.

The most profound conclusion of Newton's *Principia* was also the one most stridently rejected by Leibniz, namely, that all bodies gravitate toward one another in accordance with a new law of nature. The law of universal gravity ought to be the apotheosis of Newton's new approach to answering questions about nature, a sign of his triumph over his metaphysically-obsessed predecessors and interlocutors. And yet it is precisely here, as we will see, that Newton's vexed attitude toward metaphysics is apparent. His attempt to leave an old discipline behind was not entirely successful. Moreover, when it came time to sum-up his most cherished beliefs in the General Scholium to the *Principia* in the midst of his vociferous debate with the Leibnizians, Newton surprisingly chose to make a metaphysical argument

A. Janiak (✉)
Department of Philosophy, Duke University, Durham, NC, USA
e-mail: janiak@duke.edu

223

M. Stan, C. Smeenk (eds.), *Theory, Evidence, Data: Themes from George E. Smith*, Boston Studies in the Philosophy and History of Science 343, https://doi.org/10.1007/978-3-031-41041-3_11

about powers and substances. He embraced a view that was not out of place amidst the prominent metaphysical discussions of the early modern period. As we will see, these two aspects of the *Principia* each involve the notion of a force or power. And they each help to highlight the extent to which Newton made great strides in transforming natural philosophy even while keeping one foot firmly planted in the metaphysical past.

In the first seven propositions of Book III in the *Principia*, Newton articulates a series of increasingly general conclusions about the impressed force he had studied in the greatest depth, namely gravity. As George Smith pithily puts it, these propositions express a series of "increasingly problematic theses" (1999, 32). We proceed from considering Jupiter and its satellites, to the planets orbiting the sun, to the moon orbiting the earth, until we reach the proposition with greatest generality and the one that is most problematic. Proposition Seven reads as follows: "Gravity is in all bodies universally [*Gravitatem in corpora universa fieri*], and is proportional to the quantity of matter in each."[1] The latter half of this conclusion was the result of considerable empirical work on Newton's part.[2] His discovery that the force of gravity is proportional to the *quantitas materiae* (i.e., the mass) of a body was a key aspect of his startlingly novel conception of many phenomena in nature that were previously conceived of as disparate, from the planetary orbits to the tides to the free fall of bodies. But the conclusion did indeed pose a problem: Newton had not merely stated that gravity affects, or is present in, all bodies within the reach of current observations, or those bodies between Earth and Jupiter's solar orbit, etc. He had instead stated, *Gravitatem in corpora universa fieri*. What does that mean?

The first sign of trouble is that Newton's conclusion resists easy translation. Cohen and Whitman render it as: "Gravity exists in all bodies universally," but *fieri* is not normally rendered by *exists*, and of course for a philosopher, the latter term is not to be used lightly.[3] One might follow the first English translation of the *Principia*, Andrew Motte's work of 1728: "Gravity is in all bodies universally."[4] That rendering has the benefit of avoiding difficulties with the word 'exists,' but it is not quite literal. The most literal rendering may be: "Gravity is effected in all bodies universally." Unfortunately, these do not *seem* to be equivalent statements because the terms in question do not seem synonymous; none of the statements lacks difficulties.

---

[1] Throughout this paper, I typically use the now standard translation of *Principia mathematica* by Cohen and Whitman (Newton 1999), deviating from its familiar renderings only when necessary. The Latin original is taken from Cohen and Koyré (Newton 1972).

[2] No one has done more than George Smith to illuminate the way in which Newton marshaled evidence for his conclusions in the *Principia*. See, e.g., the now classic account in Smith (1999).

[3] Many thanks to George Smith for his expert guidance in translating Newton's sentence.

[4] However, in the Motte translation, as modified by Cajori, Proposition Seven reads that "there is a power of gravity pertaining to all bodies," which obviously raises other questions (Newton 1960, 414). In her translation, Du Châtelet has: "La gravité appartient à tous les corps" (Newton 1759, vol 2, 21).

Newton was well aware that the conclusion of his theory of gravity, expressed in Proposition Seven of the *Principia*, was confusing. So, at the height of the dispute with Leibniz and his supporters, he made significant alterations for the second edition of 1713. In that edition, published under the editorship of Roger Cotes with the support of Samuel Clarke and Richard Bentley, Newton added a now famous section called the *Regulae philosophandi*. In part, it reads as follows (Newton 1999, 796; slightly modified translation):

> Those qualities of bodies that cannot be intended and remitted and that belong to all bodies on which experiments can be made should be taken as qualities of all bodies universally . . . if it is universally established by experiments and astronomical observations that all bodies on or near the earth gravitate toward the earth, and do so in proportion to the quantity of matter in each body, and that the moon gravitates toward the earth in proportion to the quantity of its matter . . . it will have to be concluded by this third rule that all bodies gravitate toward one another . . . Yet I am by no means affirming that gravity is essential to matter. By inherent force I understand only the force of inertia. This is immutable. [*Attamen gravitatem corporibus essentialem esse minime affirmo. Per vim insitam intelligo solam vim inertiae. Haec immutabilis est.*] Gravity is diminished as bodies recede from the earth.

Newton emphasizes that he has the appropriate empirical evidence to conclude *Gravitatem in corpora universa fieri*. The kind of inductive inference undergirding this conclusion was analogous to other inductive inferences that had been employed to considerable effect by modern mechanical philosophers.

The last few lines of the quotation above, however, reflect Newton's awareness that in addition to the question about the empirical evidence required for a certain type of inductive inference, there was also the question of how to *understand* what Proposition Seven means. Newton says that *Gravitatem in corpora universa fieri* does *not* mean *Gravitatem corporibus essentialem esse*. This implies that such an interpretation of Proposition Seven might be favored by some of his readers, but also indicates something deeper: Newton himself regarded the notion of an essential property as an appropriate interpretive category. For had he thought otherwise, he presumably would have remained silent, or said something to the following effect: some philosophers have suggested that I regard gravity as essential to matter, but the notion of an essential property is irrelevant for my work. Instead, Newton took the notion of an essential property to be relevant by denying that he was affirming that gravity is essential to matter. The relevance of traditional philosophical categories is also clear from the next sentence: "By inherent force I understand only the force of inertia." He thought of the *vis inertiae* as essential to matter; why else would this be the next remark? Newton regarded the notion of an essential property—he also calls it "immutable" here—as relevant to his physics; the point is that he regards the *vis inertiae*, but not gravity, as essential to matter. That is, he thinks that bodies have an immutable or essential power to resist acceleration, but he is not saying that they have an immutable or essential power to attract.

So, metaphysical categories are relevant to interpreting Newton's claim and the task is to apply them correctly.[5] Unfortunately, Newton's remarks were not

---

[5] Many thanks to Marius Stan for several conversations on these points—his challenges helped me to sharpen my thinking in this area.

clarificatory: under what interpretation of "essential" and "immutable" would those be potential renderings of Proposition Seven? What does it mean to say, "by inherent force I understand only the force of inertia"? Is the idea that if a force is inherent, it is also essential, and an essential one is immutable? Or are these synonyms?

Newton's famously cautious approach toward traditional metaphysical topics was not the sole cause of his readers' confusion.[6] In addition, the idea that gravity is universal is dissimilar to other universal claims. For instance, the common thought, "All bodies are extended," is easily understood and can then be debated. Similarly, if a philosopher says, "All bodies move," or, "All bodies are moving," once again her readers would have no trouble understanding. Of course, they might dispute the claim, deny that we could know it to be true, or the like, but those reactions are predicated on understanding it. The problem is that in Newton's milieu, gravity was not antecedently understood to be the kind of item—such as a property— whose universal status can be contemplated. Contemporary readers may find this puzzling: for us, the notion of a universal force of nature seems straightforward. However, if we consider Newton's definitions of force in the *Principia*, we can come to recognize that our attitude and that of Newton's contemporaries are essentially inverted.

The Definitions that open the *Principia* were a section of special interest to mathematicians and philosophers.[7] For the most part, Definition Three has been lost to history: "Inherent force of matter is the power of resisting by which every body, in so far as it is able, perseveres in its state of resting or of moving uniformly straight forward." This is identical to the mass of a body except for the way in which we consider it, and therefore can also be called, Newton tells us, by "the very significant name of force of inertia." With Definition Four, Newton introduces the notion of a *vis impressa*: "Impressed force is the action exerted on a body to change its state either of resting or of moving uniformly straight forward." He then adds in Definition Five that gravity (along with magnetism) is an example of an impressed force (viz., a centripetal type). Hence gravity is "an action exerted on a body" to change its state of motion. The addendum to Definition Four is also significant: "This force consists solely in the action and does not remain in a body after the action has ceased." Now for us, whereas the first force to be defined, the so-called *vis inertiae*, is not actually a force at all, the second to be defined, the *vis impressa*, is *obviously* a force. But to Newton's contemporaries, whereas the idea of a *vis*, meaning a power (*potentia*) of body, would have been familiar, the notion of a vis, meaning an *action* (*actio*), would not have been. Philosophers had traditionally pondered the various powers characterizing bodies or substances but

---

[6] For details on Newton's skeptical attitude toward metaphysics, especially in the Scholastic tradition, see Levitin (2016). Indeed, Levitin goes so far as to say that "no major natural philosopher between 1500 and 1800 said more often than Newton that they were not doing metaphysics" (*ibid.*, 68 note 57). See Koyré (1957, 159–60) for remarks on Newton's relation to metaphysics.

[7] This is certainly true of that most important of readers, Leibniz, who took notes on the Definitions. See Bertoloni Meli (1993, 96–104, 306).

the *vis impressa* is not a power in the sense of a property, at least not in any obvious sense.

In the very same edition of the text in 1713, its editor Roger Cotes tried to solve this problem. He did not share Newton's cautiousness.[8] Cotes was one of the only intellectuals who was deeply respected by Newton (Huygens was no longer alive, and Leibniz was out of favor). He used his editor's preface to try to explain the sense in which gravity, an impressed force, is in fact a property of all bodies:

> Thus all bodies for which we have observations are heavy; and from this we conclude that all bodies universally are heavy, even those for which we do not have observations. If anyone were to say that the bodies of the fixed stars are not heavy, since their gravity has not yet been observed, then by the same argument one would be able to say that they are neither extended nor mobile nor impenetrable, since these properties of the fixed stars have not yet been observed. Need I go on? Among the primary qualities of all bodies universally, either gravity will have a place, or extension, mobility, and impenetrability will not. And the nature of things either will be correctly explained by the gravity of bodies or will not be correctly explained by the extension, mobility, and impenetrability of bodies (Newton 1999, 391-92).

Cotes declares that *gravity is a primary quality*. He thought that one then ought to ask, but how do you know that it's universal? With his discussion of the empirical evidence, he answers that we know it because we have the same kind of evidence for thinking that extension is universal, an idea few would question.

To regard gravity as a primary quality, one must obviously regard it as a property; Cotes embraces the idea that gravity is the property of being heavy. So for him, Proposition VII means that all bodies are heavy. But how does this interpretation cohere with the Definitions that open the book? If gravity is an action exerted on a body, and it is also the property of being heavy, then presumably being heavy is an action. But that thought is confusing. One might think that (e.g.) a rock is heavy, and that in virtue of having that property, the rock acts (e.g., it falls), but that is not what Cotes has claimed. Consider another primary quality that is often given a dispositional analysis, impenetrability. It is natural to think that bodies are impenetrable in that they have the power to repel other bodies. Of course, in the classic lonely corpuscle case, the particle in question does not repel anything because nothing else exists, but it retains that power. In contrast, one could analyze gravity as the quality of being heavy, as Cotes does, but then one loses the idea that gravity is an action. After all, if one were to say that impenetrability is not the power to repel, but rather the action of repelling, then it becomes difficult to see how it could be a primary quality, since lots of things are impenetrable even though they are not repelling anything at a given moment.

If we look back at Newton's discussion in the *Regulae* for clarification, we may be disappointed. Although Cotes's preface seems to track Newton's discussion of

---

[8] The Cotes preface is of considerable historical importance. For instance, no less a figure than James Clerk Maxwell said in his lecture at the Royal Institution of Great Britain that Cotes was the driving force for "the doctrine of direct action at a distance" (48), rather than Newton himself (Maxwell, 1873–5: 48–9). Thanks to Margaret Maurer for this reference.

the relevant issues in the famous passage from Rule 3 quoted above, Newton did not contend that gravity is a primary quality. He *discussed* gravity amongst the list of primary qualities, just as Cotes did, adding mass to the standard roster of extension, impenetrability, etc. But he said only that "all bodies gravitate toward one another," thereby sticking with his idea that gravity is an action and avoiding Cotes's idea that it is the *property* of being heavy. He then adds his two qualifications: he is not contending that gravity is "essential" to matter, nor that it is "immutable" (as mass is). Perhaps he added those qualifications because he thought they helped to indicate why he had decided *not* to claim that gravity is in fact a primary quality. For it would certainly be reasonable to think that a primary quality is an immutable feature of a body, and that it expresses part of the essence of matter. So Newton acknowledges that his readers might interpret him to be claiming that gravity is a primary quality, but never resolves the tension among Rule 3, Cotes's preface, and the Definitions.

To highlight the importance of the Definitions, we might contrast the approach in the *Principia* with that in Newton's famous treatment of optics, published in the *Philosophical Transactions* in 1671/2, a contrast invited by the insightful discussion of Harper and Smith (1995). In that early paper, Newton attempted to evade the prominent wave/particle debate in optics by declaring that he had empirical evidence suggesting that rays of sunlight are heterogeneous, containing within them connate colors involving an index of refrangibility. He could avoid saying whether a ray of light is a *wave* or a stream of *particles*; more importantly, he could ignore the similar question, using metaphysical categories, of whether a ray of sunlight is a *substance* or a *property*. He did in fact add a (highly controversial) paragraph arguing that rays of sunlight seem to be substances rather than properties, which suggests that his favored particle theory is correct. But this was incidental—Newton's view can be articulated in metaphysically neutral terms by referring to *rays*. That was not the case with the *Principia*: Newton could not simply refer to gravity the way that one can point to a ray of sunlight entering the window. Instead, he presented gravity as an action exerted on a body. He contrasted that idea with the notion of an *inherent* force, which he said was a power. So already in the Definitions in the *Principia*, these concepts are in play: *action, body, power, inherent*. Metaphysical concepts are there from the very beginning, but without a guide. In particular, readers were left wondering how one ought to think of forces like *gravity* that are actions, as opposed to forces like the *vis inertiae* that are powers.[9]

Empirically-minded readers, or those with little interest in the historical Newton, might reply that although these concepts play a role in the Definitions, they are nonetheless dispensable. After all, the third definition concerns the "inherent force of matter," also called the *vis inertiae*, but that's just another name for the mass of a body, and of course mass is not a force at all. So perhaps this is merely a metaphysical residue, an unfortunate turn of phrase. To embrace this attitude,

---

[9] The notions of a body, its powers, inherent features, actions, and so on, were central concepts in metaphysics for centuries—see the extensive treatment in Robert Pasnau, *Metaphysical Themes*, which tackles powers and dispositions (2011, 519–540) and much else besides.

however, is to ignore the philosophical struggles of the first half of the eighteenth century. Those struggles are neatly encapsulated in the work of the century's leading mathematician, Euler, who grappled for years with the notion of the *vis inertiae*. After much deliberation, Euler finally jettisoned the notion in the 1750s, restricting the notion of force to impressed forces. Thus, these concepts were not considered dispensable, but rather required substantial analysis. More importantly, even if one jettisons Newton's idea of the *power* to resist acceleration, which he promoted his entire life, one is still left with the notion that gravity is an action that accelerates a body. Unlike in the case of mass, *there was no other concept in Newton's repertoire to employ in its stead.*[10] That concept conflicts with Cotes's analysis of what Newton meant, and Newton himself refrained from clarifying it.[11]

One might think that this interpretive conundrum is shallow, reflecting mostly the fact that Newton and Cotes saw the second edition of the *Principia* through the press during the vociferous calculus priority dispute with Leibniz and his followers (see Bertoloni Meli 1993). Indeed, from Newton's own point of view, the controversy reflected Leibniz's unfair contention that Newton had in fact proclaimed gravity to be essential to matter.[12] But it would be a mistake to overemphasize this context, for Newton's employment of metaphysical categories, coupled with his reluctance to clarify those categories, was evident long before his dispute with Leibniz began. For instance, when Richard Bentley was preparing the text of his Boyle lectures for publication in 1693, he exchanged letters with Newton in the hopes of clarifying the philosophical and theological implications of the *Principia*.[13] The

---

[10] Euler chose to solve what he regarded as the problem of gravity by reducing its action to the presence of an ether with differential density. See especially the illuminating treatment in Wilson (1992).

[11] For his part, Clarke adopts yet another approach, explaining to Leibniz in his fifth and last letter: "It is very unreasonable to call (sec. 113) *attraction* a miracle and an unphilosophical term, after it has been so distinctly declared that by that term we do not mean to express the cause of bodies tending toward each other, but barely the effect or the phenomenon itself, and the laws or proportions of that tendency discovered by experience, what is or is not the cause of it … The phenomenon itself, the attraction, gravitation, or tendency of bodies toward each other (or whatever other name you please to call it by), and the laws or proportions of that tendency, are now sufficiently known by observations and experiments." See C 5: 110–116 for the first quotation, before the ellipsis, and C 5: 124–130 for the second one, after it. So Clarke adopts a view that conflicts with the definition of impressed force in the *Principia*, but thereby recoups the notion that gravity is a property like mobility, a tendency toward a certain action. For a discussion of the seemingly significant distance between Newton's understanding of universal gravity and Clarke's rendering, see Janiak (2008, 58–74).

[12] In his anonymous review of the Royal Society's report on the calculus priority dispute, the so-called Account of the *Commercium Epistolicum*, in 1715, Newton writes (2014, 166–167): "And yet the editors of the *Acta Eruditorum*: (a) have told the world that Mr. Newton denies that the cause of gravity is mechanical, and that if the spirit or agent by which electrical attraction is performed be not the aether or subtle matter of Descartes, it is less valuable than an hypothesis, and perhaps may be the hylarchic principle of Dr. Henry More; and Mr. Leibniz: (b) hath accused him of making gravity a natural or essential property of bodies, and an occult quality and miracle."

[13] Bentley gave the first Boyle lectures concerning Christianity and the new science in 1692—they were endowed in Robert Boyle's will—and asked Newton for his guidance in understanding some

correspondence is often discussed in the context of debates concerning Newton's apparently complex attitude toward action at distance,[14] but it also reflects Newton's thinking about the status of gravity in relation to classic metaphysical categories. In early January 1693 (by the new calendar), Newton first warns Bentley not to "ascribe" to him the idea that gravity is "essential and inherent to matter" on the grounds that "the cause of gravity is what I do not pretend to know" (Newton 2014, 126). In Bentley's reply in the middle of February, he assures Newton that he will delay the publication of his Boyle lectures to continue revising them in the light of their correspondence. He then presents two series of numbered propositions, the first with six claims and the second with four. The last clause of the second proposition from that second series reads as follows: "And again, 'tis inconceivable that inanimate brute matter should (without a divine impression) operate upon & affect other matter without mutual contact: as it must, if gravitation be essential and inherent in it" (Newton 2014, 131). One week later, Newton replies:

> The last clause of the second position I like very well. It is inconceivable that inanimate brute matter should, without the mediation of something else, which is not material, operate upon and affect other matter without mutual contact, as it must be, if gravitation in the sense of Epicurus, be essential and inherent in it. And this is one reason why I desired you would not ascribe innate gravity to me. That gravity should be innate, inherent, and essential to matter, so that one body may act upon another at a distance through a vacuum without the mediation of anything else, by and through which their action and force may be conveyed from one to the other, is to me so great an absurdity, that I believe no man who has in philosophical matters a competent faculty of thinking can ever fall into it. Gravity must be caused by an agent acting constantly according to certain laws; but whether this agent be material or immaterial, I have left to the consideration of my readers (Newton 2014, 136).

Whereas Bentley wishes to assert that gravitational interactions involve a "divine impression," Newton is more cautious, indicating only that some agent must be involved. It is true that by definition, gravity as an impressed force is an action. But why *must* an agent be involved? Does this reflect a general fact about actions requiring agents? Newton does not answer this question here. As we will see, he broaches this topic again many years later in the General Scholium.

So even when corresponding with a strong supporter, Newton regarded several classic metaphysical notions as relevant for his physics, long before his famous disputes with Leibniz. Had he thought that notions like *essential, innate, inherent* were irrelevant to his physics, or regarded them as aspects of an outmoded metaphysics that is irrelevant to the physics of forces, he could have informed

---

of the principal implications of his new science of nature. Although they were originally private, the letters were first published in the eighteenth century (Newton 1756) and have since become a major source for our understanding of Newton's interpretation of his science—see Bentley (1976).

[14] The question of action at a distance in Newton's physics has received a tremendous amount of scholarly attention. The best general account of the topic as it arises throughout the history of physics is Hesse (1961). In Janiak (2008), I argued that Newton dismissed the notion that material bodies could act on one another at a distance; that argument is questioned in Ducheyne (2011) and Schliesser (2011), although it is endorsed by Chomsky (2016, 34–35). For further details, see Kochiras (2009, 2011) and Janiak (2013).

Bentley. Instead, Newton was at pains to instruct Bentley in the proper use of these concepts when describing physics to a wide audience, noting that he abhors the notion of innate gravity. Alas, his admonishment of Bentley does not clarify the prior question, what do *innate, inherent,* and *essential* mean? Are these terms synonymous?[15] Newton was not clearer on *that* issue in 1693 than he was in 1713.

So much for the first concept of force, the *vis impressa* called gravity, which was a revolutionary idea. The second concept of force is a far more traditional notion. In the Definitions, this notion appears in the guise of the famous *vis inertiae*, the power of a body to resist acceleration, a notion eventually jettisoned from physics through the work of thinkers like Euler. Whatever the merits of the notion, at least it falls within the bounds of natural philosophy since Newton predicates the power of all bodies, and the notion of a body—by which he means, anything with the quantity of matter he calls mass—obviously plays a central role for him. However, there is a twist: in the General Scholium to the *Principia*, added to the second edition of 1713, the concept of a power appears in a distinct guise. There, Newton seems to leave the bounds of empirical natural philosophy behind by predicating the power not of *bodies*, but rather of *substances*, which is obviously a central metaphysical category. Newton never clarified the relation between the power of a body and the power of a substance. Are these the same notion in two guises, or rather two distinct concepts?

The discussion of substances and their powers in the General Scholium to the *Principia* has been interpretively occluded by the more famous discussion of space and the divine in that text. But the latter discussion helps to explain the context in which Newton introduces his argument concerning substances. He writes:

> He is eternal and infinite, omnipotent and omniscient, that is, endures from eternity to eternity, and is present from infinity to infinity; God rules all things and knows all things

---

[15] Lest one think that the Newtonians implicitly agreed on the meaning of terms like *innate, inherent* and *essential*, Cotes actually endorsed an especially strong notion of *essential property*. After Bentley, Clarke and Newton prevailed upon him to write the preface, Cotes sent Clarke a draft. Clarke's objections to the draft are now lost, but we do have Cotes's reply (25 June 1713): "I received Your very kind Letter, I return You my thanks for Your corrections of the Preface, & particularly for Your advice in relation to that place where I seem'd to assert Gravity to be Essential to Bodies. I am fully of Your mind that it would have furnish'd matter for Cavilling, & therefore I struck it out immediately upon Dr. Cannon's mentioning Your objection to me, & so it never was printed. The impression of the whole Book was finished about a week ago. My design in that passage was not to assert Gravity to be essential to Matter, but rather to assert that we are ignorant of the Essential propertys of Matter & hat in respect of our Knowledge Gravity might possibly lay as fair a claim to that Title as the other Propertys which I mention'd. For I understand by Essential propertys such propertys without which no others belonging to the same substance can exist: and I would not undertake to prove that it were impossible for any of the other Properties of Bodies to exist without even Extension." (Newton, 1959, *Correspondence*, V: 412–413; cf. *ibid.*, V: 413, note 2.) For Cotes, then, an essential property P of X is such that X would lack all of its other properties if it lacked P. Leaving aside the vexed question of whether gravity should be thought of as essential in this sense, note that if Newton's own spokesperson was operating with such a strong notion of essence, this illustrates how important it was to have Newton himself clarify how he understood essences.

that happen or can happen. God is not eternity and infinity, but eternal and infinite; he is not duration and space, but endures and is present. God endures always and is present everywhere, and by existing always and everywhere constitutes duration and space [durationem & spatium constituit]. Since each and every particle of space is *always*, and each and every indivisible moment of duration is *everywhere*, certainly the maker and lord of all things will not be *never, nowhere*.[16]

Newton presents here a long-standing, but until now largely private, conception of the divine. He clearly echoes his discussion of the emanation of space from God in his unpublished manuscript *De Gravitatione*,[17] which also rejects the notion that God could be considered to be *nowhere*. There, he writes:

Space is an affection of a being just as a being [*Spatium est entis quatenus ens affectio*]. No being exists or can exist which is not related to space in some way. God is everywhere [*Deus est ubique*], created minds are somewhere, and body is in the space that it occupies; and whatever is neither everywhere nor anywhere does not exist. And hence it follows that space is an emanative effect of the first existing being, for if any being whatsoever is posited, space is posited. (Newton, *Philosophical Writings*, 40; Newton 1962, 103)

In 1713, Newton no longer uses the word *emanation* or the phrase *emanative effect*, but he clearly still endorses the notion: for space to emanate from God is for space to exist just in case God exists (no act of will or intermediary causal agent is required). Or as he puts it here, "by existing everywhere and always" God "constitutes duration and space." That is emanation without the label. Just as importantly, since every existent being must have some relation to space, then God, who necessarily exists, must exist within space. And in fact, God is "everywhere". We shall return to that perplexing claim below.

As has long been recognized, Newton's view of God, space and emanation reflects the influence of his fellow Cantabridgian Henry More.[18] For many British

---

[16] See Newton 1972, Vol. 2: 761, my translation; cf. Cohen and Whitman, *Principia*, 941.

[17] For recent discussions of the question of dating, see Ruffner (2012) and Levitin (2021), who argues forcefully that the text was written in 1671. Of course, until direct historical or archival evidence is found, the question of dating will remain speculative. The deep connections between the metaphysical claims of *De Gravitatione* and those of the General Scholium outlined in this chapter mitigate the importance of the dating issue.

[18] Sarah Hutton argues that More was probably the most influential figure in England concerning the reception of Cartesianism (Hutton 2015, 65), a view outlined in depth in Alan Gabbey's near monograph-length piece on More and Cartesianism (Gabbey 1982, 171–72). More even coined the term "Cartesianism" in 1662 in *A Collection of Philosophical Writings* (preface, xvii). More was a famous and influential figure in Cambridge when Newton arrived there in 1660; indeed, Westfall (1980, 97 note 85) indicates that he may have been *the* foremost intellectual figure there at the time. Newton engaged substantially with More's thought already as an undergraduate at Trinity College (see McGuire and Tamny 1983; Westfall 1980, 301). Later on, Newton became an "intimate associate" of More's—as Turnbull puts it at Newton, *Correspondence*, Vol. I: 306—and in a letter of 1680, More himself reported that he and Newton had "a free converse and friendship" (More to John Sharp, August 16, 1680, in *The Conway Letters*, 479). The two clearly had a sense of mutual respect (Westfall 1980, 55, 97 note 85). While furiously writing what would become the *Principia* in 1685, Newton wrote to Francis Aston, a secretary of the Royal Society, explaining that he had invited More to join a "Philosophick Meeting" in Cambridge, and for his part, More had

philosophers from the mid-to-late seventeenth century, Henry More's criticisms of "Cartesianism," which figure prominently in his correspondence with Descartes in 1648–49, were profoundly influential, and indeed, his views mediated their understanding of Descartes's ideas.[19] The views that More developed and published between 1648 and 1662 in his correspondence with Descartes and his own publications, culminating in *A Collection of Philosophical Writings* in 1662, find numerous echoes in Newton's unpublished manuscript.[20] For instance, in Book I, Chapter VI of More's *Immortality of the Soul*, which Newton had in his personal library, Axiome XVII reads as follows: "An Emanative Effect is coexistent with the very substance of that which is said to be the cause thereof." The gloss is:

> This must needs be true, because that very Substance which is said to be the Cause, is the adequate and immediate Cause, and wants nothing to be adjoyned to its bare essence for the production of the Effect; and therefore by the same Reason the Effect is at any time, it must be at all times, or so long as that Substance does exist.[21]

---

a funeral ring sent to Newton upon his death (Hall 1990, 103 and 169–70, respectively) as one of only fifteen people named in More's will (Reid 2012, 2–3)—the will is reprinted in *The Conway Letters* (Nicholson 1992, 481–83, with Newton appearing on page 482). Newton had eleven of More's works in his personal library (Harrison 1978, 195–96). Cassirer gives a detailed analysis of More's "decisive" influence on Newton's conception of space (1953, 147–50), which follows his famous account in *Das Erkenntnisproblem* (1999, Vol. 2: 372–76).

[19] Koyré (1968, 89–90) made a detailed textual and philosophical case for More's influence, carefully outlining the relevance of his correspondence with Descartes in an extensive footnote discussion. For a helpful discussion of some relevant points, see also the excellent and detailed analysis in Gabbey, 1982, e.g., at 180, 187ff, 193. The correspondence is reprinted in More's *A Collection of Philosophical Writings* in the original Latin. There is still no complete translation into English. For a helpful discussion of More's reactions to Cartesianism, see Henry, "The Reception of Cartesianism," 129–32. As for Newton, there is a complete copy of More's correspondence with Descartes extant amongst his papers in Cambridge University Library (see Hutton 2020 for discussion), and the correspondence was printed already in 1657 in the Clerselier edition of Descartes's correspondence and then reprinted by Henry More in *A Collection* in 1662 (Gabbey 1982, 206), a copy of which Trinity College obtained in 1664 (McGuire and Tamny 1983, 59).

[20] They include at least the following: to exist is to exist in space, even in God's case; therefore, God occupies space; space is an emanative effect of the first existing being, i.e., of God; therefore, space exists just in case God exists; Descartes's conception of the indefinite is confusing or even incoherent—space should instead be understood as infinite; and, Descartes's view of motion is mistaken (More, *A Collection*, preface, xi), as is his view that the Earth remains within its vortex, and in that sense is at rest according to Descartes, even though he also claims that it moves through the heavens (Descartes 1982, AT V: 388–89, which alas isn't included in the Lewis (Descartes 1953) edition, but which is included in *A Collection*, correspondence section, 93–94; the sections of this volume have separate pagination). Indeed, Newton brilliantly develops this last point in what amounts to his deepest criticism of Cartesian natural philosophy in *De Gravitatione*, one that leads him to rework the theory of motion that finds full flower in the *Principia*. Hence, far from being a text designed to criticize Henry More's views (*pace* Levitin 2021), the manuscript in fact adopts and extends many of the central criticisms that More presents against Cartesianism in the course of articulating his own views of space, time, motion, and the divine.

[21] Despite the clear affinities between their views, More and Newton differ on whether to call space a substance: Newton rejects the label in *De Gravitatione* on the grounds that space is not active, a tacitly understood criterion for substance-hood in his view; for More's part, he thinks that an item like space, which is an emanative effect of God, can still be considered a substance if it bears properties and would continue to exist even if all matter were expunged from the universe. Newton

More also makes it clear that if all matter disappeared, space would remain because God's essence is ubiquitous and God would be entirely present everywhere (see section I of Chapter VII in the appendix to the *Antidote*). Indeed, an infinite space is necessary, in More's view, because an infinite God is necessary and he conceives of God's substance or essence as occupying all of space. As we will see, Newton's view in the General Scholium bears a striking resemblance to this conception.[22]

The emanation thesis has received lots of attention.[23] But what do More and Newton mean when they contend that God is *everywhere*? Newton privately endorsed this claim early in his career in *De Gravitatione* and then publicly years later in the General Scholium. The claim cannot be understood *per se*; its meaning shifts when interpreted from distinct points of view. For instance, if seen from a theological point of view, the contention might be anodyne. Many authors in this period—including Newton himself, in one of the three notes in the General Scholium—cited a famous passage from Acts 17:28, which speaks of God as him "in whom we live, and move, and have our being". This scriptural passage was often taken to show that God is everywhere in some unobjectionable sense, one requiring no further analysis—perhaps it means that God is all-seeing, or able to act anywhere in the world, etc. Not surprisingly, it was cited frequently by authors with theologically disparate perspectives. To take four examples from within the British Isles in this historical period: the anti-Trinitarian Newton cited the passage, as did the theologically radical John Locke, but so did the more cautious theologian Samuel Clarke, rector of St. James Westminster, and the Bishop of Cloyne, George Berkeley.[24]

However, a *theologically* safe claim, one rooted in the words of St. Paul himself, may not be *philosophically* safe. For a philosopher, the claim is ambiguous: in what specific sense is God everywhere? In particular, is God's *power* everywhere, a widely accepted notion, or God's *substance* (essence), a widely derided one? For his part, More clearly endorsed the controversial view that God is present everywhere in

---

agrees with these two claims in the case of space, and yet denies that it's a substance. So, this is partly semantics. For More, see the gloss on axiom xix in Book I, Chapter VI of *Immortality* and for Newton, *see De Gravitatione, Philosophical Writings*, 35–36.

[22] Indeed, the connection between More's views and Newton's was made long before the General Scholium appeared: it was discussed in depth, e.g., by Joseph Raphson, F.R.S., in 1702, as Koyré emphasized long ago (1957, 190ff).

[23] For instance, it was the subject of an influential debate between Ted McGuire and John Carriero in 1990: see their contributions to *Philosophical Perspectives on Newtonian Science*.

[24] For a helpful discussion of relevant issues in Newton interpretation, see Snobelen (2006). Locke (1975) cited the passage at *Essay* II.xiii.26 in the course of discussing the relation between space and body. Clarke cited the passage in his fifth and last letter to Leibniz (C 5: 33–35; Leibniz 1931, Vol. 7: 427). For his part, Berkeley apparently cited the phrase more than any other Scriptural passage; it occurs in both the *Principles of Human Knowledge* and the *Three Dialogues* (see Daniel 2021, 237 and note 1).

substance and not merely in power. Some English philosophers writing after More—for instance, Newton's former friend Locke—decided to remain neutral.[25]

What reasoning does More give for endorsing this controversial view? We find the answer already in his letter to Descartes of December 1648:

> And, indeed, I judge that the fact that God is extended in his own way follows from the fact that he is omnipresent and intimately occupies the universal machine of the world and each of its parts. For how could he have impressed motion on matter, which he did once and which you think he does even now, unless he, as it were, immediately touches the matter of the universe, or least did so once? This never could have happened unless he were everywhere and occupied every single place. Therefore, God is extended in his own way and spread out; and so God is an extended thing [*res extensa*].[26]

From More's point of view, Descartes must agree that God's power to act is omnipresent, for that view is entailed by his occasionalism. Yet how could God act on bodies to which God was not present? More takes the impossibility of this notion to entail that God must be extended and omnipresent in substance. Descartes resists this claim, arguing that it's fundamentally mistaken to conceive of the divine being as extended, for God has no parts—as extended bodies must—and cannot be imagined—as extended bodies can.

But does Newton follow More again? In what sense is God "everywhere" for him? Newton answers this question in the General Scholium by using what I take to be the second concept of a force, viz. the power of a substance:

> Every sentient soul, at different times and in different organs of sense and motions, is the same individual person. There are parts that are successive in duration and coexisting in space, but neither of these exist in the person of man or in his thinking principle, and much less in the thinking substance of God. Every man, insofar as he is a thing that has senses, is one and the same man throughout his lifetime in each and every organ of his senses. God is one and the same God always and everywhere. He is omnipresent not only in power, but in

---

[25] In his classic paper "The idea of God," Jean-Luc Marion discusses a number of English philosophers, including most prominently More, Locke, and Newton, who endorse the view that God is extended and perhaps ubiquitous in space (Marion 1998, 284–90). However, Locke seems to be a bit more cautious than the others. In the *Essay* (II.xv.3) he writes: "But yet every one easily admits, That though we make Duration boundless, as certainly it is, we cannot yet extend it beyond all being. GOD, every one easily allows, fills Eternity; and 'tis hard to find a Reason, why any one should doubt, that he likewise fills Immensity: His infinite Being is certainly as boundless one way as another; and methinks it ascribes a little too much to Matter, to say, where there is no Body, there is nothing." This does not seem to resolve the ambiguity because Locke may mean that God "fills" immensity with the divine power, and not the divine substance (and of course, Locke may have wished to avoid using the concept of substance anyway). Cf. also *Essay*, II.xvii.1 and II.xxiii.21. This connects, in turn, with Locke's decision to not determine whether that "inspired Philosopher" St. Paul's famous words in Acts 17:28 (which he cites at Essay II.xiii.26) "are to be understood in a literal sense." The theological basis is sound, for Locke, as long as one does not wade into metaphysics to find a precise meaning. He even wisely avoids saying whether the "literal sense" of the scriptural passage involves the idea of substantial omnipresence. So, he is doubly neutral.

[26] See Descartes, *Oeuvres*, AT 5: 238–39; my translation.

substance: for power cannot subsist without substance. [Omnipraesens est non per *virtutem* solam, sed etiam per *substantiam*: nam virus sine substantia non potest.][27]

Unexpectedly, Newton does not take the safe route of remaining neutral on the precise meaning of the theologically unobjectionable notion that God is everywhere. Like More, he claims instead that God is *substantially* everywhere, and he does so by invoking, for the first and only time in the entire text of the *Principia*, the notion of the power of a substance.

The controversy caused by these remarks about God was perhaps clearest in the famous Leibniz-Clarke correspondence, which began just 2 years after the publication of the General Scholium. At first, Clarke repeats Newton's view from the General Scholium in his third letter to Leibniz: "God, being Omnipresent, is really present to every thing, essentially and substantially: His Presence manifests it self indeed by its Operation, but it could not operate if it was not There."[28] After this claim does not convince his correspondent, Clarke thentried to defend the view that immensity is essential to God by citing Paul's line in Acts 17:28 in his last letter to Leibniz in 1716. Although Leibniz's death ended the correspondence before he was able to reply, it's clear that the famous lines from St. Paul were regarded as too general to convince a philosopher of a specific conception of God's relationship with extension.

Moreover, as we have seen, Newton did not rely on St. Paul to support his view, he chose instead a metaphysical route to the same destination.[29] And that route was obvious to Leibniz. In his correspondence with Clarke, Leibniz strongly criticizes the view found in the *Principia* and in the Queries to the *Opticks*. He puts the point pithily in his last letter of August 1716:

---

[27] See Newton, *Principia Mathematica*, vol. 2: 762; my translation. This exact phrase appears in all extant draft copies of the General Scholium. In two such copies, Newton then added the following phrase (more precisely, in one copy it's a phrase, in the other a full sentence), "Quod fingitur sine substantia subsistere, jam fingitur esse substantia," which we might render as "What is supposed to subsist without substance is already supposed to be substance." However, this line did not appear in any published version of the text.

[28] See C 3: 12; Leibniz 1931, Vol. 7: 370. This characterization of Newton's view leaves out the specific claim that God's presence could not manifest itself through its operation unless God was there on the general grounds that a power cannot subsist (or exist) without a substance.

[29] See C 5: 33–35; Leibniz 1931, Vol. 7: 427, where Clarke calls his understanding of God's immensity "the express assertion of Saint Paul." Intriguingly, in the very next sentence after the one last quoted in the text above, Newton paraphrases Paul's famous line in Acts, without quoting it: "In him all things are contained and move, but he does not act on them nor they on him." But even if Newton thought his view was precisely the one found in Paul, as Clarke did, he decided not to rely on it, preferring instead to use the concept of a substance and its power. That may have been wise, since philosophers with disparate attitudes toward God and space cited the passage from Paul approvingly, thereby undermining Clarke's contention.

> If infinite space is God's immensity, infinite time will be God's eternity. We must therefore say that what is in space, is in God's immensity, and consequently in his essence; and that what is in time is in the essence of God. Strange expressions, which clearly show an abuse of terms.[30]

If God is essentially (substantially) present everywhere, then created things like rocks and plants will literally be *in* God. Leibniz seemed to regard this point as a *reductio* of the Clarke-Newton view.

Leibniz also understood that Newton sought to defend his thesis with a general conception of the relation between powers and substances: generally speaking, if powers cannot subsist without substances, then it must follow that even the divine power cannot so subsist. Meeting the Newtonians on their own turf, as was his specialty, Leibniz then argued that the general relation between entities and extension rules out Newton's view. After all, if "infinite space is the immensity of God," it would have to follow that some finite space is the extension of the finite body occupying it; but if that were the case, then in changing its location, a body would be changing its own extension, which is an "absurd thing."[31] If we are going to speak univocally about substances and powers, applying our view to the divine and to the quotidian, then we ought to do so for substances and their extension as well.

Leaving aside the comparative merits of these arguments, it should be clear that Newton's controversial thesis cannot be squared with the popular conception of him as merely an empirically-minded natural philosopher. First of all, substance is a classic metaphysical concept—the most central such concept, along with being—and has obviously been significant since at least Aristotle. Perhaps that is why influential accounts of the history of metaphysics in the seventeenth century—in both the so-called Continental and Analytic traditions—regularly discuss Newton's view in the General Scholium along with the ideas of the usual cast of characters from the late Scholastics to Descartes, Spinoza and Leibniz.[32] Second of all, although many philosophers in his milieu were skeptical of the idea that God is substantially omnipresent in infinite space, Newton buttressed that controversial idea with a familiar metaphysical move concerning the general relation between substances and their powers, one found in famous Scholastic authors like Suarez.[33]

---

[30] See L 5: 44, which corresponds to Leibniz 1931, Vol. 7: 399; my translation.

[31] See L 5: 37, which corresponds to Leibniz 1931, Vol. 7: 398; my translation.

[32] For instance, both Jean-Luc Marion in his piece "The idea of God" (1998) and Robert Pasnau in his *Metaphysical Themes* (2011) discuss Newton's view of God's omnipresence in the course of discussing late seventeenth century metaphysics.

[33] Indeed, we find Suarez, perhaps the most famous of the late Scholastic metaphysicians, taking the same view that Newton endorses in the General Scholium. As Suarez argues in his *Disputationes Metaphysicae* (51.2.8), God is present to the universe not just by power or action, but also by essence or substance. As Pasnau argues, right after quoting at length from Suarez (2011, 330), Newton endorses this view of God's relation to space (see also Pasnau 2011, 352–355, including the long note appended to page 355, which quotes Newton). Thus, Newton's general point about powers and substances in the General Scholium is not part of an "anti-Scholastic" move, as Levitin (2016, 71–73) claims.

There is little doubt, then, that Newton waded into metaphysical waters in his magnum opus. However, it does not follow that Newton's metaphysical views were integrated into the text of the *Principia*. On the contrary, Newton's controversial thesis in the General Scholium effects a separation between his metaphysical views found there and the rest of the natural philosophy found in the text. The separation is both conceptual and methodological in character.

Conceptually speaking, notice that if we bracket the General Scholium, *substance* is not a concept in Newton's empirical natural philosophy—instead, the key concepts are body (or mass), impressed force, space, time, motion, etc. Newton speaks of *bodies* in motion within space and time under the causal influence of various forces, but never of *substances* and their powers. The other most famous discussion of space in the text, the Scholium following the Definitions before Book I of the *Principia*, is no exception. In the Scholium, Newton *comes close* to saying that space is a substance: "absolute space by its own nature and without relation to anything external always remains homogeneous and immovable." Indeed, some philosophers think that Newton *tacitly* treats space as a substance by contending that it has its own nature characterized by certain properties. But Newton never employed the concept of a substance in the Scholium, *not even to deny that he regards space as one*. This approach contrasts with that found in *De Gravitatione*, where Newton explicitly denies that space is a substance or a property. So, the General Scholium has a unique conceptual status in the *Principia*.

Newton's thesis of God's substantial omnipresence in infinite space is also methodologically separate from the rest of the *Principia*, where the aim is to deduce conclusions *from the phenomena*. It is here that the gulf between the two forces in my title is perhaps the widest. For, leaving aside questions of interpretation, Newton certainly had a strong argument that he had deduced universal gravity from the phenomena. Indeed, one prominent strand in eighteenth-century thought said roughly: whatever the impressed force called gravity really is, Newton deduced its mathematical features from the phenomena in brilliant fashion. The same cannot easily be said about the other notion of a force, the idea of a substance's power.

This problem did not escape Newton's notice. For his part, he had insisted publicly that his conception of the divine does not diverge methodologically from the rest of his natural philosophy before the second edition of the *Principia* was published. In Query 28 to the *Opticks*, first published in the Latin translation of 1706, Newton first explains that in contrast to his predecessors, who feign "Hypotheses for explaining all things mechanically" and then refer "other Causes to Metaphysicks," he believes instead that "the main Business of natural Philosophy is to argue from Phaenomena without feigning Hypotheses, and to deduce Causes from Effects, till we come to the very first Cause, which certainly is not mechanical" (Query 28, *Opticks*, Newton 1952, 369). Next, he describes how animals perceive objects through their "immediate presence" to the substance within them. And then he concludes:

> And these things being rightly dispatch'd, does it not appear from Phaenomena that there is a Being incorporeal, living, intelligent, omnipresent, who in infinite Space, as it were in his Sensory, sees the things themselves intimately, and thoroughly perceives them, and comprehends them wholly by their immediate presence to himself... (Query 28, *Opticks*, 370).

Newton builds up the materials for an analogy between the way that creatures perceive objects through an immediate presence to the substance of those creatures to the case of their creator. But even if one grants the analogy, this is mere rhetoric, for he does not explain *how* an analysis of the *finite* phenomena could support an inference to God's substantial omnipresence *in infinite space*. And he does not present any analogical argument for this idea in the General Scholium, relying instead on a general claim about substances. It's difficult to avoid the conclusion that Newton's understanding of God's omnipresence falls methodologically outside of his empirical natural philosophy, even by his own standards. Ironically, Newton's stalwart defender, Samuel Clarke, had argued in correspondence that no argument from the phenomena supporting God's infinity in space could be made.[34] As for Newton, he remained silent on why he thought it could be.

For generations, interpreters of Newton have regarded the lack of metaphysical commitments in his work as a reflection of his commitment to a fundamentally "empiricist" approach to answering questions about nature.[35] Newton the "empiricist" has been a powerful image to many of his readers, and it may be capable of accommodating his vexed relationship with metaphysical categories in his treatment of gravity. It also reflects Newton's own sharply negative attitude toward the metaphysical tradition.[36] But it is difficult to square with his views in the General Scholium. For in that text, as we have seen, he proclaims that there is an infinite being that substantially occupies all of infinite space, and despite his assurances that the claim is apt given his methodological commitments, he does not explain how such a view could be deduced from the phenomena. Since his most general claim based on the phenomena—viz., that gravity is effected in all bodies—is universal in scope but does not constitute a claim about the infinite, the complex methodology undergirding his revolutionary approach to natural philosophy would not appear to be of much assistance here.

What are we missing as a result of Newton's vexed relationship with the metaphysical tradition? For starters, we do not know the answer to the following obvious question: what is the relation between the *vis inertiae*, the power of a body

---

[34] In his answer to the seventh letter of Daniel Waterland, the English theologian critical of Clarke's views, Clarke writes: "In like manner, the *Infinity*, the *Immensity*, or the *Omnipresence* of God, can no otherwise be proved, than by considering *a priori* the nature of a *Necessary* or *Self-Existent* cause. The *Finite* phaenomena of nature, prove indeed demonstrably *a posteriori*, that there is a Being which has Extent of *Power* and *Wisdom sufficient*, to produce and preserve all these phaenomena. But that This *Author of Nature* is *Himself* absolutely *Immense* or *Infinite*, cannot be proved from these *finite* Phaenomena; but must be demonstrated from the intrinsick nature of *Necessary Existence*." See Clarke, 1738, *Works*, Vol. 2: 756.

[35] For differing perspectives on these issues, see Stein (2002) Domski (2010) and the papers in Biener and Schliesser, *Newton and Empiricism*, (2014).

[36] As Levitin says (2016, 68), Newton had little time for metaphysics as a discipline. That is surely correct, for Newton not only uses "Metaphysicks" and its cognates pejoratively, but also narrowly. As Levitin notes, he means a philosophy that embraces "innate ideas," which captures rationalist views of the seventeenth century, but which leaves out Aristotle. Obviously, this usage reflects a rhetorical move on Newton's part.

to resist acceleration, and the power of a substance? Are all bodies substances? Did Newton tell Bentley that there must be an agent causing gravity because *any* action must have an agent, and if so, what is the scope of these categories—do they cover all *bodies*, or something more general, such as all *substances*? These are the basic questions that one would expect any metaphysically-minded philosopher to tackle. Newton's philosophy raises these questions, but unfortunately, it simply does not answer them. The great figures of the eighteenth century who followed Newton tidied up his work and left these issues behind. But for those of us who remain interested in the historical Newton, rather than the Newton of the textbooks, we cannot help but be fascinated by the complex combination of revolutionary natural philosophy and old-fashioned metaphysics that we find in his most important text.

References

Bentley, R. 1976. *Eight Boyle Lectures on Atheism*, 1692. New York: Garland.

Bertoloni Meli, D. 1993. *Equivalence and priority: Newton vs. Leibniz*. Oxford University Press.

Biener, Z., and E. Schliesser. 2014. *Newton and empiricism*. Oxford University Press.

Carriero, J. 1990. "Newton on space and time: comments on J.E. McGuire." In *Philosophical Perspectives on Newtonian Science*, ed. Phillip Bricker and R.I.G. Hughes. Cambridge, MA: MIT Press.

Cassirer, E. 1999. Das Erkenntnisproblem in der Philosophie und Wissenschaft der neueren Zeit. In *Gesammelten Werke*, vol 2. Felix Meiner Verlag.

———. 1953. *The platonic renaissance in England*. Trans. James Pettegrove. University of Texas Press.

Clarke, S., and G.W. Leibniz. 1717. *A collection of papers, which passed between the late learned Mr. Leibnitz and Dr. Clarke, in the years of 1715 and 1716*. London.

Clarke, S. 1738. The works of Samuel Clarke, D.D. Four volumes. Knapton.

Chomsky, N. 2016. *What kind of creatures are we?* Columbia University Press.

Daniel, S. 2021. *George Berkeley and modern philosophy*. Oxford University Press.

Descartes, R. 1953. *Descartes: correspondance avec Arnauld et Morus, texte Latin et traduction*, ed. G. Lewis. Librairie philosophique Vrin.

———. 1982. *Oeuvres de Descartes*, eleven volumes, ed. Charles Adam and Paul Tannery. Vrin.

Domski, M. 2010. Newton's empiricism and metaphysics. *Philosophy Compass* 5: 525–534.

Ducheyne, S. 2011. Newton on action at a distance and the cause of gravity. *Studies in History and Philosophy of Science* 42: 154–159.

Gabbey, A. 1982. Philosophia Cartesiana Triumphata: Henry More (1646–1671). In *Problems of Cartesianism*, ed. Thomas Lennon et al. McGill-Queen's University Press.

Hall, A.R. 1990. *Henry More and the scientific revolution*. Cambridge University Press.

Harper, W. and George E. Smith. 1995. "Newton's New Way of Inquiry." In *The Creation of Ideas in Physics*, ed. Jarrett Leplin. Dordrecht: Kluwer.

Harrison, J. 1978. *The Library of Isaac Newton*. Cambridge: Cambridge University Press.

Hesse, M. 1961. *Forces and fields: The concept of action at a distance in the history of physics*. Nelson.

Hutton, S. 2015. *British philosophy in the seventeenth century*. Oxford University Press.

———. 2020. Henri More et Descartes: une copie manuscrite de leur correspondance dans le Notebook de Thomas Clerke (1654). *Bulletin cartésien* 49: 162–167.

Janiak, A. 2008. *Newton as philosopher*. Cambridge University Press.

———. 2013. Three concepts of causation in Newton. *Studies in History and Philosophy of Science* 44: 396–407.

Kochiras, H. 2009. Gravity and Newton's substance counting problem. *Studies in History and Philosophy of Science* 40: 267–280.

———. 2011. Gravity's cause and substance counting: Contextualizing the problems. *Studies in History and Philosophy of Science* 42: 167–184.

Koyré, A. 1957. *From the closed world to the infinite universe*. Johns Hopkins University Press.

———. 1968. *Newtonian studies*. Harvard University Press.

Leibniz, G.W. 1931. *Streitschriften zwischen Leibniz und Clarke. Die philosophischen Schriften von Gottfried Wilhelm Leibniz*, Volume Seven, ed. C.J. Gerhardt. Lorenz.

Levitin, D. 2016. Newton and scholastic philosophy. *British Journal for the History of Science* 49: 53–77.

———. 2021. Isaac Newton's 'De gravitatione et aequipondio fluidorum': Its purpose in historical context. *Annals of Science* 78: 133–161.

Locke, J. 1975. *An essay concerning human understanding*. Based on the fourth edition, ed. P. Nidditch. Oxford University Press.

McGuire, J.E. 1990. "Predicates of pure existence: Newton on God's space and time." In *Philosophical Perspectives on Newtonian Science*, ed. Phillip Bricker and R.I.G. Hughes. Cambridge, MA: MIT Press.

McGuire, J.E. and Tamny M. 1983. *Certain philosophical questions: Newton's Trinity notebook*. Cambridge University Press.

Marion, J.-L. 1998. The idea of God. In *The Cambridge history of seventeenth century philosophy*, ed. D. Garber et al. Cambridge University Press.

Maxwell, J. C. 1873–75. Notices of the proceedings at the meetings of the members of the Royal Institution 7.

More, H. 1662. *A collection of several philosophical writings*. James Flesher.

Newton, I. n.d. *De Gravitatione et Aequipondio Fluidorum*. Cambridge University Library.

———. 1756. *Four Letters from Sir Isaac Newton to Doctor Bentley*. London: R & J Dodsley.

———. 1759. *Principes mathématiques de la philosophie naturelle*. Translated by Madame Du Châtelet. Paris: Desaint & Saillant.

———. 1952. *Opticks*. Based on the fourth edition, ed. I.B. Cohen. Dover.

———. 1959–77. The correspondence of Isaac Newton. Edited by H.W. Turnbull, et al. Cambridge: Cambridge University Press.

———. 1960. *Mathematical Principles of Natural Philosophy*. Translated by Andrew Motte, revised by Florian Cajori. Berkeley: University of California Press.

———. 1972. *Philosophiae Naturalis principia Mathematica. The third edition with variant readings*, ed. Alexandre Koyré and I. Bernard Cohen, with Anne Whitman. Harvard University Press.

———. 1999. *The principia: Mathematical principles of natural philosophy*. Trans. I. Bernard Cohen and Anne Whitman. Berkeley: University of California Press.

———. 2014. *Newton: Philosophical writings*. 2nd ed, ed. Andrew Janiak. Cambridge: Cambridge University Press, second edition.

Nicholson, M.H. 1992. *The Conway letters: The correspondence of Anne, Viscountess Conway, Henry More, and their friends, 1642–1684*. Revised edition with an introduction and new material by Sarah Hutton. Oxford University Press.

Pasnau, R. 2011. *Metaphysical themes, 1274–1671*. Oxford University Press.

Reid, J. 2012. *The Metaphysics of Henry More*. Springer Verlag.

Ruffner, J. 2012. Newton's De gravitatione: A review and reassessement. *Archive for the History of the Exact Sciences* 66: 241–264.

Schliesser, E. 2011. Newton's substance monism, distant action, and the nature of Newton's empiricism. *Studies in History and Philosophy of Science* 42: 160–166.

Smith, G. 1999. How did Newton discover universal gravity? *St. John's Review* 45: 32–63.

Snobelen, S. 2006. To us there is but one God, the father: Antitrinitarian textual criticism in seventeenth and early-eighteenth century England. In *Scripture and scholarship in early modern England*, ed. A. Hessayon and N. Keene. Algate.

Stein, H. 2002. *Newton's metaphysics*, ed. I.B. Cohen and G. Smith. The Cambridge Companion to Newton. Cambridge University Press.

Wilson, C. 1992. "Euler on action at a distance and fundamental equations in continuum mechanics." In *The Investigation of Difficult Things*, ed. P.M. Harman and Alan Shaprio. Cambridge: Cambridge University Press.

Chapter 12
Theory-Mediated Measurement and Newtonian Methodology

Michael Friedman

My argument owes much to two contemporary philosophical scholars of the development of modern physics beginning with Newton and extending through the late nineteenth and early twentieth centuries: George Smith and Howard Stein. From the former I take a conception of theory-mediated measurement, from the latter a conception of abstract structures *in the phenomena*. The relevant notion of phenomena, as we shall see, is closely related to Newtonian scientific methodology, a methodology that is clearly embraced by both philosophers. The conception of *abstract structures*, however, is not so clear, and it is especially to be distinguished from the more recent conception of "structural realism."

The paper that gave birth to this conception was John Worrall's "Structural Realism: The Best of Both Worlds?" The two worlds in question were realism and anti-realism, and Worrall attempted to do justice to both (if only somewhat tentatively) by reviving what he took to be an earlier version of structural realism articulated by Henri Poincaré. Worrall's idea, in brief, was to find sufficient continuity in scientific change to support the legitimate claims of scientific realism while simultaneously acknowledging that the referents of central theoretical terms may radically change in the process or even disappear altogether. His main example involved the mechanical models of the luminiferous and electromagnetic aether that were explored throughout the nineteenth century—which, it is supposed, were later found to have no physical reality after all, especially after Einsteinian relativity theory. Worrall's contention, in a nutshell, was that, although these aethers themselves, conceived as material substances, were not preserved over time, the mathematical structures involved in their description—wave forms continuously propagating over space and time, electromagnetic lines of force (vector fields) in

M. Friedman (✉)
Philosophy of Science, Stanford University, Stanford, CA, USA
e-mail: mlfriedm@stanford.edu

M. Stan, C. Smeenk (eds.), *Theory, Evidence, Data: Themes
from George E. Smith*, Boston Studies in the Philosophy and History
of Science 343, https://doi.org/10.1007/978-3-031-41041-3_12

the theories of Faraday and Maxwell—were preserved nonetheless. It is here that we find the invariant *mathematical structures* that remain after the disappearance of their originally-supposed *material content*.

Howard Stein's "Yes but . . .—Some Skeptical Remarks on Realism and Anti-Realism" appeared in the same 1989 volume of *Dialectica* as Worrall's paper. Here, without referring to Worrall directly, Stein embarked on an analogous project of reconciliation between two attitudes towards scientific theories that are standardly taken to be diametrically opposed to one another—this time realism and instrumentalism. Stein argued, on the contrary, that realism and instrumentalism are fruitfully combined with one another in the work of the greatest scientists, and, in particular, in the work of Newton, Maxwell, and Einstein: realism in so far as their goal is adequately and correctly "to represent the phenomena of nature," but instrumentalism also, in so far as their project can only proceed by careful attention to an accumulation of experimental evidence over time by which we take a given theory adequately and correctly to represent the phenomena.[1] We must be able to "find" the theoretical structures in question in the phenomena themselves, in something like the way in which we "find" enduring physical objects in our everyday sensory experience.

Now this talk of representing the phenomena, and of finding theoretical structures in the phenomena, may appear unsatisfactory from a realist point of view. For one of the key moments in the development of modern science—Galileo's refusal to take up Cardinal Bellarmine's invitation to confine Copernican theory to the goal of "saving the phenomena"—appears to have involved precisely a realist commitment going well beyond this goal. And a similar dialectic, of course, appears to be present in later collisions between realism and instrumentalism such as Pierre Duhem's rejection of contemporaneous molecular-atomic theories (again on behalf of "saving the phenomena") or, in our own time, in the related defense of instrumentalism (as *opposed* to realism) by Bas van Fraassen. I confess that I was not quite satisfied with Stein's position, along precisely these lines, for a considerable period of time, especially in my earlier militantly realist phase. When I finally became committed to Kant's version of Newtonianism, however, I noticed, for the first time, that Stein himself explicitly appeals to Kant's conception of phenomena in his criticism of what he takes to be *excessive* realism.[2]

---

[1] It is interesting to note that the printed version of Worrall (1989) explicitly thanks Stein for his helpful comments. It is also interesting to note that Stein states his conclusion as follows (1989, 65): "realism—yes, but instrumentalism—yes also, no only to anti-realism." Stein's substitution of "instrumentalism" for "anti-realism" can be taken to signal an implicit disagreement with Worrall's paper. As we shall see, changes in experimental evidence over time imply that what is really at issue is the *increasing* adequacy and correctness of the conclusions in question. More precisely, Newton's non-hypothetical methodology conceives changes in these conclusions are always possible, becoming either closer to adequacy and correctness or further from it.

[2] In defending his conception of adequately and correctly representing the phenomena against a form of realism demanding something more (reference to something "behind" the phenomena), Stein begins (1989, 50): "First, there is the classical philosophical problem of the epistemology of transcendent metaphysics—the problem that Kant (we may say) posed to the ghost of Leibniz,

12.1   Kant and Newton on Force and Quantity of Matter

In the second chapter of *Metaphysical Foundations of Natural Science* (1786) Kant appears sharply to differ with Newton about gravity as an immediate action at a distance:[3]

> [I]t is clear that the offense taken by his contemporaries, and perhaps even by Newton himself, at the concept of an original attraction set him at variance with himself. For he could by no means say that the attractive forces of two planets, those of Jupiter and Saturn for example, manifested at equal distances of their satellites (whose mass is unknown), are proportional to the quantity of matter of these heavenly bodies, if he did not assume that they attracted other matter merely as matter, and thus according to a universal property of matter. (4: 515)

What is at issue is the method Newton develops in Book 3 of the *Principia* for measuring the quantities of matter (or masses) of primary bodies in the Solar System by the accelerations produced in their satellites. Kant appears to suppose that Newton must assume both the universality and the immediacy of the gravitational interaction between Jupiter and Saturn independently of any intervening matter surrounding these bodies. Kant thus appears to insist on the physical reality of immediate action at a distance for gravitational force, contrary to Newton's own more agnostic approach.

The Newtonian procedure for measuring the quantities of matter (or masses) of primary bodies in the Solar System is a paradigmatic example of theory-mediated measurements (of these masses), it is therefore well worth our while to examine this method more closely. The first point to emphasize is that these measurements enabled Newton empirically to demonstrate that the center of gravity of the Solar System lies always very close to the center of the Sun (Book 3, Proposition 12). Newton was considering, in particular, the choice between the Copernican and Tychonic systems, having already eliminated the Ptolemaic system via telescopic observations (following Galileo) of the phases of Venus and Mercury. Newton could then measure the quantity of matter of the Sun by the orbital acceleration of Venus (the same on both Copernican and Tychonic systems), the quantities of matter of Jupiter and Saturn by the orbital accelerations of their satellites (likewise), and that of the earth by the orbital acceleration of the moon (the same on all three systems). It thereby turned out that the quantity of matter (or mass) of the earth

---

and that led Kant himself to the view that all theoretical knowledge is *of* the phenomenal world: the world of experience. The nub of the matter can be put very simply: it is, How can you *know* that things are as you say they are? If the claimed 'reference' of the theory is something beyond its correctness and adequacy in representing the phenomena—if, that is, for a given theory, which (we may suppose) does represent phenomena correctly and adequately, there are still two possibilities: (a) that it is (moreover) *true*, and (b) that it is (nonetheless) *false*—then how in the world could we ever tell what the actual case is?"

[3] I cite this work by volume and page numbers in Kant (1900—). The page numbers in vol. 4 appear in the margins of my contribution to the Cambridge Edition (2004). All translations from Kant's German are my own.

is much, much smaller than that of the Sun, so that only the Copernican system (in its Keplerian formulation) remained. And it is precisely this argument that has settled the empirical truth of the Copernican system (to a very high degree of approximation) ever since.

How did Newton arrive at this method for measuring the quantities of matter of the primary bodies in the Solar System? Two propositions established earlier in Book 3 are of particular importance. Proposition 6 generalizes the (Galilean) equality of the acceleration of fall for all bodies near the surface of the earth to the gravitational fields of any "planet" (i.e., primary body), now considered as extending to arbitrary distances (1999, 806): "*All bodies gravitate toward each of the planets, and at any given distance from the center of any one planet the weight of any body whatever toward that planet is proportional to the quantity of matter which the body contains.*"[4] Here we are dealing with the singular property of gravitational force that it produces the same acceleration in any body whatsoever at equal distances from the source of that force—entirely independently, in particular, of the *accelerating body's* mass. This property does not hold for other forces such as magnetism, and it amounts to the claim that the specifically gravitational force on an attracted body (i.e., its "weight") is always proportional (at a given distance) to this body's mass, and it therefore essentially involves what we now call the equality of inertial and *passive* gravitational mass.[5]

This property, the (generalized) Galilean property of the gravitational field, is a crucial part of Newton's theory-mediated measurements of the quantities of matter of the primary bodies in the Solar System. As becomes clear in Newton's following Proposition 7, however, the Galilean property alone is not sufficient. According to Proposition 7 (1999, 810): "*Gravity exists in all bodies universally and is proportional to the quantity of matter in each.*" The proof depends on Proposition 69 of Book 1, which appeals to the Galilean property (the equality of inertial and passive gravitational mass), the universality of the attractive force in question (in this case gravitation), and, crucially, the equality of action and reaction expressed in the Third Law of Motion. Consider the system of bodies consisting of Jupiter, Saturn, and their satellites. By the Galilean property there are what Stein (1967) has helpfully called "acceleration fields" around the two planets, whose actions are independent of the mass of any body "falling" in such a field. By universality, these fields extend (in the inverse-square proportion to distance) arbitrarily far into space and, in particular, between the two planets themselves. Applying the equality of action and reaction to this interaction implies that the acceleration produced by Jupiter on Saturn, for example, is proportional (at the given distance) to the *inertial*

---

[4] All citations to the *Principia* are from Newton (1999).

[5] Inertial mass, a concept introduced by Newton for the first time in Definition 3 of the *Principia*, is a measure of the resistance of a body to motion (acceleration) produced by *any* force, encapsulated in our modern equation $F = ma$. Gravitational mass, by contrast, represents the source of the gravitational field, so that, in our modern formulation, $F_{grav} = Gm_A m_B/r^2$. If A is the attracting body and B the attracted body, we then have $a_B = Gm_A/r^2$. The *passive* gravitational mass of the attracted body, taken to be equal to its inertial mass, drops out of the equation.

mass of Jupiter (and conversely for Saturn). By measuring the acceleration of one of Jupiter's satellites in the same acceleration field, we are therefore measuring the inertial mass of Jupiter as well (and similarly for Saturn and its satellites). In other words, the equality of inertial and passive gravitational mass, in the context of the universality of gravitational interaction and the Third Law of Motion, implies the equality of inertial and *active* gravitational mass as well.[6]

To better appreciate the subtlety and complexity of the Newtonian measurement of quantity of matter we need also to consider the traditional concept of *weight* from (Archimedean) statics in relation to both the distinctively Newtonian concept of inertial mass and the concept of quantity of matter that was already in use before Newton. Indeed, what Newton makes of the concept of weight is the key to his unification of the other two concepts. (Terrestrial) weight, for Newton, is the force on a body exerted by the earth's gravitational field, while inertial mass, as we have seen, represents the resistance of a body to *any* force. Given the Galilean property that all bodies fall with (approximately) the same acceleration near the surface of the earth (the constant g in Galileo's law of fall), it follows that two falling bodies are in equilibrium (in a balance, for example) just in case their (Newtonian) inertial masses are equal. Inertial mass, in this context, represents what the weights of bodies near different sources of attraction (the earth and the Sun, for example) have in common. In one use of the concept of quantity of matter, therefore, inertial mass represents the invariant or intrinsic quantity of matter that a body has independently of its interactions with other bodies.

In the seventeenth century, however, there was also another use of the concept of an intrinsic or invariant quantity of matter defined independently of weight: the amount of "stuff" in a given volume occupied by matter, which could be made more precise in terms of the ratio between the volume of absolutely dense (incompressible) matter in the volume (without empty interstices) and the total volume. This purely geometrical definition of quantity of matter was popular among adherents of the mechanical natural philosophy, and it, too, had a well-defined mathematical meaning independently of a body's interactions and motions relative to other bodies (including its merely *potential* motions as in the concept of inertial mass). The problem was that this purely geometrical concept had no clear

---

[6] Let the acceleration field on Saturn's moons be given by $a_1 = k_S/r_1{}^2$ and the acceleration field on Jupiter's moons by $a_2 = k_J/r_2{}^2$, where k_S and k_J are constants characterizing the absolute strengths of the fields independently of distance. Since the two acceleration fields extend far beyond the regions of their satellites, we also have accelerations $a_J = k_S/r^2$ of Jupiter and $a_S = -k_J r^2$ of Saturn, where r is the distance between them. But, according to the Third Law, $m_J a_J = -m_S a_S$. Therefore, $m_S/m_J = -a_J/a_S = k_S/k_J$: the absolute strengths of the two fields—their *active* gravitational masses—are proportional to their inertial masses. The need for this second step (Proposition 7) is obscured by the modern formulation of note 5 above, for it already assumes that there is a universal constant G characteristic of all gravitational interactions such that $k_A = G m_A$, where m_A is the inertial mass of the attracting body. Given the conceptual distinction between inertial and gravitational mass, however, this is precisely what Newton needs to establish—first via the acceleration-field property for passive gravitational mass and then by the above argument for active gravitational mass.

physical or empirical meaning at the time: how in the world could the ratio between absolutely dense matter within a given volume and the total volume be measured, even in principle? In that case, then, we had a mathematically well-defined quantity that was not, at the same time, empirically well-defined (empirically measurable). This underscores and makes vivid some of the important constraints on theory-mediated measurements.

If we adopt Newtonian inertial mass as the invariant concept of quantity of matter corresponding to the measurement of weight, however, the remaining problem is to explain how these measurements can then be projected into the heavens so as to apply to bodies beyond the significant reach of the earth's gravitational field. Such bodies can certainly not be quantitatively compared by placing them in a balance, for example. Here we find one of the most interesting and subtle steps in Newton's argument, which occupies Propositions 3–4 of Book 3, including the Scholium to Proposition 4. This argument, featuring what we now call the *moon test*, follows Propositions 1 and 2 of Book 3, which have established the inverse-square proportion of acceleration for the moons of Jupiter and Saturn relative to their primary bodies, and the same proportion for the primary planets (except possibly the earth) relative to the Sun. Since there are a number of different orbiting bodies in each case, the inverse-square proportion, by Kepler's third or harmonic law, applies from orbit to orbit (not just within a single orbit). Since the harmonic law depends only on distance and orbital period (independently, crucially, of the masses of the different bodies) the acceleration-field property therefore follows.[7]

In the case of the (earth's) moon considered in Proposition 3, however, there is only a single orbiting body. So we cannot appeal directly to Kepler's harmonic law in order to derive the acceleration-field property. Newton, in Proposition 3, does not even mention the harmonic law in deriving the inverse-square proportion, but appeals, instead, to "the very slow motion of the moon's apogee" (1999, 802).[8] This derivation of the inverse-square proportion, however, holds only for the (single) lunar orbit and thus does not by itself imply the acceleration-field property. What we now call the *moon test*, which is suggested in the Corollary to Proposition 3 and then carried out in Proposition 4, therefore proceeded as follows. Beginning from the inverse-square proportion for the (single) lunar orbit, Newton compared the

---

[7] Assuming circular orbits for simplicity, and using the circumstance that here $v = 2\pi rt$, with r the radius and t the period, we have $a = v^2/r$ in accordance with the formula for centripetal (or centrifugal) acceleration. In these terms, therefore, Kepler's harmonic law states that if r, R and t, T are, respectively, the distances and periods of two concentric orbits, then $r^3/R^3 = t^2/T^2$. It follows that the two accelerations stand in the ratio $1/r^2 : 1/R^2$ and thus that the variation in acceleration depends only on distance in accordance with the inverse-square proportion.

[8] Newton had already indicated this way of deriving the inverse-square proportion, by way of anticipation, in Proposition 2, with reference to Corollary 1 of Proposition 45 of Book 1. In Proposition 2, however, Newton says that the inverse-square proportion can be "proved with the greatest exactness from the fact that the aphelia are at rest" (ibid.), while in Proposition 3 he appeals to a corresponding "very slow" motion of the lunar orbit—indicating, in this case, that the inverse-square proportion holds only to a very high degree of approximation. I shall return to this circumstance below.

(actually measured) acceleration of the orbiting moon towards the center of the earth with the acceleration the moon would have if it were deprived of its orbital motion and to descend to the surface of the earth via a continual increase of this acceleration in accordance with the inverse-square proportion. It turns out, on this supposition, that the resulting acceleration at the surface of the earth would equal the Galilean acceleration g of terrestrial gravity (as determined, for example, by the motions of pendula). This gave Newton license to conclude, in Proposition 4, that the force by which the moon is maintained in its orbit is the very same force as terrestrial gravity, and also, in the Scholium to this proposition, that the relationship between falling bodies near the surface of the earth and the moon is precisely analogous to that between the different (multiple) satellites of the (known) primary bodies in the solar system (the satellites of Jupiter, Saturn, and the Sun). We now (and only now) have the acceleration-field property of gravitation in *all* of these cases.[9]

Our discussion of the moon test has revealed a further important aspect of Newton's methodology of theory-mediated measurement in the *Principia*. This methodology essentially involves approximations that can, in principle, be made continually more precise by bringing additional relevant factors to bear using the very same methodology. Thus, in the case of establishing the inverse-square proportion for the acceleration of the moon towards the earth, Newton finds that the observed value of this proportion deviates from the inverse-square—albeit by an amount that is in fact extremely small. Newton in no way completely neglects the deviation, however, but rather suggests a plausible explanation in terms of a perturbation of the otherwise stable lunar orbit by a further (relatively small) gravitational acceleration towards the Sun; and he promises to address this perturbation later on.[10] He returns to it, most explicitly, in his long treatment of the three-body problem (for the earth, Sun, and moon) later in Book 3, which extends

---

[9] Thus the main argument of Proposition 4 concludes as follows (1999, 804): "And therefore that force by which the moon is kept in its orbit, in descending from the moon's orbit to the surface of the earth, comes out equal to the force of gravity here on earth, and so (by rules 1 and 2) is that very force which we generally call gravity." The Scholium begins (1999, 805): "If several moons were to revolve around the earth, as happens in the system of Saturn or of Jupiter, their periodic times (by the argument of induction) would observe the law which Kepler discovered for the planets [i.e., the harmonic law], and therefore their centripetal forces would be inversely as the squares of the distances from the center of the earth (by prop. 1 of this book 3)." Note that the original Galilean property (constant g) holds (to a high degree of approximation) only for falling bodies sufficiently close to the surface of the Earth; for any considerable distance from the surface it must be generalized via the inverse-square proportion to the acceleration-field property.

[10] Compare note 8 above. In Phenomenon 6 of Book 3, from which Proposition 3 is derived, Newton considers applying Kepler's area law (his second law) to the moon's orbit, from which one could derive that the acceleration relative to the center of the earth is centripetal. Newton is clear, however, that the area law (relative to this center) is not quite exact (1999, 801): "Actually, the motion of the moon is somewhat perturbed by the force of the sun, but in these phenomena I pay no attention to minute errors that are negligible." Nevertheless, Newton does not neglect this error in Proposition 3 but instead formulates the promissory note in question more precisely (1999, 803): "Now this [very slow] motion of the apogee arises from the action of the sun (as will be pointed out below) and accordingly is to be ignored here."

from Proposition 25 through Proposition 35 and its Scholium.[11] Alas, however, Newton's treatment of this problem, as is well known, was not entirely satisfactory, and it was left to later mathematical scientists such as Euler and Clairault to find the correct solution after Newton's death. Nevertheless, if Newton's own solution to the three-body problem did not survive, his method for articulating and refining such problems within the (Newtonian) theory of gravity enjoyed a considerably longer life.

This method for attacking physical problems via an open-ended sequence of ever more accurate approximations has been dubbed "the Newtonian style" by I. Bernard Cohen (1980), who emphasizes the transition, within the Newtonian theory of gravity, from one-body problems, to two-body problems, to three- and many-body problems.[12] The method has been further developed by George Smith (2002, 2012, 2014), who emphasizes the importance of theory-mediated measurement in open-ended sequences of approximate reasoning. It has perhaps not been sufficiently appreciated, however, how closely such sequences of approximate reasoning are bound up with what Newton calls "the argument [or method] of induction." For what Newton means by the method of induction—which he typically opposes to that of mere hypotheses—is circumscribed by the four "Rules for the Study of Natural Philosophy" delineated at the beginning of Book 3. Rules 1 and 2 restrict causal inferences by a version of Occam's razor, which implies that the same effects should have, as far as possible, the same causes. Rule 3, perhaps the best known of the four, licenses the extension of regularities holding generally for all bodies we have observed to all bodies universally.[13] And Rule 4, added to the third (1726) edition of the *Principia*, clarifies the opposition between what Newton means by induction and what he means by hypotheses. Thus: "*In experimental philosophy propositions gathered from phenomena by induction should be* [taken as] *either*

---

[11] Proposition 25 reads (1999, 839): "*To find the forces of the sun that perturb the orbit of the moon.*" The Scholium to Proposition 35 begins by summing up the point that Newton has been most concerned to establish (1999, 869): "I wished to show by these computations of the lunar motions that the lunar motions can be computed from their causes by the theory of gravity." (The Scholium as a whole extends through p. 874.) Further discussion of Propositions 25–35 can be found in §8.14 of *A Guide to the Principia* in (1999); see also the more general reflections on Newton's work in lunar theory by George Smith in §8.15.

[12] One can survey the successful life-span of this method by focusing not only on the development of lunar theory from Newton to Euler and Clairault in the 1740's, but also on the corrections to Kepler's orbital laws due to further gravitational perturbations of the planetary orbits—including, most spectacularly, the successfully prediction of the existence of the planet Neptune by Adams and Leverrier via perturbations of the orbit of Uranus (1846). Indeed, this sequence of successes extended all the way up to the late nineteenth and early twentieth century, when (extremely small) perturbations of the perihelion of Mercury resisted all attempts to derive them within Newtonian gravitational theory—which finally gave way to a radically new theory of gravitation in Einstein's theory of general relativity (1915). I shall return to the discovery of Neptune in considerably more detail below, together with the complementary example of the perihelion of Mercury.

[13] More precisely (1999, 795): "*Those qualities that cannot be intended and remitted* [neither increased nor diminished] *and that belong to all bodies on which experiments can be made should be taken as qualities of all bodies universally.*"

exactly or very nearly true notwithstanding any contrary hypotheses, until yet other phenomena make such propositions either more exact or liable to exceptions."[14] Newton comments (*ibid.*): "This rule should be followed so that arguments based on induction may not be nullified by hypotheses."

To illustrate how this method is supposed to work, consider again the proposed concept of quantity of matter in terms of the ratio of the absolutely dense matter in a given volume to the total volume (including empty interstices) discussed above in connection with the mechanical natural philosophy. This way of understanding quantity of matter, for Newton, was merely hypothetical at the time, since there was no way of determining the relevant ratio within a given volume by empirical measurements—no super-microscope, as it were, capable of peering into the volume and estimating the variations in density there. While Newton by no means rejected the hypothesis that bodies in fact consist of perfectly dense (or perfectly hard) matter and empty space, he rightly considered it, at the time, to be only a plausible explanation of a number of given phenomena. It was treated in the Queries to the *Opticks*, therefore, as a conjecture rather than an inductively established proposition deserving to be taken to be true or very nearly true. For it had not yet found a place in an open-ended sequence of ever more accurate approximations facilitated by empirically practicable theory-mediated measurements of the relevant causally-efficacious quantities—in this case quantity of matter itself. And it is only such a sequence, as we have seen, which, for Newton, could then amount to a well-grounded "argument of induction."

This particular example is of more general significance, for analogous micro-structural hypotheses put forward by the mechanical philosophy were the main targets of Rules 3 and 4, especially when such hypotheses conflicted with Newton's own properly inductive arguments. Rule 3 was added in the second edition of the *Principia* (1713), in fact, precisely because of opposition to Newton's theory of gravitation expressed by his contemporaries Huygens and Leibniz, while Rule 4, added to the third edition (1726), served principally to reinforce this Newtonian defense. Newton's contemporaries had rejected the idea that gravity was a non-mechanical force, operating by action at a distance rather than action by contact, and they proposed instead an aetherial vortex theory that could, in principle, explain the celestial orbital motions in a way that was mechanically acceptable. But this theory, at the time, was again merely conjectural, because there was no practicable way to measure the unobservable motions in the aetherial vortex so as to provide an empirical basis for deriving the observable celestial motions from the proposed vortex theory. In sum, Newton's theory of gravitation was grounded in a proper empirical argument from the phenomena, while the vortex theory, at least at the time, was not even close to such a grounding.

---

[14] Newton 1999, 796. I have replaced "considered" at (1999, 796) with "taken as," since the same verb [*habere*] is used in both Rule 4 and Rule 3 (compare note 13 above). I follow Smith in understanding this usage as that of a technical term of art.

Thus, in the last paragraph of his comments on Rule 3, Newton emphasizes the superior inductive warrant for his theory of gravitation and states that the force of universal gravity has considerably more inductive warrant than that of impenetrability—which, of course, is fundamental to action by contact:

> Finally, if it is universally established by experiment and astronomical observations that all bodies on or near the earth gravitate toward the earth, and do so in proportion to the quantity of matter in each body, and that the moon gravitates toward the earth in proportion to the quantity of its matter, and that our sea gravitates toward the moon, and that all planets gravitate toward one another, and that there is a similar gravity of comets toward the sun, it will have to be concluded by the third rule that all bodies gravitate toward one another. Indeed, the argument from phenomena will be even stronger for universal gravity than for the impenetrability of bodies, for which, of course, we have not a single experiment, and not even an observation, in the case of the heavenly bodies. (Newton 1999, 796)

Against this background, then, Rule 4 functions primarily to make explicit the implied contrast between the properly inductive grounding of the theory of universal gravitation and the merely hypothetical contrary proposals of the mechanical philosophy.

That Newton's "argument from phenomena" (or "argument of induction") here is paired with a complimentary reticence towards mere hypotheses (as in the comment on Rule 4) is strongly confirmed by the penultimate paragraph of the General Scholium, which had already appeared in the second edition:

> Thus far I have explained the phenomena of the heavens and of our sea by the force of gravity, but I have not yet assigned a cause to gravity. Indeed, this force arises from some cause that penetrates as far as the centers of the sun and the planets without any diminution of its power to act, and that acts not in proportion to the quantity of the *surfaces* of the particles on which it acts (as mechanical causes are wont to do) but in proportion to the quantity of *solid* matter, and whose action is extended everywhere to immense distances, always decreasing as the squares of the distances. . . . I have not yet been able to deduce from phenomena the reason for these properties of gravity and I feign no hypotheses [*hypotheses non fingo*]. For whatever is not deduced from the phenomena must be called a hypothesis, and hypotheses, whether metaphysical or physical, or based on occult qualities, or mechanical, have no place in experimental philosophy. In this experimental philosophy propositions are deduced from phenomena and are made general by induction. The impenetrability, mobility, and impetus of bodies, and the laws of motion and the law of gravity have been found by this method. And it is enough that gravity really exists and acts according to the laws that we have set forth and [suffices for] all the motions of the heavenly bodies and of our sea.[15] (Newton 1999, 943)

Thus, while Newton certainly takes himself to have discovered the law in accordance with which the force of gravity acts by a proper argument from phenomena, he declines to speculate about the underlying cause or reason for this

---

[15] In suggesting the change of wording in the final sentence I am again following Smith, who pointed out that the Latin does not say anything here about "explaining" the motions in question. The verb translated as "explain" in the first sentence [*expōnēre*] does occur in the final sentence, but there it is translated as "set forth" (applied to the relevant laws), and the last clause then says that gravity is "sufficient for" [*sufficiat*] all the motions in question.

law. For no such speculation, conjecture, or hypothesis—of which the vortex theory is paradigmatic—has yet been grounded by an argument from phenomena.[16]

We are now in a position to return to Kant's apparent criticism of Newton's conception of gravitational force with which we began this section. The relevant Kantian passage introduced above occurs in the second remark to the seventh proposition in the Dynamics chapter of the *Metaphysical Foundations*.[17] So let us now place the passage in question into its immediately preceding context. After citing Corollary 2 to Proposition 6 of Book 3 of the *Principia*, Kant continues:

> [O]ne cannot adduce this great founder of the theory of attraction as one's predecessor, if one takes the liberty of substituting an apparent attraction for the true attraction he did assert, and [one] assumes the *necessity* of an impulsion through *contact* to explain the phenomenon of approach. [Newton] rightly abstracted from all hypotheses purporting to answer the question as to the cause of the universal attraction of matter, for this question is physical or metaphysical, but not mathematical. And, even though he says in the Advertisement for the second edition of his *Opticks*, "to show that I do not take *gravity* for an *essential* property of bodies, I have added one question concerning its cause," it is clear that the offense taken by his contemporaries, and perhaps even by Newton himself, at the concept of an original attraction set him at variance with himself. For he could by no means say that the attractive forces of two planets, those of Jupiter and Saturn for example, manifested at equal distances of their satellites (whose mass is unknown), are proportional to the quantity of matter of these heavenly bodies, if he did not assume that they attracted other matter merely as matter, and thus according to a universal property of matter. (4: 515)

Considering this fuller excerpt from Kant's second remark helps us to understand his perspective on the Newtonian concept of force (especially the Newtonian concept of gravitational force) more precisely.

To begin with, the fact that Kant has just been considering Proposition 6 of Book 3 shows that he is emphasizing the peculiarity of specifically gravitational force, whose relationship to mass or quantity of matter goes far beyond what is characteristic of forces in general. Indeed, Kant has just distinguished gravity from magnetism (also an attractive force) on precisely this score, since the latter satisfies only the Third Law of Motion (like all forces) but not the (generalized) Galilean acceleration-field property. And, as we have seen, it is precisely this last property, which, together with the Third Law of Motion, then enables us to measure the quantities of matter of Jupiter and Saturn by means of the accelerations of their

---

[16] As Newton tellingly points out, the action of the force of gravity "penetrates as far as the centers of the sun and the planets" and not in proportion to their surfaces ("as mechanical causes are wont to do") but rather to their "quantity of *solid* matter"—i.e., their quantity of matter or (inertial) mass. This does not mean, however, that Newton is open to genuine action at a distance. For, as forcefully expressed in his well-known letter to Bentley (1693), he takes it to be contradictory for a body to act at a place in which it is not. Local presence of a substance to another upon which it acts is therefore metaphysically necessary, but this presence could also be that of a "spirit" or other immaterial substance—including that of God, who, according to Newton, is omnipresent throughout the whole of space. I have discussed this situation at greater length in Friedman (2012a), and I shall return to it below.

[17] The seventh proposition itself reads (4: 512): "The *attraction essential to all matter* is an immediate action of matter on other matter through empty space."

satellites in Proposition 7 of Book 3.[18] Yet what is most striking about this fuller excerpt, from our present point of view, is that Kant goes on to say that Newton "*rightly* abstracted from all hypotheses purporting to answer the question as to the cause of the universal attraction of matter" (emphasis added). Indeed, Kant makes it clear that he is focusing especially on the hypotheses proposed by the mechanical philosophy, which "assumes the *necessity* of an impulsion through *contact* to explain the phenomenon of approach" (emphasis in the original). When Kant speaks of "the offense taken by [Newton's] contemporaries," therefore, he means to refer to those mechanical philosophers (especially Huygens and Leibniz) who criticized the first edition of the *Principia* for ignoring the supposed necessity of action by contact, thereby sparking Newton's responses in the second (and third) editions.

The most important of these Newtonian responses, in this connection, is the penultimate paragraph of the General Scholium, which appeared (along with Rule 3 and its commentary) in the second edition. For it is here that we find Newton's famous *hypotheses non fingo*, which is addressed, as we have seen, to the speculations of the mechanical philosophy concerning the underlying cause of gravitational attraction. Moreover, the way in which Kant himself describes the question concerning this underlying cause—as "physical or metaphysical, but not mathematical"—echoes Newton's classification of these possible hypotheses in the General Scholium: as "metaphysical or physical, or based on occult qualities, or

---

[18] See (4: 514–15): "It is commonly supposed that Newton did not find it necessary for his system to assume an immediate attraction of matter, but, with the most rigorous abstinence of pure mathematics, allows the physicists full freedom to explain the possibility of attraction as they might see fit, without mixing his propositions with their play of hypotheses. But how could he ground the proposition that the universal attraction of bodies, which they exert at equal distances around them, is proportional to the quantity of their matter, if he did not assume that all matter, merely as matter, therefore, and through its essential property, exerts this moving force? For although between two bodies, when one attracts the other, whether their matter be similar or not, the mutual approach (in accordance with the law of equality of interaction) must always occur in inverse ratio to the quantity of matter, this law still constitutes only a principle of mechanics, but not of dynamics. Thus, it is a law of the *motions* that follow from attracting forces, not of the proportion of their *attractive forces* themselves, and it holds for all moving forces in general. Thus, if a magnet is at one time attracted by another equal magnet, and at another time by the same magnet enclosed in a wooden box of double the weight, the latter will impart more relative motion to the former in the second case than the first, even though the wood, which increases the quantity of matter of this second magnet, adds nothing at all to its attractive force, and manifests no magnetic attraction of the box." Kant appears to be first asking for the basis of the claim that the source of specifically gravitational attraction (active gravitational mass) is equal to inertial mass. Newton establishes this by appealing to Proposition 69 of Book 1, which assumes both the universality of gravitational attraction and the Third Law of Motion: compare note 6 above, together with the paragraph to which it is appended. Kant goes on to use the example of magnetism to emphasize the importance of the acceleration-field property of gravitational attraction (the equality of inertial and passive gravitational mass), which, of course, is also assumed in Proposition 69 of Book 1: compare note 5 above, together with the paragraph to which it is appended.

mechanical."[19] So there is very little doubt, in my view, that Kant is endorsing rather than disputing Newton's methodological agnosticism concerning such hypotheses. He is following rather than rejecting the empirical methodology characterized by Newton's Rules for the Study of Natural Philosophy, including the opposition between hypotheses and "experimental" arguments from phenomena made explicit in Rule 4.[20]

That Kant is following rather than rejecting Newton's empirical methodology, however, does not imply that he agrees with Newton across the board. And it is in the realm of metaphysics, in particular, that the remaining differences between them become most salient. Kant explains what he means by "true metaphysics" in the Preface to the *Metaphysical Foundations*:

> All true metaphysics is drawn from the essence of the faculty of thinking itself, and is in no way fabricated [*erdichtet*] on account of not being borrowed from experience[; r]ather, it contains the pure actions of thought, and thus a priori concepts and principles, which first bring the manifold of *empirical representations* into the law-governed connection though which it can become *empirical* **cognition**, that is, experience.[21] (4: 472).

---

[19] See again the quotation from the General Scholium (1999, 943) in the paragraph to which note 16 above is appended. The most salient difference between Kant's list of alternatives and Newton's is the latter's inclusion of "mechanical." As we have seen, however, Kant has already made it clear that he is focusing especially on mechanical hypotheses.

[20] Note that the passage quoted in note 18 above (4: 514–5) cannot be read as a Kantian rejection of Newtonian inductive methodology. In particular, Kant is not implying that Newton himself is giving "physicists full freedom to explain the possibility of attraction as they might see fit." On the contrary, the *hypotheses non fingo* passage in Newton's General Scholium (1999, 943), with which Kant was clearly familiar, places severe constraints on the action of gravitational attraction: that it "penetrates as far as the centers of the sun and the planets," "acts not in proportion to the quantity of the *surfaces* of the particles on which it acts (as mechanical causes are wont to do) but in proportion to the quantity of *solid* matter" (i.e., mass), and "whose action is extended everywhere to immense distances." These are all properties that Kant attributes to his "fundamental force of attraction"—that it be a "penetrating" rather than a "surface" force (seventh explication of the second chapter), that it act in proportion to the quantity of matter (note to this seventh explication), and that it "extends immediately throughout the universe" (eighth proposition) (4: 516–18). Nor does the fact that Newton explicitly refrains from "affirming that gravity is essential to bodies" (1999, 796; immediately following the passage quoted above from his comment on Rule 3) whereas Kant explicitly asserts this of his fundamental force of attraction (4: 508; fifth proposition of the Dynamics), imply that there is a significant conflict between them. For Newton is just as clear (in the immediately following sentence in his comment on Rule 3) that he considers (only) "the force of inertia (*vis inertiae*)" to be an "inherent force (*vis insita*)" of bodies (1999, 796). Here Newton takes *inertial mass* to be essential to matter, and since, for both Newton and Kant, inertial mass is equal to gravitational mass, the cause of gravitational force—i.e., its source—is, in this sense, indeed essential to matter for both.

[21] Kant's "fabricate [*erdichten*]" corresponds to Newton's "feign [*fingêre*]," as becomes clear in Kant's polemic against what he calls the "mechanical natural philosophy" in the General Remark to Dynamics (4: 532): "[H]ere is not the place to uncover hypotheses for particular phenomena, but only the principles in accordance with which they are all to be judged[; e]verything which relieves us of the need to resort to empty spaces is a real gain for natural science, for they give the imagination far too much freedom to make up by fabrication [*Erdichtung*] for the lack of any inner knowledge of nature." For further discussion of this terminological correspondence see Friedman (2012c) (2017).

Later in the Preface, moreover, Kant makes a stronger assertion:

[M]etaphysics has busied so many heads until now, and will continue to do so, not in order thereby to extend natural knowledge (which takes place much more easily and surely through observation, experiment, and the application of mathematics to outer appearances), but rather so as to attain cognition of that which lies beyond all boundaries of experience, of God, Freedom, and Immortality[.]" (4: 477)

Metaphysics, for Kant, is not a contribution to first-order knowledge of either nature or morals, but has rather the second-order task of explaining the character and possibility of such first-order knowledge in general.[22]

Kant's revolution in metaphysics takes place a century after Newton's revolution in natural philosophy. And, to return to the issue of action at a distance, it is not only that Kant belonged to a later generation of Newtonian philosophers for whom genuine action-at-a-distance forces represented a badge of honor rather than an embarrassment. The deeper point, in the present context, is that Newton's approach to metaphysics was much more traditional, in so far as it aimed for a single all-encompassing ontology including both material and immaterial objects. Newton, in particular, represented a modern version of neo-Platonism associated with his teacher Henry More, according to which space itself was the unitary framework within which all objects, even God, were to be located. For More, Newton, and his representative Samuel Clarke, the Divinity was omnipresent everywhere in space (and time), with no obstacle, on their view, to God's unity, simplicity, or immateriality. This conception, however, was rejected by the overwhelming majority of contemporary and later natural philosophers, including Kant. Indeed, Kant's transformation of traditional metaphysics into what he called transcendental philosophy was intended definitively to block the possibility, within natural philosophy, of any interaction at all between material and immaterial objects, with the result that a body can only act on another body via an exchange of momentum between them—even if, in the case of gravity, the action of this force is "extended everywhere to immense distances."[23]

---

[22] I say "nature or morals" because God, Freedom, and Immorality, for Kant, are not mysterious inhabitants of a supersensible (noumenal) realm belonging, together with the sensible (phenomenal) realm, to a single whole of objects of cognition. In particular, they are by no means on a par with objects in nature that we encounter (in empirical natural science) as independent of us. They are rather what Kant (in the second *Critique*) calls "Postulates of Pure Practical Reason," which are derived from the necessity of the moral law flowing from our own *practical* rather than *theoretical* reason—creatures of our own free self-legislation or rational autonomy.

[23] See again Friedman (2012a), cited in note 16 above, for the application of this point to gravity. The famous letter to Bentley explicitly leaves open the possibility that the underlying cause of gravity may be immaterial.

12.2 Theory-Mediated Measurement and the "Phenomena" in Newton and Kant

Newton was the creator of a revolutionary methodological approach based on Rules 3 and 4 of the *Principia*, about which he reflected deeply and insightfully. Nevertheless, Kant's systematic discussions often have the advantage of clarifying some of the key conceptions constitutive of Newton's approach from a philosophical point of view—especially, as we shall see, the conceptions of "phenomena" and "realism."

Howard Stein's view is that the relevant conception of "phenomena" (and of "realism") is Kant's.[24] And what this means, for Kant, is that they are objects in space and time interacting by objective causal laws: they are neither non-spatio-temporal (supersensible) objects in a purely intelligible world (noumena) nor the subjective mental states from which (on some philosophical views) we make inferences to the objective spatio-temporal order. More to the point, however, they also go considerably beyond the "Phenomena" from which Newton begins his inferences to the structure of "The System of the World" in Book 3. These Newtonian Phenomena involve the relative motions among the then observable bodies in the Solar System (all relative to the fixed stars): Jupiter with its four moons, Saturn with its five moons, the earth with its moon, the Sun with its five (remaining) planets.[25] And, in this respect, Newtonian Phenomena go well beyond those of the ancient tradition of "saving the phenomena," where phenomena in that tradition involved only the changing angles between the lines of sight drawn from our position on the earth's surface and projected from there via the relevant heavenly bodies onto the celestial sphere of the fixed stars.[26] The important point

---

[24] Cf. note 2 above: "all theoretical knowledge is *of* the phenomenal world: the world of experience."

[25] Newton is using Kepler's orbital laws—the area and harmonic laws—to describe motions of satellites relative to their primary bodies. As noted above, Newton has already ruled out the Ptolemaic system (by telescopic observations of phases and transits), and his project is then to decide between Kepler and Tycho. His (provisional) neutrality between the two is expressed in Phenomenon 4 (1999, 800): "*The periodic times of the five primary planets and of either the sun about the earth or the earth about the sun—the fixed stars being at rest—are as the 3/2 powers of their mean distances from the sun.*"

[26] Newton's Phenomena go beyond the earlier tradition of "saving the phenomena" principally via the telescopic observations of phases and transits mentioned in the previous note. See especially Phenomenon 3: "*The five primary planets—Mercury, Venus, Mars, Jupiter, and Saturn—encircle the sun.*" The argument follows: "That Mercury and Venus revolve about the sun is proved by their exhibiting phases like the moon's. When these planets are shining with a full face, they are situated beyond the sun; when half full, to one side of the sun; when horned, on this side of the sun; and they sometimes pass across the sun's disk like spots. Because Mars also shows a full face when near conjunction with the sun, and appears gibbous in the quadratures, it is certain that Mars goes around the sun. The same thing is proved also with respect to Jupiter and Saturn from their phases being always full; and in these two planets, it is manifest from the shadows that their satellites project upon them that they shine with light borrowed from the

here, however, is that Newtonian Phenomena do not involve causal or dynamical information further determining the true (versus merely apparent) motions—for it is precisely this that is to be inferred from these Phenomena by applying the Laws of Motion in accordance with Newton's methodological rules.

By contrast, Kantian phenomena—which Kant calls objects of *experience*— involve precisely such causal and dynamical information. In particular, the Analogies of Experience governing the categories of substance, causality, and community are what first characterize, for Kant, the meaning of "experience"—i.e. full-fledged *empirical cognition*.[27] More precisely, experience first arises by moving from what Kant calls "perceptions" to empirical cognition by means of the Analogies of Experience.[28] And the transition from perceptions to fully objective experience, more generally, involves a transition from what Kant calls "mathematical" principles of pure understanding to "dynamical" such principles. There are two sets of mathematical principles, the Axioms of Intuition and Anticipations of Perception.[29] The dynamical principles, aside from the Analogies of Experience, include a fourth set of principles entitled the Postulates of Empirical Thought in general. The title of this fourth (and final) set of principles makes it explicit that all of the principles

---

sun." (Newton 1999, 799) These phenomena, therefore, are based on (kinematical and optical) inferences establishing and then applying the (non-Ptolemaic) perspective common to both the Tychonic and Keplerian systems. And, because both of these systems are still (provisionally) in play, the Newtonian argument is not begging the question between geocentrism and heliocentrism. NB: *Neither* Newton's Phenomena *nor* the ancient tradition involve inferences from bare subjective sense-data—*both* involve sophisticated spherical geometry.

[27] The Analogies of Experience are the third set (out of four) of Principles of Pure Understanding in the *Critique of Pure Reason*. Kant's systematic discussion of the Analogies is by far the longest, extending from (A176/B218) through (A218/B265), although there are some significant differences between the first or 'A' edition (1781) and second or 'B' edition (1787). All citations of the *Critique* are given by this standard A/B format, which is also used in the margins of the Guyer-Wood translation (1998) in the Cambridge Edition.

[28] Preceding the three Analogies corresponding to the categories of substance, causality, and community is a single principle governing all three. This principle, in the second edition, reads (B218): "*Experience is possible only through the representation of a necessary connection of perceptions.*" The "Proof" added in the second edition begins (B218–9): "Experience is an empirical cognition, i.e., a cognition that determines an object through perceptions. It is therefore a synthesis of perceptions, which is not itself contained in perception, but rather contains the synthetic unity of the manifold [of perceptions] in a consciousness, which constitutes what is essential to a cognition of *objects* of the senses, i.e., experience (not merely the intuition or sensation of the senses)." The final sentence emphasizes the *objectivity* of the experience in questions, while the statement of the principle itself emphasizes the importance of a "necessary connection [*notwendige Verknüpfung*]" of perceptions—thereby implicating Kant's differences with Hume concerning *causal necessity*. I shall return to this last point below.

[29] The general principle of the Axioms is that all empirical intuitions have an *extensive magnitude* conforming to their spatial and/or temporal extent, while the general principle of the Anticipations states that every empirical intuition has also an *intensive magnitude* corresponding to the degree of filling (out of a continuum of possible such degrees) of the empirical content within a given spatial and/or temporal extent—e.g., the degree or intensity of illumination on a given surface, or the degree of gravity or weight with which a body presses down on a balance.

govern *empirical* intuitions and *empirical* objects, not simply pure intuitions in space and time such as geometrical figures.[30]

To properly appreciate the relationship between Kantian and Newtonian phenomena, however, it is necessary to interpose the *Metaphysical Foundations* between the *Critique of Pure Reason* and the *Principia*. For, as Kant understands it, the *Metaphysical Foundations* results from realizing the purely a priori system of categories and principles of the understanding via "the empirical concept of matter"—which, as empirical, has a quite different status from the pure concepts or categories of the understanding. In the case of the Analogies of Experience, in particular, we thereby instantiate (via the empirical concept of matter) the permanence of substance (First Analogy), general principle of causality (Second Analogy), and principle of thoroughgoing interaction or community (Third Analogy). We thereby obtain three mathematically formulated specifically *physical* principles in the third or Mechanics chapter of the *Metaphysical Foundations*: the conservation of the total quantity of matter, the law of inertia, and the equality of action and reaction.[31] As is clear, these are not identical to Newton's three Laws of Motion, and the relationship between the two sets of laws is actually rather complicated. It is also clear, however, that Kant intends his three "Laws of Mechanics" to play the same role in the determination of the true as opposed to merely apparent motions in the Solar System as Newton's three Laws of Motion play in his determination of the same true motions in Book 3. In both cases, more specifically, we arrive at the center of gravity of the Solar System, which is always very close to the center of the Sun, as determining a privileged non-rotating frame of reference (relative to the fixed stars) wherein all gravitational actions of one body on another body are precisely counterbalanced by equal and opposite such reactions of the second body on the first.[32]

---

[30] The Postulates of Empirical Thought in general govern the categories of possibility, actuality, and necessity—where the category of necessity, in particular, is bound up with Kant's differences with Hume concerning the necessity of causal laws of nature (compare note 28 above). These Postulates, moreover, involve all three of the earlier sets of principles in so far as the three categories of possibility, actuality, and necessity corresponds, respectively, to the Axioms of Intuition, the Anticipations of Perception, and the Analogies of Experience. The Postulates therefore track the movement or transition from each of the three earlier sets to the next, so that, at the end, the Analogies of Experience yield causal necessity.

[31] Kant's clearest and most intuitive statement of these principles is in the Introduction to the second (B) edition of the *Critique*. They are there presented as uncontroversial examples of "pure natural science" (B20-1n): "[O]ne need merely consider the various propositions that come forth at the outset of proper (empirical) physics, such as those of the permanence of the same quantity of matter, of inertia, of the equality of action and reaction, etc., and one will be quickly convinced that they constitute a *physica pura* (or *rationalis*), which well deserves to be separately established, as a science of its own, in its whole extent, whether narrow or wide." In general, the second edition (1787) is more closely related to the *Metaphysical Foundations*, which, of course, had already been published by then.

[32] I have reconstructed Kant's argument here, in juxtaposition with the Newtonian argument, in a number of places: in Friedman (2012d), for example, and, most fully, in Friedman (2013). In the latter I provide the most detailed reconstruction I am able of the fourth or Phenomenology chapter

In comparing Kant's notion of phenomena with Newton's, therefore, the first point worth emphasizing is that, while the Newtonian argument in Book 3 begins from what Newton terms "Phenomena," Kant's reconstruction of that argument in the Phenomenology chapter of the *Metaphysical Foundations* ends with what *he* calls "phenomena"—i.e., full-fledged objects of experience, characterized, more specifically, by precisely the causal and dynamical information that Newton leaves out in his Phenomena and arrives at only later in the main argument of Book 3.[33] The second point worth emphasizing is that it is precisely the Kantian conception of phenomena that supports the relevant conception of scientific realism—i.e., the conception of adequately and correctly representing "the phenomena of nature" in Stein's 1989 paper.[34] What makes this a conception of "realism"—applied, specifically, to the original issue of Copernicanism addressed by the tradition of "saving the phenomena"—is that Newton showed for the first time how one could *empirically demonstrate* the (approximate) truth of Copernicanism by *empirically determining* the center of gravity (center of mass) of the Solar System. The question of the correct System of the World became a methodological rather than metaphysical or meta-scientific question, in so far as it could now be answered by evidential reasoning *internal* to science itself. And it is in precisely this sense, therefore, that Stein's conception of abstract structures adequately and correcting

---

of the *Metaphysical Foundations*, which, for Kant, corresponds to the Postulates of Empirical Thought in general in the first *Critique*. The basic idea of all of my reconstructions is that those motions are merely *possible* that result from an arbitrary choice of body (e.g., the earth) as the center of motion of the system, those are *actual* that can be dynamically distinguished from a contrary rotation of the surrounding space (e.g., via centrifugal forces), and those are *necessary* for which there is always an equal and opposite contrary motion. The example for actuality is a state of (true) rotation (relative to the fixed stars); the example for necessity is the totality of gravitational accelerations relative to the center of gravity of the Solar System (taking the fixed stars to be at rest). It should be noted, however, that Kant, unlike Newton, explicitly considers larger galactic structures (beginning with the Milky Way), in which one can sequentially determine the true motions relative to the centers of gravity of these larger structures *ad infinitum*. For Kant, accordingly, there is no end to the possibility of ever larger such nested structures.

[33] Thus, at the beginning of the Phenomenology chapter Kant explains the determination of true (actual or necessary) motions as a "transformation of appearance [*Erscheinung*] into experience [*Erfahrung*]" (4: 555)—where, as Kant then makes clear, the initial (purely relative) motions with which we began can only be judged as "merely possible" (ibid.). It is also clear, however, that the Phenomena with which Newton begins his determination of the true motions cannot be characterized as "appearances [*Erscheinungen*]" in the sense of (4: 555). For the latter are mere empirical intuitions that have not yet been judged by the understanding at all, and, therefore, no categories of the understanding have yet been applied. In particular, since Newton's Phenomena have a rather rich mathematical description, involving both kinematical and optical information, they fall under Kant's "mathematical" categories and principles (but not, of course, the "dynamical" ones). So they are best described, in Kantian terminology, as *perceptions* rather than mere empirical intuitions. Compare note 26 above, together with the paragraph to which it is appended.

[34] See again note 2 above, together with the paragraph to which it is appended and the preceding paragraph.

representing the phenomena is, in fact, a version of "realism"—albeit a now unorthodox "empirical" (or methodological) version.[35]

The "structural realism" defended by Worrall, by contrast, was explicitly addressed to *meta*-scientific questions. What does it mean for theories to converge over time towards the truth, and what is our justification for believing that they do so? In addressing these questions, Worrall therefore had to address the "no miracles augment" put forward by Hilary Putnam and Richard Boyd, together with the contrary "pessimistic meta-induction" defended by Larry Lauden. And, as discussed briefly at the beginning of this paper, Worrall's basic idea was to substitute mathematical structures for physical referents as the desired locus of convergence over time. The main issue that arises for us now, therefore, is to consider the relationship between the methodological conception defended here and meta-scientific versions of realism such as Worrall's.

To apply a remark I first heard from George Smith, what is essential to such Newtonian forces is that they are physical quantities or magnitudes (what we now call vector quantities) having determinate quantitative relations to other physical magnitudes—which, according to the Laws of Motion, are the accelerations produced in a body on which the force in question is supposed to act and the mass or quantity of matter of this same body. These three central concepts—impressed force, (true) acceleration, and quantity of matter—are not concepts of pure mathematics but rather physical concepts of an emerging new approach to the *empirical* study of nature. Moreover, from a modern point of view, they are theoretical rather than observational such concepts, since there is no way directly to perceive the quantity of an impressed force or the true versus merely apparent accelerations of a body or, most especially, the mass or quantity of matter of a body. This is especially obvious when we consider applying these concepts to celestial bodies in Newtonian astronomy, which, as we have seen, included precisely such an extension of their applicability by means of the wealth of new physics articulated in the *Principia*. This physics, as we have also seen, was particularly notable for the theory-mediated measurements it made possible for all three of the quantities in question.

In what sense, however, are these quantitative physical concepts structural or abstract? Our final point of emphasis is that this way of speaking derives from the distinctively Newtonian methodology of theory-mediated measurement. For Newtonian empirical methodology, as formulated in his Rules for the Study of

---

[35] Kant represents such "internal" or "empirical" realism (in contradistinction to his own "*transcendental* idealism") in the famous Refutation of Idealism added to the second edition of the *Critique of Pure Reason* (B274–9). It follows the discussion of the second postulate (governing actuality) in the Postulates of Empirical Thought in general, which centrally involves the use of the Analogies of Experience for "proving existence mediately" from given perceptions (B272–4). The following Refutation is supposed to answer a (Cartesian) objection to this procedure, and the initial example Kant gives is "undertaking [*vornehmen*] all time-determination through the change in outer relations (motions) relative to that which persists in space (e.g., motion of the Sun with regard to objects on the earth)" (B277–8). The procedure in the fourth chapter of the *Metaphysical Foundations* for determining the true motions in the Solar System by beginning with our own position here on earth appears to be very salient here.

Natural Philosophy, involved a carefully crafted combination of bold inductive extrapolation with a complementary reticence towards so far (merely) conjectural hypotheses. Newton *abstracts from* all such hypotheses, in order precisely to avoid preemptively obstructing the progress of inductive extrapolations before there are any contrary empirical grounds for doing so.

Consider the treatment of the concept of centripetal force—a species of impressed force—in the first two Propositions of Book 1. This species of impressed force is explained in Definition 5 of the *Principia*: "*Centripetal force is the force by which bodies are drawn from all sides, are impelled, or in any way tend, toward some point as to a center*" (1999, 403). Newton does not speculate concerning the underlying causes of this "drawing," "impulsion," or "tendency" but speaks merely of its direction towards a (single) center. Then, in the first two Propositions of Book 1, he appeals only to his geometrical characterization of centripetal force and the First Law of Motion (the law of inertia) in demonstrating Kepler's area law—that a body orbiting a single point in a fixed plane is moved by a centripetal force towards this point if and only if the body traces out equal areas in that plane in equal times. The law of inertia determines equality of times by the equal distances traversed by bodies moving in straight lines acted upon by no impressed forces, and Newton's demonstration then applies Euclidean geometry to substitute equal areas for equal distances in the case of bodies acted upon by only a single centripetal force. Since bodies orbiting in (approximate) conformity with Kepler's area law are ubiquitous in the Solar System, this yields theory-mediated measurements of elapsed time with extensive empirical applicability. It also enables us to make well-grounded empirical inferences from the mathematical properties of observable motions to the mathematical properties of corresponding impressed forces—which, as explained, are theoretical quantities.[36]

Newton makes the abstraction involved in his treatment of centripetal force explicit in his comments on Definition 8 of the *Principia*:

---

[36] Proposition 1 and 2 of Book 1 thereby first set in motion the very long argument of the *Principia*, which begins here and ends with the (partially completed) determination of the true System of the World in Book 3. Since these initial inferences in Book 1 involve only the definition of centrifugal force and the law of inertia, there is nothing problematically circular in this kind of theory-mediated inference. The only theoretical assumption here is the law of inertia, which had become generally accepted among virtually all natural philosophers of the time—whether proponents of the mechanical natural philosophy or not. And this is also true of almost all of Newton's arguments along the way—e.g., the move from an initial neutrality between Tychonic and Keplerian systems to a forceful demonstration of the (approximate) truth of the latter (compare note 26 above). More generally, since the theory of universal gravitation only emerges at the end of the argument, when (almost) all of the relevant theory-mediated inferences have been concluded, this characteristically Newtonian theory is not a *presupposition* of the argument. Nevertheless, there was one particularly significant gap in the evidential reasoning that had already been confronted by Newton—in his assumption (or conclusion) that the Third Law of Motion applies equally to attractions as well as impacts and pressures. I shall return to these issues below.

I use interchangeably and indiscriminately words signifying attraction, impulse, or any sort of propensity toward a center, considering these forces not from a physical but only from a mathematical point of view[; t]herefore, let the reader beware of thinking that by words of this kind I am anywhere defining a species or mode of action or a physical cause or reason, or that I am attributing forces in a true and physical sense to centers (which are mathematical points) if I happen to say that centers attract or that centers have forces. (Newton 1999, 408)

But there is a subtlety involved in Newton's abstraction from a "physical cause or reason" here, which is clarified in his discussion in the Scholium at the end of Section 11 of Book 1.[37] In this Scholium, in particular, Newton again emphasizes the purely mathematical character of his treatment:

I use the word "attraction" here in a general sense for any endeavor whatever of bodies to approach one another, whether that endeavor occurs as a result of the action of the bodies either drawn toward one another or acting on one another by means of spirits emitted or whether it arises from the action of aether or of air or of any medium whatsoever—whether corporeal or incorporeal—in any way impelling toward one another the bodies floating therein. I use the work "impulse" in the same general sense, considering in this treatise not the species of forces and their physical qualities but their quantities and mathematical proportions, as I have explained in the definitions. (Newton 1999, 588)

Moreover, the reference back to the Definitions at the end suggests that the contrast between "mathematical" and "physical" in this passage is the same as in Definition 8. In the immediately following passage (in the same Scholium), however, Newton explains more exactly what he means by "physics":

Mathematics requires an investigation of those quantities of forces and their proportions that follow from any conditions that may be supposed. Then, coming down to physics, these proportions must be compared with the phenomena, so that it may be found out which conditions of forces apply to each kind of attracting bodies. And then, finally, it will be possible to argue more securely concerning the physical species, physical causes, and physical proportions of these forces. (Newton 1999, 588–9)

Thus, while the only conditions generally presupposed in Book 1 are the Laws of Motion themselves, in Book 3 Newton considers specific empirically given forces—especially the force of gravity—for which we are in the process of empirically determining more specific conditions or laws. This would include the inverse-square law, for example, and, more generally, the acceleration-field property peculiar to gravitational force.

When Newton says that he is abstracting from the "physical cause or reason" in Definition 8, therefore, he does not mean that the centripetal force of gravity, for example, is not itself a *cause of motion*, but only that he has not yet provided a deeper explanation or reason for the specific laws governing this force. This is why, in the General Scholium containing the *hypotheses non fingo* passage at the end of the *Principia*, Newton explains that he has "not yet assigned a cause to

---

[37] This Scholium immediately follows Proposition 69 if Book 1, which, as we have seen, plays a crucial role in Book 3 in relating the centripetal force of gravity to the active gravitational mass of the central attracting body, thereby enabling theory-mediated measurements of the quantities of matter of the primary bodies in the Solar System.

gravity" itself, but, nonetheless, "gravity really exists and acts according to the laws that we have set forth" (1999, 943).[38] So there are two distinguishable levels of abstraction here—one purely mathematical and the other methodological. The first reflects the fact that Newton, throughout the *Principia*, is articulating "mathematical principles of natural philosophy." The second depends on his Rules for the Study of Natural Philosophy, which initiate the progressive application of mathematics to given natural phenomena throughout Book 3 in a way that does not "feign" hypotheses.

In the last-quoted passage from the Scholium to Section 11 of Book 1, Newton asserts that, on the basis of his method for treating forces, both mathematically and physically, "it will be possible to argue more securely concerning the physical species, physical causes, and physical proportions of these forces" (1999, 589). The obvious question is: "more securely than what?" And the obvious answer, in the present context, is that Newton's method is considerably more secure than any merely *hypothetical* method. Indeed, even contemporary partisans of the hypothetico-deductive method for testing and confirming hypotheses will readily admit that confirmatory true predictions need to satisfy substantially stronger criteria—such as simplicity, non-adhocness, and the like—than the mere deduction of true predictions. Newton himself, in any case, wanted (at least) to balance merely hypothetical reasoning with his own version of the inductive method in accordance with his Rules for the Study of Natural Philosophy and distinctively mathematical treatment of the concept of force. And since the inductive inference, more generally, begins from observational data and infers from these to a law or regularity to which the data conform, it appears, at least at first sight, to involve a closer and more immediate relationship between the data and the inferential conclusion than the hypothetico-deductive method.

Newton's version of the inductive method, however, goes well beyond these generalities. For both the data at the basis of his inductions (Newtonian Phenomena) and the laws derived therefrom (such as the laws of gravitation) are thoroughly mathematical or quantitative. This means, in particular, that the laws derived from the data are subject, in principle, to iterated quantitative corrections as both more data are observed and the inferred regularities or putative laws become less approximate and thus increasingly precise. Nevertheless, the preceding tradition of mathematical astronomy devoted to "saving the phenomena" culminated in the work

---

[38] Note 15 above is appended to the entire relevant passage from the General Scholium. As emphasized in note 20 above, the passage in question already places more specific constraints on the force of gravitation going well beyond the Laws of Motion—and, in particular, makes it quite difficult to see how the hypothetical explanations suggested by the mechanical natural philosophers could possibly provide a genuine alternative to Newton. Note 20 also argues, on this basis, that there may be less difference than one might otherwise suppose between Kant and Newton concerning the physics (and not just mathematics) of gravitational force. As already suggested at the end of the last section, however, there is a large difference indeed between Kant and Newton from a metaphysical point of view. Most obviously, Kant holds that both Euclidean geometry and his three Laws of Mechanics are synthetic a priori truths, while Newton never suggests any such thing—neither for Euclidean geometry nor his three Laws of Motion.

of Ptolemy and Copernicus. And both Ptolemaic and Copernican astronomy shared a powerful (and non-trivial) instrument for constructing such iterated quantitative corrections in the method of *epicycles*—where, if an orbit constructed at any given stage is then shown to be only approximately in conformity with data at a later stage, one could almost always make the orbit more precise by the addition of epicycles of the appropriate sizes and circular speeds.[39] In other words, as Newton himself was surely aware, both Ptolemaic and Copernican astronomy were particularly effective instruments for quantitative iterative curve fitting. The question naturally arises, therefore, of why we should need anything more than this from our inductive method. How exactly does the "Newtonian style," in I. Bernard Cohen's sense, provide a better fit with empirical data than these earlier methods of quantitative curve fitting?[40]

The answer is that Newton's very particular approach to inductive argumentation also involves, aside from the Rules of Reasoning formulated at the beginning of Book 3 of the *Principia*, the mathematical conception of force articulated in both the Definitions and Laws of Motion preceding Book 1. This means, as already explained, that Newton's method essentially involves what we would call *theoretical* concepts—mass, force, and (true) acceleration—in addition to the *observational* concepts occurring in what Newton calls Phenomena. The Newtonian Laws of Motion then mediate an inference from quantitative *observational* data to equally quantitative *theoretical* conclusions. And it is for precisely this reason that Newton can take as confirmatory data observations and experiments not accommodated in the previous tradition of mathematical astronomy limited to "saving the phenomena."

12.3 Planetary Perturbations and the Discovery of Neptune

Planetary perturbations within Newtonian theory are good illustrations of the further confirmatory data—going well beyond traditional curve-fitting strategies—that this theory can provide. And the case of the discovery of Neptune on the basis of the perturbations of the orbit of Uranus is especially valuable. In particular, the confirmatory data in question were spread out over more than half a century, and

---

[39] It has been pointed out a number of times that the method of iterated epicycles is, from a formal point of view, subsumable under the Fourier representation of arbitrary (continuous) quasi-periodic functions: just as any (continuous) quasi-periodic function can be approximated by a finite number of Fourier harmonics, any quasi-periodic motion can be approximated by a finite number of iterated epicycles. See, for example, Hanson (1960) and Gallavotti (2001)—who also cites an earlier contribution by G. Schiaparelli from 1874.

[40] For the "Newtonian style" see the paragraph to which note 12 above is appended, together with the preceding paragraph. I am indebted to a discussion with my colleague Rosa Cao (a neuroscientist and philosopher of cognitive neuroscience) for raising this question in a particularly vivid and helpful way.

they thereby illustrate both the subtlety and complexity of such data—and, most importantly, how this sequence of confirmatory date turned out to be continuous with the eventual discovery of Einsteinian general relativity in connection with a very subtle anomaly in the motion of Mercury. This example thus provides the best illustration of how the development of Newtonian gravitational theory became a necessary and irreplaceable instrument for the discovery of new and better theories. Since the story is rather long and intricate, however, the reader may wish to skip this section on a first reading and simply take it as given at the beginning of the next section where the implications of this story are further developed.

As far as the traditional curve-fitting strategies are concerned, it would have been quite possible to arrive at a considerably improved orbit for Uranus using, for example, the epicyclic methods deployed in both Ptolemaic and Copernican astronomy. That would have yielded improved approximations for the orbit of Uranus with no mention of Neptune at all. In the Newtonian tradition, by contrast, the relevant perturbations of the orbit of Uranus were *residuals*, taken to be additional to those arising from other sources of perturbative gravitational force— such as Jupiter and Saturn, for example—that had already been taken into account. The planet Neptune, in the first instance, was a body postulated or hypothesized with the required orbit and mass to produce the additional (residual) perturbations in question via the Newtonian gravitational law. The crucial point is that, in the Newtonian method, we must still actually find or observe (in this case by telescopic means) the relevant *physical source* (the massive body) responsible for these perturbations. Thus, while epicyclic methods (or other curve-fitting methods) might very well have resulted in a correction to the orbit of Uranus that would be as accurate as (or even more accurate than) the Newtonian derivation, they would have remained entirely closed off from the very striking new evidence provided by (telescopic) observations of the planet Neptune.

Newton's use of the theoretical concepts of force, mass, and (true) acceleration is not merely a means for learning more about the causal or dynamical structure of nature that is responsible for the observed orbital motions—although it certainly is that. This same apparatus provides a powerful tool for acquiring further empirical evidence, going beyond curve fitting, for the revision of any given (optical or kinematical) description of such motions in the direction of greater precision and accuracy. How do we determine, for example, that Neptune is indeed the gravitational source of the additional (residual) perturbations of Uranus' orbit? This determination requires theory-mediated measurements of the active gravitational mass of Neptune—which depend, in general, on measuring accelerations directed towards Neptune of other bodies in its celestial neighborhood. In the case of Uranus, in particular, we need its additional (residual) accelerations to be directed towards Neptune, and proportional (at a given distance) to Neptune's gravitational mass.

It is important to appreciate, however, that it was in fact quite difficult to discern these additional (residual) perturbations of the orbit of Uranus in the first place. And here, in particular, it was necessary to apply curve fitting to the data points originally taken to correspond to the (heliocentric) orbit of Uranus before the specific perturbations attributable to Neptune could emerge. The curve fitting in question, however, did not proceed via epicycles (as in Ptolemy and Copernicus), but rather by looking for elliptic orbits (following Kepler and, of course, Newton) while also taking account of the already successful cases of (Newtonian) perturbation theory developed by Laplace—as, for example, in his theories of the moon and of Jupiter and Saturn. The required orbit of Uranus would thus need to be elliptical, subject only to the proviso that deviations from the Keplerian description would need to be explained by perturbations arising from other relevant bodies in the Solar System in conformity with Newtonian universal gravitation. The crucial difference between the case of Neptune and the earlier Laplacian cases, however, is that here the planet in question had not yet been discovered. On the contrary, it needed to be first postulated or hypothesized on the basis of our best guess as to the true orbit of Uranus, where this orbit had already been regularized via curve fitting so that only the perturbative irregularities attributable to the (hypothetical) planet Neptune remained as residual phenomena.[41]

The residual phenomenon in question, as it presented itself to Adams and Leverrier, took place between 1800 and 1846, during which interval the observed heliocentric longitude of Uranus first advanced from its predicted longitude (maintaining a constant rate of advance) from 1804–5 until approximately 1822, then began to retreat from this advance (at an increasing rate) until, in 1830–31, the predicted and observed longitudes coincided; this retreat then continued (again maintaining a constant rate) until 1846. On this basis both Adams and Leverrier hypothesize a planet with orbit encompassing that of Uranus, such that the conjunction of the two planets would take place approximately in 1822, and the hypothetical planet exerted increasing attractive force on Uranus as it approached conjunction, decreasing attractive force as it passed this point. The force exerted by the planet would thus accelerate Uranus as it approached conjunction and decelerate it afterwards—although now with continually diminishing effect. In order to obtain quantitative predictions, however, each of the two astronomers needed to hypothesize a mass for the new planet, together with a size and eccentricity for its orbit. They did the best that they could on the basis of the telescopic observations of Uranus available to them, with the result that they each (independently) predicted a heliocentric longitude of the new planet for the 23rd of September 1846 between 326 degrees (Leverrier) and a bit more than 329 degrees (Adams). Johann Galle of the Royal Observatory in Berlin, at the request of Leverrier, searched for and observed

---

[41] A quite detailed but still relatively non-technical description of this process can be found in Part Two of John Herschel's *Outlines of Astronomy*: see the reprint edition of 1902 (which seems to be based on the tenth edition of 1869), pp. 659–80, together with Plate A. Herschel's account explains that and how the theoretical work of Adams and Leverrier during the years 1845–6 was in fact addressed to an already existing residual phenomenon. I am indebted to George Smith for emphasizing the importance of Herschel's account to me. I am also indebted to a seminar covering Herschel's work on residual phenomena led by Smith and Teru Miyake in the Spring of 2018—which was also assiduously followed by my colleague Rosa Cao (see again note 40 above).

the new planet (very near the ecliptic) at just short of 327 degrees of heliocentric longitude, and the real planet Neptune was thereby discovered.[42]

Although the inevitable differences in size and eccentricity—between the orbits of Neptune hypothesized by Adams and Leverrier in 1845–6, and the eventually established orbit that we now accept—presented no substantial difficulties at the time, the discrepancies in (gravitational) mass were a different matter: Leverrier's value, for example, was more than twice as large as our modern value. In the resulting perturbative residuals in the orbit of Uranus there was thus a trade-off between the distances between the two planets during their perturbative interactions and the mass of Neptune, in such a way that the latter remained quite indeterminate. In these circumstances, therefore, it would be particularly advantageous if there were an alternative method for determining the mass of Neptune independently of its interactions with Uranus. Fortunately, however, precisely such an alternative in fact emerged quite soon. Seventeen days after Galle made the first reliable telescopic observations of Neptune in 1846, William Lassell observed its largest and nearest satellite, which was later named Triton. Multiple observations of Triton's orbit then yielded the first values of Neptune's gravitational mass of the same order of magnitude as modern values: approximately 1/19,400th of the mass of the Sun.[43]

To complete the story, however, we needed more accurate investigations of the orbits of both Neptune and Uranus. This process extended throughout the 19th and early 20th centuries, with the most important such investigations being carried out in the second half of the nineteenth century by Simon Newcomb. On the basis of more accurate estimations of the distance between the two planets and the shape of Neptune's orbit, Newcomb was able to estimate the gravitational mass of Neptune—and therefore the source of its gravitational acceleration field—as a little larger than 1/19,400th of the (gravitational) mass of the Sun in 1898.[44] This result,

---

[42] The sizes and especially eccentricities for Neptune's orbit hypothesized by Adams and Leverrier differed considerably from our modern values. While they both took the orbit to be very eccentric, with the mean distance from the Sun to lie approximately between 36 (Leverrier) and 37.25 (Adams) astronomical units, we now consider Neptune's orbit to be almost a perfect circle, with a (nearly) constant distance from the Sun of approximately 30 astronomical units. Nevertheless, this did not stand in the way of an appropriate description of the orbit in the interval between 1806 and 1846 (on either side of the conjunction of the two planets). For the relevant distances from the Sun *during this interval* varied only between 32.6 and 33 astronomical units (Leverrier) or 32.8 and approximately 34 astronomical units (Adams). More importantly, however, the heliocentric longitude for the 23rd of September 1846 observed by Galle fit quite comfortably between the values predicted by Leverrier and Adams. Indeed, the three values of heliocentric longitude in question remained relatively close to one another throughout the interval between 1806 and 1847. See Morando (1995, 215–22, including figure 28.5) for an especially clear discussion and depiction of this situation.

[43] See the very helpful historical survey by David Pierce (1991). The first decent values from the orbit of Triton were calculated by Benjamin Peirce on the basis of observations (close to the now standard value) at Harvard and by Otto Struve on the basis of observations (close to the value of 1/14,500) at Pulkovo (near Saint Petersburg). Compare the analogous discussion in Herschel (1902)—see note 41 above—p. 673.

[44] See again Pierce (1991). The denominator Newcomb arrived at—19,314—became standard for the next ninety years—until the Neptune flyby of Voyager 2 in 1989 determined the now-standard value of 19,412.249 ± 0.057.

within appropriate margins of error, was then sufficient to establish that Neptune's gravitational acceleration field is indeed the source of the residual perturbations of Uranus.[45] Observations of Neptune, at this point, provided further evidence for our revised description of the orbit of Uranus—and, even more remarkably, so did observations of Neptune's satellite Triton. Continuing observations of Triton and continuing observations of the residual perturbations of Uranus provided independent cross checks on one another.

Even if Neptune had had no satellites, however, it would have been possible to determine its gravitational mass by careful observations of Neptune's orbit in relation to the residual (perturbative) accelerations of the orbit of Uranus. Nevertheless, it would still have been necessary to use the acceleration-field property of specifically gravitational force to determine the gravitational mass of Neptune directly from these accelerations of Uranus—which, like all such gravitational accelerations, are independent of the body being acted upon. The fact that this kind of determination of the mass of Neptune converges on the very same value determined by the acceleration of Triton directed towards Neptune considerably strengthens the case, for it shows that the gravitational acceleration field of Neptune does in fact extend far beyond Triton (together with the additional satellites of Neptune discovered later) all the way to Uranus. The gravitational action of Neptune on any body in the Solar System, it appears, is governed by the same inverse-square acceleration field at virtually any distance from this planet.[46] By Newtonian parity of reasoning, moreover, the same should be true for the gravitational action of any body in the Solar System on any other. Gravitational interaction, in this sense, is universal.

12.4   Implications for the Newtonian Method

The discovery of Neptune represented a spectacular success of the Newtonian gravitational law—even though, as we have seen, it in fact took considerably longer to establish than is suggested by the conventional story focused exclusively on Adams and Leverrier. Eventually, however, there was an almost equally spectacular

---

[45] Standish (1993) argues that the above value determined by Voyager 2 should still be taken as standard despite further attempts to invoke a new hypothetical 'planet X' to account for remaining residuals in Uranus's orbit; these residuals, according to Standish, simply lack a sufficiently well-behaved (regularized) structure.

[46] The otherwise useful book by Morton Grosser (1962) ignores the role of Triton in the determination of Neptune's mass, even where Lassell's discovery is reported (1962, 148–9). Grosser thereby misses the significance of the inverse-square acceleration field surrounding Neptune, which can be derived from Kepler's harmonic law (see note 7 above). Pierce (1991, 413) is clear on this in relation to Triton (and other satellites)—although he could have been more explicit that direct determinations from the perturbations of Uranus involve the same inverse-square acceleration field.

failure of the Newtonian gravitational law (spectacular in so far as it finally led to the radically revised theory of gravitation of Einsteinian general relativity) involving repeated fruitless attempts to find an appropriate Newtonian gravitational source for the very small residual perturbations in the motion of the perihelion of Mercury. But the point I want to emphasize here is that this failure of the Newtonian gravitational law involved another spectacular success of the Newtonian empirical method, which once again proved its superiority over the methods of curve fitting associated with the tradition of "saving the phenomena." Of most importance here is the fact that the anomalous motion of Mercury's perihelion was, in fact, a *residual*: a phenomenon left over from the previously known perturbations of this perihelion that had been successfully derived from the Newtonian gravitational law. It would not only have been completely impossible to "observe" this anomalous motion independently of the Newtonian law, but we would also have been left with no useful guidance for how to proceed—namely, to envision and then find empirical evidence for a radically different kind of gravitational source (relativistic space-time curvature). We can therefore make a stronger point. Not only did the Newtonian method prove its worth once again, but the Newtonian gravitational law continued to prove its worth at the same time. For it was only on the basis of this same law that we were then led to a better theory capable of accounting for the very small residual phenomenon remaining in the Newtonian theory.[47]

This point reveals a deeper meaning in Howard Stein's attempt, considered at the very beginning of this paper, to reconcile realism and instrumentalism. For it is not only the case that we find theoretical structures in the phenomena only by very careful appeals to empirical evidence. These theoretical structures are themselves scientific instruments in so far as they aim to address the continual accumulation of empirical evidence over time—in such a way that they can, and typically do, eventually lead to better theories. The goal is not to find the completed final theory, for this could only be a theory correct to any arbitrary degree of approximation. On the contrary, the point of the Newtonian method is to exploit a series of successive approximations taken always to be capable of further refinements. Theoretical structures, in this sense, are not merely instruments for empirical prediction and control. They are also essential instruments for discovering new phenomena in the Kantian sense, which include, most especially, dynamical phenomena involving theoretical (rather than merely observational) objects of experience. Among such phenomena, in particular, we find the Newtonian resolution of the ancient dispute

---

[47] The anomaly in the advance of Mercury's perihelion is given by a very small number (a bit more than 43 arc seconds per century), which, more importantly, is not even well defined independently of the Newtonian gravitational law. It is interesting to note, moreover, that Simon Newcomb had extensively investigated this anomaly in 1895, and he thus played a central role in both the success of Newton's theory in accounting for the perturbations of Uranus (see the sentence to which note 44 above is appended) and its failure exhaustively to account for the perturbations of Mercury. As we have just seen, however, both the Newtonian method and Newtonian gravitational theory proved their worth during the Einsteinian revolution that resulted in the eventual replacement of Newtonian theory.

between geocentric and heliocentric astronomy in terms of the center of gravity of the Solar System, the discovery of the planet Neptune within this same Newtonian theory, and (most dramatically) the further discovery of the precise limits of Newtonian theory within Einsteinian gravitational theory.

When we speak of Newtonian theory-mediated measurements, therefore, we do not understand the concept of "theory" in the logical sense that has dominated recent philosophy of science—as either a fixed and completed set of sentences or a fixed and completed class of models. As applied to what we now call the Newtonian theory of universal gravitation, for example, it would include both the three Laws of Motion and the law of universal gravitation: the assertion that there is an attractive force, inversely proportional to the square of the distance and directly proportional to the product of their two masses, between any two pieces of matter in the universe.[48] If we take Newton's gravitational theory to be fixed and complete in this sense, it seems that Newtonian theory-mediated measurements must already *presuppose* the completed theory of universal gravitation. And it then becomes extremely difficult to see how Newton could have applied such measurements to establish his theory of gravitation in the first place.[49]

The first point I want to make, in this connection, is that if anything is presupposed in Newton's argument for universal gravitation it is the Axioms or Laws of Motion that precede Book 1 of the *Principia*, together with the Rules for the Study of Natural Philosophy that precede Book 3. Moreover, Newton gives rather extensive arguments for the Laws of Motion (and their corollaries) in the following Scholium, which contains both appeals to the previous (and now well-established) work of Galileo, Wren, Wallis, and Huygens, together with Newton's own very careful experiments with pendula.[50] And, as far as the Rules for the Study of Natural Philosophy are concerned, Newton provides comments upon, and considerations in their favor for, all of them—at greatest length, as I have already explained, for Rule 3. More generally, as I have said, we can take both the Laws of Motion and the Rules as definitive of Newton's particular version of the inductive method.

---

[48] As explained in notes 5 and 6 above, our modern formulation of this theory also includes the well-known universal constant of proportionality $G = 6.67408 \times 10^{-11}$ m^3 kg^{-1} s^{-2}. Note 6 already suggests, however, that this is quite misleading in terms of Newton's own argument for his theory. I shall return to this point below.

[49] Clark Glymour (1980) develops a "bootstrapping" conception of theory confirmation applying a logical approach of precisely this kind: "bootstrap confirmation" involves formulas relating values of observed parameters to inferred values of theoretical parameters that are deducible from (or implied by) the theory in question. Bas van Fraassen (2009) appeals to Glymour's conception in arguing that it is misleading to take Jean Perrin's celebrated experimental investigations of Brownian motion to be evidence for the existence or reality of molecules—since these investigations, according to van Fraassen, are carried out wholly within Maxwell-Boltzmann molecular theory and therefore *presuppose* the existence of molecules.

[50] These pendulum experiments, as Newton explicitly says, pertain to the application of the Third Law of Motion only "insofar as it relates to impacts and reflections" (1999, 427). Newton also attempts briefly to extend the Third Law to attractions in the two following paragraphs, using an argument that was controversial then and remains controversial now. I return to this point below.

This method, as suggested, may indeed be reasonably said to be presupposed by Newton's argument for universal gravitation, but the theory of universal gravitation itself is not so presupposed.

Consider, for example, the acceleration-field property of gravitational force, which, as we have seen, plays a central role in the application of Newtonian method in both the *Principia* and later gravitational investigations extending over the next two centuries and beyond. This property results from a complex (but still direct) inductive extrapolation in accordance with Newton's Rules 3 and 4. The induction begins in the work of Galileo as a constant acceleration for all falling bodies near the surface of the earth and continues in the work of Kepler as his third or harmonic law for first the planets and then (in the hands of contemporary astronomers and Newton himself) the satellites of Jupiter and Saturn. Newton also extends it, via the moon test, to an inverse-square proportionality governing both the (now almost) constant accelerations for bodies falling near the surface of the earth towards its center and the acceleration of the moon towards this same center. The extrapolation is continued, finally, all the way to inverse-square acceleration fields surrounding any gravitating body—fields extending, in accordance with Rule 4, potentially to infinity.

This inductive extrapolation includes a further important property of gravitational acceleration fields. The fields of different bodies in the Solar System vary markedly in their strengths, in so far as they involve very different accelerations generated at any given distance: the Sun's field is much stronger than the earth's and, more generally, than that of any other body in the System; Jupiter's acceleration field is significantly stronger than those of all the other planets; and so on. This means that the fields of gravitational force surrounding each body vary, from body to body, in direct proportion to the accelerations thereby generated at any given distance: if we fix the gravitating body in question, the resulting accelerations vary only with the distance from that body in accordance with the inverse-square law. Every gravitating body, therefore, is associated with a distinct measurable physical quantity of (absolute) gravitational field strength, which we now call active gravitational mass. And it remains an open question, as far as direct inductive extrapolation is concerned, whether active gravitational mass is or is not identical with (or proportional to) inertial mass. Nevertheless, the direct inductive extrapolation already suffices, by itself, to complete the Newtonian resolution of the dispute between geocentrism and heliocentrism: the center of active gravitational mass in the Solar System, i.e., the center of accelerated motion in the System, lies always very close to the center of the Sun.[51]

---

[51] The acceleration-field property, involving the concept of active gravitational mass, suffices equally well for all the astronomical applications we have considered so far—including the discovery of Neptune, the discovery of the anomaly in Mercury's perihelion resisting all attempts to derive it within Newtonian theory, and even the eventual (and unexpected) discovery of the general relativistic explanation of this anomaly (see again note 47 above, together with the paragraph to which is appended). Hence, the inference Newton made (and Kant applauded) to the final identification of active gravitational and inertial mass, although certainly plausible at the time, is

Therefore, while the full law of universal gravitation in its modern formulation certainly implies the acceleration-field property, it is by no means necessary to derive the latter from the former. Indeed, the full law of universal gravitation, including, in particular, the identification of active gravitational and inertial mass, was considerably more hypothetical in Newton's time (and extending many years beyond) than the acceleration-field property itself. For there was little or no direct empirical evidence, during much of the relevant period, supporting the application of the Third Law of Motion to specifically gravitational interactions.[52] Thus the full law of universal gravitation remained largely hypothetical in Newton's sense, while the acceleration-field property, by contrast, remained a paradigmatic product of the proper application of Newton's method of induction. This does not mean that the full law of universal gravitation was implausible, however, nor that it was not far superior to the competing proposals (involving various vortex models and the like)

---

not actually needed for the astronomical applications we have considered—including the treatment of Jupiter and Saturn introduced, and extensively discussed, throughout the first section of my discussion. In particular, if we return again to notes 5 and 6 above, we now see that the constants k characterizing the absolute strengths of the relevant gravitational acceleration fields already suffice for the Newtonian treatment of Jupiter, Saturn, and their perturbations: there is no need to factor the constants k into products of the gravitational constant G and the relevant *inertial* masses m of the gravitational sources, and we can then identify the *active* gravitational mass in general with the relevant constant k. It also follows, more generally, that we do not need to apply the Third Law of Motion (governing *inertial* mass) in order successfully to apply the acceleration-field property. I am indebted to George Smith for this fundamental insight.

[52] Newton presents substantial evidence for the Third Law of Motion in the Scholium to the Laws of Motion. In particular, he provides a detailed discussion, covering three full pages, of experiments he has undertaken with collisions of spherical bodies in pendular motion. The aim was to test the conservation of quantity of motion (what we now call momentum) in all such cases, taking due consideration of both the resistance of the air and the degree of elasticity of the bodies. These experiments are impressive, but, as Newton himself emphasizes, they pertain to the application of the Third Law of Motion only "insofar as it relates to impacts and reflections" (see note 52 above). Newton then devotes the next two paragraphs to "attractions." In the first paragraph (1999, 427–8) he discusses magnetic attractions and refers to an experiment with a lodestone and piece of iron floating next to one another in still water on different vessels, which are found to attain a state of rest in equilibrium (in accordance with the conservation of momentum). In the second paragraph (1999, 428) Newton then turns to the case of the earth and its various parts, now assumed to attract one another mutually and in proportion to their "weights" towards one another (thus also remaining in equilibrium in accordance with the conservation of momentum). This second paragraph, however, presents a thought experiment rather than a real one, and the first has the disadvantage that magnetic attractions (as Newton well knew) obey quite different laws than gravitational attractions: they do not, tellingly, exhibit the acceleration-field property. Newton also, however, envisioned experiments on the conservation of momentum applied to gravitational interactions between terrestrial bodies—experiments that were eventually successfully carried out much later. But these later experiments (including the Cavendish experiment) were far too sensitive and complicated (due to the weakness of gravitation compared with other forces) to be practicable in Newton's time. In the case of astronomical evidence for conservation of momentum applied to gravitational interactions, finally, the acceleration-field property itself presents a formidable obstacle, although conservation of angular momentum between the earth and the moon plays a necessary role in explaining the tides. I am here once again indebted to Smith.

of the mechanical natural philosophy. Indeed, it eventually turned out to be well supported by direct empirical evidence after all, but only at about the same time that it finally decisively failed—due to quite different and completely unexpected problems involving relativistic phenomena.

These relativistic phenomena, I hasten to add, also affected the status of the Newtonian formulation of the acceleration-field property via an inverse-square law. In particular, the anomaly in the perihelion of Mercury called into question both the inverse-square law and, a fortiori, the law of universal gravitation at the same time. Moreover, both laws are equally theoretical, in the modern sense, since both govern Newtonian impressed forces, which are by no means directly observable. Yet the quantities of these forces were nonetheless empirically determinable by well-behaved theory-mediated measurements. Both the inverse-square law and the full law of universal gravitation governed phenomena in the Kantian sense involving dynamical, rather than merely kinematical or optical, properties of bodies: objects of experience rather than mere perceptions. The difference between the two is only that the full law of universal gravitation remained a work in progress throughout almost all of the development of Newtonian theory. The Newtonian formulation of the acceleration-field property via an inverse-square law, by contrast, was an early, and entirely non-hypothetical, product of the Newtonian inductive method itself—quite independently of the difficulties afflicting the application of the Third Law of Motion to gravitational interactions.

In any case, however, Rule 4 of the Newtonian inductive method undermines any conception of empirical laws or theories as capable of eventually attaining a fixed and completed form in which no failures of exactness can remain. This rule, as we have seen, expresses a delicate balance between inductions and hypotheses, and I repeat it here for convenience: "*In experimental philosophy propositions gathered from phenomena by induction should be taken as either exactly or very nearly true notwithstanding any contrary hypotheses, until yet other phenomena make such propositions either more exact or liable to exceptions*" (Newton 1999, 796). He there comments: "This rule should be followed so that arguments based on induction may not be nullified by hypotheses." Rule 4 is not part of a theory of confirmation, in the modern sense, governing degrees of rational belief or conditions under which a proposition is likely to be either true or very nearly true. Newton is rather formulating a procedural rule for "taking" a proposition to be true or very nearly true notwithstanding any contrary hypothesis. Such "taking" is encouraged, moreover, only until certain later new phenomena are discovered, and it thereby acquires a crucially important temporal dimension. As the new phenomena are sequentially discovered over time, the proposition in question is then "taken" to be either more exact or liable to exception (in which case the claim to exactness is revised).

There is no suggestion here of completeness or finality even if, at a given stage, the proposition in question is "taken" to be more exact than before. For the whole point of Rule 4, as Newton reemphasizes in his comment, is to prevent proper inductions from being "nullified" by hypotheses. Only observed phenomena (in the Newtonian sense) belonging to a properly inductive argument can "nullify" such a proposition, but never a mere hypothesis not yet based directly on phenomena that we have already observed. In the end, therefore, Newton's version of the inductive method is committed to the continuing possibility of sequential corrections by new observed phenomena, with no finality, no end in sight. For, if there were such an end, there would remain no *continuing* sequential corrections on the basis of which alone empirical science can proceed: empirical science, as Newton understands it, would simply cease to exist. Newtonian method is necessarily and intrinsically open-ended.[53]

12.5   Kant and the Newtonians

Kant's conception of natural science was by no means as open-ended as Newton's. On the contrary, as we have seen, he took both the Euclidean geometry of space and the Newtonian Axioms or Laws of Motion (as reformulated in Kant's Three Laws of Mechanics) to be synthetic a priori principles. And, although Kant certainly had room for the fine-grained consideration of planetary perturbations characteristic of celestial mechanics after Newton, he did not have room for the relativistic revolution that displaced both the Newtonian Laws of Motion and the Euclidean geometry of space.[54] Nevertheless, there remain two central respects in which Kant and Newton are in substantial agreement.

---

[53] Newton's approach is so open-ended that even the principles definitive of his methodology are themselves subject to reconsideration. These principles include his conceptions of mass, force, and true acceleration, as characterized by the Laws of Motion, together with his Rules for the Study of Natural Philosophy. In the Preface to the first edition of the *Principia* Newton outlines his mathematical treatment of forces, as adumbrated in Books 1 and 2, followed by his application of this treatment to nature in Book 3: "[I]n book 3, by means of propositions demonstrated mathematically in books 1 and 2, we derive from celestial phenomena the gravitational forces by which bodies tend toward the sun and toward the individual planets[; t]hen the motions of the planets, the comets, the moon, and the sea are deduced from these forces by propositions that are also mathematical" (1999, 382). The Preface then concludes as follows: "I hope that the principles set down here will shed light on either this method of philosophizing or some truer one" (1999, 383). Stein in effect makes this same point when he emphasizes that conclusions "reached on the basis of a comprehensive consideration of what has been discovered from phenomena are in the nature of the case open to *re*-consideration when *more* things have been learned" (2002, 291).

[54] Towards the end of the section "On the regulative use of the ideas of pure reason" in the Appendix to the Transcendental Dialectic in the first *Critique*, Kant is considering the three regulative principles of manifoldness, affinity, and unity (A662/B690): "The affinity of the manifold, without detriment to its variety, under a principle of unity, concerns not merely the things, but, much more, the properties and powers [*Kräfte*] of things." Kant proceeds to illustrate the use of these principles by a progression in the orbits of the planets from circular to elliptic to parabolic (three cases allowed by the inverse-square law). He next turns to the particular moving force—gravitation—responsible for these orbits and then alludes to the planetary perturbations (A663/B691): "Thus, under the guidance of these principles, we arrive at unity of the genera of these paths according to their form; and we thereby further arrive, however, at unity of the cause of all the laws of their motion (i.e., gravitation). From there we afterwards extend our conquests

The first point of agreement is that both Kant and Newton attribute radical open-endedness to specifically *empirical* natural science. Kant does not extend such open-endedness to what he takes to be the synthetic a priori parts of Newtonian theory—the Laws of Motion and the Euclidean geometry of space, which together constitute what Kant calls *pure* natural science. Nevertheless, he does extend Newtonian open-endedness to the more inclusive *proper* natural science (containing both synthetic a priori and empirical propositions) discussed in the *Metaphysical Foundation*. In particular, Kant explicitly extends such open-endedness to what he takes to be the *empirical* law of universal gravitation.[55] Indeed, Kant calls attention to the empirical limitations of the law of universal gravitational at the end of the passage from the Appendix to the Dialectic in connection with the planetary perturbations cited in note 54 above. Immediately after the sentence where he alludes to the planetary perturbations, Kant completes his train of thought as follows (A663/B691): "Finally, we even add still more to this than experience can ever confirm—namely, in accordance with the rule of affinity, we even imagine [*sich denken*] hyperbolic cometary orbits, in which these bodies leave our solar system entirely, and, proceeding from sun to sun, unify in their orbits the more distant parts of a for us unlimited cosmic system, which is interconnected through one and the same moving force [*bewegende Kraft*]."

This final move, Kant emphasizes, involves more than what experience can ever confirm: the hyperbolic orbits in question leave our Solar System and proceed from sun to sun (and therefore from system to system) without limit.[56] The necessary background here is Kant's speculative cosmology in his pre-critical *Universal Natural History and Theory of the Heavens* (1755), according to which the cosmos is structured by a never-ending sequence of nested galactic systems, beginning with our Solar System and then proceeding to the Milky Way galaxy, a rotating system of such galaxies, and so on. This conception is transposed, from an epistemological point of view, into Kant's critical conception of Absolute Space as a regulative idea of reason (the ideal limit of such a never-ending sequence) in the fourth or Phenomenology chapter of the *Metaphysical Foundations*.[57] And, in accordance

---

further, seeking also to explain all variations and apparent deviations from these rules from the same principle."

[55] See the following striking assertion towards the end of the General Remark to Dynamics in the second chapter of the *Metaphysical Foundations*: "[N]o law of either attractive or repulsive force may be risked on a priori conjectures. Rather, everything, even universal attraction as the cause of weight [*Schwere*], must be inferred, together with its laws, from data of experience[.]" (4: 534)

[56] The hyperbolic case (also allowed by the inverse-square) was treated as essentially different from the three previous such cases (circular, elliptic, and parabolic) by both Kant, in his pre-critical speculative cosmology (next sentence), and by his friend and correspondent Johann Heinrich Lambert in his *Cosmological Letters* (1761).

[57] The Phenomenology chapter, in particular, makes clear that (and why) the sentence just quoted above (A663/B691) does not contradict itself—by, for example, also asserting as an established fact (not a conjecture or speculation) that the moving force in question (i.e., gravitation) unifies the entire (seemingly infinite) cosmos. For Kant, however, this idea is rather a regulative ideal for inquiring into the cosmos sequentially in a systematically structured way

with this now purely regulative conception, although it is always possible to confirm the law of universal gravitation (to some or another degree of exactness) within any given (finite) galactic system, there is no way to confirm in advance the structure of even the next (still unexamined) system—not to mention that of the entire (potentially infinite) sequence of such systems.

There is a deeper and more general point of agreement between Kant and Newton in connection with theory-mediated measurements and the relationship, more generally, between observation (perception) and theory (conception). In the *Critique of Pure Reason* Kant considers this relationship abstractly in terms of his central distinction between the faculties of sensibility and understanding. The former is a passive faculty of intuition for receiving sensory impressions, the latter an active faculty of thought for conceptualizing such impressions. Kant sums up the matter in a famous dictum (A51/B75): "Thoughts without content are empty, intuitions without concepts are blind." Although this may appear to be nothing but plain common-sense, it implicates an elaborate discussion of what Kant calls *schemata*, which systematically relate concepts—pure mathematical concepts (like 'triangle'), pure concepts of the understanding (the categories), and empirical concepts (like 'weight')—to sensible (spatio-temporal) intuition. Thus, for example, the three categories of relation, substance, causality, and community, are schematized in terms of temporal relations: substance in terms of permanence in time, causality in terms of succession in time, community in terms of simultaneity in time. What matters here, however, is that, in the *Metaphysical Foundations*, Kant realizes or instantiates the three relational categories via the empirical concept of matter to obtain three empirical (but still theoretical) concepts of mass, force, and interaction. Moreover, this same instantiation of the relational categories by the three corresponding empirical (but still theoretical) concepts of mass, force, and interaction then yields Kant's three Laws of Mechanics—which, in turn, realize or instantiate the three Analogies of Experience.

The instantiation of the Analogies of Experience by Kant's three Laws of Mechanics is *quantitative*. For these Laws involve, respectively, the precise conservation of the total quantity of matter (mass), the maintenance of precisely the same direction and speed for a body affected by no external forces (inertia), and a precise quantitative equality between action and reaction. Measurement of mass or quantity of matter, which is paradigmatic of theory-mediated measurement in general, is thus of fundamental importance, and it is exemplified, for both Newton and Kant, by systems of attractive relationships involving the Sun, the planets (including the earth), and their moons. Moreover, this particular system of theory-mediated measurements, for both Newton and Kant, enables us to settle the dispute between geocentrism and heliocentrism by a straightforward empirical determination of the center of mass (center of gravity) of the Solar System—thereby implementing what Kant would call *empirical* realism.[58]

---

[58] Note 35 suggests a connection between the Refutation of Idealism in the first *Critique* (as an argument for *empirical* realism) and the Phenomenology chapter of the *Metaphysical Foundations*.

Thus, in the realm of (empirical) natural science the schemata of the empirical (but still theoretical) concepts of mass, force, and interaction are just the procedures of theory-mediated measurement employed by Newton in the *Principia*. We know, however, that Newton himself also had an important place for much more general changes in fundamental principles, including the Laws of Motion.[59] For, as we have seen in the example of the history of planetary perturbations, culminating in the discovery of the "true" general relativistic cause of the perturbations of Mercury, it is still possible to apply Rule 4 in a rigorous and meaningful way. Newton's original method, in this sense, can be subject to approximative generalization in virtually all of its fundamental principles. In so far as Kant's interpretation of Newton's original method can be subject to analogous generalization, it can be re-interpreted within the contemporary interpretation of Newtonian methodology embraced by both Howard Stein and George Smith. And this, I believe, does credit to all four of these great methodologists.

References

Cohen, I.B. 1980. *The Newtonian revolution.* Cambridge University Press.

Cohen, I.B., and G.E. Smith, eds. 2002. *The Cambridge companion to Newton.* Cambridge University Press.

Friedman, M. 2012a. Newton and Kant on absolute space: From theology to transcendental philosophy. In *Interpreting Newton: Critical essays*, ed. A. Janiak & E. Schliesser, 342–359. Cambridge University Press.

———. 2012b. Carnap's philosophical neutrality between realism and instrumentalism. In *Analysis and interpretation in the exact sciences*, ed. M. Frappier, D. H. Brown, and R. DiSalle. Springer.

———. 2012c. Newton and Kant: Quantity of matter in the metaphysical foundations of natural science. *Southern Journal of Philosophy* 50: 482–503.

———. 2012d. The *prolegomena* and natural science, Kants *prolegomena*. Ein kooperativer Kommentar, ed. H. Lyre and O. Schliemann. Klostermann.

———. 2013. Kant's construction of nature: A Reading of the *metaphysical foundations of natural science.* Cambridge University Press.

Friedman, M. 2017. *Kant's conception of causal necessity and its legacy. Kant and the Laws of nature,* ed. M. Massimi and A. Breitenbach. Cambridge University Press.

Gallavotti, G. 2001. Quasi periodic motions from Hipparchus to Kolmogorov. *Rend Mat Acc Linei* 12: 125–152.

Glymour, C. 1980. *Theory and evidence.* Princeton University Press.

Grosser, M. 1962. *The discovery of neptune.* Harvard University Press.

Hanson, N.R. 1960. The mathematical power of Epicyclic astronomy. *Isis* 51: 150–158.

Herschel, J.F.W. 1902. Outlines of astronomy: Part two. *BiblioLife* (historical reproduction).

Janiak, A., and E. Schliesser, eds. 2012. *Interpreting Newton: Critical essays.* Cambridge University Press.

Kant, I. 1900–. Kant's gesammelte Schriften, Georg Reimer, later Walter de Gruyter.

———. 1998. *Critique of pure reason.* Trans. P. Guyer and A. Wood. Cambridge University Press.

---

[59] See note 53 above, together with the paragraph to which it is appended and the preceding paragraph.

————. 2004. *Metaphysical foundations of natural science*. Trans. M. Friedman. Cambridge University Press.

Morando, B. 1995. The *golden age of celestial mechanics. Planetary astronomy from the renaissance to the rise of astrophysics part B: The eighteenth and nineteenth centuries*, ed. R. Taton and C. Wilson. Cambridge University Press.

Newton I. 1999. The *Principia*: Mathematical principles of natural philosophy. Trans. I.B. Cohen and A.W. Whitman, and ed. assisted by J. Budenz, preceded by I. B. Cohen, a guide to Newton's principia. University of California Press.

Pierce, D.A. 1991. The mass of Neptune. *Icarus* 94: 413–419.

Smith, G. E. 2002. The methodology of the *principia. The Cambridge companion to Newton.*, ed. I. B. Cohen and G. E. Smith, 138–173. Cambridge University Press.

————. 2012. How Newton's *principia* changed physics. Interpreting Newton: Critical essays, ed. A. Janiak and E. Schliesser, 360–395. Cambridge University Press.

Smith, G.E. 2014. Closing the loop: Newtonian gravity, then and now. In *Newton and empiricism*, ed. Z. Biener and E. Schliesser. Oxford University Press.

Standish, E.M. 1993. Planet X: No dynamical evidence in the optical observations. *The Astronomical Journal* 105: 2000–2006.

Stein, H. 1967. Newtonian space-time. *Texas Quarterly* 10: 174–200.

————. 1989. Yes but ... – Some skeptical remarks on realism and anti-realism. *Dialectica* 43: 47–66.

————. 2016. *Newton's metaphysics. The Cambridge companion to Newton*, ed. I. B. Cohen and G. E. Smith, 321–381. Cambridge University Press.

van Fraassen, B. 2009. The perils of Perrin, in the hands of philosophers. *Philosophical Studies* 143: 5–24.

Worrall, J. 1989. Structural realism: The best of both worlds? *Dialectica* 43: 99–124.

Chapter 13
Immediacy of Attraction and Equality of Interaction in Kant's "Dynamics"

Katherine Dunlop

13.1 Introduction

Kant's *Metaphysical Foundations of Natural Science* (*MFNS*), published in 1786, has proved difficult to situate in the context of eighteenth-century responses to Newton. One point beyond dispute is that Kant is not satisfied with the "metaphysical foundations" thus far proffered by Newton and his followers. He echoes some familiar Leibnizian criticisms (such as those concerning absolute space) and, in a passage we will examine closely, insists that rejecting "the concept of an original attraction" would put Newton "at variance with himself" (4:515). In light of Kant's robust defense of the immediacy and universality of gravitational attraction, the overall similarity between his three "mechanical laws" and Newton's "Axioms, or Laws of Motion" has made it the "standard" view that Kant "intends to justify Newtonian science in general and Newtonian mechanics in particular as stated in" Newton's laws of motion (Watkins 2019, 89). But it is unclear whether the alternative "foundations" he proposes are for Newton's theory of motion or something more akin to Leibniz's physics. Recent scholarship has documented the Leibnizian provenance of Kant's three laws,[1] and argued for the relevance of later figures strongly influenced by Leibniz.[2]

This paper does not attempt to settle whether the science Kant seeks to ground is identical to Newton's theory of motion. But it does argue that Kant can be seen

---

[1] See Stan 2013 and Watkins 2019.

[2] See, for instance, Dunlop (forthcoming), Dunlop 2022, and Stan 2014.

K. Dunlop (✉)
University of Texas at Austin, Austin, TX, USA
e-mail: kdunlop@utexas.edu

© The Author(s), under exclusive license to Springer Nature Switzerland AG 2023   281
M. Stan, C. Smeenk (eds.), *Theory, Evidence, Data: Themes from George E. Smith*, Boston Studies in the Philosophy and History of Science 343, https://doi.org/10.1007/978-3-031-41041-3_13

to make contact with Newton's thought, even when his arguments are couched in more Leibnizian terms.[3] As I explain in §13.2, I follow Michael Friedman in taking Kant to express the same worry as Roger Cotes: that Newton's application of the Third Law presupposes that bodies attract immediately and at a distance. While Friedman's reading demonstrates the importance of Kant's comment on Newton for the overall project of *MFNS*, it fails to respect Kant's distinction between the "dynamical" and "mechanical" phases of his project, as I argue in §13.3. In this paper, I seek to understand Kant's claim that Newton is committed to "original," universal attraction within the context of Kant's dynamics, where he himself situates it; and also as a demand for explicit acknowledgement rather than justification, which marks a difference with Cotes.

Kant's claim is challenging to understand in the context of his dynamics for several reasons. First, his "dynamical"/"mechanical" distinction is rooted in the *vis viva* controversy (as I explain in §13.4), which originated in Leibniz's criticism of Descartes, continued to engage mainly Continental thinkers, and is widely held to have ended in the 1750s.[4] So Kant's demand that Newton acknowledge a "dynamical" principle of (immediate and universal) interaction can seem not only gratuitous, but antiquarian. Second, Kant makes explicit that matter is first brought under quantitative concepts only in "mechanics," which makes it difficult to even formulate a principle of equality of interaction at Kant's "dynamical" level, let alone relate it to Newton's thought. My main aim in §13.6 is to explain how Kant understands the dynamical principle that he claims to be presupposed both by his own mechanical law of equality of interaction, and by Newton.

My aim in §13.9 is to show how Newton can be seen to presuppose equality of interaction at a level of consideration that counts as "dynamical" for Kant. Specifically, I claim Newton assumes equality of interaction with respect to forces that are attributed to bodies without regard to the bodies' states of motion, and are not compared in terms of quantity of matter. We find forces treated in such a manner in Section 11 of Book I. (In pointing us toward this early application of the Third Law, Kant's response further differs from Cotes'). In the Conclusion, I suggest how acknowledging a "dynamical" principle of interaction in this context might have furthered Newton's own aims, different as they are from Kant's. Newton insists that in Section 11 he treats forces "mathematically" rather than "physically." Casting his reasoning as "dynamics" in Kant's sense—which involves causal considerations, but prescinds from what Newton would call "physical" factors—would go some way toward showing how "mathematical" reasoning can have causal significance, as is required for it to establish the conclusions of Book III.

---

[3] This point is similar to that made in Warren 2001.

[4] In a widely cited paper that is mainly concerned to argue that the controversy was not settled by d'Alembert in 1743, L. Laudan nonetheless concedes that "by about 1755 or 1760 most prominent natural philosophers felt that the controversy ... was effectively over" (1968, 135). But see also Smith 2006.

13.2   Kant and Cotes on Newton's Application of the Third Law

In *MFNS* the statements of matter's properties and of laws governing motion and force are "Theorems" [*Lehrsätze*] deduced from "Explications" [*Erklärungen*]. The latter function as definitions, in that they either express features that can be seen immediately (without any need for inference) to belong to the concept of matter, or to a concept of its activity or powers; or else serve merely to distinguish different concepts. Kant takes himself to prove, from such a basis, that the essence of matter includes attractive force which acts immediately and "through empty space" (Theorem 7, 4:512). In a "Remark" following this proposition,[5] Kant argues that Newton himself has to accept immediate action at a distance, as an essential feature of matter, in order to "ground the proposition that the universal attraction of bodies, which they exert at equal distances around them, is proportional to their quantity of matter" (4:514).

I regard it as an important insight of Michael Friedman's (which appears as early as Friedman 1992, 156–7) that Kant is making the same "deep" point as Newton's contemporary Roger Cotes. The point is that Newton's argument that gravitational attraction is proportional, at a given distance, to the attracting body's mass "involves a key application of the Third Law of Motion to the interactions among primary bodies in the solar system," an application which "must take place between any pair of bodies directly, as if no intervening matter were (significantly) involved in the interaction" (2013, 220). To understand how Kant's response is distinctive, we should first set out the overall agreement between Kant and Cotes.

Kant argues that Newton applies "the law of equality of interaction" in a manner that does not entitle him to his conclusion: "For although between two bodies, when one attracts the other, ... the mutual approach (in accordance with the law of equality of interaction) must always occur in inverse ratio to quantity of matter, this ... is [only] a law of the *motions* that follow from attracting forces, not of the proportion of the *attractive forces* themselves" (4:514–5). Kant goes on to claim that for Newton's argument to be valid, the law must be applied to bodies "merely as matter," or in virtue of an "essential" property.

Cotes claims that, "in the train of reasoning by which I would make out" Proposition 7 of Book III, he is "persuaded" that the phrase "*Et cum attraction omnis mutua sit*" is "true when the Attraction may be properly so call'd", but "otherwise ... may be false."

> You will understand my meaning by an Example. Suppose two Globes A & B placed at a distance from each other upon a Table, & that whilst A remains at rest B is moved toward it by an invisible Hand. A by-stander who observes this motion but not the cause of it, will say that B does certainly tend to the centre of A, & thereupon he may call the force of the invisible Hand the Centripetal force of B & the attraction of A. (Newton 1959-1977, vol. V, 392)

---

[5] For simplicity, I call the passage in question "the Remark" following Theorem 7, although it is the second of two Remarks.

Cotes distinguishes such a force from "a proper and real Attraction of A." He argues that such a bystander "cannot conclude," from the law of the equality of action and reaction, "contrary to his Sense & Observation that the Globe A does also move toward the Globe B & will meet it at the common centre of Gravity of both Bodies." While this conclusion would follow from the "Supposition that the Attractive force resides in the Central Body," that is A, Cotes charges that Newton only "tacitly makes" this assumption, without securing it.[6]

In contrast to Cotes, Kant does not give an example to show how centripetal attraction could be mediated. But he does claim Newton is committed to "true attraction," "which takes place without the mediation of repulsive forces," rather than "apparent attraction," in which the "striving to approach" is brought about by "forces of pressure and impact" directed oppositely to matter's repulsive force (4:515).

A more important difference is that only Cotes suggests that Newton lacks justification for the manner in which he applies the Third Law: he writes to Newton that "'till this objection be cleared, I would not undertake to answer one who should assert that You do *Hypothesim fingere*" (*op. cit.*). Kant does not construe Newton's assumption of the immediacy of attraction as a hypothesis in need of support, but rather as integral to (though implicit within) Newton's conception of matter. So while Kant similarly claims Newton risks being "at variance with himself," he nowhere suggests that Newton infringes on his own ban on hypotheses[7]; his point is rather that to "take offense" at "the concept of an original attraction" would leave Newton denying something to which he is, at least tacitly, committed (4:515).

13.3 Quantity of Matter and Mechanics in *MFNS*

Since Friedman has significantly advanced our understanding of the Remark following Proposition 7 by drawing attention to how Newton applies the Third Law, I should now explain why I think there is room to advance further.

Friedman holds that Kant's reference to "the Newtonian argument for the proportionality of universal gravitation to matter" at 4:514 is closely related to the so-called "balancing argument," given earlier in the same chapter. The balancing argument

---

[6] On Zvi Biener and Chris Smeenk's illuminating account, Cotes should be taken, or at least was taken by Newton, to "suggest that *two* invisible hands are acting jointly to move the globes in a way identical with the predictions of Newton's theory" (2012, 120). Cotes thus shows, on this reading, that the reaction force can be taken to originate in something that pushes B and to act on something that—in the best case—pushes A.

[7] On the contrary, Kant suggests that it would infringe Newton's ban on hypotheses if one "substitute[d] an apparent attraction for the true attraction he did assert, and assume[d] the *necessity* of an impulsion through *impact* to explain the necessity of approach". According to Kant, Newton himself "rightly abstracted from all hypotheses purporting to answer the question as to the cause of the universal attraction of matter" (4:515).

consists of two theorems. The first asserts that matter must have attractive force as well as the repulsive force which Kant has already claimed is necessary for matter's possibility, and the second asserts that "no matter is possible through mere attractive force without repulsion" (4:510). (So while Kant has independent grounds for asserting the necessity of repulsive force, he thinks its necessity can also be shown by considering attractive force "on its own", to see what this force "could achieve in isolation for the presentation of matter" (Remark to Theorem 6, Dynamics, 4:511). Friedman takes the argument to "hinge" on "the interplay between three different concepts of quantity of matter at work in [*MFNS*]" (2013, 207). We will see that in the following chapter of the book, "Mechanics," Kant distinguishes a new concept of quantity *of motion* from one that is already in play; but in that chapter he defines "quantity of matter" univocally, without distinguishing it from a previously introduced concept.[8] So it is Friedman's innovation to distinguish this "mechanical" concept (defined as "the aggregate of the movable in a given space") from two "dynamical" ones, each linked to one of the fundamental forces (attraction and repulsion). On Friedman's interpretation, the balancing argument aims specifically to show that the mechanical concept is connected to each of the dynamical concepts (2013, 191), and in the Remark to Theorem 7, Kant is likewise concerned to link a mechanical concept of quantity of matter with a dynamical one.

Finding such a concern in this passage explains a striking difference between Kant's allusion to Newton here and a related allusion in the pre-Critical *Physical Monadology*.[9] And by highlighting "the particularly intimate connection between

---

[8] Friedman contends that in "Dynamics", Kant "introduces the concept of quantity of matter ... in terms of a 'measure of intensive filling of space'" [*i.e.* density] (2013, 189–90). In *Kant's Natural Philosophy*, Marius Stan likewise distinguishes between "dynamical" and "mechanical" varieties of quantity of matter, characterizing dynamical quantity of matter as a parameter that determines the strength of a body's force to repel "any object trying to occupy the volume" that it takes up (*ms.*, 43–4; see also Smith 2013). For Stan at least, it is clear that the role of this quantity is to explain impenetrability rather than any kind of agency that "bodies, qua determinate volumes of matter, exert" (*ms.*, 45); so it could as well be attributed to a field. I have argued (in Dunlop forthcoming) that the only concept in *MFNS* apt to express bodies' quantity of matter is the concept of "*die Quantität der Materie*" defined in "Mechanics", and prior to its introduction, references to the "*Grösse*" of matter should be taken to designate matter's size (*i.e.* volume), and translated "magnitude". In the passage in which Friedman takes Kant to introduce "the concept of quantity of matter [as a] measure of intensive filling of space" (4:521), Kant speaks of "matter filling its space to a determinate degree [*Grad*]", not of the quantity [*Quantität*] or magnitude [*Grösse*] of matter itself. I would argue similarly that here "*Grad*" denotes the "degree" of force rather than "quantity" of matter.

[9] In what Friedman describes as "the corresponding section of" *Physical Monadology*, Kant alludes to Corollary 3 to Proposition 6, Book III, construing it as a proof that "free motion is not possible in a medium which is completely filled" by elements all equal in density to one another (1:486). Friedman stresses that in this work, the concept of mass or quantity of matter is introduced "as a measure of a body's force of inertia." Because Kant here conceives of the force of inertia as a third fundamental force, distinct from and independent of the other two, the concept of mass is not connected to the forces of attraction or repulsion. So what we have, from the point of view of *MFNS*, is "only a mechanical concept of the quantity of matter that is not yet linked with any dynamical concept. In particular, the concept of mass or force of inertia has no intrinsic connection

gravitational attraction (as immediate action at a distance) and the concept of mass or quantity of matter" (Friedman 2013, 215), Friedman is able to explain the importance of the Remark for *MFNS* as a whole. The Remark ends with the contention that Newton must assume that the planets attract "other matter ... according to a universal property of matter" in order to conclude that "the attractive force of two planets, Jupiter and Saturn for example, manifested at equal distances of their satellites (whose mass is unknown)" are proportional to the planets' quantity of matter (4:515). Friedman is surely correct to understand this as an allusion to Corollaries 1 and 2 of Proposition 8, Book III, in which Newton infers the ratios of the masses of the primary bodies (Earth, Saturn, Jupiter and the sun) from the values of their satellites' centripetal accelerations. For Friedman, this procedure for empirically determining the planets' quantity of matter is central to *MFNS* as a whole. Kant claims his project is to "make possible the application of mathematics to the doctrine of body" by supplying principles for concepts that "belong to the possibility of" the "concept of matter in general" (4:472). Friedman takes Kant to explain how the concepts in question (such as movability and inertia) can have "quantitative structure" (2013, 32).[10] Crucial to this account is the "transition from the terrestrial to the celestial realm, whereby we articulate a generally applicable concept of quantity of matter" (2013, 78). Specifically, we "extend" the quantitative structure exhibited by terrestrial weighing to all bodies in the universe, however distant from the earth, and thereby "embed the parochial comparison of quantities of matter in terms of weights relative to the earth's surface within a more general system of comparisons based on (more general) equilibria between changes of momenta involving gravitational accelerations" (2013, 568).

I do not have space to further discuss the details of Friedman's interpretation and its advantages. Despite its lucidity and plausibility, it does not completely accord with the organization of *MFNS* as a whole, nor with details of the text. I will highlight these features in order to show, beginning in §13.4, how they give us another way to understand Kant's criticism of Newton's argument for the proportionality of attractive force to quantity of matter.

To begin with a detail, Kant rephrases Corollary 2 to Proposition 6, Book III (which is the only passage of *Principia* cited in the Remark). Kant renders this Corollary as: "if the aether or any other body were without weight, it could, since it differs from every other matter only by its form, be transformed successively by gradual change of this form into a matter of the same kind as those which on earth have the most weight; and so the latter, conversely, by gradual change of their

---

with the fundamental force of attraction, and, accordingly, Kant's pre-Critical version of the dynamical theory of matter contains ... no proof that the fundamental force of attraction must act immediately at a distance" (Friedman 2013, 216–7). Conversely (I take it), Kant's inclusion of such a proof in *MFNS* shows that there is a concept of quantity of matter closely associated with the dynamical notion of the fundamental attractive force.

[10] The possibility of quantitative structure is explained by exhibiting an addition operation by which combined amounts of, *e.g.*, motion or static pressure can be found equal to a single instance. See Friedman 2013, 54.

form, could lose all their weight, which is contrary to experience, *etc.*" (4:515). In *Principia*, the argument that for the weights of bodies "to be altered with the forms" would be "entirely contrary to experience" is given in the preceding Corollary 1. The relevant portion of Corollary 2 reads: "if the aether or any other matter were entirely *devoid of gravity or gravitated less in proportion to its quantity of matter*, then, since it does not differ from other bodies except in the form of its matter, it could by a change of its form be transmuted by degrees into a body of the same condition *as those which gravitate the most in proportion to their quantity of matter*; and on the other hand, the heaviest bodies, through taking on by degrees the form of the other body, could by degrees lose their gravity" (Newton 1999, 809; emphasis added). It seems significant that Kant rewords the italicized portions of Newton's text, for by replacing Newton's references to gravity with references to weight, Kant suppresses mention of the proportionality of attractive force to matter.[11]

Another detail incongruous with Friedman's interpretation is that Kant raises the question of how Newton could ground the proportionality of attractive force to quantity of matter about one-third of the way through the Remark, but does not allude to Newton's procedure for finding the planets' quantity of matter until the very last sentence. So it is odd to construe the latter as the passage's main concern. The placement of the allusion suggests that Newton's procedure is important to Kant, not for its own sake, but as a context in which the assumption of universal, immediate gravity is put to use.

The overall organization of *MFNS* gives Kant a reason to defer consideration of the proportionality of attractive force to quantity of matter until the following chapter, "Mechanics." As Friedman himself remarks, it is only in "Mechanics" that "the concept of mass or quantity of matter is first treated as a magnitude" (2013, 183). This is not coincidental; Kant in fact argues that quantity of matter can be "estimated" only "mechanically" (4:541).

Kant introduces the subject of mechanics, and contrasts it with dynamics, in terms of how these disciplines deal with "moving forces": the "original" forces of attraction and repulsion (treated in dynamics) serve to "impart" motion, while "[i]n mechanics, by contrast, the force of a matter set in motion is considered as *communicating* this motion to another." Hence, while in dynamics matter is considered merely as filling space, in mechanics it must be considered as "itself moved" (4:536), and indeed as "something that has moving force *through its motion*" (4:538).

Kant concludes that the estimation of quantity of matter must be mechanical because to find "the quantity of matter, which is moved with a given speed, and has moving force," one must find the body, or quantity (*sc.* volume) of matter that "acts by [its] own inherent motion" to bring about the effect in question"

---

[11] Three pages later, Kant defines weight as "the tendency to move in the direction of greater gravitation" and defines gravitation as an action of immediate and universal attraction, without specifying any measure of its quantity. This explication of weight corresponds to what Ernan McMullin distinguishes as "passive dispositional" ("PD-") gravity (1978, 59–60), which contrasts with the "active" variety in being less suggestive of agency, thus less obviously linked to mass.

(4:540). In the course of this complex argument, Kant indicates why the features of matter considered in the first chapter, "Phoronomy," are not a sufficient basis for determining the quantity of matter. In a preliminary "Explication," he defines "quantity of matter" as "the aggregate of the movable in a determinate space"; he proceeds to argue that, first, the only "generally valid measure" of matter's quantity is the quantity of its motion (4:538),[12] and, second, that the only measure of (quantity of) motion that can serve to determine quantity of matter is mechanical. Kant's argument for this second claim turns on a distinction between "mechanical" estimation of quantity of motion, in terms of "the quantity of the moved matter and its speed together," and "phoronomical" estimation, by the quantity of speed alone (4:537). He argues that the phoronomical measure does not suffice to represent "composition of many motions equivalent to one another," which is required for quantitative representation of motion.[13] So quantity of matter can be found only through the mechanical measure of (quantity of) motion (and because the concept of quantity of matter depends on that of quantity of motion only with respect to "its application to experience," not in the order of conceptual explication, circularity is avoided.[14])

---

[12] Kant contends that the quantity of matter "cannot be immediately determined *by an aggregate* of its parts", first, because matter is infinitely divisible (4:537); and second, because "even if this occurs in comparing the given matter with another of the same kind, in which case the quantity of matter is proportional to the size of the volume, it is still contrary to the requirement" that matter's quantity is "to be estimated in comparison with" every other matter whatsoever. Hence we must estimate the quantity of matter in terms of "its own inherent motion" (4:538).

[13] Kant argues, in particular, that "in phoronomy it is not appropriate to represent a motion as composed of many motions *external to one another*, since the movable [is] here represented as devoid of moving force" and thus can differ in quantity from the "several of its kind", with which it enters into "composition", only with respect to speed. (It is in this context that Kant criticizes the distinction between dead and living force in terms of velocity (4:539), as I discuss in the main text below.) Kant's claim in the following "Remark" that "only the aggregate of the moved can yield, at the same speed, a difference in quantity of motion" (4:540) may clarify his point here: the elements entering into composition are all assumed to move at the same speed, as parts of one "aggregate", and thus in phoronomy they cannot be represented as differing in quantity from, *i.e.* as "external to" one another. (See Stan *ms.*, 50 for a lucid account of this assumption and its consequences.) It follows that the "seemingly phoronomical concept of quantity of motion, as composed of many motions of movable points, external to one another yet united in a whole" must give place to the "mechanical concept" on which the points "are thought as something that has moving force *through its motion*" (4:538).

[14] As we just saw, the first claim in Kant's argument is that quantity of matter must be determined in terms of its motion (rather than "immediately ... *by an aggregate* of its parts"), and the apparent circle arises because what the second claim shows to be the only applicable notion of quantity of motion already incorporates quantity of matter. Thus, as Kant says, "the quantity of matter must be estimated by quantity of motion at a given speed, but ... the quantity of motion must, at the same speed, be estimated by the quantity of matter moved". He concedes that "this alleged circle would be an actual one, if it were a reciprocal derivation of two identical concepts from one another", but contends that "it contains only the explication of a concept, on the one hand, and its application to experience, on the other" (4:540).

In the "Remark" following this argument, Kant argues that the features of matter treated in the "Dynamics" chapter are also insufficient to determine the quantity of matter. He explains at the beginning of the "Remark" that the necessity of representing the quantity of matter as "the aggregate of movables (external to one another)" implies that the quantity of matter is extensive. Hence, it has "no *degree* of moving force at a given speed that would be independent of this aggregate, and could be considered merely as intensive magnitude" (4:539).[15] While Kant is most concerned to draw the further implication that matter cannot consist of monads (4:540), it also follows that matter's quantity cannot be determined through its inherent attractive and repulsive forces, since their quantity is intensive (measured in degrees). Later in the "Remark," Kant makes explicit that the quantity of matter can be exhibited only "mechanically," "by the quantity of its own inherent motion, and not dynamically, by that of the original moving forces" (4:541).

Since Kant purposefully delays quantitative assessment of matter until the "Mechanics" chapter, it is hard to see how the empirical procedure for determining quantity of matter can be important for understanding the Remark to Theorem 7 of "Dynamics" (or the "Dynamics" chapter more generally). This gives us reason to seek beyond Friedman's interpretation.

13.4   Interpretive Challenges Raised by Kant's Dynamical/Mechanical Distinction

We have seen, first, that Kant differs from Cotes in demanding explicit acknowledgement of the assumption of the immediacy of attraction, rather than justification; and second, that Kant insists on separating "dynamical" from "mechanical" notions. In particular, Kant objects that the only principle Newton gives to justify his inference (to the proportionality of quantity of matter to attractive force) is "a law of the *motions* that follow from attracting forces," hence "a principle of mechanics," when what the inference requires is a principle "of dynamics," namely of "the proportion of the *attractive forces* themselves" (4:515). But it is difficult to understand how Kant can mount his criticism in the context of his dynamical/mechanical distinction. We are challenged to explain, first of all, how Kant can formulate the principle of equality of interaction on which he takes Newton to rely, using only the limited conceptual resources at his disposal in the "Dynamics" chapter. Relatedly, it is incumbent on us to explain what meaning phrases such as "the universal attraction of bodies . . . is proportional to their quantity of matter" (4:514) can have for Kant here, as presumably they do (even if, as I have suggested, he is not mainly concerned

---

[15] Kant characterizes "extensive" magnitude in terms of the priority of the parts over the whole (A162/B203), and "intensive" magnitude as that "which can only be apprehended as a unity" (A168/B210), thus not as comprised of antecedently given parts. For thorough discussion see Sutherland 2022, 87–90.

with assigning quantity to matter). Second, we must explain how Kant can take Newton to be committed to a principle formulated in terms so different from his own.

A main reason that Kant's dynamical/mechanical distinction seems so foreign to Newton's thought is that it is rooted in the Cartesian-Leibnizian controversy concerning the measure of force, that is, whether the quantity conserved in interaction is the product of "bulk" and speed (as Descartes and his followers maintained), or the product of quantity of matter and the square of the velocity (as Huygens, Leibniz, and others argued).

Kant's initial contrast between "mechanics" and "dynamics," in terms of whether the forces they treat belong to bodies "through their motion" or not, puts his "mechanical" moving forces into correspondence with the "dead" and "living" forces distinguished by Leibnizian thinkers. For when Leibniz first introduces the term "*vis viva*," in *Specimen dynamicum*, he characterizes it as "ordinary force, joined with actual motion" (1989, 121).[16] In a later passage, Kant makes this correspondence explicit, while simultaneously calling into question another formulation of the distinction between living and dead forces. As Konstantin Pollok has made especially clear, the formulation Kant criticizes is in terms of velocity, such that dead forces confer on bodies only an infinitely small speed, and living forces, the finite speed attained in "actual motion" (Pollok 2001, 399).[17] Kant contends that in these terms "there can be no difference between living and dead forces" in the context of mechanics, where moving forces are considered "as those which bodies have insofar as they themselves are moved," whether with finite or infinitely small speed (4:539). He suggests that, if the terminology "still deserves to be retained," it would be "much more appropriate to call dead forces those, such as the original moving forces of dynamics, whereby one matter acts on another, even when we abstract completely from its own inherent motion, and also even from its striving to move"; conversely, "all mechanical moving forces, that is, those moving by inherent motion," could be called "living," regardless of their degree of speed (which "may even be infinitely small"; 4:539). Kant's disparagement of the

---

[16] Closer to Kant's time, Émilie Du Châtelet distinguishes "dead force," as a "simple tendency to motion," from "living force," as "the tendency a body has when it is actually in motion" (1740, §519). Du Châtelet reckons dead force as the product of a body's mass [*masse*] and the speed [*vitesse*] with which it would move in the first moment (§§536–9, §561).

[17] For Kant and his contemporaries, this difference in velocity corresponds to a difference in how the forces act, whether by pressure [*Druck*] or impact [*Stoss*]. In a paper presented to the Berlin Academy in 1744, Euler writes that "Leibniz, and those who have followed … call pressures dead forces and percussions living forces" (1746, 27). Kant alludes to this version of the distinction in his 1749 "True Estimation": the "observation" that "an actual motion, a blow or push, for example, always carries with it more power than a dead pressure … was perhaps the seed of an idea that could not remain unfruitful in the hands of Herr von Leibniz" (1:15). The view that these differences make it necessary to introduce distinct measures for the two forces is undermined by showing that the measure used for pressure can also be used for impact, for which it is necessary to argue that the exchange of forces in impact is not instantaneous, as Kant does at 4:552. See Pollok 2001, 395–8, and Hepburn 2010 on Euler's use of this strategy.

Leibnizian distinction between dead and living forces thus calls into question his own separation between dynamics and mechanics (*cf.* Pollok 2001, 399–400); but given the use he makes of the latter distinction, he is clearly committed to drawing it in more acceptable terms.

With respect to Newton, in particular, it is far from obvious that he assumes or presupposes equality of interaction with respect to forces that are "dynamical" on Kant's understanding: forces attributed to bodies that can as well be regarded as rest, and without presupposing that matter "acts by its own inherent motion in a mass," through a simultaneous motion of its parts, which Kant illustrates by a phenomenon of impact (the "water hammer," 4:540). For both the documentary record of Newton's work leading up to *Principia*, and the justification Newton gives there (in the Scholium to the "Axioms" section) for the Third Law, indicate that the evidence for the Law comes from collisions.[18] So as we move forward, a main question will be whether we can find a body of results or level of discourse in *Principia* in which forces are attributed to bodies that can as well be regarded as at rest, and are not characterized in terms of quantity of matter. I argue for the affirmative in §13.9.

13.5 Kant's Official "Dynamical" Law of Equal Interaction

Kant holds that the "law of equality of interaction" is not by itself sufficient to ground Newton's claim that attractive force is proportional to quantity of matter, unless this law is understood to apply directly to the interacting bodies, without taking account of any intervening matter, and thus as presupposing the equality of interaction. So to understand Kant's view of Newton's reasoning, we must see how there can be a principle of equality governing the interaction of the forces themselves. Kant claims that in addition to the "mechanical" law of equality of interaction that pertains to the "the *motions* that follow from attracting forces" (4:514), there is also a "dynamical" principle of equality of interaction which applies to the forces themselves (4:515). This gives us an obvious candidate for a principle justifying Newton's claim. But, as I will explain, it does not suffice for that purpose. Understanding its insufficiency will motivate the identification of a genuinely "dynamical" principle of equality of interaction, in §13.6.

We must first consider why the "mechanical" law is not a sufficient basis for asserting the proportionality of attractive force to quantity of matter. In the "Dynamics" chapter, Kant claims that the "law of the equality of interaction" requires that the "mutual approach" between bodies that attract one another "must

---

[18] Recently Mark Wilson, following Michael Spivak, has called attention to the "audacious generalization" by which Newton moves from the "experimental fact" of the conservation of momentum in collisions to a general law for all forces (2020, *n.p.*). For additional discussion see Jammer 1957, 125–7.

always occur in inverse ratio to the quantity of matter." Such a distribution of speeds is indeed required by Kant's "Third Law of Mechanics," which states that "In all communication of motion, action and reaction are always equal to one another" (4:544). In "Dynamics", Kant gives an example to show that this law, understood as "a law of the *motions* that follow from attracting forces," in fact "holds for all moving forces in general": A magnet will impart more motion to an equal magnet when the attracting magnet is "enclosed in a wooden box of double the weight," "even though the wood, which increases the quantity of matter of this ... magnet, adds nothing at all to its magnetic force, and manifests no magnetic attraction of the box" (4:515). This interaction conforms with Kant's mechanical law, since the motion imparted is increased in proportion to the quantity of matter of the attracting mass (that is, the box and magnet together). But since in this case there is no corresponding increase in attractive (magnetic) force, proportionality of moving force to quantity of matter cannot be assumed for all such interactions.

In "Dynamics," Kant indicates that a "dynamical" principle (of equality of interaction) would be a law "of the proportion of the *attractive forces* themselves" (4:515); in "Mechanics," he formulates a principle that seems to fit this description. Corresponding to the "mechanical" law, which pertains to communication of motion, there is another, "namely, a *dynamical*," law of equality of interaction "among matters," which holds insofar as "one matter *imparts* [its] motion originally to" another (rather than communicating it) and, "at the same time, produces [the same] in itself" (4:548). Kant's designation of these laws as "mechanical" and "dynamical," respectively, corresponds to the powers of matter thematized in "Mechanics" and "Dynamics."

Kant claims the dynamical law of equality of action and reaction is "easily shown in a similar manner" to the mechanical law. This "manner" is, specifically, by showing that the bodies' motions can be described in such a way that each body has an equal share of the interaction. I will first outline Kant's argument, then consider how he characterizes each body's part in the interaction.

Kant's argument for the dynamical law is as follows (with numbers inserted by me):

[I]f matter A exerts traction on matter B, then it *compels* the latter to *approach* it, or [(1)] equivalently, it *resists* the force with which the latter might strive to *remove* itself. But since [(2)] it is all the same whether B removes itself from A, or A from B, this resistance is, at the same time, a resistance exerted by body B against body A, insofar as the latter may be striving to remove itself from the former; and so traction and counter-traction are equal to one another. [(3)] In just the same way, if A repels matter B, then A resists the *approach* of B. But, since [(4)] it is all the same whether B approaches A, or A approaches B, B also resists the approach of A to precisely the same extent; so pressure and counter-pressure are also always equal to one another. (4:548-9)

Steps (1) and (3) of the argument are warranted by equivalences in Kant's definitions of "attractive force" and "repulsive force" in "Dynamics."[19] Steps (2)

---

[19] Specifically, step (1) is warranted by the first clause of Explication 2 of "Dynamics", which defines "attractive force" as the moving force "by which a matter can be the cause of the approach

and (4) derive their warrant from the impossibility of representing any motion with respect to "absolute" space, and the corresponding relativity of all motion descriptions. Kant claims absolute space is "*in itself* nothing, and no object at all", but rather "signifies only any other [more inclusive] relative space", *i.e.* material reference frame, with respect to which any given "relative space" can be represented as movable. This more inclusive frame earns the designation of "absolute" because I abstract from "the matter that designates it" (of which I know nothing), and thus represent it "as a pure, nonempirical, and absolute space"; and because this "enlarged space always counts as immovable" for purposes of representing the enclosed relative space as movable (4:482).

To understand how these relativity considerations function in the proof of the dynamical law, it is useful to consider how they are deployed in the proof of the mechanical law. The proof is a so-called "construction" of "the action in the community of the two bodies," which apportions the speed of approach[20] between the two in inverse proportion to their masses. Kant begins by remarking that all changes of relation in space, "insofar as they may be *causes* of certain effects" (which I take to mean, insofar as the interaction arises through moving forces that belong to the bodies only in virtue of their motion), "must always be represented as mutual." He explains that the motion of a body cannot be thought "in relation to another *absolutely at rest* that is thereby also to be set in motion"; rather, the latter body "must be represented as only *relatively at rest* with respect to the space that we relate it to, but as moved, together with this space, in the opposite direction, with precisely the same quantity of motion in absolute space as the moved body there has toward it" (4:545). It is the relativity considerations that require us to think of the latter body, together with the "relative space" with respect to which it is (represented as) at rest, as moving in a more inclusive space. In the proof of the first case of the dynamical law, the relativity considerations make it just as legitimate to represent A (and the relative space with respect to which it is at rest) as moving away from B, as to represent B as removing itself from A (taken to be at rest); similarly for the second case.

The relativity considerations thus ensure, in the proof of the dynamical law, that an action giving rise to a certain motion can be redescribed in terms of an oppositely directed motion. But this is not yet to show that the force exercised in the action is met with an equal and opposite force, so Kant's so-called dynamical law seems to face the same difficulty as his mechanical law. As Friedman remarks, "Kant says here that the equal and opposite contrary motion A experiences from the mechanical resistance of B is an effect of the force 'originally' exerted by A itself," without saying "that B also 'originally imparts' a motion to A." So, just as

---

of others to it (or, what is the same, by which it resists the removals of others from it)". Step (3) is warranted by the second clause, which defines "repulsive force" as the moving force "by which a matter can be the cause of others removing themselves from it (or, what is the same, by which it resists the approach of others to it)" (4:498).

[20] Specifically, the velocity of body A as it approaches body B, calculated "with respect to the relative space [material reference frame] in which B is *at rest*" (4:546).

the action of the magnet enclosed in a wooden box counted as greater according to the mechanical law, without a corresponding increase in attractive force, the reaction of a non-magnetized piece of iron would count as equal according to the dynamical law, without any exercise of attractive force (2013, 356).

Kant's failure to say that "B also 'originally imparts' a motion to A" manifests a more general similarity between Kant's mechanical and dynamical laws. As Marius Stan has shown in a number of papers, in the Leibnizian tradition "reaction" was understood as resistance on the part of a *passive* body, and equated with the force expended by the *active* body, so that what Leibnizians called "the law of [equality of] action and reaction" was in fact "a principle of one-way causation, not mutual interaction" (forthcoming). Stan argues persuasively that Kant's mechanical law of equality of action and reaction belongs in this tradition, albeit as a "repair" or "correction" of the paradigm (2013, 504), and his dynamical law presumably also belongs.[21] Since on this view motion is produced in body A only through B's resistance, not through an attractive force "original" to B, by Kant's lights his dynamical law is not sufficient to ground Newton's claim that attractive force is proportional to quantity of matter.

13.6 Kant's Formulation of the Presupposed Principle of Interaction

I have just argued that Kant's dynamical law of equality of action and reaction is insufficient to ground Newton's claim because, as Kant says of the mechanical law, it holds "of the *motions* that follow from attracting forces, not of the proportion of the *attractive forces* themselves" (4:515; *cf.* Friedman 2013, 356). In a footnote to his "Proof" of the mechanical law, however, Kant supplies another principle that can serve to validate Newton's inference.

In the footnote, Kant invokes the distinction between the mechanical and phoronomical concepts of quantity of motion. His point here is that according to phoronomical estimation (that is, merely in terms of speed), it is arbitrary whether to ascribe a motion to a body or an equal and opposite one to the "relative space" in which the body moves. (We have seen that every relative space can be represented as moving with respect to a more inclusive one.) According to mechanical consideration, however, "it is no longer arbitrary, but rather *necessary*, to assume each of the two bodies as moved, and indeed, with equal quantity of motion in the opposite direction—but if one is at rest relatively with respect to the space, to ascribe the required motion to it, *together with the space*" (4:547n.).

Kant's immediate concern is, presumably, to show that in the "construction" that proves the (mechanical) law, it is not merely permissible, but rather obligatory to represent the reacting body as "moved, together with the [relative] space [we relate

---

[21] Stan himself explicitly excludes the dynamical law from consideration.

it to], in the opposite direction, with precisely the same quantity of motion" as the acting body has toward it (4:545). Since Kant's purpose in contrasting phoronomical and mechanical consideration is to establish the necessity of representing bodies this way *in mechanics*, our question is whether such representation becomes necessary only in mechanics, or whether it is already required in dynamics (which comes between phoronomy and mechanics in Kant's scheme).

Kant contends that while in phoronomy "the motion of a body was considered merely with respect to space, as change of relation *in space*," in mechanics "a body is considered in motion relative to another, ... to which, through its motion, it has a causal relation" (4:547*n*.) What makes it necessary to represent the reacting body as moving together with its relative space (in an opposite direction and with equal quantity of motion) is specifically that the mechanical concept of quantity of motion brings "the quantity of substance (as moving cause) ... into the calculation at the same time [*zugleich*]." Now while the two concepts of quantity of motion belong to phoronomy and mechanics, respectively, and consideration of motion as change in spatial relation belongs to phoronomy, what is specific to mechanics are causal relations that bodies have "through their motion"; causal considerations in general are first introduced in dynamics.[22] Since causal considerations are already incorporated, together with the notion of "quantity of substance," into the mode of consideration that Kant here designates as "mechanical," Kant may not have thought it necessary to formulate his argument at a separate, dynamical level. And when we examine Kant's argument in detail, we see that the reasoning can in fact be formulated at the level of dynamics—that is, using resources available in the "Dynamics" chapter—though it also makes acute the problem of how quantities of motion can be represented as equal.

To explain why it is "*necessary* to assume each of the two bodies as" moved with the same quantity of motion (in opposite directions), and hence to ascribe motion to the relative space with respect to which one body is at rest, Kant first indicates that "communication" of motion is possible only through the activity of the "original" moving forces treated in "Dynamics".[23] Kant then argues that when motion is imparted by these forces, an equal and opposite reaction always follows.

> Now since both forces *always act mutually and equally in opposite directions* [*beiderseitig in entgegengesetzten Richtungen und gleich*], no body can act by means of them on another body through its motion, without just as much reaction from the other with the same [*mit gleicher*] quantity of motion. Hence no body can impart motion to an *absolutely resting* body through its motion; rather, the latter must be moved precisely with the same [*mit derselben*] quantity of motion (together with the space) in the opposite direction as that which it is supposed to receive through the motion, and in the direction, of the first. (4:547*n*., emphasis added)

---

[22] Kant begins "Dynamics" by telling us that the "dynamical explication of the concept of matter" adds, to the phoronomical explication, "a property relating as cause to an effect" (4:496).

[23] "For one body cannot act on another through its own inherent motion, except either in approach by means of repulsive force, or in withdrawal by means of attraction" (4:457*n*.).

Kant can thus be seen to argue that equality of interaction necessarily holds at the "mechanical" level because it necessarily holds at the "dynamical" level.

As I understand Kant, he holds (in contrast to Cotes) that Newton does not need to justify his view that bodies mutually attract immediately and at a distance, but rather to own up to this commitment. Thus to fill the lacuna Kant finds, he needs to supply an articulation of, rather than an argument for, the view. This makes acute the question of how equality of interaction can be asserted using the resources available in the "Dynamics" chapter, where the mechanical notion of quantity of matter is not yet available (and the phoronomical notion of quantity of motion is clearly inadequate).

I think we can make headway by attending to the broader meaning of the German word "*gleich*" (and its nominalization, "*Gleichheit*") in comparison to the English "equal" (and "equality"). While "equal" means sameness of quantity, and is used only in an extended sense to express other relations (as in "equal treatment before the law"), the Kant scholar Daniel Sutherland has observed that "*gleich*" can mean at least four other distinct relations of sameness.[24] I suggest that when Kant is operating at the level of (*i.e.* limiting himself to the resources available in) the "Dynamics" chapter, he exploits the open-endedness of "*gleich*" to indicate that action and reaction are the same with respect to some *as yet unspecified* property or measure. To be clear, I do not take Kant to assert that action and reaction are the same in a non-quantitative sense, but rather to leave unspecified in what way they are the same (which turns out to be, of course, with respect to quantity of motion, as that notion is understood in "Mechanics").

13.7   Newton's Measures of Centripetal Force

I have identified, within Kant's argument that equality of interaction holds at the "mechanical" level because it holds at the "dynamical" level, a principle that can ground the reasoning by which Newton proves the proportionality of attraction to quantity of matter. This principle is not formulated in terms of familiar "mechanical" notions, such as mass and momentum, that feature in (for instance) Friedman's lucid account. Since Newton justifies (and seems to have first conceived) the Third Law in these "mechanical" terms, it is not clear why he should be committed to Kant's principle. My final goal in this paper is to sketch a line of reasoning that shows Newton to be committed to equality of interaction with respect to what Kant would call "dynamical" moving forces: those attributed to bodies without regard to their state of motion, and not compared with respect to quantity of matter.

---

[24] These relations of sameness are: with respect to a (non-quantitative) property or circumstance; with respect to "all readily noticeable intrinsic properties"; "with regard to all intrinsic properties"; or strict (numerical) identity (Sutherland 2008, 148).

The first point to observe is that in the "Definitions" that precede the "Axioms, or Laws of Motion," Newton distinguishes several "measures" of force: specifically, of "centripetal force", which Newton defines as that by which "bodies are drawn from all sides, are impelled, or in any way tend, toward some point as to a center" (1999, 405). He distinguishes three "kinds of quantity" of such a force, which he also designates as "measures" of it. The *absolute* quantity is "greater or less in proportion to the efficacy of the cause propagating it from a given center through the surrounding regions" (1999, 406). The *accelerative* quantity is "proportional to the velocity which it generates in a given time." The *motive* quantity is "proportional to the motion which it generates in a given time" (1999, 407).

Before laying down these measures, Newton defines "quantity of motion" as "the measure of motion that arises from the velocity and the quantity of matter conjointly," *i.e.* as their product (1999, 404). Since we are here to understand the motion as generated by a force in a given time, during which it acts to alter velocity, the motive quantity of a centripetal force is a function of its accelerative quantity as well as of the moved body's quantity of matter, *i.e.* its mass (which is assumed to be unaffected by the force). The measures are therefore interrelated, despite quantifying different aspects of (centripetal) force.

In the actual cases of attractive force that most interest Newton, the accelerative quantity will be the same for any body at a given distance from a given center. Newton says of the accelerative quantity of force that it is referred "to the place of a body as a certain efficacy diffused from the center through each of the surrounding places in order to move the bodies that are in those places," whereas the absolute quantity is referred to the center itself (1999, 407). This links the absolute quantity to the motive and accelerative, since the "efficacy" with which different central bodies attract, and thus the absolute quantity of their forces, can be determined in terms of their accelerative quantities (at equal distances).

To illustrate these notions, Newton gives two examples of centripetal force. One is "the force that produces [terrestrial] gravity." At places sufficiently "near the surface of the earth" the "accelerative gravity" is "the same in all bodies universally," so the "motive gravity, or weight, is as the body" (*i.e.* proportional to its mass), but "in an ascent" to more remote places, the "accelerative gravity becomes less" and the weight decreases proportionately (1999, 408). The "absolute" quantity of gravitational force comes into consideration only as Newton shows that gravity is universal.[25] So Newton instead uses the example of magnetic force, "which is greater in one lodestone and less in another, in proportion to the bulk or potency of the lodestone," to illustrate the absolute quantity of force.

---

[25] Thus in Corollaries 1 and 2 to Proposition 8, Book III, Newton compares the "weight force experienced by a body of given mass ... on the surface of the earth, the moon, the sun, or any of the planets" and then computes "the values of the gravitational forces at equal distances from the centers of the sun, Jupiter, Saturn, and the earth". The resulting distribution of values can be described in terms of the greater efficacy of the sun's production of gravitational force, so that the absolute quantity of its gravitational force is greater (Cohen 1999, 105).

By explicitly distinguishing the absolute and, in particular, the accelerative quantity of (centripetal) force from the motive quantity, Newton can be seen as supplying the resources to relate forces to motions without having to take account in every case of quantity of matter. From a bird's-eye vantage, it is easy to see how this serves Newton's purposes in *Principia*, for in Book III he infers the quantity of matter of the planets from the motions of their satellites (as is particularly emphasized in Friedman 2013). I now want to consider how distinguishing these measures is useful to Newton in Book I.

13.8 Introduction of the Third Law in Section 11

Newton's distinction among measures would be easy to overlook, for Newton uses the notion of motive force to formulate his Laws of Motion. According to the Second Law, a change of motion is "proportional to the motive force impressed" as well as "along the straight line in which that force is impressed" (Newton 1999, 416). In contrast, the Third Law speaks of the equality of "actions" of bodies, without indicating which measure applies; but in the following discussion, it is explicit that such actions produce "equal changes in the motions, not in the velocities" (417). Yet with a few notable exceptions (Corollaries 3, 5, and 6 of the Laws of Motion), in the first 10 sections of Book I the Second Law is used only to determine the direction of changes in motion. (Corollary 1 of the Laws and Proposition 2 are illustrative examples.) So the motive measure of force comes under consideration at the same stage that the Third Law is first applied[26]: Section 11 of Book I.

Newton begins Section 11 by remarking that he has, "up to this point, ... been setting forth the motions of bodies attracted toward an immovable center". But such motion "hardly exists in the natural world", since "attractions are always directed toward bodies, and—by the third law—the actions of attracted and attracting bodies are always mutual and equal". Hence, the motions now to be "set forth" are "of bodies that attract one another" (1999, 561). The main point of this introductory passage, and of the concluding "Scholium", is that the section consists of "mathematics" rather than "physics," meaning that it considers the "quantities and mathematical proportions" of forces rather than their "species and physical qualities" (588). Newton seems to have issued this disclaimer because the results of the section so clearly anticipate the comparison of these "mathematical proportions" with "phenomena," which constitutes "physics."[27] Not only does Newton highlight

---

[26] In his *Guide*, Cohen observes that in general, "whenever Newton introduces 'force' without any qualification, he intends accelerative measure", and this is most strictly true for these sections of Book I. A search of the *Principia*'s text reveals that after the "Definitions" and "Axioms, or Laws" sections, the first occurrence of the term "motive" is in Section 11 of Book I.

[27] Newton claims that mathematics investigates the quantities of forces and their proportions, under "any conditions that may be supposed"; then, "coming down to physics, these proportions must be compared with the phenomena, so that it may be found out which conditions of forces apply to each kind of attracting bodies" (1999, 588–9).

the applicability of the present treatment of bodies (in contrast to considering attraction toward "immovable centers"), but he says in the section's concluding Scholium that its results direct us to consider "the analogy between centripetal forces and the bodies toward which those forces tend," and on whose "natures" the forces depend (*ibid*.). Since Newton is to this extent taking account of the causes of kinematic—or in Kant's classification, "phoronomical"—phenomena, we may ask whether this discussion is situated at what Kant considers the "dynamical" level (where Kant himself first considers their causes) or whether the forces are in every instance related to the motion of the central bodies and characterized in terms of quantity of matter (as in Kant's "mechanics").

For purposes of understanding Kant's comment on Newton, this question is particularly relevant, because the section already contains an application of the Third Law that presupposes the immediacy of attraction. In the section's final theorem, Proposition 69, the Third Law is applied directly to two bodies, A and B. A and B are each assumed to attract every body belonging to the system with an "accelerative force that [is] inversely as the square of the distance from the attracting body," which thus equally affects all bodies at equal distances. Since the distance between A and B equals the distance between B and A, the "absolute" quantity of centripetal force towards A is in the same ratio to the "absolute" quantity of centripetal force towards B as the "accelerative attraction" of each body towards A is to the "accelerative attraction" of each body towards A is to the "accelerative attraction" of each body towards B. As William Harper notes, it is the application of the Third Law to A and B, which sets the "motive" force by which A attracts B equal to that by which B attracts A, that "makes" the ratio of their accelerative attractions equal to that of their masses; it is thus also the ratio of their absolute attractive force (Harper 2016, 248). This theorem underwrites the reasoning in Book III (Propositions 7 and 8) in which attraction is shown to be universal, proportional to quantity of matter, and found by means of accelerative measures.

13.9 Newton's Implicit Use of the Third Law and Motive Measure

I will now argue that although the Third Law is in effect, as Newton notes, from the beginning of Sect. 11, its application remains implicit, as does the use of the motive measure of force, in the first seven propositions. And when the Third Law and quantity of motive force come to be explicitly considered, in the eighth proposition (Proposition 64), the attractive force is treated without regard to the motion of attracting bodies, thus not as a "mechanical" force in Kant's sense.

The first seven propositions of Section 11 treat systems in which two bodies attract one another. In determining the trajectories of these bodies, Newton exploits a constraint imposed by the application of the Third Law, or more specifically by the fourth Corollary of the Laws of Motion. He contends in the beginning of the section that if there are two bodies, "neither the attracting nor the attracted body can be at rest, but both (by Cor. 4 of the Laws) revolve around a common center of gravity"; and several bodies that either all attract one another or "are all attracted by or attract a single body," "must move with respect to one another in such a way that the common center of gravity either is at rest or moves uniformly straight forward" (Newton 1999, 561, translation modified slightly). Thus the trajectories arising from the bodies' interaction can be described in terms of their motion relative to their common center of gravity. More importantly, the distance between two bodies must be apportioned the same way, relative to the center, at each moment in the evolution of the system (since the product of the first body's mass and its distance from the center equals the product of the second body's mass and its distance, and the two distances lie in a straight line). Hence each body's accelerative tendency toward the center can be defined, given its accelerative tendency relative to the other body.[28] The problem of determining each body's motion is thereby reduced to a pair of "one-body" problems, of the kind Newton has already solved in preceding sections of Book I.[29] Accordingly, Newton claims that for two bodies, "attracting each other with any kind of forces and not otherwise acted upon or impeded," their motions ("in any way whatever") "will be the same as if they were not attracting each other but were each being attracted with the same forces by a third body set in their common center of gravity" (1999, 565). By the time Newton asserts this, in the fifth proposition (Proposition 61) of Section 11, he has already made use of the proportional relationship and opposite direction of the bodies' distance from the center.

I say that the application of the Third Law remains implicit throughout this reasoning because Newton nowhere appeals to the general fact of the equality of the bodies' actions on each other, not even to derive the proportionality and opposite direction of the bodies' distance from the center. Rather, in the proof of the first proposition (Proposition 57) he asserts these properties as evident,[30] in virtue of the way the problem has been framed at the section's opening, namely as a mutual attraction between bodies. By the same token, the quantity of motive force is not explicitly considered in this reasoning. Once Newton has asserted that the two bodies' distances from their common center of gravity are in inverse proportion to their masses, he gives no more consideration to the masses, but uses only kinematic

---

[28] This explanation comes directly from notes of George E. Smith's lecture course on *Principia*.

[29] In his lectures Smith explains, further, that the modern approach to determining the two-body problem is to write down two simultaneous equations of motion, one for each body, and then use the constraint to reduce to a pair of one-body problems. S. Chandrasekhar gives a detailed account of Newton as proceeding in this way in Proposition 57 of Book I (1995, 208).

[30] The proof of Proposition 57 begins: "For the distances of these bodies from their common center of gravity are inversely proportional to the masses of the bodies and therefore in a given ratio to each other and ... in a given ratio to the total distance between the bodies. These distances, moreover, rotate around their common end-point with an equal angular motion because, since they always lie in the same straight line, they do not change their inclination toward each other ... " (Newton 1999, 561).

information about the bodies' motions with respect to this center and each other. There is, then, a stage of reasoning in which interactions are construed as equal, but the measure with respect to which they are equal—the motive quantity of force—is presupposed rather than explicitly applied. In that respect, it corresponds to Kant's proleptic assertion of the equality of the action of original moving forces.

Beginning in Proposition 64, Newton considers systems of more than two interacting bodies. By the time he wrote *Principia*, Newton seems to have been acutely aware that the constraint imposed by Corollary 4 of the Laws of Motion does not generally make it possible to determine the motion of bodies in such systems, because they may be drawn in a direction different from that of the center of gravity.[31] Accordingly, in *Principia* he begins by considering the special case of a system in which the centripetal force is in the direction of the center of gravity, so he can follow the same strategy as in the two-body cases.

Proposition 64 considers how to "find the motions of more than two bodies in relation to one another" when "the forces with which bodies attract one another increase in the simple ratio [*i.e.* as a constant multiple] of the distances from the centers" (Newton 1999, 566). Newton proceeds by induction, first concluding from earlier results that the first pair of bodies (*T* and *L*) describe ellipses around their common center of gravity (*D*), then relating each additional body's motions to those found so far. When Newton adds a third body (*S*) to the system, he uses the parallelogram rule to decompose the forces between it and each of the first two bodies into a component directed along line *SD* and one directed along the line between *T* and *L* (on which *D* lies). Our interest is in his treatment of the accelerative forces directed along lines parallel to *SD*; Newton claims that by attracting *T* and *L* "with *motive actions SD x T and SD x L (which are as the bodies)*," these forces "do not at all change the situations of those bodies in relation to one another, but make them equally" move along the parallels to *SD* (Newton 1999, 566–7, emphasis added). In his treatment of the movement of body *S*, Newton argues that it must move in the way required ("to revolve in orbit with just the right velocity" around

---

[31] In late 1684, Newton composed the now well-studied tract *De Motu*, the first draft of which asserts that "the major planets revolve in ellipses having a focus in the center of the sun, and radii drawn [from the planets] to the sun describe areas proportional to the times, entirely as Kepler supposed" (quoted in Cohen 1984, 234). I. Bernard Cohen has observed that in this draft, Newton fails to take account of the equality of the planets' reaction forces to the force with which the sun attracts them, in consequence of which the sun as well as the planets must revolve around their common center of gravity, so that this center and not the sun is the focus of the planets' orbits. Newton corrects this oversight in the so-called "Copernican" Scholium, added to a later draft of *De Motu*: "By reason of the deviation of the Sun from the center of gravity, the centripetal force does not always tend to that immobile center, and hence the planets neither move exactly in ellipses nor revolve twice in the same orbit." Since every planet has "as many orbits as it has revolutions", and "the orbit of any one planet depends on the combined motion of all the planets, not to mention the action of all these on each other," Newton draws a pessimistic conclusion: "to consider simultaneously all these causes of motion and to define these motions by exact laws admitting of easy calculation exceeds, if I am not mistaken, the force of any human mind" (Hall and Hall 1962, 280).

C, the center of gravity common to S and the T-L system) because "the sum of the *motive forces SD x T and SD x L*, which are proportional to the distance CS, tends toward the center C" (Newton 1999, 567, emphasis added).

Here, in contrast to the first seven propositions, the motive quantity of force is explicitly brought into consideration (in the italicized text). But in determining the motions of body S and point D, Newton treats the centripetal forces as directed toward a fixed center of gravity, as the constraint permitted him to do in the two-body case. Insofar as the deviation of the bodies (and the center of gravity of the T-L system) from this center is disregarded in the reasoning, the attractive force is considered without regard to the motion of the body (or center) to which it is attributed, thus as a "dynamical" force in Kant's sense.

13.10 Conclusion

On my reading Newton presupposes (instead of explicitly asserting) equality of interaction with respect to motive force, and attributes attractive forces to bodies without regard to their motion, in Propositions 57 through 64 of Book I. I do not claim he explicitly countenances a level of discourse corresponding to Kant's "dynamics". It is widely recognized that Newton proceeds by successively introducing "real-world" complications into abstract or idealized scenarios,[32] and it would be rash to assume every stage of this reasoning is significant from a foundational point of view. I would even concede that Newton's idealizing or abstracting away from the motion of sources of attractive force, and their quantity of matter, could be merely strategic. As regards Proposition 64, Book I, the difficulty involved in taking account of the "deviation" of sources of attractive force from the center of gravity is sufficient to explain why Newton delays consideration of it. Regarding the earlier Propositions of Section 11, the usefulness of the constraint imposed by the Third Law's application, specifically for reducing two-body problems to one-body problems, suffices to explain why Newton only presupposes equality with respect to motive force. And looking ahead, the use of Book I's results in Book III to derive ratios of masses from kinematic phenomena explains why in Book I, Newton bases his results on kinematic information as far as possible.

So Newton's selective consideration of what Kant calls "dynamical" features of matter may have no further importance, within his own framework, than enabling or facilitating certain results. I think Kant's articulation of how equality of interaction

---

[32] I. Bernard Cohen introduced the term "Newtonian style" for this manner of reasoning. Here I mean to emphasize what Cohen, in his *Guide to Newton's* Principia, states as one "advantage" of the style: that it allows Newton "to begin by exploring the simplest system, and then to add various degrees of complexity one by one, rather than have to face … multiple interrelated problems all at once" (1999, 149). I think Cohen is correct to describe the idealized or abstract stages of this reasoning as "mathematical," but his further characterization of them as "mental constructs" is highly contentious.

is presupposed, within what he calls "dynamics," is still valuable for understanding Newton.

To begin, attention to the question whether Newton assumes equality of interaction at the dynamical level brings to light his direct application of the Third Law to bodies A and B in the proof of Proposition 69 (Book I), and how it presupposes immediacy of attraction. This shows that immediacy of attraction is assumed at an even more basic stage of Newton's reasoning than that discussed by Cotes. So Cotes's insight is complemented by Kant's observation that Newton rests his assertion of the proportionality of "the universal attraction of bodies" to quantity of matter on a principle of equality of interaction that belongs, specifically, to dynamics.

We have seen that Kant's own "dynamical" principle of equality of interaction is in a sense proleptic, for the sameness (of quantity of motion) that it asserts is given precise quantitative significance only in mechanics. Raising the problem of how the actions of bodies can be treated as equal at the dynamical level leads us to notice what I have described as Newton's "implicit" application of the Third Law, earlier in Section 11, and how Newton's formulation of the principle lends itself to this application. As we saw, the Third Law contrasts with the Second in that Newton does not indicate which measure of force is in use. Now as George E. Smith has shown, Newton initially considered other candidates for the *Principia*'s third Law of Motion, including the statement of conservation of linear momentum that appears as Corollary 3 to the Laws.[33] Just as the Second Law is formulated in terms of motive force, Corollary 3 is formulated in terms of quantity of motion. We may speculate that Newton ultimately chose a formulation devoid of these quantitative notions in order to facilitate the way he uses the principle at the dynamical level.

On my reading, Kant's response to Newton differs from Cotes' both in locating the assumption of immediacy of interaction earlier, and in demanding explicit articulation rather than justification. It is worth reflecting on how it could have furthered Newton's purposes to acknowledge that he takes bodies to interact mutually and immediately, even when he is abstracting from bodies' quantities of matter and states of motion. I do not think Newton would accept Kant's charge that he must acknowledge this in order to show the proportionality of attractive force to quantity of matter. But I do think that an acknowledgement would further understanding of Newton's treatment of force in Section 11 of Book I. As Andrew Janiak has observed, Newton's characterization of his method as "mathematical" suggested to many, especially eighteenth-century, readers that "he aims to avoid action at a distance by denying that gravity really exists, construing it perhaps as a mere calculating device"[34]; meanwhile, those who took seriously

---

[33] "The quantity of motion, which is determined by adding the motions made in one direction and subtracting the motions made in the opposite direction, is not changed by the actions of bodies on one another" (Newton 1999, 420). See Smith 2008.

[34] Janiak 2009, 54. A prominent example of this kind of reading of Newton is the anonymous 1688 review of *Principia* in *Journal des sçavans*.

Newton's claim to have shown in *Principia* that "gravity really exists" and "suffices for" celestial and terrestrial motions[35] were left to wonder how "mathematical" reasoning could establish such a conclusion. In this regard, Kant's separation of "dynamics" from "phoronomy" and "mechanics" has the advantage of giving entrée to causal considerations, without yet taking account of what Newton characterizes as the "physical sites and causes"[36] of forces, thereby clarifying how reasoning that abstracts from the "physical" can pertain to causes.

Acknowledgements I wish to thank the editors of this volume for their very helpful comments; thanks especially to Marius Stan for sharing his unpublished work. George E. Smith's meticulous and incisive scholarship is a model for history of philosophy, and I am greatly indebted to George for what understanding I have of Newton's *Principia*. I hope this chapter shows how better understanding of Newton can also illuminate the thought of more canonical figures.

References

Biener, Z., and C. Smeenk. 2012. Cotes's queries: Newton's empiricism and conceptions of matter. In *Interpreting Newton*, ed. A. Janiak and E. Schliesser. Cambridge University Press.

Chandrasekhar, S. 1995. *Newton's Principia for the common reader*. Oxford University Press.

Cohen, I. 1984. Newton's steps to universal gravity. *Birth of a New Physics*. W. W. Norton.

Du Châtelet, É. 1740. Institutions de Physique. Prault fils.

Dunlop, K. 2022. The significance of Du Châtelet's proof of the parallelogram of forces. *Époque Émilienne*, ed. R. Hagengruber, 85–112. Springer.

———. forthcoming. Physics. *The Kantian mind*, ed. S. Baiasu and M. Timmons. Routledge.

Euler, L. 1746. De la force de percussion et de sa véritable mesure. *Mémoires de l'Academie des Sciences de Berlin* 1: 21–53.

Friedman, M. 1992. *Kant and the exact sciences*. Harvard University Press.

———. 2013. *Kant's construction of nature*. Cambridge University Press.

Hall, A.R., and M.B. Hall. 1962. *Unpublished scientific papers of Isaac Newton*. Cambridge University Press.

Harper, W. 2016. Newton's argument for universal gravitation. *Cambridge companion to Newton*, 2nd ed., ed. R. Iliffe and G. E. Smith. Cambridge University Press.

Janiak, A. 2009. *Newton as philosopher*. Cambridge University Press.

Leibniz, G. W. 1989. *Philosophical essays*. Ed. and Trans. R. Ariew and D. Garber. Indianapolis: Hackett.

---

[35] In the *Principia*'s concluding General Scholium (Newton 1999, 943). I have followed Janiak's translation (2009, 55).

[36] In the paragraph introducing the three measures of centripetal force, Newton says of the concept of "absolute force" (which is "referred to the center of the forces", in contrast to "bodies seeking a center" or those bodies' places) that it is "purely mathematical, for I am not now considering the physical causes and sites of forces" (Newton 1999, 407). I take it that Kant's attribution of "mechanical" moving force to matter that "acts by its own inherent motion as a mass" (4:540) counts as consideration of a "physical" cause or site of force. Newton himself progresses to consideration of "physical causes" of *Principia* insofar as he claims gravity causes the "phenomena of the heavens and our sea", though he famously prescinds from "assign[ing] a cause to gravity" (1999, 943).

McMullin, E. 1978. *Newton on matter and activity*. University of Notre Dame Press.
Newton, I. 1959–1977. *The correspondence of Isaac Newton*, ed. H.W. Turnbull et al. Cambridge University Press.
———. 1999. *The* Principia: *Mathematical principles of natural philosophy*. Trans. and Ed. I. B. Cohen and A. Whitman. University of California Press.
Pollok, K. 2001. Kant's "Metaphysische Anfangsgründe der Naturwissenschaft": Ein Kritischer Kommentar. Felix Meiner Verlag.
Smith, G.E. 2006. The vis viva dispute: A controversy at the dawn of dynamics. *Physics Today* 59: 31–36.
———. 2008. Isaac Newton. *The Stanford encyclopedia of philosophy* (Fall 2008 edition), ed. Edward N. Zalta. https://plato.stanford.edu/archives/fall2008/entries/newton/
Smith, S. 2013. Does Kant have a pre-Newtonian picture of force in the balance argument? *Studies in History and Philosophy of Science* 44: 470–480.
Stan, M. 2013. Kant's third law of mechanics: The long shadow of Leibniz. *Studies in History and Philosophy of Science* 44: 493–504.
———. 2014. Once more unto the breach: Kant and Newton. *Metascience* 23: 233–242.
———. Forthcoming. Newtonianism and the physics of du Châtelet's Institutions de Physique. *Collected wisdom of the early modern scholar: Essays in honor of Mordechai Feingold*, ed. Gideon Manning and Anna Marie Roos. Springer.
Stan, M. Manuscript. Kant's Natural Philosophy.
Sutherland, D. 2008. From Kant to Frege: Numbers, pure units, and the limits of conceptual representation. *Royal Institute of Philosophy* Supplement: 63.
———. 2022. *Kant's mathematical world*. Cambridge University Press.
Warren, D. 2001. Kant's dynamics. *Kant and the sciences*, ed. Eric Watkins. Oxford University Press.
Watkins, E. 2019. *Kant on laws*. Cambridge University Press.
Wilson, M. 2020. Newton in the pool hall: Subtleties of the third law. *The Oxford handbook of Newton*, ed. Eric Schliesser and Chris Smeenk, online edition. https://doi.org/10.1093/oxfordhb/9780199930418.013.6

Chapter 14
Remarks on J. L. Lagrange's *Méchanique analitique*

Sandro Caparrini

> *A discovery of Newton was of a two-fold character—he made it, and then others had to find out that he had made it.*
> A. de Morgan, *Essays on the Life and Work of Newton (1914, 38).*

There is no shortage of books and papers citing Lagrange's *Méchanique analitique* (1788)[1] Yet, though armed with a thorough knowledge of secondary literature, the modern reader is likely to struggle with every page. Even the most dedicated Lagrangian scholars are not hardy enough to follow the book all the way through. Like many classics of the exact sciences, the *Méchanique analitique* is a labyrinthine microcosm.

Indeed, this is a book that begs for a critical edition with an extensive introduction and copious footnotes. Already by the middle of the nineteenth century, when the *Méchanique analitique* was required reading for any aspiring mathematician, the need was felt for a corrected reprint, with multiple appendices and a few explanatory notes (Lagrange 1853–1855). Nowadays, after two centuries of tumultuous changes in science, the bare text is closed to the unskilled in higher mathematics and to anyone who is not well-versed in the byways of eighteenth-century mechanics. (And while we are speaking of desiderata, we also need a serious edition of all of Lagrange's surviving correspondence which is rigorously faithful to the original texts.)[2]

Leaving aside the problem of a critical edition, I offer here a number of disjointed remarks, or a collection of footnotes, to selected aspects of the *Méchanique analitique*. They aim to explore some less-traveled paths in the history of mathematics

---

[1] A core bibliography of scholarship is appended to Pulte (2005).

[2] Previously unknown letters by Lagrange keep being discovered (Caparrini 2007).

S. Caparrini (✉)
Politecnico di Torino, Torino, Italy

© The Author(s), under exclusive license to Springer Nature Switzerland AG 2023
M. Stan, C. Smeenk (eds.), *Theory, Evidence, Data: Themes from George E. Smith*, Boston Studies in the Philosophy and History of Science 343, https://doi.org/10.1007/978-3-031-41041-3_14

and mathematical physics at the end of the eighteenth century. These notes are, of necessity, quite technical; to understand the role of the *Méchanique analitique* in the development of the exact sciences, we need to go through the text line by line and equation by equation.[3]

A quarter century after the *Méchanique analitique*, Lagrange published a second edition, retitled *Mécanique analytique* (1811–1815). The two editions differ in many ways, and should be considered separately; in what follows they will be referred to by their respective titles. Quotations from the *Méchanique analitique* are from the 1788 edition; for the other works of Lagrange I have relied upon his collected *Oeuvres* (1867–1892). Most of the translations from the French are taken from the English version of the *Mécanique analytique* published in 1997 by A. Boissonnade and V. N. Vagliente; any remaining ones are my own. For the dates of publication of Lagrange's works I follow the chronology established by Taton 1974. For the sake of clarity, the formatting of notations and the layout of the equations has sometimes been slightly modernized.

14.1 Why Is the *Méchanique analitique* So Important?

There are a number of detailed descriptions of the *Méchanique analitique* (Barroso-Filho 1994; Pulte 2005). Our purpose here is not to repeat these overall studies but rather to survey quickly those points that will be needed later on.

At the time of its publication, the book was widely considered the zenith of eighteenth-century mechanics. This is hardly surprising, considering that the book was conceived as a general unification of all branches and results of mechanics, especially those obtained in the preceding half-century. Lagrange was certainly successful in regard to systems with a finite number of degrees of freedom, but perhaps not equally so with continuum mechanics. By and large, the *Méchanique analitique* set the stage for the treatises of the following century.

The great mathematical physicists of the generations following Newton increasingly developed mechanics using the language of algebra rather than geometry. By the 1760s in higher mechanics diagrams were mainly used as a basis for setting up general equations. The reason behind this major theoretical and methodological shift was an increasing desire for generality in results and conciseness in proofs. Lagrange took this trend to a new level, for there are no diagrams at all in the *Méchanique analitique*. After Lagrange, most textbooks and research papers in mechanics no longer relied much on synthetic geometry. The influence of Lagrange's purely algebraic approach to mathematics was felt even in analytic geometry (the very adjective 'analytic' was taken over from the *Méchanique analitique*).

---

[3] For a recent evaluation of the place of Lagrange in eighteenth-century mechanics, see Caparrini and Fraser (2013).

The book is neatly divided into Statics and Dynamics. This division reflects Lagrange's new approach to higher mechanics.

Statics is entirely founded on the *principle of virtual work*. The essence of the principle had been known since Hellenistic times: in force-multiplying devices (lever, pulley, wheel and axle, inclined plane, screw, and wedge, plus any combinations of them) what is gained in force is lost in velocity (Capecchi 2012). Lagrange elevated this rule of experience to the rank of abstract first principle ("it can be considered as a kind of axiom of mechanics", p. 12): Given a system of mass points connected in any way, there will be equilibrium if

$$Pp + Qq + Rr + \cdots = 0,$$

where $P, Q, R, \dots$ are the applied forces and p, q, r, etc. are any small displacements of the points consistent with the constraints and estimated in the directions of the forces (Lagrange 1788, 22).

Dynamics is reduced to Statics by adding the so-called 'inertial forces' $m\frac{d^2x}{dt^2}$, $m\frac{d^2y}{dt^2}$, $m\frac{d^2z}{dt^2}$ (*d'Alembert's principle*), thus obtaining

$$\sum m \left(-\frac{d^2x}{dt^2}\delta x - \frac{d^2y}{dt^2}\delta y - \frac{d^2z}{dt^2}\delta z + P\delta p + Q\delta q + R\delta r + \dots \right) = 0,$$

where Σ denotes a summation over all particles and all forces [1788, p. 195].[4] The physical meaning of this formula is best understood in component form

$$\sum_i m_i \left[\left(X_i - \frac{d^2x_i}{dt^2} \right)\delta x_i + \left(Y_i - \frac{d^2y_i}{dt^2} \right)\delta y_i + \left(Z_i - \frac{d^2z_i}{dt^2} \right)\delta z_i \right] = 0,$$

where X_i, Y_i, Z_i are the Cartesian components of the resultant force per unit mass on the ith particle and we have introduced the indicial notation for summing over the particles (1788, 200). From this modified principle of virtual work, Lagrange first obtains the general theorems of dynamics, then lays the foundations of the main branches of mechanics: rigid bodies, vibrations, mutually gravitating mass points, and fluid motion.

Today anyone who opens the pages of the *Méchanique analitique* will surely look for the Lagrange equations

$$\frac{d}{dt}\left(\frac{\partial L}{\partial \dot{q}_i} \right) - \frac{\partial L}{\partial q_i} = 0, \ (i = 1, 2, 3 \dots),$$

where L is the difference between the kinetic and the potential energies, q_1, q_2, q_3, etc. are generalized coordinates and $\dot{q}_1, \dot{q}_2, \dot{q}_3, \dots$ the corresponding generalized

---

[4] Eighteenth-century mathematical notations do not always coincide with our own.

velocities. Lagrange obtains them, as a direct consequence of the principle of virtual work, in a form slightly different (though basically equivalent) from the one which is known today (1788, 226).[5] They had made their first appearance a few years earlier (in 1782), but it is here that they began to play a central role in general dynamics.

14.2 A Stratified Text

The composition of *Méchanique analitique* has a complex history.[6] In 1756 the young Lagrange submitted an article on the principle of least action for publication in the *Mémoires* of the Berlin Academy of Sciences; unfortunately, for reasons unknown, the paper never appeared, and is now lost. He then worked for about three years on a book about his calculus of variations and its applications to mechanics, but this too never materialized. In the end, he only published a succinct overview of the application of the principle of least action to selected topics of general mechanics. Evidently, all of these works may be interpreted as intermediate stages in the evolution of the final treatise.

In addition to the *Méchanique analitique*, Lagrange also published several papers on fundamental branches of mechanics. They always begin with a few methodological remarks, followed by a historical introduction and a section on general principles. Only after these preliminaries does Lagrange proceed to the special question he has set himself. My readers need no reminding that this is also the structure of every main subdivision of the *Méchanique analitique*. Historians of science are thus confronted with the problem of determining how these early works influenced the final treatise. Moreover, since Lagrange began planning the book quite early in his mathematical career, this is in effect a question of *mutual* influences.

A quarter of a century after the first edition debuted, Lagrange published an updated second edition (1811–1815). The many revisions and additions nearly doubled the size of the book. It was this greatly enlarged treatise that was read throughout the nineteenth century, and led to a new mathematical discipline called *analytic mechanics*. Surprisingly, the differences between the two editions have rarely been explored.[7]

---

[5] For a more recent version of Lagrange's derivation, see Goldstein (1950).

[6] See Barroso-Filho (1994); Galletto (1991); Caparrini (2014).

[7] Steps in this direction have been taken by Pulte (2005).

14.3 "No Figures Will Be Found in This Book ... "

"... The methods I present require neither constructions nor geometrical or mechanical arguments, but solely algebraic operations subject to a regular and uniform
procedure." These famous words from Lagrange's preface to the *Mécanique
analitique* have been handed down from one generation to the next, by way of
popular histories of mathematics and historical footnotes in technical treatises. It is a
measure of Lagrange's influence that, to a present-day expert in analytic mechanics,
this dictum states the obvious and does not require any further comment.

However, the true extent of Lagrange's elimination of the visual aspects of
mechanics is sometimes overlooked. For his revulsion is not limited to the
Mécanique analitique: nowhere in Lagrange's *Oeuvres* is there a single illustration.
Some rough sketches can only be found in a few early texts that were not meant
for publication: his letters to Euler of 1755 (cf. *Oeuvres* **14**) and the lectures on
calculus given in the late 1750s at Turin's Royal School of Artillery (Borgato and
Pepe 1987).

Clearly, there are serious issues at stake here, which go to the heart of what
Lagrange understood by mathematics. It recalls the time, during the 1960s, when
Bourbaki's followers clamored, "Down with triangles!," and tried to replace
classical geometry with linear algebra, even in secondary schools. Indeed there are
similarities between Lagrange and Bourbaki, for Lagrange's entire work could be
considered a vast foundational program that sought to algebraize all of mathematics.
The geometric interpretation and the physical meaning of the algebra was left to the
reader. Yet an over-reliance on the algebraic language does not necessarily imply
the absence of geometry. Bourbaki and Lagrange did not eliminate triangles: put
simply, they did not care about trifling details.

Since for Lagrange algebra was the medium but not the message, it could be
rewarding to investigate the role of geometry in his mathematics. In fact, there are
new ideas about geometry hiding behind formulae in the *Mécanique analitique*.

14.4 The Dot Product of Two Vectors

A note of caution is needed here. Since the word *vector* was formally adopted
into mathematical physics by Sir William Rowan Hamilton in 1844 to denote the
imaginary part of a quaternion, purists may question its use in connection with
works published in the eighteenth century. However, we note that, as early as the
seventeenth century, the Latin locution *radius vector* made its way into astronomy.
We are thus justified in calling *vector* any quantity represented by directed line
segments subject to the parallelogram rule. In other words, our vectors are those
encountered in physics textbooks for high school.

We have before us quite an interesting state of affairs. While the abstract
concept of vector belongs to the second half of the nineteenth century, most of

the fundamental theorems on the composition of directed line segments had been discovered earlier. How much earlier? Surprisingly, some important work on the systems of vectors was done by Lagrange as far back as 1775. These results then turned up in the *Méchanique analitique*.

Consider, for example, the section of the *Méchanique analitique* concerning the kinematics of a rigid body rotating about a fixed point O. Given a set of orthogonal Cartesian axes $(O; \xi, \chi, \zeta)$ fixed in space, Lagrange defines a co-moving set of axes by taking three co-moving points $P'(\xi', \chi', \zeta')$, $P''(\xi'', \chi'', \zeta'')$, $P'''(\xi''', \chi''', \zeta''')$ such that the directed line segments OP', OP'', OP''' are mutually orthogonal:

> It is clear that ξ', χ', ζ' are nothing more than the coordinates of an arbitrary point of the system; and that similarly, ξ'', χ'', ζ'', and ξ''', χ''', ζ''' are the coordinates of two other arbitrary points of the system. [. . .] To abbreviate these expressions, let us assume $\xi'\xi'' + \chi'\chi'' + \zeta'\zeta'' = G'$, $\xi''\xi''' + \chi''\chi''' + \zeta''\zeta''' = G''$, $\xi'\xi''' + \chi'\chi''' + \zeta'\zeta''' = G'''$. If the three points of the system, which we have assumed as given, are disposed in such a manner that they form three right triangles with the origin of the axes as their common vertex, that is, if they are on three straight lines passing through the origin and making right angles with one another, it is obvious that we will have $G' = 0$, $G'' = 0$, $G''' = 0$. (Lagrange 1788, 340–42; my translation)

In vector terminology, to achieve orthogonality Lagrange sets $OP' \cdot OP'' = 0$, $OP'' \cdot OP''' = 0$, $OP' \cdot OP''' = 0$. Undeniably, he knew the geometric meaning of what we call the *dot product* of two vectors.

Lagrange had already worked out the elements for a theory of systems of three radii vectores about ten years before, in the first part of his *New solution to the problem of the rotational motion of a body of arbitrary shape and that is not acted on by any force* (1775a). There, among other theorems about directed line segments, he proves the formula

$$x'x'' + y'y'' + z'z'' = \left(\sqrt{a'a''}\right) \cos \, \varepsilon,$$

where $a' = x'^2 + y'^2 + z'^2$, $a// = x''^2 + y''^2 + z''^2$, and ε is the angle between the radii vectores $\sqrt{a'}$ and $\sqrt{a''}$ (see *Oeuvres* 3, 585). Except for notational details, this is our dot product. This formula also appears in his *Analytic solutions to some problems on tetrahedra* (1775b), written about the same time as the *New solution* (*Oeuvres* 3, 668).

After the publication of Lagrange's two papers, the 'dot product' was taken up by Euler in his memoir *A new method for determining the motion of rigid bodies* (1776), where it appears in the form

$$\cos \, \theta = \cos \, \alpha \, \cos \, \alpha' + \cos \, \beta \, \cos \, \beta' + \cos \, \gamma \, \cos \, \gamma',$$

which expresses the angle between two lines in space in terms of their direction cosines (*L. Euleri Opera omnia* (II)**9**, 105]). In this new incarnation, it earned a place in every textbook of analytic geometry and higher mechanics of the nineteenth century, whereas Lagrange's original formulation lay dormant for almost a century.

14.5 Cross Product of Two Vectors

The first part of the *New solution* is a theory of systems of radii vectores of three moving points, and the *Analytic solutions* is a study of tetrahedra considered as the portion of space enclosed by three radii vectores. Given the subject matter of these two papers, it should come as no surprise that Lagrange discovered a number of properties of vectors long before the abstract concept 'vector' had been formulated. In the same way we recognize that, during the seventeenth century, Fermat and Barrow knew some theorems of differential and integral calculus even before the general theory was born.

Let us return to the section on the 'General formulae relative to rotational motion' in *Méchanique*. After the passage cited above, Lagrange takes up anew the problem of defining a system of Cartesian axes fixed within the rotating rigid body:

> Although the preceding analysis is very direct, it is possible to obtain the same results more naturally from this geometric consideration: that the position of a generic point in space is completely determined by its distance from three given points (Lagrange 1788, 343; my translation).

This time he considers two points P', P'' moving with the rotating body, and defines a system of co-moving axes with the origin at the fixed point O and two axes aligned along the segments OP' and OP'':

> Now, since the positions of these points in the body is arbitrary, let us suppose for simplicity that their distances from the center of the body is equal to 1, and, moreover, that the straight lines drawn from the center to these two points are at right angles. [...] From which it follows that a, b, c are nothing more that the rectangular coordinates of a generic point of the body referred to three axes passing through the center and fixed within the body, one of which passes through the point with coordinates ξ', χ', ζ', and the other through the point represented by ξ'', χ'', ζ'', while the third axis is perpendicular to the other two (Lagrange 1788, 349; my translation).

In short, Lagrange refers the position of any point $P(\xi, \chi, \zeta)$ to a new set of axes by the formulae

$$\xi = a\xi' + b\xi'' + c\left(\chi'\zeta'' - \zeta'\chi''\right),$$

$$\chi = a\chi' + b\chi' + c\left(\zeta'\xi'' - \xi'\zeta''\right),$$

$$\zeta = a\zeta' + b\zeta'' + c\left(\xi'\chi'' - \chi'\xi''\right),$$

where the points $P'(\xi', \chi', \zeta')$ and $P''(\xi'', \chi'', \zeta'')$ define the position of two axes (1788, 350f). In the language of vectors, this means $OP = aOP' + bOP'' + cOP' \times OP''$ fixed, where $P(x, y, z)$ is any point, and a, b, c are real coefficients. Then he adds the conditions

$$\xi'^2 + \chi'^2 + \zeta'^2 = 1,$$

$$\xi''^2 + \chi''^2 + \zeta''^2 = 1,$$

$$\xi'\xi'' + \chi'\chi'' + \zeta'\zeta'' = 0,$$

that is, $OP' \cdot OP'' = 0$, $\| OP' \| = \| OP'' \| = 1$, and then Lagrange defines the position of a fourth co-moving point $P'''(x''', y''', z''')$ by the formulae

$$\xi''' = \chi'\zeta'' - \zeta'\chi'',$$

$$\chi''' = \zeta'\xi'' - \xi'\zeta'',$$

$$\zeta''' = \xi'\chi'' - \chi'\xi'',$$

or $OP = OP' \times OP''$, observing that

$$\xi'''^2 + \chi'''^2 + \zeta'''^2 = 1,$$

$$\xi'\xi'' + \chi'\chi'' + \zeta'\zeta'' = 0,$$

$$\xi''\xi''' + \chi''\chi''' + \zeta''\zeta''' = 0,$$

that is, $OP' \cdot OP''' = OP'' \cdot OP''' = 0$, and $\| OP \| = 1$. Briefly, OP', OP'', $OP''' = OP' \times OP''$ is an orthonormal basis. Thus we see that Lagrange was in possession of the formulae for the *cross product* of two vectors.

This part of *Méchanique* too comes from the introductory sections of his *New solution*. In reading the source, it is plain that Lagrange was aware of most of the geometric interpretations of these formulae. For instance, Lagrange knew the interpretation of his expression for what we call the cross product $\mathbf{u} \times \mathbf{v}$ as the (oriented) area of the parallelogram spanned by the vectors $\mathbf{u}$ and $\mathbf{v}$ (*Oeuvres* **3**, 586).

14.6   Scalar Triple Product of Three Vectors

In the *Méchanique analitique* there occurs the algebraic expression.

$$\xi'\chi''\zeta''' - \chi'\xi''\zeta''' + \zeta'\xi''\chi''' - \xi'\zeta''\chi''' + \chi'\zeta''\xi''' - \zeta'\chi''\xi''',$$

where ξ', χ', ζ', ξ'', χ'', ζ'', ξ''', χ''', ζ''' are the direction cosines of three directed unit segments drawn from the same origin. However, he said nothing of its geometrical interpretation (Lagrange 1788, 354).

To understand what Lagrange had in mind, let us consider the chapter on "The determination of the orbits of comets" in the second edition. Here Lagrange marks with the letters C, C', C'' three positions of a comet (i.e., any moving point) on the surface of the unit sphere centered on the observer:

> Having reduced the problem of comets to final equations in one unknown, it remains to examine the quantities which are assumed to be known. [. . .] Let us begin with the quantity $G = [lm'n'' + l''mn' + l'm''n - l''mn' - lm''n' - l'mn'']$ from which the others are only derived. [. . .] It is easy to prove that the quantity G is nothing more than the volume, taken six times, of the triangular pyramid for which the apex is at the center of the sphere whose radius is assumed equal to unity and which rests on the spherical triangle CC' C'', that is, which has for base the rectilinear triangle made by the chords of the three arcs CC', CC'', $C'C''$ (Lagrange 1997 ; *Oeuvres* **12**, 56-8; trans. Boissonnade and Vagliente).

Put differently, G is the volume of the parallelepiped spanned by the unit vectors $OC = (l, m, n)$, $OC' = (l', m', n')$ and $OC'' = (l'', m'', n'')$. It is impossible to escape the conclusion that Lagrange had the formula for the *scalar triple product* of vectors. He also invented the concise notation $(C \; C' \; C'')$ for these expressions.

Here too Lagrange recycled old ideas. The interpretation as volumes of these expressions goes back to the *New solution* (*Oeuvres* **3**, 585) and the *Analytical solutions* (ibid., 669). The chapter on comets in the *Méchanique analitique* is modeled on his paper *On the problem of the determination of the orbits of comets* (1785).

These quantities are now best viewed as determinants, that is

$$G = \begin{vmatrix} l & m & n \\ l' & m' & n' \\ l'' & m'' & n'' \end{vmatrix}.$$

While Lagrange does not say anything about this interpretation, he could not have missed it, for the first general theories of determinants—then called *résultantes*—were published independently by Laplace and Vandermonde about the same time as Lagrange's two papers of 1775 (cf. Muir 1906, 15 ff).

14.7 Cauchy's (So-Called) Inequality

Incidentally, it is worth noting that, in the course of his research on systems of radii vectores, Lagrange discovered the formula.

$$(a\alpha + b\beta + c\gamma)^2 = \left(a^2 + b^2 + c^2\right)\left(\alpha^2 + \beta^2 + \gamma^2\right) - \\ - (a\beta - b\alpha)^2 - (b\gamma - c\beta)^2 - (c\alpha - a\gamma)^2,$$

now called *Lagrange's identity*; it appears in the *New solution* as $\xi^2 + \eta^2 + \zeta^2$ $= a'a'' - b^2$ (cf.)*Oeuvres* **3**, 581]. In terms of linear algebra, it takes the form.

$$\|a\|^2 \|b\|^2 - (a \cdot b)^2 = \|a \times b\|^2,$$

where *a* and *b* are 3-dimensional Euclidean vectors with the respective lengths $\|a\|$ and $\|b\|$.

A few years later, in a paper entitled *General remarks on the motion of several bodies mutually attracting in the inverse ratio of the square of the distances* (1779) Lagrange obtained the inequality.

$$a\alpha + b\beta + c\gamma < \sqrt{(a^2 + b^2 + c^2)}\sqrt{(\alpha^2 + \beta^2 + \gamma^2)},$$

which is an easy consequence of the above result (*Oeuvres* **4**, 412). Mathematicians will recognize it as the *Cauchy-Schwarz inequality*, also known as the *Cauchy–Bunyakovsky–Schwarz inequality*. It was given in *n* dimensions in Cauchy's *Analyse algébrique* (1821; see *Oeuvres de Cauchy* (II)**3**, 373). Much later, it reappeared in the works of H.A. Schwarz and V. Y. Buniakovsky. Perhaps it should be renamed the *Lagrange-Cauchy inequality*, in honor of its real discoverers.

14.8   The Mechanics of Systems of Bodies

While the statement in the Preface of the *Méchanique analitique* about the absence of (synthetic) geometry has been elevated to iconic status, the preceding paragraph is much less cited:

> In addition, this work will have another use. The various principles presently available will be assembled and presented from a single point of view, in order to facilitate the solution of the problems of mechanics. Moreover, it will also show their interdependence and mutual dependence, and will permit the evaluation of their validity and scope (Lagrange 1788, v; trans. Boissonnade and Vagliente).

The great achievement of eighteenth century mechanics was the creation of formally coherent, nearly definitive theories for several different *systems* of bodies. The backbone of this success was a series of general principles, the most important of which were the conservation of *vis viva*, the principle of momentum, the principle of moment of momentum, the principle of virtual work, the principle of least action, and the principle of the conservation of motion of the center of gravity. Before Lagrange, it was generally acknowledged that they could be derived from some set of first principles, but in effect they had been demonstrated (if at all) only in the context of specific theories, starting from a number of different principles, and using a variety of mathematical methods.

Lagrange put an end to the discussions by proving these theorems for most mechanical systems, starting from unobjectionable first principles, via uniform analytic methods.[8] Thus he accomplished in general what Euler had done for single points, rigid bodies, and perfect fluids. With respect to the architecture of the general theorems, all nineteenth-century textbooks and treatises are direct descendants of the *Méchanique analitique*.

14.9   A Notation for Partial Derivatives

While partial derivatives appeared de facto as early as the beginning of the eighteenth century, the recognition of their importance as new mathematical entities worthy of independent research began only in the 1740s (Lützen 1994). Then came the problem of notation. In the 1750s, Euler denoted by $\left(\frac{dP}{dy}\right)$ the derivative of a function $P(x, y)$ holding x constant; this notation found wide acceptance for half a century, but there were a few contenders. The 'round delta' letter, ∂, briefly appeared in an article by Legendre in 1786, and the notation $\frac{\delta V}{dx}$ was used by Monge about 1770.

Lagrange's *Méchanique analitique* may have played a role in this period of notational indecision, for Lagrange denotes by $\frac{\delta V}{\delta \xi}$ the partial derivative of the function V with respect to ξ (1788, 226). Yet Cajori's *History of Mathematical Notations* (1929), the standard text in this area, does not cite this important step toward our, modern notation.

14.10   Spatial Symmetries and Conservation Laws

Historical consensus has it that Lagrange was not concerned with the problem of space in physics. However, a close examination of his work shows that he opened up new vistas on space and dynamics. While Newton and Euler had studied the structure of space and time in relation to the principle of inertia, Lagrange explored the connection between spatial invariance and conservation laws.

In the *Méchanique analitique*, Lagrange first applies his invariance technique to the problem of finding the equilibrium of forces and momenta:

> Let us consider an arbitrary system or assemblage of bodies or mass points mutually in equilibrium to which are applied various forces. If, for an instant, the action of these forces ceased to be mutually equilibrated, the system would begin to move and whatever its

---

[8] In the *Méchanique analitique*, Lagrange essentially deals with systems of mass points subject to workless holonomic constraints independent of time and acted on by central forces. That said, we are oversimplifying things here. Different sections of the *Méchanique analitique* have different degrees of generality, and in some cases the physical assumptions are not stated.

motion, it could always be considered as composed of (1) A translational motion common to all bodies; (2) A rotational motion about an arbitrary point; (3) A relative motion of the bodies expressing their change of position and their distance from one another. But, if they are to be in equilibrium, the bodies cannot have any of the motions cited above. However, it is obvious that the relative motions depend on the manner in which the bodies are arranged with respect to one another. Consequently, the conditions required to preclude these motions must be specifically fit to each system. Furthermore, the motions of translation and rotation can be independent of the configuration of the system and they can take place without changing the relative position of the bodies composing the system.

Thus the consideration of these two types of motion must furnish the general conditions or properties for equilibrium. These conditions and general properties are what we shall investigate (Lagrange 1788, 25; trans. Boissonnade and Vagliente).

Lagrange's argument is now familiar, though in our textbooks it is formulated in terms of the Lagrangian function rather than through the principle of virtual work. The basic idea is quite simple. Let us imagine that a system of n particles in equilibrium has been given an arbitrary translation

$$x_i \to x_i + a, \quad y_i \to y_i + b, \quad z_i \to z_i + c.$$

It is natural to assume that the mechanical properties of the system do not change (provided forces and mutual distances stay the same). Formally, the virtual work formula should be independent of a, b, c, that is, the coefficients of a, b, c should be zero. This translational invariance yields the equilibrium of forces

$$\mathbf{F}_1 + \mathbf{F}_2 + \mathbf{F}_3 + \cdots = \mathbf{0},$$

where $\mathbf{F}_i$ is the total force on the ith particle (Lagrange 1788, 28).

Likewise, an arbitrary rotation of the system about the z-axis yields the equilibrium of the moments of forces about this axis. Formally, using cylindrical coordinates $x = \rho cos\varphi$, $y = \rho sin\varphi$, the virtual work formula should not change under the substitution

$$\varphi \to \varphi + \mathrm{d}\varphi,$$

that is, the coefficient of $\mathrm{d}\varphi$ should be zero. The same rotational invariance plainly holds true for the other axes as well. Together, the three formulae are equivalent to the vector equilibrium of torques

$$\mathbf{M}_1 + \mathbf{M}_2 + \mathbf{M}_3 + \cdots = \mathbf{0},$$

where $\mathbf{M}_i = \mathbf{r}_i \times \mathbf{F}_i$ is the torque on the ith particle (Lagrange 1788, 29).

Later in the book, Lagrange applies the same method to dynamics (1788, 198–203). As is to be expected, in this more general context translational invariance leads to the balance of linear momentum

$$\sum_i F_i = \sum_i m_i \frac{d^2 r_i}{dt^2},$$

while rotational invariance implies the balance of moment of momentum

$$\sum_i r_i \times F_i = \sum_i m_i r_i \times \frac{d^2 r_i}{dt^2}.$$

As always with the *Méchanique analitique*, the main argument was not new. The idea of applying spatial invariance so as to obtain the balance of momentum and moment of momentum went back to Lagrange's very first paper in dynamics, his *Application of the preceding method to the solution of different problems in dynamics* of 1762. In this case, the invariance argument was used in conjunction with the principle of least action (*Oeuvres* **1**, 380). Later still, in his *General remarks* (1779), Lagrange found the balance laws for an isolated system of mutually gravitating points by infinitesimally varying the gravitational potential (*Oeuvres* **4**, 403–6).[9]

These ideas were entirely original and threw much light on the significance of the conservation principles. We see here the first step on the way which, more than a century later, led Emmy Noether to her fundamental theorems.[10] Unfortunately Lagrange stuck to the algebra and did not elaborate these concepts in relation to the formal properties of physical space. He might well have said something about space being homogeneous and isotropic. Even more unfortunately, at that time philosophers did not take notice of these mathematical results. While Newton's absolute space ignited the thinking of Clarke, Leibniz, Berkeley and Kant, Lagrange's space invariance had to wait the twentieth century to get the recognition it deserved.

14.11   From the First to the Second Edition

The second edition of Lagrange's treatise (1811–1815), with the title now spelled *Mécanique analytique*, was published nearly a quarter of a century after the first. By

---

[9] More precisely, Lagrange considered an unnamed function

$$\sum_{i \neq j} = M_i M_j / \sqrt{\left[(x_i - x_j)^2 + (y_i - y_j)^2 + (z_i - z_j)^2 \right]},$$

where M_i are the masses of the particles and x_i, y_i, z_i their positions referred to fixed rectangular Cartesian axes.

[10] This story has been reconstructed in Kosmann-Schwarzbach (2011).

then, a new generation of mathematicians, physicists and engineers were redefining mathematical physics. In his new Preface, Lagrange took notice of the changing landscape:

> The present edition is in many respects a new work based on the same outline but augmented. I have further developed the principles and general formulas and I have introduced numerous additional applications in which the solutions to the major problems in the domain of mechanics will be found. (*Mécanique analytique*, *Oeuvres de Lagrange* **11**, xiv; trans. Boissonnade and Vagliente]

How should we interpret his claim that the second edition is "a new work"? Certainly, not in the sense that the foundations had been consolidated, for the initial chapters are unchanged, except for minor improvements. Nor are we to look for mere exercises or examples, as there are no such things in Lagrange. In fact, the "further [developments of] the principles and [the] general formulae" reveal Lagrange's reactions to new trends in mathematical physics: geometric mechanics,[11] questions of invariance, and conservation laws.

As for the "numerous additional applications," a comparison with the mathematics literature up to 1810 shows that Lagrange had inserted into his treatise many of the results that had been discovered since the appearance of the first edition. In a way, he was showing the mathematicians of the new generation that their efforts were but corollaries of his general theory. Some of these "additional applications" are easily found: Carnot's theorems on the loss of *vis viva* during impact (*Oeuvres* **11**, 310), Laplace's invariable plane (*Oeuvres* **11**, 285), and Poisson's development of the theory of the variation of arbitrary constants (*Oeuvres* **12**, 176). Others are more recondite.

14.12   Projections of Plane Surfaces

Among the additions to the second edition is the proof that the sum of the squared products of the mass of every particle and the sectorial area described by its radius vector on each coordinate plane per unit time, that is,

$$
\left[\sum_i m_i \left(y_i \frac{dz_i}{dt} - z_i \frac{dy_i}{dt} \right) \right]^2 + \left[\sum_i m_i \left(z_i \frac{dx_i}{dt} - x_i \frac{dz_i}{dt} \right) \right]^2
$$
$$
+ \left[\sum_i m_i \left(x_i \frac{dy_i}{dt} - y_i \frac{dx_i}{dt} \right) \right]^2,
$$

---

[11] In a sense, mechanics has alway been based on geometry. Here, by the locution "geometrical mechanics" I refer to the line of research initiated mainly by Poinsot and Möbius at the beginning of the nineteenth century. For details, see Ziegler (1985).

is invariant under a rotation of the coordinate axes (*Oeuvres* **11**, 284). Vectorially speaking, this is the squared magnitude of the total angular momentum of the system, and thus Lagrange has nearly shown that angular momentum is a vector.

In pursuing this argument Lagrange was covering the same ground as a then-recent paper by Poisson on the representation of moments by means of plane surfaces (1808). Poisson considered a system of plane surfaces and studied the sum of their projections on the coordinate planes of an orthogonal Cartesian system. This amounts in effect to a geometric interpretation of the theory of moments. Poisson's results became quite well-known and had a place in every textbook of mechanics and analytic geometry up to the end of the nineteenth century (Caparrini 2002).

14.13   Internal Forces and Torques

During the eighteenth century the theorem that, in a system of mass points, the net resultant and their total torque are both equal to zero was not yet known. Only for some special mechanical systems, like rigid bodies, did it seem intuitively true (Truesdell 1968). Therefore, when in the *Méchanique analitique* Lagrange wrote the principles of linear momentum and of moment of momentum, he did not mention the distinction between internal and external forces.

As far as we know, the first proof that the total internal forces and torques both vanish are due to Laplace, who published it in *Traité de Mécanique céleste* (1798b; *Oeuvres de Laplace* **1**, 62–4). He also remarked that, on account of these lemmas, internal forces and torques do not appear in the principles of linear momentum and of moment of momentum.

Lagrange took note of all this in the second edition of his treatise. His proof of Laplace's lemmas is a simple corollary of translational and rotational invariance: evidently, internal forces and torques are not affected by rigid translations or rotations of the whole system. These lemmas were first demonstrated for statics, then taken over into dynamics (*Oeuvres* **11**, 47–50; 274, 279).

It is worth noting that in those same years Laplace's lemmas also appeared in the works of Binet, Cauchy and Poisson (Caparrini 1999 and 2002). This is further evidence, if any were needed, that, while the *Méchanique analitique* belongs entirely to the eighteenth century, the *Mécanique analytique* has to be evaluated in the context of early nineteenth century mechanics.

14.14   The Vector Representation of Infinitesimal Rotations

The vectorial composition of infinitesimal rotations was discovered by P. Frisi in a 1759 book on the precession of the equinoxes; but it went unnoticed until Lagrange republished it (Caparrini 2002).

Lagrange did not immediately recognize the vector nature of his theorems. In the *Méchanique analitique*, he demonstrates that any infinitesimal rotation of a rigid body with a fixed point O can be obtained as a sequence of three rotations about three orthogonal axes intersecting at O:

> It follows in general from all this that arbitrary rotations $d\psi$, $d\omega$, $d\phi$ about three rectangular axes meeting at some point are composed into one rotation $d\theta = \sqrt{d\psi^2 + d\omega^2 + d\varphi^2}$ about an axis passing through the said point of intersection and making angles λ, μ, ν with these three axes such that $cos\lambda = d\psi/d\theta$, $cos\mu = d\omega/d\theta$, and $cos\nu = d\varphi/d\theta$; and conversely, that an arbitrary rotation $d\theta$ about a given axis can be resolved into three partial rotations expressed by $d\theta \cdot cos\lambda$, $d\theta \cdot cos\mu$, and $d\theta \cdot cos\nu$ about three axes intersecting at right angles on a point of the given axis, and which make with this axis the angles λ, μ, ν. This gives, as is easily seen, a very simple way to compose and resolve rotational motions (Lagrange 1788, 33; trans. Boissonnade and Vagliente (revised)).

It is easy to see how close Lagrange comes to establishing the vectorial character of infinitesimal rotations, yet he fails to do so.

Lagrange considers again the same problem in the second edition. This time, after the passage reported above, he adds a completely new analysis of the problem. He demonstrates that the partial rotations transform as the components of a force, that is, he succeeds at last in formulating the vectorial composition of infinitesimal rotations;

> It is clear from this development that the composition and resolution of rotational motions are entirely analogous to rectilinear motions. Indeed, if on the three axes of rotation $d\psi$, $d\omega$, $d\phi$ one takes from their point of intersection lines proportional respectively to $d\psi$, $d\omega$, $d\phi$, and if one draws on these three lines a rectangular parallelepiped, it is easy to see that the diagonal of this parallelepiped will be the axis of the composite rotation $d\theta$ and will be at the same time proportional to this rotation $d\theta$. From this result — and from the fact that rotations about the same axis can be added or subtracted (depending on whether they are in the same or opposite directions), exactly as the motions that are in the same or opposite directions, we must draw the general conclusion that the composition and resolution of rotational motions is done in the same manner and by the same laws that the composition or resolution of rectilinear motions, by substituting for rotational motions rectilinear motions along the direction of the axes of rotation (*Oeuvres* 11, 61; trans. Boissonnade and Vagliente (revised)).

It is quite likely that Lagrange took this result from Frisi, for he knew Frisi's work well. He probably decided to highlight the geometrical meaning of his results for a new generation of more geometrically-minded mathematicians. However, his contemporaries attributed to him all the merits of the discovery. It is through the *Mécanique analytique* that the vector theory of angular velocity made its way into our treatises.

14.15   The Geometric Theory of Moments

At the beginning of the nineteenth century, thanks to the work of Euler, Laplace, Poinsot, Prony, Poisson, Binet and Cauchy, it became apparent that moments of forces can be represented by directed line segments and composed according to the parallelogram rule (Caparrini 2002).

Lagrange took notice of this discovery in the second edition of his treatise. As we have seen, he had demonstrated that infinitesimal rotations admit a vector representation. Given this result, his proof of the vector composition of moments follows immediately from the apparently natural hypothesis that the moment of a force about an axis is proportional to the corresponding rotation in a given infinitesimal time (*Oeuvres* **11**, 60). Unfortunately, Lagrange's proof is not correct, because this assumption does not take into account the role of the moments of inertia.

14.16 Formal Invariance

As mentioned above, in the *Méchanique analitique* the entire dynamics is founded on the formula

$$\sum m \left(-\frac{d^2x}{dt^2}\delta x - \frac{d^2y}{dt^2}\delta y - \frac{d^2z}{dt^2}\delta z + P\delta p + Q\delta q + R\delta r + \ldots \right) = 0.$$

This visibly leads to an invariance problem: while the force terms $\sum m(P\delta p + Q\delta q + R\delta r + \cdots)$ are independent of the coordinates, the inertial terms $\sum m \left(\frac{d^2x}{dt^2}\delta x + \frac{d^2x}{dt^2}\delta x + \frac{d^2x}{dt^2}\delta x \right)$ apparently depend on the axes. Lagrange tackled the question in the second edition:

> There is an important addition in Section II [of Part II: Dynamics]. It is shown for which cases the general formula of dynamics and consequently, the equations which result for the motion of a system of bodies, is independent of the position of the coordinate axes in space. This demonstration gives a means of completing a solution by the introduction of three new arbitrary constants where some constants would have otherwise been assumed to be equal to zero (*Mécanique analytique, Oeuvres* **11**, xvi; trans. Boissonnade and Vagliente).

The proof is by direct substitution (*Oeuvres* **11**, 269). Lagrange shows that the virtual work per unit mass and unit time

$$d^2x\delta x + d^2y\delta y + d^2z\delta z$$

is not affected by an orthogonal transformation of coordinates.

This is one of the first demonstrations of coordinate invariance. A few years earlier, Laplace had studied the effect of an orthogonal transformation of coordinates on the three sums of the projections on the coordinates planes of the areas swept out by the radii vectores of a system of particles (1798a). Then Poisson in 1808 shed light on the geometric background of this theorem, demonstrating implicitly the independence of the solution from the choice of axes.

Apparently, the idea spread slowly. In 1834 Gabriel Lamé demonstrated the invariance of the Laplacian and other differential operators, and as late as 1841 Cauchy devoted a paper to the invariance of what we call the dot product, the triple scalar product and the magnitude of the cross product.

14.17 The Lagrangian Function

Lagrange's equations in the *Méchanique analitique* look somewhat unfamiliar at first sight, for they are expressed in the form

$$d\frac{\delta T}{\delta d\xi} - \frac{\delta T}{\delta \xi} + \frac{\delta V}{\delta \xi} = 0,$$

where, in Lagrange's terminology, $2T$ is the *vis viva*, ξ is a coordinate, δ denotes the partial derivative (as I explained above), and V a function of the coordinates (Lagrange 1788, 226).[12] Apart from notational nuisances, they differ from the formulae currently in use mainly because of the absence of the Lagrangian function $L = T - V$. A search through the relevant literature shows that Lagrange's equations were mostly written this way in the nineteenth century.

To find the combination $T - V$ we must turn to Lagrange's first paper on the variation of arbitrary constants in mechanics (1809), his last burst of creative genius. There he introduces the function $R = T - V$, and a perturbation function Ω that takes into account, for example, moving centers of force. The equations of motion then take the form

$$d\frac{dR}{dr'} - \frac{dR}{dr}dt = \frac{d\Omega}{dr}dt,$$

which is close enough to that reported in our textbooks (*Oeuvres* **6**, 778). The general theory of the variation of arbitrary constants in mechanics was inserted into the second edition, where the function $T - V$ was denoted by Z (*Oeuvres* **11**, 347). The new function was then taken up, for example, by Poisson (1809) and Binet (1841).

Let us fast forward to the second half of the century. The function $T - V$ was called *Lagrangian* in E. J. Routh's *Treatise on the Stability of a Given State of Motion* (1877, 47). Ten years later, in his celebrated paper "On the physical meaning of the principle of least action," Helmholtz introduced the *kinetische Potential* $H = V - T$ as a kind of "world function" encompassing different physical phenomena Von Helmholtz (1887). Finally, the terminology was standardized in E. T. Whittaker's *Treatise on the Analytical Dynamics of Particles and Rigid Bodies* (1904, 38).

Acknowledgments Many thanks to Chris Smeenk and Marius Stan for inviting me to contribute to this *Festschrift*. Marius helped tremendously in clarifying my arguments. I am very grateful to Marta Serrano for her help with English.

---

[12] While Lagrange in the *Méchanique analitique* does not say much about the physical meaning of the function V (but see 1788, 208), readers would have recognized the force function introduced by d'Alembert, Daniel Bernoulli, Maupertuis and Euler more than thirty years before.

I spent the academic year 2005/06 as visiting Fellow at the Dibner Institute for the History of Science and Technology in Cambridge, Mass. George Smith was then the acting director. While steering the Institute through the rough seas of academic politics, he found time to take good care of a group of international scholars and to write innovative essays on Newton. He helped me in many ways, and I welcome this opportunity to thank him.

Bibliography

Barroso-Filho, W. 1994. *La mécanique de Lagrange. Principes et méthodes*. Paris: Karthala.

Binet, M.J. 1841. Mémoire sur la théorie générale de la variation des constantes arbitraires, dans les formules générales de la dynamique et dans un système d'équations analogues plus étendues. *Journal de l'Ecole Polytechnique* XVII (cahier 28): 1–94.

Borgato, M.T., and L. Pepe. 1987. Lagrange a Torino (1750-1759) et le sue lezioni inedite nella Reale Scuola di Artiglieria. *Bollettino di Storia delle Scienze Matematiche* 7: 3–43.

Cajori, F. 1929. *A history of mathematical notations*. Vol. 2. Chicago: Open Court.

Caparrini, S. 1999. On the history of the principle of moment of momentum. *Sciences et Techniques en Perspective* 3: 47–56.

———. 2002. The discovery of the vector representation of moments and angular velocity. *Archive for History of Exact Sciences* 56: 151–181.

———. 2007. An unpublished letter by Lagrange concerning the Turin academy of science. *Atti della Accademia delle Scienze di Torino; Classe di Scienze Fisiche. Matematiche e Naturali* 141: 45–52.

———. 2014. The history of the *Méchanique analitique*. *Lettera Matematica: International Edition* 1–2: 47–54.

Caparrini, S., and C. Fraser. 2013. Mechanics in the eighteenth century. In *The Oxford handbook of the history of physics*, ed. J.Z. Buchwald and R. Fox, 358–405. Boston: Elsevier.

Capecchi, D. 2012. *A history of virtual work Laws*. Milan: Springer.

Cauchy, A.-L. 1821. Cours d'analyse de l'Ecole Polytechnique Royale Ire Partie. Analyse algébrique. Paris: de Bure. Reprinted in *Oeuvres complètes de Cauchy*, series II 3.

———. 1841. Mémoire sur divers théorèmes relatifs à la transformation des coordonnées rectangulaire. *Exercices d'Analyse et de Physique Mathématique* 2: 273–286. Reprinted in *Oeuvres complètes de Cauchy*, series II, vol. 12, 310–325.

de Morgan, A. 1914. In *Essays on the life and work of Newton*, ed. P.E.B. Jourdain. Chicago, London: Open Court.

Euler, L. 1776. Nova methodus motum corporum rigidorum determinandi. *Novi Commentarii academiae scientiarum Petropolitanae* 1775: 208–238. Reprinted in *L. Euleri Opera omnia*, ser. II, vol. 9, 99–125.

Galletto, D. 1991. Lagrange e la *Mécanique Analytique*. *Memorie dell'Istituto Lombardo, Accademia di Scienze e Lettere, Classe di Scienze Matematiche e Naturali* 29 (3): 78–179.

Goldstein, H. 1950. *Classical mechanics*. Cambridge, MA: Addison-Wesley.

Kosmann-Schwarzbach, Y. 2011. *The Noether theorems: Invariance and conservation Laws in the twentieth century*. New York: Springer.

Lagrange, J.L. 1762. Application de la méthode précédente à la solution des différens Problèmes de Dynamique. *Miscellanea Taurinensia. Mélanges de philosophie et de mathématiques de la Societé Royale de Turin pour les années 1760–1761*: 196–268. Reprinted in *Oeuvres* 1: 365–468.

———. 1775a. Nouvelle solution du problème du movement de rotation d'un corps de figure quelconque qui n'est animé par aucune force accélératrice. *Nouveaux Mémoires de l'Académie des Sciences et Belles-Lettres, année 1773*: 85–120. Reprinted in *Oeuvres* 3: 579–616.

———. 1775b. Solution analytique de quelques problèmes sur les pyramides triangulaires. *Nouveaux Mémoires de l'Académie des Sciences et Belles-Lettres, année 1773*: 149–176. Reprinted in *Oeuvres* 3: 661–92.

————. 1779. Remarques générales sur le movement de plusieurs corps qui s'attirent mutuelle-ment en raison inverse des carrés des distances. *Nouveaux Mémoires de l'Académie des Sciences et Belles-Lettres, année 1777*: 155-172. Reprinted in *Oeuvres* 4: 401–18.

————. 1782. Théorie de la libration de la Lune & des autres phénomènes qui dépendent de la figure non sphérique de cette planète. *Nouveaux Mémoires de l'Académie des Sciences et Belles-Lettres, année 1780*: 203–309. Reprinted in *Oeuvres* 5: 5–122.

————. 1785. Sur le problème de la détermination des orbites des Comètes d'après trois observations (Troisième Mémoire dans lequel on donne une solution directe & générale de ce problème). *Nouveaux Mémoires de l'Académie des Sciences et Belles-Lettres, année 1783*: 296–332. Reprinted in *Oeuvres* 4: 496–532.

————. 1788. *Méchanique analitique*. Paris: Desaint.

————. 1809. Mémoire sur la théorie générale de la variation des constantes arbitraires, dans tous les problèmes de mécanique. *Nouveaux Mémoires de l'Institut national, classe des Sciences mathématiques et physiques, année 1808*: 257–302. Reprinted in *Oeuvres* 6: 771–804.

————. 1811–1815. *Mécanique analytique. Nouvelle édition, revue et augmentée par l'auteur.* Vol. 2. Paris: Courcier.

————. 1853–1855. *Mécanique analytique. 3ᵉédition, revue, corrigée et annotée par J. Bertrand.* Vol. 2. Paris: Mallet-Bachelier.

————. 1867–1892. In *Oeuvres de Lagrange*, ed. J.-A. Serret, vol. 14. Paris: Gauthier-Villars.

————. 1997. Analytical mechanics, trans. A. Boissonnade and V.N. Vagliente. Dordrecht: Kluwer.

Lamé, G. 1834. Sur les lois de l'équilibre du fluide éthéré. *Journal de l'Ecole Polytechnique* XIV, cahier XXIII: 191–288.

Laplace, P.S. 1798a. Mémoire sur la détermination d'un plan qui reste toujours parallèle a lui même, dans le mouvement d'un système de corps agissant d'une manière quelconque les unes sur les autres et libres de toute action étrangère. *Journal de l'Ecole Polytechnique*, t. II, cahier V: 155-159. *Oeuvres de Laplace* 14: 3–7.

————. 1798b. *Traité de mécanique céleste. Tome premier.* Vol. 1. Paris: Duprat. Reprinted in *Oeuvres de Laplace*.

Lützen, J. 1994. Partial Differential Equations. In *Companion Encyclopedia of the history and philosophy of the mathematical sciences*, ed. I. Grattan-Guinness, vol. 1, 452–469. London: Routledge.

Muir, T. 1906. *The theory of determinants in the historical order of development*. Vol. 1. London: Macmillan.

Poisson, S.D. 1808. Note sur différentes propriétés des projections. *Correspondance sur l'Ecole Polytechnique* 1: 389–394.

————. 1809. Mémoire sur la théorie générale de la variation des constantes arbitraires dans les questions de mécanique. *Journal de l'Ecole Polytechnique, t.* VIII, cahier XV: 266–344.

Pulte, H. 2005. Joseph Louis Lagrange, *Méchanique analitique*, first edition (1788). In *Landmark writings in Western mathematics 1640–1940*, ed. I. Grattan-Guinness, 208–224. Boston: Elsevier.

Routh, E.J. 1877. *A treatise on the stability of a given state of motion*. London: Macmillan.

Taton, R. 1974. Inventaire chronologique de l'œuvre de Lagrange. *Revue d'histoire des sciences* 17: 3–36.

Truesdell, C. 1968. Whence the law of moment of momentum? In *Essays in the history of mechanics*, 239–271. Berlin: Springer.

Von Helmholtz, H. 1887. Über die physikalische Bedeutung des Princips der kleinsten Wirkung. *Journal für die reine und angewandte Mathematik* 100: 137–166. Reprinted in *Wissenschaftliche Abhandlungen* 3: 203–48.

Whittaker, E.T. 1904. *A treatise on the analytical dynamics of particles and rigid bodies.* Cambridge: University Press.

Ziegler, R. 1985. *Die Geschichte der geometrischen Mechanik im 19. Jahrhundert* ed. Stuttgart: Steiner.

Chapter 15
Ptolemy's Scientific Cosmology

N. M. Swerdlow

The purpose of this essay is to show that there was one person, perhaps only one, who developed a rigorously scientific cosmology nearly two thousand years ago. Cosmology is the largest of all subjects, with a long history, and the cosmology considered here is the one that endured for the longest part, nearly three-quarters, of that history. By cosmology I mean a description of the universe as a whole and of the arrangement of its principal parts. But by scientific cosmology, I mean something more. For a cosmology to be considered scientific, three criteria must be met: it must be quantitative, physical, and empirical. Quantitative means that it must assign some scale of distances and sizes to its parts. Physical means that it must rest upon some kind of physical or mechanical principles, causes, for the ordering and motion of its parts. Finally, empirical means that it must either be derived from observation or, if derived theoretically, as from physical or mechanical principles, it must be confirmed by observation. I set out these criteria, which may not be exclusive or exhaustive but are surely within reason, in order to define our subject and exclude mythological or theological cosmologies, which generally meet none of the criteria, and philosophical cosmologies, as those of Plato, Aristotle, and Descartes, which may meet one or two in some way, although not rigorously, but surely do not meet all three. I am not concerned with whether the cosmology or the criteria upon which it is based are correct. Compared to our present understanding, the scale may be off by orders of magnitude, the physics may be entirely mistaken, and the observations may be inaccurate. But if all three are present I would call the cosmology scientific, and if not I would not call the cosmology scientific.

If now we apply these criteria, the earliest scientific cosmology of which we have any knowledge is that of Ptolemy, Claudius Ptolemaeus, who lived in Alexandria

N. M. Swerdlow (✉)
University of Chicago, Chicago, IL, USA

California Institute of Technology, Pasadena, CA, USA

© The Author(s), under exclusive license to Springer Nature Switzerland AG 2023   327
M. Stan, C. Smeenk (eds.), *Theory, Evidence, Data: Themes from George E. Smith*, Boston Studies in the Philosophy and History of Science 343, https://doi.org/10.1007/978-3-031-41041-3_15

in the second century, perhaps from about 100 to 175, and wrote the definitive treatises of antiquity on nearly every branch of applied mathematics, astronomy, astrology, optics, harmonics, and cartography, omitting only mechanics. If he was not the greatest scientist of antiquity, which I believe he was, he was one of the two or three greatest, and in the entire history of science he belongs with those we place in the next rank after Newton. His most important work in astronomy was called in Greek the *Mathematical Collection* or *Mathematical Treatise*, but all but pedants call it by the Arabic contraction of its title, the *Almagest*, and so shall we. It is a comprehensive treatise on every aspect of mathematical astronomy, that is, spherical astronomy, the theory of the sun, moon, and planets, eclipses, the fixed stars, and the phenomena, risings and settings, of the stars and planets. He catalogs the coordinates of over one thousand stars down to the sixth magnitude, which meant apparent size (*megethos*), not brightness (*lamprotes*) as we us the term today, a change of definition that occurred only around the beginning of the nineteenth century. He gives an extremely detailed description of the Milky Way, in which he considers the locations of all its twists and turns, where it is denser and where rarer, but never speculates on what it could be, except that it belongs among the stars. Among his other works are the *Handy Tables*, intended for practical computation, and the *Planetary Hypotheses*, of which we shall be concerned with the parts on physical models, or mechanisms, of motions in the heavens and on distances and sizes of bodies within the planetary system.

The method of Ptolemy's astronomy and cosmology may be characterized simply: it is rigorously empirical and rigorously mathematical, indeed, it is the most strictly empirical of all cosmologies. Every 'hypothesis'—a technical term which means a model, for every motion of every body—is either derived or confirmed by observation, although not all the observations are presented in detail, and every numerical parameter is derived from observation by strict mathematical procedures. There is, however, a large range of precision in his observations, from positions and times measured to within a few minutes for the derivation of parameters— although their accuracy is more variable and there are systematic errors—to rough, even qualitative observations for demonstrating the properties of models. It is principally those rough, qualitative observations that concern us. Paradoxically, since they do not depend upon precise measurement, there can be no doubt about the conclusions drawn from them, as small inaccuracies in the observations would make no difference. I will begin with the demonstrations in the *Almagest* and then take up the physical models and theory of distances and sizes of bodies in the *Planetary Hypotheses*, which depend as much upon physical principles as upon observation.

Ptolemy's first demonstration is that the heavens are spherical in form and move as a rotating sphere. The evidence is entirely empirical. It can be seen that the sun, moon, and the stars that rise and set, which include the planets, are carried from east to west along arcs of parallel circles, and that after setting rise again at the same place except for the independent motions of a few bodies. Likewise, those stars that never set describe parallel circles around the celestial pole, stars closer to the pole describing smaller circles, stars farther from the pole describing larger circles. All of these phenomena, appearances, correspond to motion of a rotating sphere. Other possibilities are excluded as contrary to fact, meaning contrary to observation. That the stars do not move in straight lines from and to infinity is shown by their sizes not changing as they pass from the horizon to the meridian and by the fact that they reappear in the same place each time they rise. That they are not lighted as they rise and extinguished as they set is clear because they appear the same to all observers even though their times of rising and setting differ at different locations on the earth, both to the east and west and to the north and south, for some stars are always visible to observers to the north but rise and set to observers to the south. And any motion other than the rotation of a sphere would produce a variation in the apparent sizes and separations of stars, as their distance from observers would change, but such a variation is not seen to occur except that bodies appear somewhat larger near the horizon, an effect that Ptolemy here attributes to moisture in the air although in his later *Optics* he explains it as a perceptual illusion, for measurements with instruments show no change in size. This is the first more or less correct description of the 'moon illusion' as it is called, that the moon looks larger near the horizon than at higher elevations, and that it is only an illusion of perception.

Ptolemy next demonstrates that the earth taken as a whole is sensibly spherical. That it has curvature to the east and west, in longitude, is shown by the fact that stars rise and set for observers to the east before they do so for observers to the west, and more specifically that lunar eclipses, which occur at a unique time, when the moon is opposite the sun in the shadow of the earth, as shown in Fig. 15.1a, are seen later

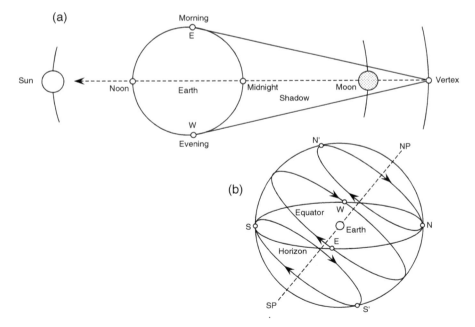

Fig. 15.1 Caption

in the night by observers to the east and earlier by observers to the west, and the differences in time are proportional to the distances between the observers. In the later *Geography* 1.4 he cites a famous lunar eclipse eleven days before the Battle of Arbela in 331 BC that was observed in Arbela in Persia at the fifth hour of night and at Carthage in Africa three hours earlier at the second hour, and uses this to establish a longitudinal distance of 45°, which is too large as correctly the distance is $33^3/4°$. That the earth has curvature to the north and south, in latitude, is shown by more stars that are circumpolar and fewer stars that rise and set as one travels away from the equator toward the poles, and the contrary as one travels toward the equator. In Fig. 15.1b, the celestial equator and stars between *NN'* and *SS'* rise and set obliquely to the horizon; at northern latitudes stars north of *NN'* are circumpolar and always visible, stars south of *SS'* are circumpolar and never visible, which is opposite at southern latitudes, and changes in the number of stars that rise and set or are always or never visible, correspond to the distances to the north or south of the equator. Since the earth is uniformly curved in both perpendicular directions, it is sensibly spherical. Were it flat, stars would rise and set for all observers at the same time; were it concave, stars would appear to rise earlier for observers in the west than in the east, were it cylindrical with observers on the curved surface, all stars visible at all would rise and set and none would be circumpolar. Hence, these contrary to fact conditions may be excluded.

The most complex demonstration is that the earth is in the middle of the heavens. It is a proof by contradiction: were the earth not in the middle, what *would* be seen is not *in fact* seen. Ptolemy considers two possibilities, shown in Fig. 15.2, off axis but equidistant from the poles, on axis but removed toward one pole. In the first case (a), with the earth off axis in the plane of the celestial equator, were the observer at the earth's equator, the horizon, shown by the vertical broken line, would divide the heavens into two unequal parts and the sun would spend unequal times above and below the horizon, contrary to the observation that day and night are always equal at the equator. For any other location, with the horizon shown by a sloping solid line, equal day and night would not occur at the equinoxes (viz. the intersections of the ecliptic and the equator, shown by the signs for Aries and Libra), where in fact they occur. Instead, they would occur only where—and if—the horizon bisects some circle parallel to the equator. Further, it is known from observation that the equinoxes and the solstices, shown by the signs for Cancer and Capricorn, are separated exactly by quadrants, but the angles measured from the earth off axis in the figure are manifestly unequal. In the second case (b), with the earth on axis but removed from the center toward one pole, the horizon, shown by the sloping line, would divide the heavens into two unequal parts, and these would differ with the latitude of the observer: no inequality at the equator, the greatest inequality at the pole, and different unequal divisions in between. The horizon would thus divide the zodiac, constellations along the great circle of the ecliptic, unequally, as in the figure fewer than four constellations are visible above the horizon and eight invisible below. But this is contrary to the observation that everywhere and at all times six signs of the zodiac are visible and six invisible, showing that the horizon does indeed bisect the zodiac for all observers. Since now the cases of off axis but equidistant

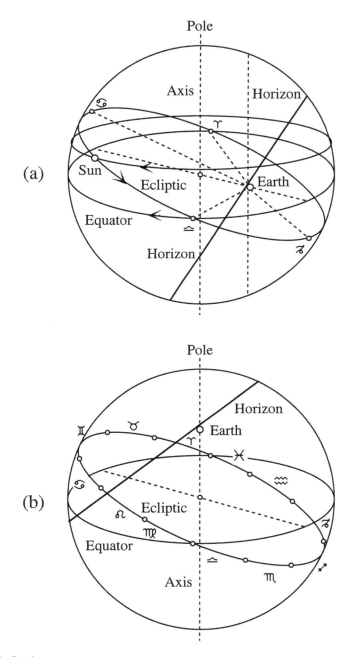

Fig. 15.2 Caption

from the poles and on axis but removed toward one pole have been eliminated, the combination off axis and removed toward one pole, that is, an arbitrary position, is also eliminated, and the earth can only be in the center of the heavens. Ptolemy has yet more arguments for both cases, of which these are a selection.

This important demonstration allows Ptolemy to dispose of two further proofs rapidly. First, the earth has the ratio of a point compared to the heavens because were it of any appreciable magnitude, the very effects just excluded would be seen from its surface and they are not. Hence, its surface is sensibly also at the center of the heavens and its entire body has the ratio of a point to the heavens. Second, the earth does not move from place to place away from the center because, once again, the excluded effects would be seen at times if it did, and they are not ever seen. He also has a physical argument that if it did so move, on account of its bulk it would move the fastest and leave behind less heavy bodies adjacent to the earth and air, if all heavy bodies tend toward the same limit, which is never seen and is absurd to contemplate. What Ptolemy has shown by this series of demonstrations is the applicability of the model we call the celestial sphere, a sphere of infinite or indefinite or unit radius with the observer at the center, the fundamental model of the heavens in spherical astronomy. That it may not describe the true universe is possible if the distance of the earth from the center is itself insensible compared to the distance of the heavens. But empirical demonstrations cannot by their very principle show what is not sensible, and we shall see that Ptolemy has yet another reason for placing a limit on the distance of the heavens.

For there is still one possible motion of the earth: that it rotates in place while the heavens remain motionless—which cannot be decided by observing the heavens, since all appearances would be the same whether the earth or the heavens made one rotation each day. However, Ptolemy answers this with two counterfactual arguments that depend on observing objects in the air. If the earth rotated to the east and the air did not, objects in the air, such as birds, clouds, and projectiles, would be left behind to the west, which manifestly they are not. And if the air moved with the rotating earth, the same objects would either be left behind by the rotation of both the earth and air—which they are not. Or if they were carried around, as it were, joined to the air, they could have no independent motion against such a rapid motion (as the air would necessarily have). It is one thing to move independently against winds such as we find them near the surface of the earth, but quite something else to move independently against the motion of the entire air, reaching as much as a thousand miles an hour at the equator, that would be produced by the rotation of the air together with the earth. (Cannot a fish in a river swim upstream or downstream? Not if the river is flowing at hundreds of miles per hour.) Since birds, clouds, and projectiles do have independent motions, and are not left behind to the west, the obvious conclusion is that neither the earth nor the air rotate and the earth is entirely at rest. And since it is the heavens and not the earth that perform the daily rotation, the heavens cannot be all that large, and thus any departure of the earth from the center would have noticeable effects, as shown in the earlier demonstrations, of which none are observed. See how one demonstration reinforces another.

So much for the establishment of the general description of the universe on an empirical foundation. In setting out the models for the motions of the sun, moon, and planets, Ptolemy uses observations to derive the properties of each model in the same way he did for the general description of the heavens and earth. But the observations and demonstrations are specific to the apparent motion of each body. We shall pass over the sun and moon to consider the more complex case of the planets, which uses some of the methods also used for the sun and moon. First we consider the complete model, shown in Fig. 15.3, so that we shall know the object

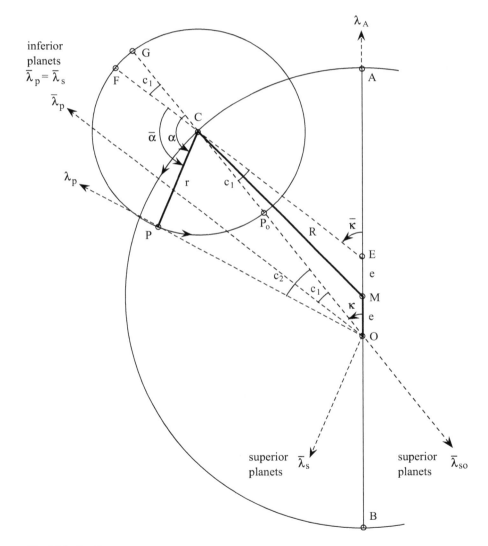

Fig. 15.3 Caption

of our demonstrations. The earth is at O and at an eccentricity e a point M is the center of an eccentric circle of radius R on which the center C of an epicycle moves through the mean eccentric anomaly $\bar{\kappa}$ measured from the direction of the apogee of the eccentric A, uniformly with respect to a point E, later called the equant point, at twice the eccentricity $2e$ from O. The line EC is parallel to the direction $O\bar{\lambda}p$ from the earth to the mean longitude of the planet. The planet P moves on an epicycle of radius r through the mean anomaly $\bar{\alpha}$ measured with respect to the extension of EC to the mean apogee F of the epicycle. The parameters e, $2e$, and r are stated proportionally in the measure of R, e.g. for Mars where $R = 60$, $e = 6$, $2e = 12$, $r = 39;30$. The position of the planet as seen from the earth, the direction OP, is found from two corrections or equations: $c1$, the equation of center, which corrects direction EC to OC, the mean eccentric anomaly $\bar{\kappa}$ to the true eccentric anomaly $\kappa = \bar{\kappa} \pm c_1$, and the mean apogee of the epicycle F to the true apogee G, correcting the mean anomaly $\bar{\alpha}$ to the true anomaly $\alpha = \bar{\alpha} \pm c_1$; and $c2$, the equation of the anomaly, which corrects direction OC to OP. The true longitude of the planet is then $\lambda_p = \bar{\lambda}_p \pm c_1 \pm c_2$.

There is an important difference between the inferior and superior planets— a consequence of the fact that their models are transformations of underlying heliocentric models (though of course this was not known). For the inferior planets—Mercury and Venus, which have a limited elongation on either side of the sun—the mean longitude $\bar{\lambda}_p$ always lies in the direction $O\bar{\lambda}_s$ of the mean longitude of the sun, $\bar{\lambda}_p = \bar{\lambda}_s$, the mean sun, which moves uniformly with the annual motion of the sun, from which the direction of the true sun, which moves nonuniformly, differs by at most about $\pm 2°$. In the case of the superior planets, Mars, Jupiter, Saturn, which reach opposition to the sun, the radius of the epicycle CP is always parallel to the direction from the earth to the mean longitude of the sun $O\bar{\lambda}_s$. As a consequence, when the planet is in opposition to the mean sun, separated by an elongation of $180°$, it lies at the perigee of the epicycle on OC at Po with the mean sun in the direction $O\bar{\lambda}_{so}$, and thus an observation of the direction of the planet Po gives the direction of the center of the epicycle C, which is otherwise unobservable.

Ptolemy explains qualitatively the observational confirmation of each part of this model: (1) the motion of the planet on an epicycle and the direction of the motion, (2) the motion of the center of the epicycle on an eccentric, (3) the constant distance of the epicycle's center C from the eccentric point M that lies at half the distance from O to E, about which the motion of C is uniform, that is, $OM = \frac{1}{2}OE$. This idea is known as the bisection of the eccentricity, and it is the most important innovation in Ptolemy's planetary theory. He gives only the principles of these demonstrations without specific details, but they can be reconstructed and, interestingly, each one depends upon a different planet and is then extended to all the planets since, although they cannot be proved for each individually, neither can they be disproved or contradicted. We shall consider each separately, although the last only in outline.

The proof of motion on an epicycle can only be done for an inferior planet, since the position of the sun gives the approximate direction of the center of the epicycle

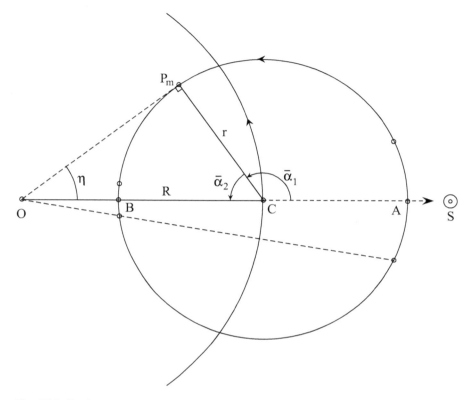

Fig. 15.4 Caption

at all times, which is not known for a superior planet, and three locations of the planet on the epicycle can be determined by observation with sufficient accuracy for the demonstration. As Mercury is usually invisible, the demonstration can only practically be done for Venus. Ptolemy says that a characteristic of motion on an epicycle (with the forward direction at apogee) is that the time from greatest speed to mean speed is greater than the time from mean speed to least speed. In Fig. 15.4, which is drawn for Venus, greatest speed occurs when the planet is at A, superior conjunction with the mean sun, and least speed, really greatest retrograde motion, at B when it is at inferior conjunction. Mean speed occurs when the line from the earth O to the planet Pm is tangent to the epicycle, at greatest elongation η from the mean sun. The question is, does this model apply to Venus, is the time from A to Pm greater than the time from Pm to B? Neither A nor B can be directly observed since they occur in the arcs of invisibility close to the sun, in the parts of the epicycle between the broken lines, but their time can be estimated by dividing the time from last to first visibility on either side of the sun. The planet can easily be observed at Pm, although it is necessary to do so for several nights and then interpolate for the time of greatest elongation from the computed position of the mean sun. If we now

Fig. 15.5 Caption

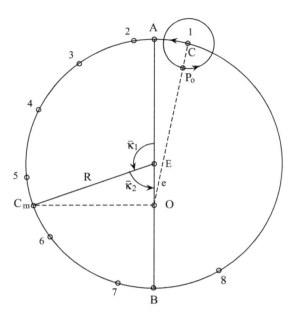

represent the time by the angle $\bar{\alpha}$ of the mean anomaly, which increases uniformly, we find that, to the nearest ten days,

$$A \rightarrow P_\mathrm{m} \ \bar{\alpha}_1 = 220^\mathrm{d}, \ P_\mathrm{m} \rightarrow B \ \bar{\alpha}_2 = 70^\mathrm{d}, \ \bar{\alpha}_1 - \bar{\alpha}_2 = 150^\mathrm{d},$$

a very great difference. Thus, even the roughest observations are adequate to show that the time from A to Pm is greater than the time from Pm to B, which proves that an epicycle with forward motion at apogee is applicable to Venus and, by extension, to all the planets.

Ptolemy's criterion for motion on an eccentric is that the time from least speed to mean speed is greater than the time from mean speed to greatest speed, just the opposite of motion on an epicycle in the forward direction at apogee. The best planet for this demonstration is Jupiter—because Saturn has a rather long period; and the highly irregular motion of Mars makes interpolation uncertain. In Fig. 15.5, showing an eccentric with center E about which uniform motion takes place, the least speed occurs at apogee A, the greatest speed at perigee B, and the mean speed at Cm, a quadrant from A and B as seen from the earth O, where CmO is perpendicular to the apsidal line AB. It is necessary to have a provisional direction for the apsidal line, which Ptolemy could obtain from earlier Greek or Babylonian planetary theory or estimate from the very series of observations used to confirm the model. We have noted that when a superior planet is at opposition it is at the perigee of the epicycle—hence, an observation of the planet Po gives the direction of the center of the epicycle C. Thus, from a series of, say, eight oppositions of Jupiter from before A to after B, one can interpolate for the time when the center of the epicycle is at A (where the motion is slowest), B (where it is fastest), and Cm, which is 90° from both A and B. Now letting the mean eccentric anomaly $\bar{\kappa}$, which increases uniformly, represent the time, we find to the nearest ten days,

$$A \to C_m \ \bar{\kappa}_1 = 1140^d, \ C_m \to B \ \bar{\kappa}_2 = 1020^d, \ \bar{\kappa}_1 - \bar{\kappa}_2 = 120^d,$$

once again a difference so large that even the roughest observations and estimates of the locations and times of A, B, and Cm are adequate to show that the center of the epicycle moves on an eccentric.

We have, however, assumed that E, the point about which the angular motion of the center of the epicycle is uniform, is the center of the eccentric. Ptolemy next specifies (Fig. 15.3) that the center of the eccentric M is at a point lying halfway between O and E, thus at one-half the eccentricity of E, a discovery known as the bisection of the eccentricity. As he later describes the eccentricities the other way around, 'using rough estimation, the eccentricity one finds from the greatest equation of ecliptic anomaly turns out to be about twice that derived from the size of the retrograde arcs at greatest and least distances of the epicycle.' What this means is that the eccentricity $OE = 2e$ that determines the equation of center $c1$ and the direction of the center of the epicycle OC is twice the eccentricity $OM = e$ that determines its distance OC, which also affects the direction and distance of the planet OP. *The purpose of the bisected eccentricity and equant motion is the distinction of direction and distance.* The demonstration is too complex to present here, but we note that it is based upon the variation of the length of the retrograde arc of Mars, from about 20° at apogee of the eccentric to about 10° at perigee, an unmistakable difference, while the length of the retrograde arcs of Saturn and Jupiter hardly vary at all. The principle of the demonstration is that from the variation of the retrograde arc Ptolemy finds an eccentricity that determines the distance of the center of the epicycle independently of the eccentricity found from oppositions that determines its direction, finding that the eccentricity that determines distance is about half the eccentricity that determines direction. Hence, the eccentricity found from oppositions must be bisected, and the center of the eccentric lies at half the eccentricity from the earth of the equant point about which the angular motion of the center of the epicycle is uniform. But this demonstration establishes only the bisection, not the accurate value of the eccentricity, which must be found more rigorously.

The bisected eccentricity greatly complicates the derivation of the parameters of the model. In fact, the eccentricity and direction of the apsidal line requires the most complex and extended series of computations—not only in the *Almagest*, but in all of ancient mathematics pure and applied, a procedure Ptolemy surely worked out himself that is well worth summarizing. The principle of the derivation is shown in Fig. 15.6. Because the radius of the epicycle is always parallel to the direction from the earth to the mean sun, as shown earlier, at opposition the planet P is at the perigee of the epicycle, in line with the earth O and the mean sun $\bar{S}$, and thus lies in the direction of the center of the epicycle C, which can therefore be found from the direction OP. Observation of P at three oppositions, found by interpolation for when

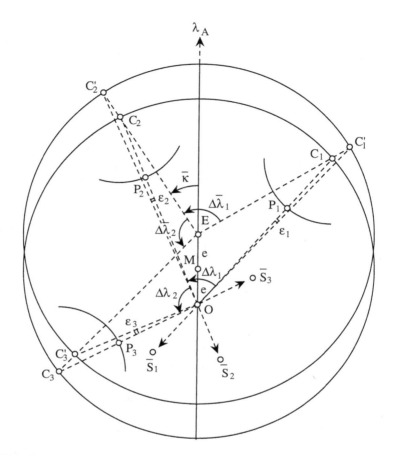

Fig. 15.6 Caption

$\lambda_p - \overline{\lambda}_s = 180°$, define two angles $\Delta\lambda 1$ and $\Delta\lambda 2$ seen from O of apparent motion of C in the circle about M. Further, the intervals of time $\Delta t1$ and $\Delta t2$ between the oppositions define two angles $\Delta\overline{\lambda}_1$ and $\Delta\overline{\lambda}_2$ of mean motion about the equant E, and thus two arcs for three locations C' in the circle about E. Note that the angles of apparent motion are to points C on the circle about M and the arcs of mean motion are between points C' in the circle about E. It is this separation—of the mean and apparent motion to two different circles—that makes the derivation so complicated. Ptolemy's solution is iterative. He lets C coincide with C' in the circle about E, assumes falsely that $\Delta\lambda_1$ and $\Delta\lambda_2$ apply to C' in the circle about E as seen from O, and solves for the double eccentricity $OE = 2e$ and the mean eccentric anomaly $\overline{\kappa}$ at each opposition. He then bisects OE so that $OM = e$, places C on the circle about M, and tests whether $2e$, e, and $\overline{\kappa}$ at each opposition can reproduce the observed $\Delta\lambda_1$ and $\Delta\lambda_2$. This not being so, he finds the small correction angles ε, which are added algebraically to $\Delta\lambda_1$ and $\Delta\lambda_2$, i.e. $\Delta\lambda'_1 = \varepsilon_1 + \Delta\lambda_1 - \varepsilon_2$,

$\Delta\lambda'_2 = \varepsilon_2 + \Delta\lambda_2 - \varepsilon_3$, to give a better estimate of the correct angles between points C' in the circle about E as seen from O. The procedure is repeated, using the *corrected* $\Delta\lambda'_1$ and $\Delta\lambda'_2$ and the *original* $\Delta\lambda_1$ and $\Delta\lambda_2$, to find a new solution for $OE = 2e$ and $\overline{\kappa}$, again bisecting OE to $OM = e$, and testing $2e$, e, and $\overline{\kappa}$ against the observed $\Delta\lambda_1$ and $\Delta\lambda_2$. Then, if necessary, finding new values of ε, correcting the *original* $\Delta\lambda_1$ and $\Delta\lambda_2$ (*not* $\Delta\lambda'_1$ and $\Delta\lambda'_2$) to the *second corrected* $\Delta\lambda''_1$ and $\Delta\lambda''_2$, and again using the *original* $\Delta\overline{\lambda}_1$ and $\Delta\overline{\lambda}_2$ solving for $2e$ and $\overline{\kappa}$ — testing $2e$, e, and $\overline{\kappa}$, as many times as necessary, until it is possible to compute $\Delta\lambda_1$ and $\Delta\lambda_2$ with the model and parameters in agreement with the observed $\Delta\lambda_1$ and $\Delta\lambda_2$. Each solution requires well over a hundred computational steps, but because the eccentricities are small, convergence to minutes is fairly rapid, and in Ptolemy's exposition two solutions suffice for Saturn and Jupiter and three for Mars, although he probably made adjustments to produce rapid convergence. The subtraction of the true eccentric anomaly κ (not shown) from the longitude of one opposition then gives the longitude of the apogee, $\lambda_A = \lambda_p - \kappa$.

The result of this bisection is the best approximation to Kepler's first two laws of motion without knowing Kepler's laws. I will show that it is so, because of its great importance to the history of astronomy. Kepler's model is shown in Fig. 15.7a, in which the planet P moves in an ellipse with center M and the sun S at one focus such that the line SP joining the sun to the planet describes an area ASP proportional to time. Now, the line EP joining the second, empty focus of the ellipse to the planet describes an angle AEP that happens to be very nearly proportional to time, and this is the principle upon which Ptolemy's model works. For in Fig. 15.7b we superimpose Kepler's model of an ellipse and Ptolemy's model of a circle, now shown heliocentrically, and we see that the circle is in fact the major auxiliary circle of the ellipse, the circle constructed on the major axis. And the equant E, about which the planet P' moves uniformly through the mean eccentric anomaly $\overline{\kappa}$, is the empty focus of the ellipse, about which the motion of the planet P is very nearly uniform. Next, we draw the ordinate PN to the apsidal line and extend it to meet the circle at $\overline{P}$ (the position the planet would occupy were it describing area $AS\overline{P}$ in the major auxiliary circle), which was Kepler's earlier model before he discovered the ellipse. We see that the direction SP lies between $S\overline{P}$ and SP', and in the case of Mars with a large eccentricity of about 0.1, the differences of the directions on either side of SP amount to at most about ± 8 minutes of arc, which is the greatest *theoretical* difference between Ptolemy's model and Kepler's model with the same parameters. If the models were drawn to scale for Mars, the circle, ellipse and directions would be nearly indistinguishable. In fact, Ptolemy's model has other problems, as errors in the parameters, and errors in the second inequality greater still. But the theoretical difference between the models for the first inequality is this small, and Kepler spent years of hard work eliminating that error of just 8 minutes of arc.

We have seen that Ptolemy's planetary theory, like his general description of the universe, is based upon the strictest empiricism, for every part of his model has been derived or confirmed by observation, using observations so unequivocal, differences in time of more than 100 days, differences in arc of $10°$, that there can

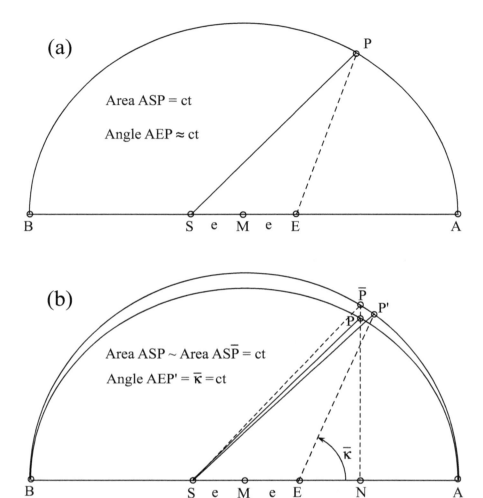

Fig. 15.7 Caption

be no doubt of the observations or of a model based upon them, and far more precise
observations for the derivation of parameters. Yet, Ptolemy's strict empiricism could
also lead to errors due to faulty observations, but these errors show how rigorously
empirical his method was. Mercury has a particularly complicated model motivated
by observations indicating that the planet reaches a greater sum of elongations on
both sides of the mean sun at $\pm 120°$ from apogee than at $180°$, where the maximum
sum of elongations would be expected. The problem was that greatest elongations
on both sides of the sun could not be observed near apogee and perigee. Because
Mercury is frequently out of sight, these elongations are visible just on one side (of
the Sun). Thus Ptolemy doubled the one side he could observe, and drew an incorrect

conclusion that the greatest elongations at 180° from apogee were less than those at
±120°. Still, his procedure was strictly empirical, using as much information as the
observations provided. Likewise, the models for latitude are quite complicated, with
variable inclinations of the planes of the epicycles due to rather rough conventional
values of extreme latitudes to the north and south that, if taken strictly, as he did,
required the variable inclinations. In this case, Ptolemy later corrected his models
through improved observations, for in the *Planetary Hypotheses* the inclinations of
the planes of the epicycles are fixed, as indeed they should be, and the results are in
excellent agreement with correct planetary latitudes as they are actually seen.

But the *Planetary Hypotheses* is of importance for yet more reasons, for it is
there that Ptolemy sets out his complete cosmology, of physical models for the
motion of bodies in the heavens and of their distances and sizes. The *Planetary
Hypotheses*, which successive changes show to be later than the *Almagest* and
the *Handy Tables*, is in two books. Book 1, most of which survives in Greek,
concerns the construction of instruments, analogue computers of the kind later
called *equatoria* for the computation of positions of the sun, moon, and planets
in longitude and latitude, for which there were once tables of mean motions. The
remainder of Book 1, concerning distances and sizes, and Book 2, descriptions of
the physical models, do not survive in Greek, but the entire work was translated
into Arabic in the ninth century before the Greek text of these parts was lost. What
Ptolemy established entirely empirically in the *Almagest* is the arrangement of the
heavens and earth, the foundation of his cosmology, and models, each with with its
own parameters, for the motions of the sun, moon, and planets. But the models,
as in Fig. 15.3 for the planets, are purely geometrical, not physical. And, there
is no unified scale of distances from the center of the world or between different
planets. So, two of our three criteria for scientific cosmology—physical in the sense
of physical or mechanical causes of motion and quantitative in the sense of a scale
of distances—have not been met.

Corresponding to the geometrical models in the *Almagest* are physical models
in Book 2 the *Hypotheses* composed of spheres, the mechanisms in the heavens
that bring about the motion of the individual bodies, although the source of motion
for the spheres is provided by the heavenly body itself in the same way that the
mind of an animal directs the motion of all its parts. The models can be composed
of complete spheres or of equatorial cross-sections of spheres the thickness of the
planets' latitudinal deviations from the ecliptic. The spherical model for all the
planets except Mercury is shown in Fig. 15.8 in a plane equatorial cross-section.
The planet P is carried by, and directs the rotation through the mean anomaly $\bar{\alpha}$, of
an epicyclic sphere with center C, which in turn is carried by an eccentric sphere,
the width of and entirely containing the epicyclic sphere, with center M that rotates
uniformly through the mean eccentric anomaly $\bar{\kappa}$ about the equant point E—this
motion likewise directed by the planet. Just how the rigid spherical body of the
eccentric sphere (for the spheres rotate as rigid bodies) can do this (viz. can rotate
uniformly about the equant E, which is not on the diametral axis passing through its
center M); and how the epicycle carrying the planet rotate uniformly with respect
to the line EC—these were raised as physical objections to Ptolemy's model by a

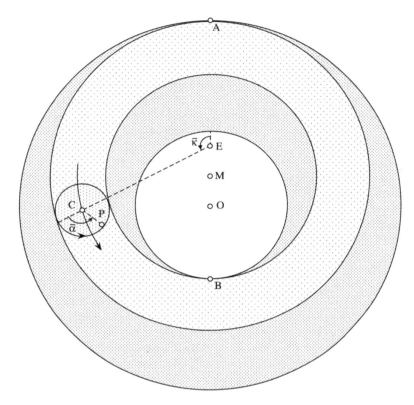

Fig. 15.8 Caption

number of Arabic astronomers. And, they were the original impetus to Copernicus's planetary theory, which relied heavily upon his Arabic predecessors. But that is a history that goes beyond our subject.

Note that in the figure there are two additional spherical bodies within and without the eccentric sphere, each with one surface touching the eccentric sphere and concentric to M and another surface—the inner surface of the inner sphere and the outer surface of the outer sphere—concentric to the earth at O, both surfaces of the inner sphere passing through the perigee B and both surfaces of the outer sphere passing through the apogee A. The result is that the inner and outer surfaces of each spherical model are concentric to the earth, to the center of the world, so that the spherical model of an inner planet can be placed inside and touching the surface passing through B and the spherical model of an outer planet can be placed outside and touching the surface passing through A. That is, the spheres are contiguous, they *nest* with no empty space between them. 'This arrangement is most plausible,' Ptolemy says, 'for it is not conceivable that there be in Nature a vacuum, or any meaningless or useless thing.' We must therefore imagine seven of these nested together, extending upward from the moon to Saturn, with additional spheres for the moon and Mercury and fewer spheres for the sun.

The consequence of this assumption, or physical principle, of contiguous spheres is that the least distance of any planet is equal to the greatest distance of the planet directly beneath it, and its greatest distance is equal to the least distance of the planet directly above it. Which is highly significant for cosmology, because it offers a way of establishing a scale of distances of the entire planetary system. Since for each planet the variation of relative distance is given by the parameters in the *Almagest*, the eccentricity e and radius of the epicycle r in units of the radius of the eccentric R, the absolute distance of any one planet allows the absolute distance of every planet to be computed. Thus, from the *Almagest* we know the least relative distance $d' = R - r - e$, greatest relative distance $D' = R + r + e$, and thus the ratio of greatest to least distance D'/d'. Hence, if we are given the least absolute distance of a planet d, it greatest absolute distance D is given by $D = (D'/d')d$. And from the principle of nested spheres, the least distance of the next higher planet $d_{n+1} = D_n$. In the same way we may compute its greatest distance D_{n+1} and proceed outward through the entire planetary system.

In the *Almagest* Ptolemy had found the absolute distances of two bodies in units of the radius of the earth r_e. Namely, the least and greatest distances of the moon (here rounded to $33r_e$ and $64r_e$), computed from a direct measure of parallax and the parameters of its model. And, he found the mean distance of the sun, viz. $1210r_e$, by a demonstration from the apparent radius of the earth's shadow at the distance the moon passes through it in an eclipse when equal to the apparent radius of the sun. These distances were found and used for computation of eclipses. From the mean distance of the sun and its eccentricity of 1/24, its least distance is $1160r_e$ and greatest distance $1260r_e$. Now from the greatest distance of the moon $64r_e$, he finds the distances of Mercury and Venus, noting that they alone can fit between the moon and the sun, even leaving a space between the greatest distance of Venus and the least distances of the sun, which he accepts, remarking that he cannot account for it although he is aware that the derivation of the distance of the sun, which is very sensitive to small changes in the distance of the moon, can be adjusted by slightly increasing the moon's distance, so as to reduce the distance of the sun and close up the space. Then, from the greatest distance of the sun $1260r_e$, he computes the distances of Mars, Jupiter, and Saturn, and concludes that the sphere of the fixed stars is to be placed just beyond the greatest distance of Saturn at about $20,000r_e$. The result is a scale of distances in radii of the earth for all the bodies in the heavens, as shown in the table.

Planet	D'/d'	Distance re			Diameter		Volume ve
		Least	Greatest	Mean	Apparent s'_p	True se	
Moon	–	33	64	48	$1^1/3$	$^1/4 + {}^1/24$	$^1/40$
Mercury	88/34	64	166	115	$^1/15$	$^1/27$	$^1/19683$
Venus	104/16	166	1079	$622^1/2$	$^1/10$	$^1/4 + {}^1/30$	$^1/44$
Sun	–	1160	1260	1210	1	$5^1/2$	$166^1/3$
Mars	7/1	1260	8820	5040	$^1/20$	$1^1/7$	$1^1/2$
Jupiter	37/23	8820	14,189	11,504	$^1/12$	$4^1/3 + {}^1/40$	$82^1/2 + {}^1/4 + {}^1/20$
Saturn	7/5	14,189	19,865	17,026	$^1/18$	$4^1/4 + {}^1/20$	$79^1/2$
Stars	–	20,000	20,000	20,000	1st $^1/20$	$4^1/2 + {}^1/20$	$94^1/6 + {}^1/8$

Ptolemy also converts the distances to stades—there are eight stades to a mile—where the radius of the earth is taken as $2 + {}^1/2 + {}^1/3 + {}^1/30$ myriad stades (a myriad is ten thousand) or $28,666^2/3$ stades, from an equatorial circumference of 180,000 stades, also given in *Geography* 7.5. To give the most impressive number, Saturn's maximum distance is five myriad myriad and $6946^1/3$ myriad stades, that is, $569,463,333^1/3$ stades or $71,182,916^2/3$ miles. The fixed stars would then be a bit farther, at about $71,666,666^2/3$ miles, which is not so small a universe.

The entire system is shown to scale with distances in radii of the earth in Fig. 15.9, in which the range is so great that we have separated the central spheres of the earth, the moon, and Mercury, and drawn them at ten times the scale of the rest of the figure. The earth drawn to scale would be a point less than one-thirtieth the least distance of the moon.

But he can do yet more, namely, compute the true sizes of the bodies—likewise empirically—because he has measurements of the apparent diameters of the planets and first-magnitude stars, which he takes to apply to mean distance, as fractions of the apparent diameter of the sun. They are all too large, spurious images of naked eye observation due to the limited resolving power of the eye, but planets and stars really do appear to have some size depending upon their brightness, and these values were accepted as canonical until Galileo, using the telescope, showed them to be incorrect. If we call the mean distance of the sun r_s and of the planet r_p, the true diameter of the earth se and of the sun $s_s = 5^1/2 se$ in the *Almagest*, and the apparent diameter of the planet s'_p where the apparent diameter of the sun $s'_s = 1$, the true diameter of the planet $s_p = (s'_p \cdot r_p)/(r_s/s_s)$. Since $r_s/s_s = 1210/5 = 220$ is constant, taking Mars as an example, $s_p = (^1/20 \cdot 5040)/220 \approx 1^1/7 se$. The apparent and true diameters are shown in the table. To convert s'_p to minutes multiply by $30'$, from which it is obvious that all are too large, especially $1^1/3 \cdot 30' = 40'$ for the moon. The diameter of the sun, $5^1/2 se$, is too small by a factor of twenty, because the solar distance is too small by a factor of twenty. Note that Jupiter, Saturn, and the first-magnitude stars are all more than four times the diameter of the earth, of interest in deciding just what these things could be, concerning which Ptolemy says nothing, although he must have wondered about it. By cubing the diameters, he finds the

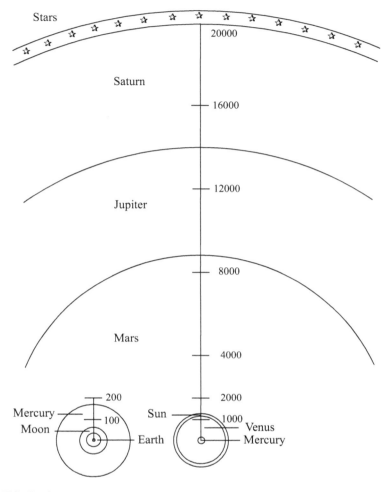

Fig. 15.9 Caption

volumes of all the bodies in units of the volume of the earth, *ve*, also shown in the table, in which it can be seen that Mercury is very small and the large bodies are very large, so again one might wonder what they could be. Finally, he concludes that if the distances have been correctly given, then these are the correct volumes. But if the distances are greater, meaning if there are spaces between the spheres, then these are the minimum values possible. Thus, Ptolemy established lower limits for the distances and sizes of all the bodies in the planetary system and the largest stars.

Ptolemy's cosmology had a long life and was still alive and well in the age of Copernicus, Kepler, and Galileo. Copernicus himself held to the physics of contiguous spheres, which were only doubted in the late sixteenth century, as by Tycho, Kepler, and Galileo, and the heliocentric theory itself, with its profound effect on cosmology, was not established until the first half of the 17th. The very completeness and specificity of Ptolemy's cosmology made it particularly attractive, for who could not admire knowing the exact distance and size of a remote planet or star? From the Arabic translation of the *Planetary Hypotheses* the physical, spherical models, or something like them, were described in detail in Ibn al-Haytham's *Configuration of the World*, translated into Latin in different versions in the twelfth and thirteenth centuries, which appears to be the source of the models described in Peurbach's *Theoricae novae planetarum*, the most well-known text on planetary theory through the late 15th and 16th centuries. There were any number of recomputations of the distances and sizes with minor, and in a few cases major, variations, of which two were translated in the twelfth century. One was by al-Farghānī, in what became a popular introduction to astronomy; and another by al-Battānī, in a more advanced work—and from these, the numbers passed to writers of every kind in Europe, who also seemed delighted to know the universe in such detail. Regiomontanus, in his *Oration on the Mathematical Sciences* delivered in Padua in 1464 as introduction to a series of lectures on al-Farghānī, said in praise of astronomy:

> You measure the depth of the earth, you ascend the heights of the heavens. You show that the sun is one hundred and sixty-six times greater than the earth, and the moon is equal to nearly a fortieth part of the earth. You compare all the stars to the volume of the earth according to accurate proportions. You investigate the dimensions of the celestial spheres, you promise to measure the size and distance from the earth of the earthly vapor set on fire in the highest region of the air, which is called a comet.

To give an amusing earlier example, Roger Bacon, who also took his distances from al-Farghānī, says that if one were to walk from the earth to the least distance of the moon, a modest $105,787^1/2$ miles, at the rate of 20 miles a day, it would take 14 years, 5 months of 30 days, 1 month of 29 days, and there would still be $7^1/2$ miles left to go.

Another writer who knew this cosmology and refers to it was Dante, on whose astronomy, drawn principally from al-Farghānī, there are whole volumes, including an edition of Gerard of Cremona's translation of the latter figure. In the *Paradiso* Dante, guided by Beatrice, traverses the spheres of the successive planets, and then in Canto 22, in crossing from the sphere of Saturn to the sphere of the fixed stars, Beatrice bids him look back through 'bright sphere upon sphere,' through the seven spheres to the tiny earth. 'And all the seven heavens,' he says, 'showed to me their magnitudes, their speeds, and the distances of each from each.' These are doubtless the numbers from Farghānī. There are also two very specific references. One is in *Convivio* 2.13, in which each planet is compared to a liberal art, and the explanation of why Mercury is compared to dialectic is straight out of our subject:

And the heaven of Mercury may be compared to Dialectic on account of two properties: (1) that Mercury is the smallest star in the heavens, for the quantity of its diameter is not more than 232 miles according to what al-Farghānī assumes, who says it to be 1/28 part of the diameter of the earth, which is 6500 miles; (2) the other property is that it is concealed by the rays of the sun more than any other star. And these two properties are in Dialectic: for (1) Dialectic is smaller in its corpus than any other science, since it is perfectly compiled and completed in as much text as is found in the *Ars vetus* and *nova*, and (2) it is more concealed than any science in that it proceeds with more sophistical and probable arguments than any other.

The *Ars vetus* and *nova* are Aristotle's logical works. The other reference is more interesting and ingenious. In *Paradiso* 9 Bishop Folquet of Marseilles explains that when he was young he devoted himself to Venus but now he dwells, along with other reformed *innamorati*, as Rahab of Jericho, in her sphere, 'that heaven in which the shadow cast by your world reaches a point.' What could this mean? It is really quite simple if you know where to look. The perigee of Venus is at a distance, with Farghānī's number, of 167 terrestrial radii, and the apogee at 1120 terrestrial radii—again Farghānī's number. But the length of the shadow of the earth is shown by Ptolemy to be 268 terrestrial radii, a number also given by Farghānī, and hence about 100 terrestrial radii of the length of the shadow cone and its vertex lie within the sphere of Venus, which is thus not quite perfect. And that is why, if you read the *Paradiso* carefully, you will find that up to the sphere of Venus there are any number of former sinners, as repentant soldiers and lovers, but from the sun on there are only the exceedingly virtuous, as doctors and soldiers of the Church. It is all a matter of knowing your Ptolemaic cosmology as well as Dante.

Sources and Further Reading

Ptolemy's cosmology was discovered as recently as 1964 by Willy Hartner, when he proposed that distances of the sun, moon, and planets contained in numerous, principally Arabic, astronomical works originated in Ptolemy's *Planetary Hypotheses*, known in Arabic as *Kitāb al-Manshūrāt*, 'Book of the sawn-off pieces', referring to models in the form of equatorial cross-sections of spheres the thickness of the planets' latitudinal deviations from the ecliptic. Willy Hartner's publication is 'Medieval Views on Cosmic Dimensions and Ptolemy's Kitāb al-Manshūrāt,' *Mélanges Alexandre Koyré*, Paris, 1964, 1.254–82, reprinted in Hartner, *Oriens-Occidens*, 319–48 (Hildesheim: G. Olms, 1968). The Arabic text had not been published and no section on distances was contained in the surviving Greek text of Book 1 or in the German translation of the Arabic text of Book 1 by Ludwig Nix and of Book 2, begun by Nix and completed after his death by Francis Buhl and Paul Heegaard, published in J. L. Heiberg's 1907 edition of Ptolemy's *Opera astronomica minora*. It appeared as though any section about distances was hopelessly lost. But Hartner could not have been more correct. In 1966 Bernard Goldstein examined a manuscript of the Hebrew translation made from the Arabic, and there at the end of Book 1 was the section on distances and also sizes, which he then found in the two known Arabic manuscripts. Goldstein published a translation of the section concerning distances and sizes with an edition of the Arabic text consisting of a facsimile of the London, British Museum manuscript with variants from the Leiden, Golius manuscript in 'The Arabic Version of Ptolemy's *Planetary Hypotheses*,' *Transactions of the American Philosophical Society* N.S. 57.4, Philadelphia, 1967. It appears that Nix had not translated the last part of Book 1, and Buhl and Heegaard, who completed Nix's translation of Book 2, did not know that the translation of Book 1 was incomplete. In this way Ptolemy's cosmology was lost, then discovered, and has since become very well known.

The best and standard translation of the *Almagest* is by G. J. Toomer, *Ptolemy's Almagest* (Springer, 1984; reprinted Princeton University Press, 1998). The most comprehensive studies of the *Almagest* and Ptolemy's other astronomical works are O. Pedersen, *A Survey of the Almagest*, Odense University Press, 1974 (second ed. Springer, 2011), and O. Neugebauer, *A History of Ancient Mathematical Astronomy* (Springer, 1975). The general description of the world in the *Almagest* has been treated in detail by L. Taub, *Ptolemy's Universe, The Natural and Ethical Foundations of Ptolemy's Astronomy* (Open Court, Chicago, 1993). There are two accounts of Ptolemy's derivation of the bisection of the eccentricity from the retrogradations of Mars: J. Evans, 'On the Function and Probable Origin of Ptolemy's Equant,' *American Journal of Physics* 52: 1080–9 (1984); N. M. Swerdlow, 'The Empirical Foundation of Ptolemy's Planetary Theory,' *Journal for the History of Astronomy* 35: 249–71(2004). Other derivations have been described using other methods. There is an evaluation of the published texts and translations of the *Planetary Hypotheses* by G. J. Toomer, *Isis* 81: 757–8 (1990). For Book 1, the edition of the Greek text and Latin translation by John Bainbridge, *Procli sphaera, Ptolemaei de Hypothesibus Planetarum liber singularis* (London, 1620), is excellent; and there is a recent English translation of the Greek text by E. Hamm, *Ptolemy's Planetary Theory: An English Translation of Book One, Part A of the Planetary Hypotheses with Introduction and Commentary*, PhD Dissertation, University of Toronto, 2011. For Book 2, one must still rely upon the German translation in Heiberg's edition, although one hopes that will soon change. The spherical models in Book 2 have been described most completely by A. Murschel, 'The Structure and Function of Ptolemy's Physical Hypotheses of Planetary Motion,' *Journal for the History of Astronomy* 26: 33–61 (1995). Many of these subjects have recently been considered within Ptolemy's entire work in J. Feke, *Ptolemy's Philosophy. Mathematics as a Way of Life* (Princeton University Press, 2018). The most comprehensive study of distances and sizes of bodies in the heavens, of cosmology, from antiquity through the end of the seventeenth century is A. van Helden, *Measuring the Universe, Cosmic Dimensions from Aristarchus to Halley* (University of Chicago Press, 1985).

Chapter 16
Revisiting Accepted Science: The Indispensability of the History of Science

George E. Smith

16.1 Theses

My theses are synopsized in my title, so let me begin by expanding on it, starting with the subtitle. The main point of the paper is to argue for the indispensability of the history of science to the philosophy of science, yet by the end I hope to make clear how the converse holds as well. Because both history and philosophy of science involve diverse pursuits, my claim of their mutual indispensability applies only insofar as each of them concerns itself with the nature and scope of the "knowledge" achieved in modern science (The shudder quotes serve not to foreclose from the outset on those who question whether the word is strictly applicable to it). Although adopting Kant's phrasing exaggerates the situation, in some respects, I am going to argue, philosophy of science without history of science is empty, and history of science without philosophy of science is blind. My hope for some time has been a rapprochement between the two. The crucial step, nevertheless, is for philosophers of science to recognize a need for a certain sort of history of science. The principal task of the paper will be to spell out just what sort that is.

Turning to the main title, the key term is 'revisiting.' I cannot explain it, however, until I first clarify 'accepted,' which I use a little differently from most others. The underlying thought is common with the usual use of the term: a claim is accepted when questions to which it is an answer are taken, at least provisionally, to have been settled. Here, nevertheless, I intend the term to apply more narrowly: a claim has become accepted within a scientific community when the community begins to presuppose it as a *constitutive* element in their ongoing research. If that research is challenged on the grounds that there is not yet sufficient warrant for presupposing

G. E. Smith (✉)
Tufts University, Medford, MA, USA
e-mail: george.smith@tufts.edu

© The Author(s), under exclusive license to Springer Nature Switzerland AG 2023
M. Stan, C. Smeenk (eds.), *Theory, Evidence, Data: Themes from George E. Smith*, Boston Studies in the Philosophy and History of Science 343, https://doi.org/10.1007/978-3-031-41041-3_16

the claim, then it has not yet been accepted. Equally, if difficulties emerge in the research, then the presupposed claim has not yet been accepted unless it is *prima facie* immune from being viewed as responsible for them. In analogy with the presumption of innocence until proved guilty, a scientific community has accepted a claim only when they grant it a presumption of truth until ongoing research gives clear evidence to the contrary.

Being accepted, as I use the term, has nothing as such to do with what individual scientists happen to believe. To offer a trifling example, once the galvanometer became the standard means for measuring electric current in the 1830s and 1840s, Ampère's law had become accepted even if those using the instrument were entirely unaware of its presupposing that Ampère's law holds to very high precision. Similarly, from the 1660s to the 1960s, the period and length of pendulums served to determine the strength of surface gravity. I wonder how many who used the resulting values gave any thought to the fact that those values were predicated on Galilean gravity—that is, uniform gravity acting along parallel lines—in direct contradiction with Newtonian centrally directed inverse-square gravity, the strength of which is what they thought was being measured. Being accepted, as I use the term here, has only to do with what the community is prepared to take for granted, perhaps tacitly, in ongoing research, and not with how individuals in the community have thought about the matter.

The truly narrow aspect of my use of 'accepted' derives from the requirement that it be a *constitutive* element in ongoing research, and not merely *heuristic*. Heuristic elements, no matter how much they are taken for granted and how heavily they are relied on to guide research, can subsequently be discarded without jeopardizing the results obtained from the research; to discard a constitutive element, however, *prima facie* undercuts the results that presuppose it.[1]

An example I have discussed at length elsewhere[2] involves the turn-of-the-century measurements J. J. Thomson and others made of the mass-to-charge ratio of the negatively charged constituents of cathode rays, of thermionic and photoelectric emissions, and of the discharge from uranium—that is, of what subsequently became known as the electron. Those measurements constitutively presupposed that the negatively charged constituents in question satisfy certain laws that theretofore had been shown to hold for charged particles in motion. Everyone making the measurements at the time surely thought of those constituents as particles, thereby precluding their exhibiting such wave-like behavior as diffraction. That this was a discardable heuristic emerged only in the 1920s when electrons were experimentally

---

[1] My distinction between constitutive and heuristic elements somewhat parallels the one Pierre Duhem draws between the discardable "explanatory" elements of physical theory and the "representational" elements. See Duhem 1991, Part I, especially Chapter 3.

[2] See Smith 2001, and "Getting Started: Building Theories from Working Hypotheses," the second lecture in the series, *Turning Data Into Evidence: Three Lectures on the Role of Theory in Science*, given at Stanford University in 2007 and available online at http://www.stanford.edu/dept/cisst/events0506.html.

confirmed to undergo diffraction, yet none of the measurements of their mass-to-charge ratio made over the thirty prior years were affected.[3]

A more nuanced example is the nineteenth-century ethereal substrate of Fresnel's transverse waves in optics. Mary Somerville in 1840 remarked, "The existence of an ethereal fluid is now proved,"[4] and went on to speculate on its long-term effects on the motions of orbiting bodies. The "proof" at the time consisted merely of an inability to conceptualize the transverse waves without a medium to support them—this quite independently of the subsequent efforts to treat electromagnetism in terms of non-uniformities in the ether. Yet virtually all of the results in wave optics during the nineteenth century remained intact after the ether was abandoned at the end of the century, for the ether itself was entering only heuristically into those results, and not constitutively. By contrast, it entered constitutively into the Michelson-Morley experiment, which is why the null results of that experiment gave reason to reconsider it.

The ether example highlights my reason for confining "accepted" to claims that enter constitutively into ongoing research: that research puts them to a test in ways that it does not put heuristic elements. More of that in a moment, however. For now let me just concede that the distinction between constitutive and heuristic elements is not at all straightforward in practice. Whether an element is entering ongoing research constitutively or heuristically has to be an historical question not only because it is a question about *ongoing* research, and hence its answer can change over time, but even more so because its answer can vary from one piece of research to another. It is also a philosophical question because it involves a retrospective "rational reconstruction" of earlier evidential reasoning, as illustrated by the wave-like features of the electron and the Michelson-Morley experiment, even though in those cases, and most others in which something has gone wrong in research, the reconstruction was carried out by scientists. Because the rest of this paper will be about how to combine historical and logical analyses of evidence within extended research traditions, and the constitutive-heuristic distinction requires just such analyses to resolve it on a case by case basis, I am not going to elaborate further on it now.[5]

Anyway, the most important aspect of accepted results for my purposes does not depend on that distinction. If the last fifty years of studies in the history of science have taught us nothing else, they have taught us that a great many fundamental claims of science first became accepted, in my sense, on the basis of remarkably little evidence. Examples abound: the law of inertia and Boyle's law in the seventeenth century; more recently Einstein's general theory of relativity, which initially had only the small anomaly in the precession of the perihelion of Mercury and a few problematic measurements of the bending of light in support of it; and

---

[3] Ironically, J. J. Thomson's son, George Thomson, shared the Nobel Prize in Physics in 1937 for his experimental efforts establishing the diffraction of electrons.

[4] Somerville 1840, 27.

[5] For more, see my lecture "Getting Started" cited in note 2.

Bohr's model of the atom, the only empirical test of which at the time it became accepted in my sense was a theoretical value of Rydberg's constant that differed from the measured value by six percent. Steven Shapin and Simon Schaffer made the limitations of evidence behind accepted results of experiments the central theme of their *Leviathan and the Air-Pump*. Several other historians of science, following their lead, have appealed to the limited evidence scientific communities require to accept not just experimental results, but all sorts of other fundamental claims, to argue that modern science does not merit the lofty epistemic standing that scientists have long claimed for it.

Some philosophers have reacted to that view by trying to show, in the face of mounting historical evidence to the contrary, that science really does have strict canons of acceptance that set it apart from all other forms of inquiry and give it a unique claim to achieving knowledge. I am not going to take that tack here. I agree with historians and sociologists of science that the evidence available at the time many of the most fundamental principles of science, and many results of experiments as well, became accepted in my sense fell far short of showing, with any finality, that those principles and results were true.

I have worded the scope of my agreement with historians and sociologists of science carefully. I am not granting them their customary further claim that acceptance of results in science is dominated by political, sociological, and other extra-scientific considerations that are orthogonal to the goal of acquiring knowledge. To the contrary, my view is that, more often than not, the claims they say are accepted in the face of inadequate evidence are in fact accepted on proper evidential warrant. The warrant in question is not, however, for the final truth of those claims, or even for a high probability of final truth. Rather, the evidence gives warrant for accepting the claims as constitutive working hypotheses enabling further research. That is, (1) the evidence gives *prima facie* support for the claim; (2) predicating further research on the claim has promise for markedly increasing the ability to marshal evidence within that research; and (3) the research in question promises to include ways of quickly exposing shortcomings in the working hypothesis that would otherwise threaten to lead down a long garden path—that is, an extended body of research that ultimately has to be thrown out because it was predicated on a fundamentally mistaken working hypothesis. Accepting a claim that meets these three requirements has everything to do with the goal of acquiring knowledge, though admittedly whether what emerges is actually knowledge depends overwhelmingly on the future research itself, and not on the grounds for accepting the working hypothesis on which it is predicated.

Which brings me to the key word of my title, 'revisit.' Everyone knows that accepted science sometimes gets revisited, for Einstein's theory of gravity replaced Newton's, the ideal gas law was found not to hold precisely of any real gas, and the classical principles presupposed in the Rayleigh-Jeans radiation law gave way to radical new principles with Planck's law. My point is broader than that. If some accepted principle is constitutively presupposed in ongoing research, then in some respect or other it is generally being tested in that research. Granted, the aim of the research is rarely to test that principle, for it has become accepted. Rather, the test is indirect, *en passant*, as it were. No one usually even notices that the accepted principle is being tested unless some problem emerges, an anomaly that turns out to be sufficiently recalcitrant that the assumptions underlying the research come to be reconsidered.

But what about when no problem or anomaly emerges? To some extent or other the presupposed principle has passed a test, and, just as interest accrues to money in a bank, further evidence has accrued supporting it. The further evidence, like the interest I am now getting from my bank, may amount to next to nothing. Millions of galvanometers are out there measuring electric current, but they are scarcely making much difference to the body of evidence supporting Ampère's law. I am going to argue, however, that ongoing research can subject accepted science presupposed in it to far more stringent *en passant* tests than is obvious at first glance. The only way to tell is to examine the details of that research.

So far I have said little beyond what Pierre Duhem pointed out a century ago: principles of accepted science continue to be tested to the extent that they enter into ongoing research, and hence the evidence for them typically grows with time. As he noted, the evidence for Newton's laws of motion in 1787, when Laplace thought he had removed the last orbital discrepancy with his explanation of the secular acceleration of the Moon, was far greater than when the *Principia* first appeared in 1687, and it was far less than the evidence for them a century later in 1887, at the time of the Michelson-Morley experiment. I want to say more than that:

> Whatever claim any of the sciences have to being epistemically different from other areas of inquiry, it lies at least as much (usually far more) in the way in which they have constantly revisited previously accepted claims, testing them anew, often much more stringently than those claims had been tested when they first came to be accepted, as it lies in any canons of initial acceptance—and this even when the testing has been only *en passant*.

But then philosophers of science who want to assess the nature of the knowledge achieved in any area of research are going to have to examine the history of that research, determining how evidence for various claims has or has not grown over time during the course of it.

That is the sense in which I say that history of science—more narrowly, the history of evidence—is indispensable to philosophy of science. But a straightforward history of research in an area is generally by itself not going to tell us how stringently various claims have continued to be tested, for the research is not focused on testing those claims, but instead on discovering new things about the world. Careful critical analysis of the details of the research is needed to reveal the ways in which accepted claims are being tested in the process and the stringency of those tests. Simply put, history of science is indispensable to the philosophy of science because the question, "How has evidence for a claim grown over the course of continuing research?" is an historical question; and philosophy of science is indispensable to history of science because critical analysis of the sort philosophers typically engage in is needed to assess the evidence, especially when continued testing is *en passant*. That is why we need a rapprochement between history and philosophy of science.

I am going to defend these theses in two steps in the rest of the paper. First I am going to appeal to my studies of the history of Newtonian orbital mechanics to demonstrate how rewarding a history of evidence can turn out to be. Then I am going to offer a few examples involving other areas of science that have raised important philosophical questions which, I claim, can be answered only through similar historical studies.

16.2 Revisiting Newton's Theory of Gravity[6]

No one would deny that Newton's theory of gravity continued to be tested as calculations based on it were made of orbital motions in our solar system and compared with the observed motions. Although such calculations began in Book 3 of the *Principia*, they became central to orbital astronomy only after the Euler-Mayer lunar theory of the early 1750s and Laplace's breakthroughs on the Jupiter-Saturn interaction and the secular acceleration of the Moon in the mid-1780s. They have remained central ever since. Two moments in that history are usually singled out: the discrepancy between the calculated and observed motions of Uranus that led to the discovery of Neptune in the mid-1840s—which Norwood Russell Hanson called the "zenith" of Newtonian gravity; and the 43 arc-second per century residual discrepancy between the calculated and observed precession of the perihelion of Mercury—Hanson's "nadir" of Newtonian gravity.[7] Other than these two, the view, at least among philosophers of science, seems to be that the only advance from the first two centuries of comparing calculated with observed orbits was increasingly closer agreement as the calculations were refined and the observations became more precise. In other words, the exactness of the theory was being tested, in the process markedly tightening the bands of accuracy with which it could legitimately be said to hold, but nothing more.

A decade ago I began looking at the history of post-Newtonian orbital research, asking if that was the only respect in which his theory was tested over its first two hundred years. I came away with seven surprises that, as a matter of autobiographical fact, are what convinced me of the value of the detailed study of the history of evidence in individual areas of research.

My first surprise was that the "test-question" was not, "Do the calculated motions agree with the observed motions?," but instead "Can robust physical sources compatible with Newtonian theory be found for each clear, systematic discrepancy between the calculated and the observed motions?" By "robust" I mean a source

---

[6] [Note revised in 2022] This section summarizes the conclusions from two long essays of mine, Smith 2014 and "Pending Tests to the Contrary: The Question of Mass in Newton's Law of Gravity" (in preparation). A shorter lecture version of the former is available online at http://www.stanford.edu/dept/cisst/events0506.html. [Added in 2022: A recently published essay of mine (Smith 2019) summarizes the points made in the long, yet-to-be published essay.] I do not have space to repeat the arguments for my conclusions in this paper. Those who find what I say here excessively *ex cathedra* should turn to those published essays.

[7] See Hanson 1962.

that has other observable and hence confirmable consequences besides accounting for the discrepancy. I should have realized that this was the test-question all along just from comparing the Neptune and Mercury-perihelion cases. What finally drove it home was the realization that there was at least one widely recognized discrepancy between calculated and observed orbits from 1687 until 1993, save for a quarter century beginning in 1787 during which Laplace's breakthroughs were being assimilated and the comparisons that were being made were more anticipatory than meaningful. The small anomaly in the precession of the perihelion of Mercury proved to be the "nadir" of Newtonian gravitation only after the failure of a half-century effort to find a source for it compatible with Newtonian theory. All the other discrepancies ended up revealing some detail of our planetary system, the least subtle of which was Neptune, that theretofore had not been taken fully into account in the calculations.

The reason why the test-question cannot simply concern agreement with observation is worth noting. The calculations based on Newtonian theory not only require a good deal of empirical information, like the masses of the planets. They also include a proviso, to use Carl Hempel's term, that no other forces need to be taken into account.[8] Agreement with observation would have consequently been evidence at least as much for this proviso as for the accepted theory. The different test-question puts those engaged in the recondite work of calculating orbits in a different light. They were not just re-testing Newtonian theory again and again, in a manner that Thomas Kuhn tended to dismiss when writing about "normal science." They were pursuing theretofore unnoted details of our planetary system that make a difference in the orbital motions and the subtle, but nonetheless detectable, differences those details make.

My second surprise concerned what was being tested. Take for example the anomaly in the motion of Uranus that led to the discovery of Neptune. It would have been masked if the significantly larger gravitational effects of Saturn on Uranus had not been included in the calculation first. So, the discovery of Neptune provided evidence not only for Newton's theory, but also for the specific aspects of Saturn that entered into calculating its effects on Uranus, for these were no less presupposed in the anomaly that emerged than Newton's theory was. The point generalizes. Each time a discrepancy emerges and a robust physical source for it is found, that source is incorporated into the new calculations, and the process is repeated, typically with still smaller discrepancies emerging that were often theretofore masked in the calculations. So, what was being tested each time when a new discrepancy emerged and a physical source for it was being sought was not only Newtonian theory, but also all the previously identified details that make a difference and the differences they were said to make without which the further systematic discrepancy would not have emerged.

Notice that as the loop shown in Fig. 16.1 is repeated with more and more details incorporated into the calculations, and smaller and smaller discrepancies

---

[8] See Hempel 1988.

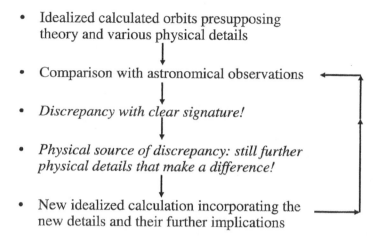

- Idealized calculated orbits presupposing
  theory and various physical details

- Comparison with astronomical observations

- *Discrepancy with clear signature!*

- *Physical source of discrepancy: still further
  physical details that make a difference!*

- New idealized calculation incorporating the
  new details and their further implications

Fig. 16.1 "Closing the Loop": the logic of testing in orbital mechanics

emerging, that tighter and tighter constraints are being placed on finding robust physical sources for new discrepancies. They had become tight enough in the case of Mercury's perihelion that no source for the residual anomaly in it could be found that was compatible with Newtonian theory.[9] Notice further what it is about the theory of gravity that was being tested in this process. It was not merely how well the theory represents observations—the sort of question one asks of curve-fits. It was whether the theory can serve to establish which physical features make detectable differences in orbital motions and what differences they make. As the discrepancies became smaller, that became an increasingly stringent test of Newtonian theory.

This shows that increasingly strong evidence was accruing to Newtonian theory over the first two hundred years of orbital research based on it. My third surprise was just how strong that evidence was. To illustrate this I am going to use an example that was of paramount importance in the history of evidence in orbital research, but has been entirely missed by philosophers of science. As most any textbook on orbital astronomy tells us, the first fully successful account of lunar motion was the Hill-Brown theory of 1919. That theory contains some 1400 perturbational terms, the vast majority of them derived from Newtonian gravity. What textbooks rarely tell us is that it also included a fudge, what Brown called "the Great Empirical Term," shown in Fig. 16.2 as the solid line. That there was some recalcitrant discrepancy in the motion of the Moon Newcomb had picked up in 1870, but it took the 1400 terms of Hill-Brown theory for it to emerge with the clear signature displayed in

---

[9] In particular, a proposal Simon Newcomb had made in 1895 of changing the value of the exponent in Newton's inverse-square law from -2 to -2.0000001612 had become untenable once Ernst Brown had shown in 1903 that such a change was much too large to reconcile with the motion of the perigee of the Moon. See, respectively, Newcomb 1895 and Brown 1903.

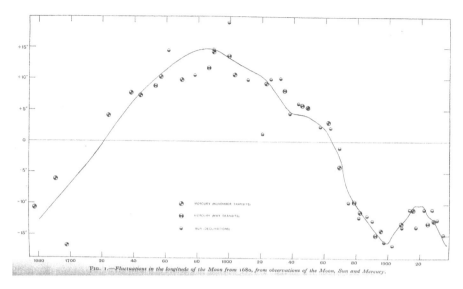

Fig. 1.—*Fluctuations in the longitude of the Moon from 1680, from observations of the Moon, Sun and Mercury.*

Fig. 16.2   Residual discrepancies in the longitudes of the Moon, Mercury, and the Sun in 1939

the figure. The question it posed was whether some still further force was acting on the Moon or whether instead the rotation of the Earth was fluctuating slightly over a roughly 200 year period.

Harold Spencer Jones resolved this question in 1939 when he established that a parallel discrepancy was showing up in the motions of Mercury, Venus, and the Earth once their mean-motions were renormalized to allow comparison with the Moon's.[10] Same-effect-same-cause reasoning implied a fluctuation in the rotation of the Earth. That has turned out to be robust, as demonstrated by lunar-laser-ranging following space-shots to the Moon and the subsequent discovery of lesser fluctuations in the Earth's rotation that the fluctuation shown in the figure had masked.[11] This detail that makes a difference led in 1950 to the end of the two thousand year tradition of taking sidereal time as our ultimate standard of time.

This example brings out more than just how stringent continued testing of Newtonian theory became over the course of the history of orbital research. It also underscores that not all the discoveries in that research concern gravity. The robust fluctuation in the Earth's rotation corroborated 1400 some terms concerning details in our planetary system that make a detectable difference in the Moon's motion.

---

[10] Spencer Jones 1939. Figure 16.2 of the present paper is Fig. 16.1, p. 551, of that paper. The dots on the figure display the comparable discrepancies for Mercury and the Sun, once Spencer Jones renormalized their mean-motions to be on the same basis with the mean-motion of the Moon, a renormalization that is needed in order to compare commensurately the contributions made by a fluctuation in the rotation of the Earth to their respective geocentric longitude discrepancies.

[11] See Lambeck 1980.

Most of those terms involve gravitational forces, but not all of them—most notably, terms involving viscous forces in our oceans that are slowly transferring angular momentum from the Earth to the Moon.[12]

There is a still more important respect in which the strength of the evidence coming out of this example surprised me. One might naturally think the evidence issuing from any comparison of calculation with observation is greatest when they match one another within observational accuracy. This example shows that that is not automatically true. Least-squares statistical methods are used in orbital mechanics to set the values of such parameters as Keplerian orbital elements and masses. As a consequence, when orbital calculations agree with observation, some ambiguity always remains about the extent to which the agreement is arising from mere curve-fitting versus the extent to which it confirms gravitation theory. Just such ambiguity would have been present had Hill-Brown theory achieved agreement with observation from the outset and Brown had not needed his Great Empirical term. That discrepancy, however, together with Spencer Jones' success in identifying what has turned out to be a robust physical source for it, removed the ambiguity in that case. For, the evidence that the discrepancy itself is physical became evidence that all the elements entering into the calculation from which it emerged are physical too. *The evidence for physicality was thus stronger with the discrepancy than it would have been without it!* That was my fourth surprise.

My fifth and sixth surprises came from the residual anomaly in the precession of Mercury's perihelion and what the transition to Einstein's theory of gravity has told us about claims to knowledge coming out of two centuries of research based on Newtonian gravity. As is well known, Einstein required Newtonian gravitation to hold in an asymptotic limit as he developed his new theory of gravity—specifically in a static, weak-field limit. That he did so was just as well because the 43 arc-seconds per century anomaly in the perihelion of Mercury that was initially the sole evidence for his theory presupposes Newtonian gravity. The 43 arc-seconds is not something anyone can observe. It is what one gets after subtracting the 531 arc-seconds per century precession produced by Newtonian gravitational effects of the other planets on Mercury from the observed 574 arc-seconds per century—that is, the 574 arc-seconds that remain once one introduces a roughly 5600 arc-seconds correction of the raw observations to compensate for the wobble of the Earth.[13] So, the 43 arc-seconds could count as evidence for Einstein's new theory only if that theory left intact, at least to the requisite level of precision, all the Newtonian contributions to the 531 arc-seconds that yielded the 43 arc-seconds.

Einstein's immediate goal in requiring that Newtonian theory hold in the static, weak-field limit was most likely to assure that all the evidence that had accumulated

---

[12] [Note added in 2022: The 1919 version of the Hill-Brown theory did not include the full effect of the tidal forces causing a secular acceleration of the Moon and a corresponding deceleration of the Earth's rotation. This was added only after the 1925 confirmation of the full magnitude of the tidal forces, including those of the liquid outer core].

[13] To appreciate the complexity of extracting the residual 43 arc-seconds per century in the motion of the perihelion of Mercury, see Clemence 1947.

in support of that theory carry over immediately as evidence for his new theory. The stringency with which Newtonian theory had been tested gave him good reason to require continuity of evidence in the transition to the new theory.[14] None of this, however, seemed especially surprising. What did surprise me was what Einstein's asymptotic limit did for the hundreds of details that Newtonian theory had singled out as making detectable differences in orbital motions in our planetary system. The conclusion just reached about the details contributing to the 531 arc-seconds per century generalizes. As a matter of historical fact, all of the details singled out as making detectable differences during the two centuries of prior research carried over intact into post-Einstein orbital mechanics. Save for some qualifications concerning levels of precision, the same details are still making the same differences as before. The only substantive change is that new details that make a difference have subsequently been added to them, details that do not invariably make the differences within Newtonian theory that Einstein's theory says they make.

A natural conclusion to draw from this is that the difference-making details identified during the two hundred years of orbital research predicated on Newtonian theory had, and still have, more of a claim to knowledge—to being permanent—than the theory itself ever had. But that cannot be correct. In an experiment one can confirm that the change in the value of some variable makes a difference simply by changing that value and seeing what happens. But orbital mechanics in those two centuries included no experiments. All conclusions about what changes in variables make what differences came out of Newtonian theory. We are still getting such conclusions out of Newtonian theory, the most prominent being the existence of dark matter. So, Newtonian theory must still have some sort of claim to being knowledge. Of course, it cannot have any claim to being the final word. But all along and still now it has had a claim to holding, projectably, *to sufficiently high accuracy over the domain of our solar system to establish, with seeming finality, a huge number of detailed features that make detectable differences in orbital motions in our planetary system.* The causal aspect of this claim makes it very different from saying merely that Newtonian theory yielded a representation of those motions to a certain level of accuracy—something a curve-fit could have done. That was my sixth surprise, the specific strong respect in which Newtonian theory continues to have claim to permanence.[15]

One final surprise. I have all along been speaking of the stringency with which Newtonian theory was tested during the course of orbital research. Now I have to add a surprising qualification. A fundamental part of Newtonian theory is that the strength of the gravitational attraction toward any body varies as the mass of that body. The first two centuries of Newtonian orbital research never tested that

---

[14] See Buchwald and Smith 2002.

[15] See "The Revolution That Didn't Happen," in Weinberg 2001, 187–206; and "T. S. Kuhn's Non-Revolution: An Exchange," *ibid.*, 207–9. I propose that the remarks made in the paragraph that ends with this note, properly expanded as in my Smith 2014, spell out grounds for the claims Weinberg makes in the latter of these two.

claim in any way at all. Throughout its history the masses of celestial bodies have been inferred from the strengths of the centripetal acceleration fields surrounding them. But then the only way in which those masses entered into any of the orbital calculations was as the product *GM*, the strengths of the fields. In other words, those masses were all along mere placeholders for the field strengths from which they were inferred and which they were then used to calculate. The thesis that gravitational attraction varies as the mass of attracting bodies thus got a free ride over the history of orbital research because it was *not* entering constitutively into that research.[16]

Lest anyone jump to a conclusion that those engaged in Newtonian gravitational research were unaware that claims about the masses of attracting celestial bodies were not being tested during that research, I best add something about how the claim that the gravitational field around any body is proportional to its mass was tested.[17] The second volume of Laplace's *Mécanique Céleste* (1799) was largely devoted to confirming that claim not from orbits, but from what we now call physical geodesy, in particular the strict relationship between the non-spherical flattening of the Earth and the variation of surface gravity with latitude implied by Clairaut's theorem for an Earth in hydrostatic equilibrium under Newtonian universal gravity.[18] Uncertainties in the geodetic measurements of the flattening and worries about the effects of surface irregularities on surface gravity limited Laplace to concluding that the data were generally supportive of Newton's universal gravity, "even though it is not so strictly verified in this case, as in the motion of

---

[16] The one factor seemingly involving the masses of celestial bodies entering into orbital calculations was the two-body correction to Kepler's 3/2 power rule, used to infer the mean distance of planets from their periods. Newton himself presents the correction in terms of the masses, but in doing so he assumed that the two bodies are orbiting about their common center of gravity. The two-body correction can be derived without the assumption that the neutral point the two bodies are orbiting is their common center of gravity; so derived, it can be stated directly in terms of the respective strengths of the centripetal acceleration fields without any occasion to mention masses. Moreover, observations were not precise enough to confirm the need for the two-body correction, even for Jupiter, until well into the twentieth century. For details, see my "Pending Tests to the Contrary: The Question of Mass in Newton's Law of Gravity."

[17] My "Pending Tests to the Contrary" discusses the history of evidence for this claim in detail. The remarks here merely summarize the conclusions of that essay. [Added in 2002: A summary of the history of the numerator in Newton's law of gravity can be found in my Smith 2019, cited in note 6 above.]

[18] Clairaut's equation for a body in hydrostatic equilibrium held together by Newton's inverse-square universal—that is, particle-to-particle—gravity states that there is a fixed relation, given the ratio of the centrifugal force to pure gravity at the equator *m*, between the variation of the measured acceleration of gravity with latitude, *g*, and the flattening of the Earth *f*:

$$g/g_{\text{equator}} = 1 + [(5/2)\, m - f]\, \sin^2 \varphi$$

This equation, which is here given only to first-order in *m*, can be rewritten to replace g/g_e on the left with the ratio between the lengths of a one-second pendulum at latitude and at the equator.

the planets."[19] Discrepancies between the measured values of the flattening and the variation of surface gravity with latitude remained throughout the nineteenth century, largely because of non-uniformities in the latter from one meridian to another, thereby continuing to limit the conclusiveness of the evidence from physical geodesy for the claim that the gravitational field around a body is proportional to its mass.

If Cavendish had varied the masses and distances between them in his famous experiment of 1798 in which he was, in effect, determining the value of the gravitational constant G, he would have provided direct evidence for the claim along lines that Newton himself had originally proposed.[20] But Cavendish did not vary the masses; he instead simply assumed Newton's law in order to infer the mean density of the Earth from his experiments, employing the same masses in all his trials. Over the course of the second half of the nineteenth century, however, Cavendish-like experiments were carried out by a number of individuals in order to determine the value of G, and the attracting masses (as well as their shapes) did vary substantially from most of those experiments to the others. Though this was not their recognized objective, for they were taking Newton's law of gravity for granted just as Cavendish had, those individuals, through the convergence of their measured values of G across their several experiments, produced the strongest evidence as of the beginning of the twentieth century that gravity is proportional to the mass of the attracting body. Strikingly, a physicist, Jonathan Zenneck, reached exactly that conclusion in a review article of 1901.[21]

There is a crucial lesson to learn from the fact that claims about the masses of celestial attracting bodies got a free ride during the two centuries of Newtonian research on orbital motions. The only way to tell just what aspects of a theory were being tested during ongoing research that presupposed it, and how stringent those tests were, is to examine the history of that research in detail, analyzing the logic of just what was and what was not being tested at each stage. This is just what Laplace

---

[19] Laplace 1966, 931. The strict verification that Laplace refers to in the case of the motion of the planets is of the inverse-square centripetal acceleration field around each body.

[20] Cavendish 1798. Although it has gone largely unnoticed, in Book 1, Proposition 92 of his *Principia* Newton proposed a sequence of *experimenta crucis* first with spheres and then with bodies of different shapes to confirm his inverse-square particle-to-particle gravity.

[21] Zenneck 1903–1921; reprinted (in English) in Renn and Schemmel 2007, 77–112. In particular, Zenneck concludes (p. 88):

> The G-determinations of Poynting and of Richarz and Kriger-Menzel are of special value in relation to the question of how far the proportionality of the attractive force to the mass is guaranteed for masses of the same material. Both experimenters used unobjectionable laboratory methods carried out with the greatest care. Both determinations employed the same material (lead) and the same method of measurement, but masses of very different magnitudes (154 or 100,000 kg). Even though in one case the mass was 650 times greater than the other, the results agree to approximately 0.2%.

[Added in 2022: It was Christopher Smeenk who originally called my attention to Zenneck's article].

did in turning from orbital mechanics to physical geodesy with Volume 2 of his *Mécanique Céleste* and Zenneck did a century later in preparing his review article. The importance of examining the history of evidence in gravity research was already clear, I hope, from the discussions of the first six of my surprises. Regardless, my seventh surprise really drives the point home that, when I say the history of science is indispensable, I am talking about fine details of that history.

16.3   Some Further Examples Requiring Study of the History of Evidence

A decade of effort on the three-hundred-year history of evidence in orbital mechanics is what convinced me of the indispensability of studying the history of evidence in highly specific areas of research in order to answer questions about knowledge achieved in those areas. I do not see why a piece of my autobiography, however, should be all that compelling to anyone else, especially to anyone who has to be taking my word for what they are going to find in that history. So, let me turn to three examples of philosophic issues raised in work done by others for which I do not presently know what a detailed investigation of the history of evidence is going to reveal. Even without knowing what the outcome of such an investigation will be, I am convinced that the philosophic issues raised in each case can be resolved only by means of it. For, in each case prior results have been revisited during ongoing research, and the resolution of the issues turns entirely on how those results came to be tested further, *en passant* or otherwise, as they were revisited.

My *first* example derives from Shapin and Schaffer's *Leviathan and the Air-Pump*, which I am told has sold more copies over the last 50 years than any other book in the history of science except Kuhn's *Structure of Scientific Revolutions*. The historiographical purpose of their book, they say late in it, was "to show the inadequacy of the method which regards experimentally produced matters of fact as self-evident and self-explanatory."[22] The three matters of fact they employ to that end are the Torricellian vacuum, the effectiveness of the air-pump in producing a vacuum, and Boyle's spring of the air. All three are presented as outcomes of experiments in one of the Harvard Case Studies,[23] where the experiments are presented not only as straightforwardly establishing their scientific results in the face of "non-scientific" resistance to them, but also as illustrating to students of humanities the nature of experimental knowledge. In contrast to most science textbooks, that Harvard case study was careful to include the disputes at the time over the vacuum, the difficulties with the newly invented air-pump, and the vagaries in the data from which Boyle extracted his law. Shapin and Schaffer did not dismiss it as sloppy history.

---

[22] Shapin and Schaffer 1985, 225.
[23] Conant 1957.

Their complaint was that the case study presented the experimental results as if they settled matters when the step from them to the Royal Society's endorsing the vacuum, the air-pump, and Boyle's so-called law was anything but straightforward. However much I might wish in other contexts to dispute their analysis of what was involved in settling those matters during the seventeenth century, that is not my concern here. I am perfectly willing to grant that the experimental results in question fell short of establishing the vacuum at the closed end of Torricellian tubes, the effectiveness of the air-pump, or the relationship between the pressure and density of air expressed by the Boyle-Mariotte law. My complaint with Shapin and Schaffer is, why stop the discussion with 1680? Three and a half centuries of research have subsequently been carried out on those three topics. Are we supposed to take it as "self-evident and self-explanatory" that any lacunae in their initial acceptance have never been revisited?

Consider the air-pump, which has continued to be developed over those three centuries, not merely as a laboratory instrument, but for a wide range of practical applications as well. Until flat-screen displays became the vogue, we all sat for many hours a day in front of cathode-ray tubes that had been evacuated to a measured pressure of around a millionth of an atmosphere. Many experiments are presently conducted in so-called ultra-vacuums of a billionth of an atmosphere, though it is worth noting that science now tells us that the number of molecules remaining in a single cubic centimeter at room temperature in an ultra-vacuum far exceeds a billion. The word 'vacuum' has come to have rather complex reference compared to what it had in the 1660s.

The Boyle-Mariotte law was revisited as early as Newton's *Principia* of 1687, but far more intensively during the research of Gay-Lussac and others early in the nineteenth century, leading to its expansion into what, following Regnault's research of the 1840s, came to be known as the ideal gas law. Figure 16.3 shows a table of Regnault's results that he stated "prove that the law assumed correct by physicists, viz. that air expands the same fraction of its volume at 0°, whatever its density, is not accurate."[24] A philosophically intriguing aspect of Regnault's work is that the instrument he used to measure temperature was a constant-volume air-thermometer, which presupposes the ideal gas law—this in experiments in which he was measuring systematic deviations from that law. During the twentieth century the virial expansion emerged as the rigorous replacement for the ideal gas law. When I, working as an engineer, need really accurate values for the compressibility of air, I turn to tables calculated, to four significant figures, from the experimentally determined first three terms of the virial expansion.[25]

These examples scarcely begin to cover the richness of the three hundred fifty years of research in pneumatics since Torricelli proposed his vacuum and Boyle conducted his experiments on the spring of the air. If we are going to reach conclusions about the nature and scope of the knowledge achieved in pneumatic

---

[24] Regnault 1902, 135.

[25] Sychev et al. 1987.

Pressure at 0°.	Pressure at 100°	Density of air at 0°, that of air at 0° under a pressure of 760 mm. being taken as unity.	1 + 100 a
109.72	149.31	0.1444	1.36482
174.36	237.17	0.2294	1.36513
266.06	395.07	0.3501	1.36542
374.67	510.35	0.4930	1.36587
375.23	510.97	0.4937	1.36572
760.00	....	1.0000	1.36650
1678.40	2286.09	2.2084	1.36760
1692.53	2306.23	2.2270	1.36800
2144.18	2924.04	2.8213	1.36894
3655.56	4992.09	4.8100	1.37091

The third column of the table comprises the densities of the gas at the temperature of melting ice ; it will be seen that these vary from 0.1444 up to 4.8100, that is, from 1 to 33.3, and for so great a change in density, the coefficient of expansion of the gas changes only from 0.3648 to 0.3709.

The above determinations therefore prove that the law assumed correct by physicists, viz., that air expands [on heating] the same fraction of its volume at 0°, whatever its density, is not accurate ; air expands, between the same temperature-limits, by amounts which are greater in proportion as the density of the gas is higher, or, in other words, as the molecules are closer together.

Fig. 16.3 Regnault's "Proof" that the Ideal Gas Law is not exact

research, we cannot restrict our attention to its first 25 years. We need to adopt a long vista to see how claims that had become accepted by 1680 have been revisited during the subsequent 330 years.

Understand, however, that just pointing to those 330 years of further research does not answer the complaints of Shapin and Schaffer. To do that we are going to have to subject the various stages of continuing research over those centuries to the same critical scrutiny to which they subjected the first decades. I know a fair amount about the conceptual history of the theory of gasses, but I have little idea of what surprises anyone is going to find if they look with that kind of scrutiny at the three and a half centuries of experimental research in pneumatics. Perhaps some fundamental tenets of pneumatics have gotten the same sort of free-ride in those experiments that Newton's claim that attraction varies as the mass of the attracting body got over two hundred years of orbital research. Philosophers of science need to appreciate that surprise is the ultimate redeeming feature of the labor that goes into historical research!

Early in their book Shapin and Schaffer remark:

> The Harvard history has itself acquired a canonical status: through its justified place in the teaching of history of science it has provided a concrete exemplar of how to do research in the discipline, what sorts of historical questions are pertinent to ask, what kinds of historical materials are relevant to the inquiry, what sorts are not germane, and what the general form of historical narrative and explanation ought to be. *Yet it is now time to move on* from the methods, assumptions, and the historical programme embedded in the Harvard case history and other studies like it. ... We became increasingly convinced that the questions we wished to have answered had not been systematically posed by previous writers. (Shapin and Schaffer 1985, 4; italics added)

I do not want to deny the indispensability of local studies of "making knowledge" with their characteristically short vistas for addressing questions of the sort Shapin and Schaffer raised, and questions about initial acceptance generally. When, however, the questions concern the nature and scope of knowledge achieved in specific areas of modern science, with pneumatics as one among many examples, short vistas cannot suffice. This gives reason to say anew that it is now time to move on beyond such studies.

My *second* example comes out of Hasok Chang's more recent book on thermometry, *Inventing Temperature*.[26] Chang, in keeping with Shapin and Schaffer, emphasizes the experimental—and social—complications of the century of effort that went into devising the two fixed points and unit interval between them that formed our temperature scale. He, however, did not stop with 1780.[27] His book continues through most of the next century, examining how the scale initially specified in terms of the expansion of mercury was revisited, giving way first to gas thermometers and then to the Kelvin-Joule theoretical characterization of absolute temperature in terms of its relation to energy and entropy, and as a by-product of this a temperature scale tied, as it continued to be, to the ideal gas law.

Chang's book thus ends up striking the same theme I am pressing here, that whatever is epistemically distinctive about science lies more in the process through which accepted results are revisited and refined than in their initial acceptance. If Chang had stopped with 1780, in the manner in which Shapin and Schaffer stopped with 1680, readers would have been left with the sense that temperature as we know it is an artificially devised quantity pieced together through a complex social process. That would have left them wondering whether temperature as we know it could have so little claim to being a real physical quantity. By going on, Chang indicated how it came to have a century later much more of such a claim.

---

[26] Chang 2004.

[27] My choice of the year 1780 is slightly arbitrary. The "Report of the Committee Appointed by the Royal Society to Consider the Best Method of Adjusting the Fixed Points of Thermometers; and of the Precautions Necessary to Be Used in Making Experiments with Those Instruments" was published in 1777, and Jean-André de Luc's *Essay on Pyrometry and Aerometry, and on Physical Measures in General*, published in 1779, first appeared in *Philosophical Transactions of the Royal Society* in 1778.

Absolute temperature, minus 273.7°	Air-thermometer temperature[b]
0	0
20	20 + 0.0298
40	40 + 0.0403
60	60 + 0.0366
80	80 + 0.0223
100	100
120	120 − 0.0284
140	140 − 0.0615
160	160 − 0.0983
180	180 − 0.1382
200	200 − 0.1796
220	220 − 0.2232
240	240 − 0.2663
260	260 − 0.3141
280	280 − 0.3610
300	300 − 0.4085

Fig. 16.4 Joule-Thomson comparison of absolute and air-thermometer temperatures

Let me explain what I mean by that. The table displayed in Fig. 16.4, taken from Chang's book,[28] compares the Kelvin-Joule ideal-gas absolute temperature scale with the constant-volume air-thermometer scale. I am less interested in the pattern of correction between the two than in the fact that the two fixed points—refined conceptions of the freezing and boiling points of water[29]—were carried over into the ideal-gas-based absolute temperature scale. One might think this a mere matter of convention were it not for the fact that the absolute scale includes a third salient point besides those two, absolute zero. As such, it provides a test of the physicality, in contrast to artificiality, of the two fixed-points. If the absolute zero extrapolated from the two fixed-points by Thomson and Joule had turned out not to be physically characterizable, there would have been reason to reconsider and perhaps reconstitute the fixed-points as well as the unit between them that specifies the quantity of temperature employed in physics.[30] As we noted, that is precisely what happened

---

[28] Chang 2004, 196. The table originally appeared in Joule and Thomson 1854, the second of four articles with that title, the others appearing in *Phil. Trans.* in 1853, 1860, and 1862.

[29] See Chang 2004, Chapter 1 or the papers cited in note 28.

[30] Such a revision, though not one undercutting my main point with this example, was made in 1954, when "the Tenth General Conference on Weights and Measures defined the thermodynamic temperature scale by selecting the triple point of water as the fundamental fixed point and assigning to it the temperature of exactly 273.16 °K. This new absolute thermodynamic scale, called the Kelvin scale, does not differ from the thermodynamic 'centigrade' scale by an amount which can be determined experimentally at this time." (Wilson and Arnold 1958) Other refinements of the temperature scale are discussed in that chapter.

in the middle of the twentieth century when the sidereal day was abandoned as the fundamental unit of time after Spencer Jones provided clear evidence that it, as such a unit, is parochial in a systematically misleading way. The history of thermometry thus offers an answer to a philosophically interesting question: what claim does temperature, defined as we know it in terms of its two classic, socially constructed fixed-points, have to being a physically real quantity?

That leads me to a further question that Chang's book does not answer. The history of thermometry did not end with Kelvin, Joule, and Rankine. The discovery of the Seebeck electromotive force, tying an electric potential to temperature, ultimately led to the development of thermocouples as instruments for measuring temperature, and the discovery of the relationship between temperature and phenomena of radiation similarly led to the development of pyrometers. Temperature occurs linearly in conjunction with the universal gas constant R in the ideal gas law:

$$p = \left(\frac{n}{V}\right) RT$$

In the Stefan-Boltzmann law for radiation energy, by contrast, temperature occurs to the fourth power, in conjunction with Boltzmann's constant k:

$$\overline{E_{Rad}} = \frac{8\pi^5 V}{15c^3 h^3} k^4 T^4$$

And in Planck's black-body radiation law, the preferred basis for pyrometry, temperature together with Boltzmann's constant occurs in an exponential:

$$E_\nu = \frac{8\pi h \nu^3}{c^3} \times \frac{1}{\exp(h\nu/kT) - 1}$$

I can now pose my question: to what extent has radiation research provided still more stringent tests of our temperature scale and its fixed points—or, alternatively, of the now preferred[31] fixed-point for the absolute thermodynamic scale given by the value at which ice, liquid water, and water vapor are in equilibrium? Some scales we invent are clearly artificial. My favorite example is the Richter scale, which derives from a correlation between a specific measurement on a specific type of seismometer and damage done by earthquakes to buildings in California. Useful though it may be for engineering purposes, it has been of little use in earthquake research, where instead the quantity called the *moment magnitude* is now used.[32] Orbital research revisited our fundamental scale for time, ultimately replacing it. But the two fixed-points of the temperature scale remain in place—or at least the one of the two singled out at the beginning of this paragraph. Is this because they have withstood increasingly stringent tests?

---

[31] See the preceding note.

[32] Aki and Richards 2002, 48 and 655.

As Newton indicated[33] and Duhem subsequently emphasized, questions about the physicality versus the conventionality and artificiality of any scientific quantity turn on the range of well-behaved lawlike relationships that quantity bears to other quantities. Few, if any, scientific quantities are more ubiquitous than temperature, entering as it does into relationships governing phenomena far removed from those that served to determine its original fixed-points and fundamental interval between them. If there are shortcomings in temperature as we have constituted it, they are more likely to show up in relations in which it occurs in the fourth power and exponentially than when it occurs linearly. So, has the last century of radiation research given us added reason to regard the classic two fixed-points of temperature to have picked out a real physical quantity? This depends on whether that research has really tested, *en passant* or otherwise, those points. I don't know the answer to the question, for while I am familiar with what the *Handbook of Physics* says about the matter, I have not gone back through the history of radiation research, critically examining the data and what they showed. And that is my point: it can be answered only through a detailed study of the history of evidence in radiation research.

My *third* example involves measurement of such fundamental quantities as the universal gas constant R and Boltzmann's constant k, which is just R divided by Avogadro's number. One way in which the physicality of the thermodynamic temperature scale might have been revisited was through precise measurements mediated by radiation theory of k and comparison of the value with other values obtained for k or with values obtained for R and Avogadro's number.[34]

Such fundamental constants as the speed of light in a vacuum and the aberration-of-light constant have been central to astronomy for a long time. Full appreciation, nevertheless, of the evidence that can be obtained from precise measurement of fundamental constants seems to have emerged only in the period from 1897 through 1913, starting with J. J. Thomson's measurement of the mass-to-charge ratio m/e of what we now call the electron and continuing through the precise measurements of its charge e, Planck's constant h, Avogadro's number, Faraday's constant from electrolysis, and Rydberg's constant, that is, the wavelength of the principal line in the spectrum of hydrogen.[35] Jean Perrin won the Nobel Prize in 1926 for his marshaling those measurements into a "proof," so to speak, of the discontinuous nature of matter—what philosophers call the "reality" of molecules and atoms.

Several philosophers have recognized the importance of the measurement of fundamental constants to philosophic disputes concerning realism, especially in

---

[33] I specifically have in mind Newton's remarks about the measurement of time in his Scholium on space and time; see Newton 1999, 410.

[34] [Added in 2022: As noted in the Postscript to this paper, in the new SI units of 2018 the values of k, R, and Avogadro's number have now been specified as exact, with the value of k in effect re-specifying the unit of temperature.]

[35] See my Smith 2001, cited in note 2 above, especially pp. 60–64; that paper stops short of including the seminal article by Mosely 1913–1914, which ties atomic number directly to x-ray spectra and indirectly to the constants I discussed.

microphysics;[36] but neither philosophers nor historians, so far as I know, have given attention to the subsequent history of the measurement of fundamental constants. Raymond Birge took it upon himself to publish preferred values for the six constants cited above and some twenty-five others in a 73-page article in 1929. He followed that a dozen years later with a second article listing preferred values, after which Richard Cohen and Jesse DuMond issued a series of further articles over the next quarter century, as well as a 1957 book they dedicated to Birge, *Fundamental Constants of Physics*, surveying the entire topic and its history.[37]

Throughout that period the task of identifying and publishing the preferred values remained the domain of a handful of otherwise less prominent physicists. The number of fundamental constants for which they tabulated values grew from fewer than 40 in Birge's initial publication of 1929 to more than 50 in 1965, and of course the values listed gained some in precision.

All this began to change in 1969 when the international physics community decided that the preferred values of the fundamental constants are important enough for the newly created committee, CODATA (the Committee on Data for Science and Technology) to form a sub-committee on fundamental constants. It has been responsible for the "official" values since then. Table 16.1 lists a sequence of articles putting forward recommended values, starting with Birge's 1929 article.[38] One thing to notice is that the committee is now publishing new sets of preferred values every four or so years, in memoranda of a hundred or so pages. A more important thing to notice is that the number of "fundamental" constants has grown, from the fewer than 40 with which Birge began to now more than 130 (Because of the theoretical inter-relationships among them, the precise number of *separate* constants in each of the publications is open to interpretation; the full table in the 2006 CODATA report has 222 entries, but what I deem to be 137 separately measurable constants). The overwhelming majority of the new constants, typical of which is the fine-structure constant, pertain in one way or another to microphysics.

As with the six constants I singled out from the 1897 to 1913 period—all of which are included in the "abbreviated list" from the 2006 CODATA report shown in Fig. 16.5—the majority of the ninety added constants are related to others, within physical theory, often in more than one way. Measurements of almost all of them, needless to say, are theory-mediated, but thanks to their multiple theoretical inter-

---

[36] See, for example, Salmon 1984, 213–29; Harper and Smith 1995; Harper 1997; Achinstein 2001; my lecture, "Getting Started," cited in note 2 above; and, most recently, van Fraassen 2009. [Added in 2022: Smith and Seth 2020, prompted by van Fraassen's paper, provides a detailed history of efforts to determine the values of fundamental constants from 1900 to 1913].

[37] Cohen et al. 1957. In 1983 an entire issue of *Philosophical Transactions of the Royal Society* (Series A, vol. 310, no. 1512) was devoted to the fundamental constants and their values, and two years later B. W. Petley published a book (Petley 1985) on them and why their measurement is important.

[38] [Added in 2022: The 1969 entry in the Table was not in the 2009 version of this paper. This was a mistake on my part, at the time not appreciating the extent to which the 1969 paper by Taylor et al. was a watershed moment in the history of the precision measurement of fundamental constants].

Table 16.1 History of publication of preferred values of fundamental constants

Title, Author	Journal	Number of constants
"Probable Values of the General Physical Constants," Raymond T. Birge	*The Physical Review Supplement*, 1929, pp. 1–73	<40
"The General Physical Constants, as of August 1941 with details on the velocity of light only," Raymond T. Birge	*Reports on Progress in Physics*, 1941, pp. 90–134	<60
"Least-Squares Adjustment of the Atomic Constants, 1952," Jesse W. M. DuMond and E. Richard Cohen	*Reviews of Modern Physics*, 5, 1953, pp. 691–708	<70
"Analysis of Variance of the 1952 Data on the Atomic Constants and a New Adjustment, 1955," E. Richard Cohen, Jesse M. W. DuMond, Thomas Layton, and John S. Rollett	*Reviews of Modern Physics*, 27, 1955, pp. 363–390	See preceding
"Our Knowledge of the Fundamental Constants of Physics and Chemistry in 1965," E. Richard Cohen and Jesse M. W. DuMond	*Reviews of Modern Physics*, 37, 1965, pp. 537–594	30 primary constants
The Fundamental Constants and Quantum Electrodynamics, B. N. Taylor, W. H. Parker, and D. N. Langenberg	A *Reviews of Modern Physics Monograph*, Academic Press, 1969	>30
"The Least-Squares Adjustment of the Fundamental Constants," E. Richard Cohen and Barry N. Taylor, under the auspices of CODATA	*Journal of Physical and Chemical Reference Data*, 2, 1973 pp. 663–734	>50
"The 1986 Adjustment of the Fundamental Physical Constants," E. Richard Cohen and Barry N. Taylor, representing CODATA	*Reviews of Modern Physics*, 59, 1987, pp. 1121–1148	>100
"CODATA Recommended Values of the Fundamental Constants, 1998," Peter J. Mohr and Barry N. Taylor	*Reviews of Modern Physics*, 72, 2000, pp. 351–495	>120
"CODATA Recommended Values of the Fundamental Constants, 2002," Peter J. Mohr and Barry N. Taylor	*Reviews of Modern Physics*, 77, 2005, pp. 1–107	>120
"CODATA Recommended Values of the Fundamental Constants, 2006," Peter J. Mohr, Barry N. Taylor, and David B. Newell	*Reviews of Modern Physics*, 80, 2008, pp. 633–730	>130

TABLE XLIX An abbreviated list of the CODATA recommended values of the fundamental constants of physics and chemistry based on the 2006 adjustment.

Quantity	Symbol	Numerical value	Unit	Relative std. uncert. u_r
speed of light in vacuum	c, c_0	299 792 458	m s^{-1}	(exact)
magnetic constant	μ_0	$4\pi \times 10^{-7}$	N A^{-2}	
		$= 12.566\,370\,614... \times 10^{-7}$	N A^{-2}	(exact)
electric constant $1/\mu_0 c^2$	ϵ_0	$8.854\,187\,817... \times 10^{-12}$	F m^{-1}	(exact)
Newtonian constant of gravitation	G	$6.674\,28(67) \times 10^{-11}$	m^3 kg^{-1} s^{-2}	1.0×10^{-4}
Planck constant	h	$6.626\,068\,96(33) \times 10^{-34}$	J s	5.0×10^{-8}
$h/2\pi$	$\hbar$	$1.054\,571\,628(53) \times 10^{-34}$	J s	5.0×10^{-8}
elementary charge	e	$1.602\,176\,487(40) \times 10^{-19}$	C	2.5×10^{-8}
magnetic flux quantum $h/2e$	Φ_0	$2.067\,833\,667(52) \times 10^{-15}$	Wb	2.5×10^{-8}
conductance quantum $2e^2/h$	G_0	$7.748\,091\,7004(53) \times 10^{-5}$	S	6.8×10^{-10}
electron mass	m_e	$9.109\,382\,15(45) \times 10^{-31}$	kg	5.0×10^{-8}
proton mass	m_p	$1.672\,621\,637(83) \times 10^{-27}$	kg	5.0×10^{-8}
proton-electron mass ratio	m_p/m_e	1836.152 672 47(80)		4.3×10^{-10}
fine-structure constant $e^2/4\pi\epsilon_0\hbar c$	α	$7.297\,352\,5376(50) \times 10^{-3}$		6.8×10^{-10}
inverse fine-structure constant	α^{-1}	137.035 999 679(94)		6.8×10^{-10}
Rydberg constant $\alpha^2 m_e c/2h$	R_∞	10 973 731.568 527(73)	m^{-1}	6.6×10^{-12}
Avogadro constant	N_A, L	$6.022\,141\,79(30) \times 10^{23}$	mol^{-1}	5.0×10^{-8}
Faraday constant $N_A e$	F	96 485.3399(24)	C mol^{-1}	2.5×10^{-8}
molar gas constant	R	8.314 472(15)	J mol^{-1} K^{-1}	1.7×10^{-6}
Boltzmann constant R/N_A	k	$1.380\,6504(24) \times 10^{-23}$	J K^{-1}	1.7×10^{-6}
Stefan-Boltzmann constant $(\pi^2/60)k^4/\hbar^3 c^2$	σ	$5.670\,400(40) \times 10^{-8}$	W m^{-2} K^{-4}	7.0×10^{-6}
Non-SI units accepted for use with the SI				
electron volt: (e/C) J	eV	$1.602\,176\,487(40) \times 10^{-19}$	J	2.5×10^{-8}
(unified) atomic mass unit $1\,u = m_u = \frac{1}{12}m(^{12}C)$ $= 10^{-3}$ kg mol$^{-1}/N_A$	u	$1.660\,538\,782(83) \times 10^{-27}$	kg	5.0×10^{-8}

Fig. 16.5 "Abbreviated list" of the recommended values of fundamental constants, 2006

relations, there are generally multiple ways of measuring them, involving different theoretical presuppositions. Much of the committee's task is to decide which ways of measuring each of them are currently yielding the most accurate value, and that requires comparisons not only of values determined in different ways, but of the different ways themselves.

The "abbreviated list" indicates how markedly the number of significant figures has increased from the time of Perrin; it is continuing to increase. In the case of the converging measures of the universal gravitation constant, G, the number of significant figures gives us error bands on the precision to which the gravitational field strength varies with the mass of the attracting body.[39] What the increased precision is giving us in the case of the other 130 some constants is the key question.

---

[39] Specifically, the value listed in the 2006 CODATA report implies that the dependence of gravitational force on the mass of the attracting body holds to within 1 part in 10 thousand. It was only 2 parts in a thousand in 1910, while Eötvös by then had shown the dependence of the forces on the mass of the attracted body holds to within 5 parts in 100 million and Brown had shown that the exponent of distance is -2.0 to within 2 parts in 100 million. That explains why the bounds on the precision of G, a constant of proportionality if Newton's law holds exactly, have long since become bounds on the precision with which gravitational forces are proportional to the mass of the attracting body.

One curiosity worth mentioning is that far more frequently than one might expect, new officially recommended values have fallen outside the published error-bands or uncertainty of their immediate predecessors. Not surprisingly, that has made the committee now chary about publishing error-bands.[40]

An international committee that decides every few years which eight- or ten-significant-figure measurements are the most accurate is right up the alley of Shapin and Schaffer, not to mention their colleague in Edinburgh, Donald MacKenzie. How does this international institution that persists over time with changing members, most of whom remain virtually unknown to those outside the scientific community, reach their decisions?

Philosophers of science have still better reason to look at the continuing history of the values of the fundamental constants. The evidence for much of what physics claims to know about microphysical structure, from the level of molecules to that of electron-photon interactions, lies within those measurements. How stable is each of the different ways of measuring each constant, and what does that stability tell us about the specific theoretical principles entering that way of measuring it? To what extent do different ways of measuring the same constant, with their different theoretical assumptions, converge on the same value, and what conclusions about those assumptions can we draw from the degree of convergence?

The stringency with which the theoretical assumptions are being tested in the measurement of any one constant depends first on the precision of the measurement and second on the number of cross-checks provided by the relationships between the constant in question and other independently measured constants. Issues about which parts of microphysics have been more stringently tested than others can be answered only by looking carefully at the history of the measurement of each inter-related group of constants. Physicists are not going to worry about a question like this until problems surface in ongoing research. Philosophers need not wait for such problems to surface.

Hopefully, my three examples suffice to show that the importance of the history of evidence to philosophy of science is not limited to gravity research. I could have chosen any number of other examples. I chose the three I did because they are especially pertinent to philosophic issues about the knowledge gained in microphysics. Had I more space, the one other example I would have included is the initial half-century inability to reconcile the theoretical and measured ratios of the specific heats of gases. As Maxwell pointed out in 1860,[41] this conflict between

---

[40] Kenneth G. Wilson, personal communication. The obvious problem with bounding the error in theory-mediated measurements of fundamental constants is unrecognized sources of systematic error in the measurements, something the spread in the measured data from any one measurement cannot expose. Wilson also reminded me of a central point Donald MacKenzie makes in his 1990 book: error-bands are themselves socially constructed. For just this reason we should not expect them invariably to be monotonically decreasing as they are revisited during ongoing research.

[41] Maxwell first raised this objection, somewhat crudely, in 1860 in a report of a meeting of the British Association for the Advancement of Science; see Document III, 8 in Garber et al. 1986, 320f. The objection is elaborated more carefully in his "On the Dynamical Evidence of the

theory and observation gave clear evidence that one of the most fundamental premises of statistical mechanics, the equipartition of energy, is not true. The first proposal that began to get anywhere on the problem was at the initial Solvay Conference of 1911,[42] and even then it took fifteen more years until the last discrepancy was eliminated and some quite stunning microstructural conclusions could be drawn from the measured values.[43] What do fifty or more years in the face of clear evidence that one of the most fundamental assumptions of that research is false tell us about modern science?

16.4 Toward a Rapprochement Between History and Philosophy of Science

I proposed at the outset that, whatever claim any science has to lofty epistemic standing, it is generally going to lie in the extent to which fundamental principles of that science have been tested not only again and again over the historical course of research in it, but more importantly ever more stringently. My summary of the way in which such further testing has happened over the course of the history of gravity research has supported my proposal for that science. The further examples I have offered pose questions that suggest similar support for my proposal may be coming out of them. What we need are studies in the history of evidence in individual, narrow areas of research, demarcated by the extent to which the principal research-concerns in them have remained the same. The fact that any continued testing of fundamental principles has likely been largely *en passant* means that philosophers and historians have some hard work ahead of them, digging critically into little known publications and even reconstructing past experiments. I can attest from my efforts on the history of evidence in gravity research that such hard work can be highly rewarding. I am sure that Hasok Chang would say the same about his efforts on thermometry.

I am asking a good deal of philosophers of science in proposing that they undertake such studies. For one thing, I am asking them to put to one side, at least for the moment, many of the issues that have dominated the field over the last century—confirmation theory, issues of realism versus, say, constructive empiricism, and abstract concerns about theory-change and the semantics of theories. If I am right, we are still woefully unprepared to address such issues, for we do not really know what the evidence consists of in most areas of research. That is what I mean when I

Molecular Constitution of Bodies," of 1875 (*ibid.*, Document II, 24), where he calls the ratio of the specific heats "the greatest difficulty which the kinetic theory of gases has yet encountered" (p. 229).

[42] See Kormos Barkan 1999, especially Chapter 11.

[43] See Fowler 1929, Chapter III, 50–71; and Partington and Schilling 1924, Chapter 6, especially 229–41.

say that so much of philosophy of science is empty, for it is about a stick-figure, and at times even a fantasy, version of science, not the real thing. I am asking philosophers of science to start reading primary source material not simply of the greats, like Newton and Einstein, but of such less heralded figures as Simon Newcomb and Spencer Jones, in my case, and Jean-André De Luc and Victor Regnault, in Chang's. The hard work is not finding and reading such sources, but learning to read them in their historical context, as others at the time and immediately afterwards read them.

Philosophers do bring one special training to this task, their skill in exposing and critically evaluating hidden assumptions in evidential reasoning and thereby making explicit what the logic of that reasoning was. Scientists of course have that skill too, but they rarely bring it to bear on the past unless something has gone wrong in current research *or* they are charged with writing a review article. What I am asking of philosophers of science in reviewing the history of evidence in any field is to put themselves in a position to write a sequence of review articles from the perspective of key moments in that history, critically assessing what the evidence had shown to that moment, its limitations, and the relationship between these and the chief questions with which research at that time was grappling. The retrospective stance of history will generally provide some insight into shortcomings in the evidence that those at the time would have had trouble formulating. But that is as much a problem as a blessing insofar as care is needed not to become too omniscient in viewing matters from the perspective of the time. When and how insights about shortcomings in the evidence that are clear to us first emerged is part of the history of evidence.

Surprisingly, I am asking much less of historians of science. Admittedly, much of history of science has concerned how conceptual frameworks have developed over time, usually through novel shifts made by the greats, and more recently with how the social context of scientific research in specific locales has affected it. There are, nonetheless, some well-known studies in the history of evidence, such as the one that got me started on the history of gravity research some twenty-five years ago, Curtis Wilson's "From Kepler's Laws, So-called, to Universal Gravitation: Empirical Factors."[44] Shapin and Schaffer, for that matter, were themselves engaged in a study of the history of evidence, though granted more with the goal of showing how scientists at the time and philosophers and historians of science more recently have misconceived that evidence. My chief request of historians of science is simply not to rule out as illegitimate, even before the work begins, studies involving such long, narrow vistas as the history of comparisons made by specialists of calculated versus observed motions of the planets, say from Ptolemy to today, exploring exactly why increasingly greater agreement has been achieved.

Fashion has a way of turning into dogma. *Leviathan and the Air-Pump* produced a fashion in the history of science of concentrating on the local making of knowledge, local both geographically and temporally. That fashion has led to

---

[44] See Wilson 1970.

doubts about the legitimacy of questions stretching across multiple generations of specialists in many different countries in the history of science. Shapin and Schaffer accused the Harvard historians of being blind to what was involved in establishing matters of fact in seventeenth century experiments in pneumatics. When I say that much recent history of science is blind, what I mean is that it has, like the drunk looking for his car keys where the street is lighted, been looking myopically in the wrong place if the question is whether the sciences have any extraordinary claim to knowledge. One of the wonderful things about history is that one can legitimately ask so many different questions. The chief thing I am requesting of historians of science is not to dogmatically reject certain questions that have recently been out of fashion.

Why do I think that this is important? Early in his book, *Making Natural Knowledge*, is an extraordinarily astute remark by Jan Golinski: "The history of science has had a long struggle to free itself from science's own view of the past."[45] This, on my view, has been the signal achievement of the last fifty years of the history of science, a goal Tom Kuhn was aiming for when he published *The Structure of Scientific Revolutions*, initiating a still continuing conflict between historians and philosophers of science.

The fact that a struggle was needed for history of science to free itself indicates that the shortcomings of science's own view of its past are too elaborate to cover here. A key unfortunate aspect of that view especially relevant to this paper can be summarized in terms akin to those used by Shapin and Schaffer: *once one of the greats proposed a breakthrough together with a modicum of reasons to take it seriously, anyone properly trained in science should have seen almost immediately that it had to be true—almost as if the proposal, once put forward, were self-evident.* Subsequent efforts to test it therefore amounted to little more than confirming the obvious in order to safeguard against any slight possibility of a mistake. That attitude, expressed succinctly in Kuhn's famous remark, "Mopping-up operations are what engage most scientists throughout their careers," grossly undervalues the contribution the vast majority of scientists over the centuries have made to the development of knowledge in their field.[46]

The mistake lying behind this attitude is one of conflating a distinction that Newton went to pains to make in his "Rules of Reasoning" in the *Principia*: it is one thing to have sufficient evidence to take some claim to be true for purposes of ongoing research, and quite another for it be true.[47] The mistake is understandable,

---

[45] Golinski 1998, 2.

[46] Kuhn 1970, 24.

[47] Newton's precise phrasing in his third Rule of Reasoning, introduced in the second edition of the *Principia*, is, "Those qualities of bodies that cannot be intended and remitted and that belong to all bodies on which experiments can be made *should be taken* to be qualities of all bodies universally;" and in the fourth Rule, added in the third edition, "In experimental philosophy, propositions gathered from phenomena by induction *should be taken* to be true, either *accuratè* or *quamproximè*, notwithstanding any contrary hypotheses, until yet other phenomena make such propositions either more exact or liable to exception." In both cases the Latin verb translated with my emphasis added

for once scientists take something to be true and cease questioning it in their ongoing practice, they lose sight of the distinction between that and its simply being true—that is, they do so until some unexpected result gives reason to reconsider it. One nevertheless has to wonder how so many scientists can put so little value on their own work. In highlighting that value I am not denying the genius of the insights of the greats, even though I do think that scientists under-appreciate the extent to which prior developments in a science create the opportunity for such moments of genius. All I am denying is that those groundbreaking insights are what give modern science its claim to special epistemic standing. That claim, when it is valid, lies instead in how stringently the groundbreaking proposals came to be tested over subsequent years, decades, and even centuries. In glossing over the subsequent history, science's view of its past has done at least as much to create the "science wars" as historians and sociologists of science like Kuhn, Shapin, and Schaffer. Indeed, we should applaud them for enabling us all to free ourselves from that view of the past.

Historians of science have freed us from it by showing time and again how claims that scientists and philosophers of science have made about how science is different from other areas of inquiry do not withstand critical scrutiny. *But it is now time to move on.* Kuhn's master project was not to undermine all claim that the sciences have to special epistemic standing, but to lay out a proper understanding of the respects in which the sciences represent humanity's greatest epistemic achievement. Kuhn, on my view, deflected that project down a dead-end path not merely by being so dismissive of normal science, but also by overemphasizing aspects in which the sciences do not differ from other areas of inquiry. It is time we return, free of the classic misconceptions, to questions about how the sciences really *are* different from other intellectual pursuits. It is because these questions will require a concerted effort from both historians and philosophers of science that I urge a rapprochement between the two. We might not merely answer questions we want answered about the nature and limits of knowledge produced in the sciences. We might provide an understanding of science that will be of value to scientists and the community at large.

16.5   Postscript, 2022

This paper began as an invited talk for a 2009 conference in Bergamo, Italy on "the philosophical history of science" organized by Niccolò Guicciardini. The paper then appeared in an issue of *The Monist* (vol. 93 (2010): 545–579) edited by him with the same title. I have made minor corrections to the original in the present version,

---

is a form of *habere*. Newton's initial formulation of the third Rule, handwritten in his annotated copy of the first edition, and also in John Locke's copy, was, "The laws and properties of all bodies on which experiments can be made *are* the laws and properties of bodies universally." (Cohen 1971, 25) So, Newton came to emphasize the distinction between "take to be" and "are" only subsequently, but then took the care much later to phrase the fourth rule correspondingly.

most of them indicated in the endnotes. None of these corrections in any way alters any of the claims or proposals made in the original.

Even more so than its plea for a rapprochement between history and philosophy of science, the rhetorical tone of the original paper reflects my half decade (2001–2006) as Acting Director of the Dibner Institute for the History of Science and Technology, where I was surrounded every day by top-flight senior and post-doctoral historians. I had come to see the conflict between philosophers and historians and sociologists on the epistemic standing of science as arising largely from their focusing on different aspects of science and hence talking at cross-purposes. The proposal to focus on the long-term development of evidence in narrow areas of specialty was intended to put them on a common page, so to speak. At the time I saw challenges to the epistemic standing of science as largely an academic concern. In the years since then it has come increasingly to impinge on public policy, giving all the more reason for those addressing the question of that status to pursue at least the beginnings of an answer that we can all support. I have nevertheless done nothing in the present version to alter its somewhat dated rhetorical tone.

The corrected version remains dated in one important further respect. As of 2009 I had just begun looking in detail into the history of precision measurements of the fundamental constants in collaboration with the physicist Kenneth G. Wilson. At the time we were focusing on the CODATA publications of recommended values of 1998, 2002, and 2006, asking what the fact that some later recommended values fell outside the error-bands of earlier ones was revealing about the status of the announced values. In the period between 2009 and 2022 CODATA published three more sets of recommended values, in 2010, 2014, and 2018. The last of these incorporates the new SI units in which the values for Planck's constant, Avogadro's number, the elementary charge, and Boltzmann's constant have been taken as exact in defining the new units for the joule, the mole, the coulomb, and the degree Kelvin.

In originally proposing a philosophical study of the history of the precision measurement of the fundamental constants as illustrating the process of revisiting prior results in physics I had most in mind a conclusion put forward in CODATA 2006, that the agreement to so many significant figures of the most precise value for the fine structure constant strongly linked to QED with the most precise value at most weakly linked to QED had provided "a truly impressive confirmation of QED theory." The new SI units and philosophical issues that have been raised about what they amount to give a whole new reason for such a project, now focusing on questions about what bearing the revisiting of those values time and again over almost a century has on these issues. Fortunately, one such study of that history specifically in relation to the new SI units has recently been published, Nadine de Courtenay's 2022 paper. It illustrates forcefully how important it is to consider recent developments in physics as a product of a history of revisiting prior results.

References

Achinstein, P. 2001. *The book of evidence*. Oxford University Press.

Aki, K., and P.G. Richards. 2002. *Quantitative seismology*. 2nd ed. Sausalito: University Science Books.

Brown, E.W. 1903. On the verification of the Newtonian law. *Monthly Notices of the Royal Astronomical Society* 64: 396–397.

Buchwald, J.Z., and G.E. Smith. 2002. Incommensurability and discontinuity of evidence. *Perspectives on Science* 9: 463–498.

Cavendish, H. 1798. Experiments to determine the density of the Earth. *Philosophical Transactions of the Royal Society* 88: 469–528.

Chang, H. 2004. *Inventing temperature: Measurement and scientific progress*. Oxford University Press.

Clemence, G.M. 1947. The relativity effect in planetary motions. *Reviews of Modern Physics* 19: 361–364.

Cohen, I.B. 1971. *Introduction to Newton's 'Principia'*. Harvard University Press.

Cohen, E.R., K.M. Crowe, and J.W.M. DuMond. 1957. *Fundamental constants of physics*. New York: Interscience Publishers.

Conant, J.B. 1957. Robert Boyle's experiments in pneumatics. In *Harvard case studies in experimental science*, Vol. 1, Case 1, ed. J.B. Conant, 1–63. Harvard University Press.

de Courtenay, N. 2022. On the philosophical significance of the reform of the international system of units (SI): A double-adjustment account of scientific enquiry. *Perspectives on Science* 30: 549–620.

Duhem, P. 1991. *The aim and structure of physical theory*. Trans. Ph. Wiener. Princeton University Press.

Fowler, R.H. 1929. *Statistical mechanics: The theory of the properties of matter in equilibrium*. Cambridge University Press.

Garber, E., S.G. Brush, and C.W.F. Everitt. 1986. *Maxwell on molecules and gases*. Cambridge: MIT Press.

Golinski, J. 1998. *Making natural knowledge: Constructivism and the history of science*. Cambridge University Press.

Hanson, N.R. 1962. Leverrier: The zenith and nadir of Newtonian mechanics. *Isis* 53: 359–378.

Harper, W. 1997. Isaac Newton on empirical success and scientific method. In *The cosmos of science: Essays of exploration*, 55–86. University of Pittsburgh Press.

Harper, W., and G.E. Smith. 1995. Newton's new way of inquiry. In *The creation of ideas in physics*, ed. J. Leplin, 113–166. Dordrecht: Kluwer.

Hempel, C.G. 1988. Provisos: A problem concerning the inferential function of scientific theories. In *The limitations of deductivism*, ed. A. Grünbaum and W. Salmon, 19–36. Berkeley: University of California Press.

Joule, J.P., and W. Thomson. 1854. On the thermal effects of fluids in motion. Part II. *Philosophical Transactions of the Royal Society* 144: 321–364.

Kormos Barkan, D. 1999. *Walther Nernst and the transition to modern physical science*. Cambridge University Press.

Kuhn, Th. 1970. *The structure of scientific revolutions*. 2nd ed. University of Chicago Press.

Lambeck, K. 1980. *The Earth's variable rotation: Geophysical causes and consequences*. Cambridge University Press.

Laplace, P.S. 1966. *Celestial mechanics*. Trans. N. Bowditch, Vol. II (1799). Bronx: Chelsea Publishing.

MacKenzie, D. 1990. *Inventing accuracy: A historical sociology of nuclear missile guidance*. Cambridge: MIT Press.

Mosely, H.G.J. 1913–1914. The high-frequency spectra of the elements. *Philosophical Magazine* 27: 703–713 and 1024–1034.

Newcomb, S. 1895. The elements of the four inner planets and the fundamental constants of astronomy. In *Supplement to the American ephemeris for 1897*. Washington, DC: Government Printing Office.

Newton, I. 1999. *The* Principia: *Mathematical principles of natural philosophy*. Trans. I.B. Cohen and A. Whitman. Berkeley: University of California Press.

Partington, J.R., and W.G. Schilling. 1924. *The specific heats of gases*. New York: Van Nostrand.

Petley, B.W. 1985. *The fundamental constants and the frontier of measurement*. Bristol: Adam Hilger.

Regnault, V. 1902. Researches upon the expansion of gases. Second memoir. In *The expansion of gases by heat: Memoirs by Dalton, Gay-Lussac, Regnault, and Chappuis*, ed. and trans. W.W. Randall, 121–150. New York: American Book Company.

Renn, J., and M. Schemmel, eds. 2007. *Gravitation in the twilight of classical physics. Vol. 3 of the genesis of general relativity*. Dordrecht: Springer.

Salmon, W. 1984. *Scientific explanation and the causal structure of the world*. Princeton University Press.

Shapin, S., and S. Schaffer. 1985. *Leviathan and the air-pump: Hobbes, Boyle, and the experimental life*. Princeton University Press.

Smith, G.E. 2001. J. J. Thomson and the electron: 1897–1899. In *Histories of the electron: The birth of microphysics*, ed. J.Z. Buchwald and A. Warwick, 21–76. Cambridge: MIT Press.

———. 2014. Closing the loop: Testing Newtonian gravity, then and now. In *Newton and empiricism*, ed. Z. Biener and E. Schliesser, 262–351. Oxford University Press.

———. 2019. Newton's numerator in 1685: A year of gestation. *Studies in History and Philosophy of Modern Physics* 68: 163–177.

Smith, G.E., and R. Seth. 2020. *Brownian motion and molecular reality. A study in theory-mediated measurement*. Oxford University Press.

Somerville, M. 1840. *On the connection of the physical sciences*, Facsimile edition, 2005. London: Elibron Classics.

Spencer Jones, H. 1939. The rotation of the Earth, and the secular acceleration of the Sun, Moon, and planets. *Monthly Notices of the Royal Astronomical Society* 99: 541–558.

Sychev, V.V., et al. 1987. *Thermodynamic properties of air*. National Standard Reference Data Service of the USSR, English language, ed. Th. B. Selover, Jr. Washington, DC: Hemisphere Publishing Corporation.

van Fraassen, B.C. 2009. The perils of Perrin, in the hands of philosophers. *Philosophical Studies* 143: 5–24.

Weinberg, S. 2001. *Facing up: Science and its cultural adversaries*. Harvard University Press.

Wilson, C. 1970. From Kepler's laws, so-called, to universal gravitation: Empirical factors. *Archive for History of Exact Sciences* 6: 89–170.

Wilson, R.E., and R.D. Arnold. 1958. Thermometry and pyrometry. In *Handbook of physics*, ed. E.U. Condon and H. Odishaw, 5–30. New York: McGraw-Hill.

Zenneck, J. 1903–1921. Gravitation. In *Encyclopädie der mathematischen Wissenschaften*, Vol. 5.1, ed. A. Sommerfeld, 25–67. Leipzig: B. G. Teubner.